TROPICAL DELTAS OF SOUTHEAST ASIA—SEDIMENTOLOGY, STRATIGRAPHY, AND PETROLEUM GEOLOGY

Edited by:

F. HASAN SIDI
Jason Geosystems Asia, Mulia Tower, Suite 2414, Gatot Subroto Kav 9-11, Jakarta 12930, Indonesia

DAG NUMMEDAL
Institute for Energy Research, Department of Geology and Geophysics, University of Wyoming - Laramie, Laramie, Wyoming 82071-4068, U.S.A.

PATRICE IMBERT
TotalFinaElf E & P, Suite 700, 800 Gessner, Houston, Texas 77024, U.S.A.

HERMAN DARMAN
Brunei Shell Petroleum Co. Sdn. Bhd., Brunei Darussalam

HENRY W. POSAMENTIER
Anadarko Canada Corp., 425 1st. Street S.W., Calgary, Alberta, T2P 4V4, Canada

Laura J. Crossey, Editor of Special Publications
SEPM Special Publication Number 76

Tulsa, Oklahoma, U.S.A. *September, 2003*

SEPM and the authors are grateful to the following for their generous contribution to the cost of publishing

Tropical Deltas of Southeast Asia—Sedimentology, Stratigraphy, and Petroleum Geology

Institute for Energy Research, University of Wyoming

Institute of Marine Geology and Chemistry, National Sun Yat-sen University

Natural Sciences and Engineering Research Council of Canada

Unocal Corporation

Yoshiki Saito

Contributions were applied to the cost of production, which reduced the purchase price, making the volume available to a wide audience

---

SEPM (Society for Sedimentary Geology) is an international not-for-profit Society based in Tulsa, Oklahoma. Through its network of international members, the Society is dedicated to the dissemination of scientific information on sedimentology, stratigraphy, paleontology, environmental sciences, marine geology, hydrogeology, and many additional related specialties.

The Society supports members in their professional objectives by publication of two major scientific journals, the Journal of Sedimentary Research (JSR) and PALAIOS, in addition to producing technical conferences, short courses, and Special Publications. Through SEPM's Continuing Education, Publications, Meetings, and other programs, members can both gain and exchange information pertinent to their geologic specialties.

For more information about SEPM, please visit **www.sepm.org.**

ISBN 1-56576-086-7

6128 E. 38th Street, Suite 308
Tulsa, OK 74135-5814, U.S.A.

Printed in the United States of America

TROPICAL DELTAS OF SOUTHEAST ASIA—SEDIMENTOLOGY, STRATIGRAPHY, AND PETROLEUM GEOLOGY

F. Hasan Sidi, Dag Nummedal, Patrice Imbert,
Herman Darman, and Henry W. Posamentier, Editors

CONTENTS

Introduction

Quaternary Systems

Tertiary Systems

Introduction

DR. GEORGE P. ALLEN

SPECIAL PUBLICATION IN THE HONOUR OF THE LATE DR. GEORGE P. ALLEN:

MEMORIAL COMMENTS

BERNARD C. DUVAL
Institut Français du Pétrole, Rueil Malmaison, France

Its is a privilege for me to be invited to write memorial comments in honor of George ALLEN, a personal friend and appreciated colleague of many years. I should not say "personal", because everyone who knew George and was fortunate enough to work with him closely felt rapidly and profoundly that a direct, natural trustworthy relationship was born.

This man was exceptionally gifted with a sense of clarity in his oral and written expression, of generosity in his passion for sharing his own experience and ideas with others. He listened and accepted all questions and gave straightforward, convincing answers. Within our company, his communication skills were even made more than perfect by his subtle mastery of French.

When we discussed his future before he started a new career with the University of Queensland, I could not but agree that this was a great opening, fitting very well the purpose of his life, his attraction for research and teaching. He was a great expert, advisor (he continued consulting), and instructor in this sphere of work, a man of the field, respectful of observations, with an eye for the telling detail and also for the larger structure. In this sense, sequence stratigraphy was the ideal field of application for his skills because he knew how to articulate between sedimentary building blocks of different orders and scales.

I can mention three domains where his impact was particularly remarkable:

Firstly his teaching skills come to mind. He gave innumerable courses on applied sequence stratigraphy, both on cores and in the field. He created a real school through which almost all geologists and geophysicists from TOTAL acquired the basic know-how and practice needed for their specific trades, a common language, and understanding of the subtle phenomenona involved in sequence stratigraphic applications. Of course we had wonderful operating models which gave substance to his theoretical views. The Mahakam Delta owes him a large part of his stardom status in the world of sedimentology!

Secondly, his expertise was instrumental in practical applications to reservoir prediction and characterization. His name appears in many studies of this type, sorting out which parts of the various sequences to be encountered in a particular play were more sand-prone, and therefore contributing to more efficient seismic interpretation. He was a practical man and established easily applicable rules to relate the geometry of a reservoir to its environmental and sedimentary characters, always referring to the field reality (*"return to the rocks"*, he said so many times!). The results were convincing enough to serve as a basis to reserve certification. One famous accomplishment was the recognition that field models of tidal mouth-bars from the modern delta, only a few meters thick, but with huge areal extension, one hundred km^2 or more, were also applicable to older sequences of Miocene age. Together with pressure studies, this helped estimate much higher ranges of gas reserves than would have been predicted from the wells in the Peciko and Tunu giant fields.

Finally, he was a world-renowned authority in the profession. His numerous articles, books, papers given on many occasions (particularly at AAPG and SEPM conferences) bear witness to his outstandingly brilliant image. He has always been a welcome invited speaker on sedimentology and sequence stratigraphy in general and deltaic environments in particular.

I saw him several times participate in panels, without ostentation and with his natural charm, compose in real time without preparation a few magic "cartoons" that fitted so well the purpose and were so clearly and brilliantly described that everyone agreed that nothing else had to be said, and that was it!

He contributed greatly to the company's image of technical excellence throughout the world of exploration. We are indebted to him for having facilitated through training and advice, successful cooperation programs in several countries such as Indonesia and Venezuela that were truly strategic for TOTAL. Thus he paved the way for excellent relationships with partners, national companies and academic institutions.

George, our Friend and Master, we miss you, but in many ways, you are still with us!

Bernard C. Duval

Tropical Deltas of Southeast Asia—Sedimentology, Stratigraphy, and Petroleum Geology
SEPM Special Publication No. 76, Copyright © 2003
SEPM (Society for Sedimentary Geology), ISBN 1-56576-086-7, p. 3.

ACKNOWLEDGMENTS

This special publication was firstly initiated by FOSI (Indonesian Sedimentologists Forum) and is dedicated to the late Dr. George Allen (1942–1998), whose contribution to deltaic research in Southeast Asia, especially the Mahakam in Indonesia, has been so monumental. George passed away much too early in his life due to a heart attack, but he probably died the way he would have wanted to ... at the head of a field trip in the Spanish Pyrenees.

Based on the nature of the results presented in this volume, most of the papers were reviewed by sedimentologists working in the fluvial to coastal environments and several petroleum practitioners whose reservoirs are in this fascinating setting. We gratefully acknowledge the individuals listed alphabetically below:

- Allen W. Archer
- Harry Doust
- Joan Esterle
- Howard D. Johnson
- Simon Lang
- Sandy Phillips
- John Suter
- Denis Tan
- Michael Tesson
- Robert Tye
- Kenrick van Noord
- Michael Wilkes
- Brian J. Willis

The enthusiasm of these reviewers and their willingness to help has really improved the content of this publication.

F. Hasan Sidi – Jakarta
Dag Nummedal – Wyoming
Patrice Imbert – Houston
Herman Darman – Brunei
Henry Posamentier – Calgary

Tropical Deltas of Southeast Asia—Sedimentology, Stratigraphy, and Petroleum Geology
SEPM Special Publication No. 76, Copyright © 2003
SEPM (Society for Sedimentary Geology), ISBN 1-56576-086-7, p. 4.

A FRAMEWORK FOR DELTAS IN SOUTHEAST ASIA

DAG NUMMEDAL

Institute for Energy Research, Department of Geology and Geophysics,
University of Wyoming, Laramie, Wyoming 82071-4068, U.S.A.

F. HASAN SIDI

ConocoPhillips Indonesia, Mulia Tower, 4th floor, Gatot Subroto Kav 9-11, Jakarta 12930, Indonesia
present address: Jason Geosystems Asia, Mulia Tower, Suite 2414, Gatot Subroto Kav 9-11, Jakarta 12930, Indonesia
e-mail: hsidi@jasongeo.com

AND

HENRY W. POSAMENTIER

Anadarko Canada Corp., 425 1st. Street S.W., Calgary, Alberta, T2P 4V4, Canada

INTRODUCTION

The early years of process–response sedimentological research were strongly focused on deltas, for a variety of reasons. They were important for accumulation of oil, gas, and coal; they were environmentally sensitive; they were readily accessible to academic communities; and consequently they were intensely studied. This characterized the decades of the 1960s and 1970s. With the shift in emphasis towards predictive stratigraphy and the stacking of depositional systems, and away from the stratigraphic architecture of depositional systems themselves, research on deltas reached a plateau in the 1980s and early 1990s. Today, however, the widespread use of shallow- as well as deep-penetration seismic data, cores from subsurface reservoirs, vibracores from modern environments, sophisticated oceanographic tools, and numerical modeling has resulted in a rejuvenation in delta research. In addition, a global province that hitherto had received relatively little attention increasingly became a focus for research—the equatorial zone of Southeast Asia. It is the objective of this volume to bring to the fore a category of deltas with which many sedimentologists and stratigraphers are, at best, vaguely familiar. It is expected that this volume also will stimulate new research on tropical deltas by highlighting how their facies and stratigraphic architectures differ from mid- and high-latitude ones, by emphasizing their significance to the global sediment budget, and by stressing their uniqueness within a petroleum systems framework.

This special publication emphasizes the need for models intrinsic to tropical deltas of Southeast Asia to supplement the more conventional general models currently in vogue, based on past studies of deltas such as the Mississippi, the Rhône, and many other large and small mid-latitude deltas. In this introductory paper, we briefly review the framework within which tropical deltas of Southeast Asia are deposited and the significance of each of the papers in the volume, evaluate trends in the sedimentology and stratigraphy of deltaic systems, and suggest a possible path for future research and applications, relevant to Southeast Asia.

WHY FOCUS ON SOUTHEAST ASIA?

The tropics are different. They differ from the mid-latitude climate belts, where most models for deltaic sedimentation were developed, in almost all the fundamental physical factors that shape depositional systems. The climate is warm and equable, so weathering rates and sediment yields are high; rainfall is abundant, so rivers have high discharge. In Southeast Asia especially, this high discharge is readily conveyed to the sea because of abundant small and steep drainage basins. Low-latitude tides are strong and play an effective role in the exchange of sediment from continental to marine environments. Once delivered to the marine realm, the strong Indonesian throughflow and deepwater tides control effective sediment dispersal across shelves and through deep-water straits. Storms, which are remarkably effective agents for marine sediment dispersal in many mid-latitude regions, are hardly present in equatorial zones because of the weak Coriolis effect. Because of low storm intensity combined with the limited wind fetch within the confined seas that characterizes this region of peninsulas and archipelagos, Southeast Asian seas have the lowest wave energies on earth.

The papers in this book explore how the combination of these complex factors has shaped deltas in this region. In many instances their morphology and stratigraphic architecture differs significantly from the (mid-latitude) norm in part because of the intimate linkage of carbonates and clastics in this climate zone. Sedimentological surprises such as distributary channels floored by thick accumulations of fluid mud lend a bit of "mystery" to tropical deltas. We hope that, rather than being merely a summary of tropical deltas, this book may open the door to a new and active phase of sedimentological and stratigraphic research in tropical environments across the globe. Almost all of our continents, at one geologic time or another, have migrated through the earth's tropical zone. Consequently, the active pursuit of research on modern tropical depositional environments has the potential to improve our interpretation of ancient rocks beyond the limits of today's tropics.

Sediment Discharge

In a comprehensive evaluation of data on sediment discharge for all large rivers in the world, Milliman and Meade (1983) concluded that more than 70% of the terrigenous sediment entering the world ocean comes from rivers in southern Asia and Oceania (Fig. 1). High and episodic rainfall, steep topographic relief, and relatively young and easily erodible rocks are the factors that drive this exceptional sediment discharge. Also, the generally small drainage basins provide little storage capacity and greater response to episodic events, such as floods and landslides (Milliman and Syvitski, 1992). In a recent study focused on Indonesian rivers, Milliman et al. (1999) concluded that rivers draining the six major islands of Indonesia are responsible for 20–25% of the total sediment discharge to the world ocean (Fig. 2), even though the land area that they drain is only about 2% of the world total.

Tropical Deltas of Southeast Asia—Sedimentology, Stratigraphy, and Petroleum Geology
SEPM Special Publication No. 76, Copyright © 2003
SEPM (Society for Sedimentary Geology), ISBN 1-56576-086-7, p. 5–17.

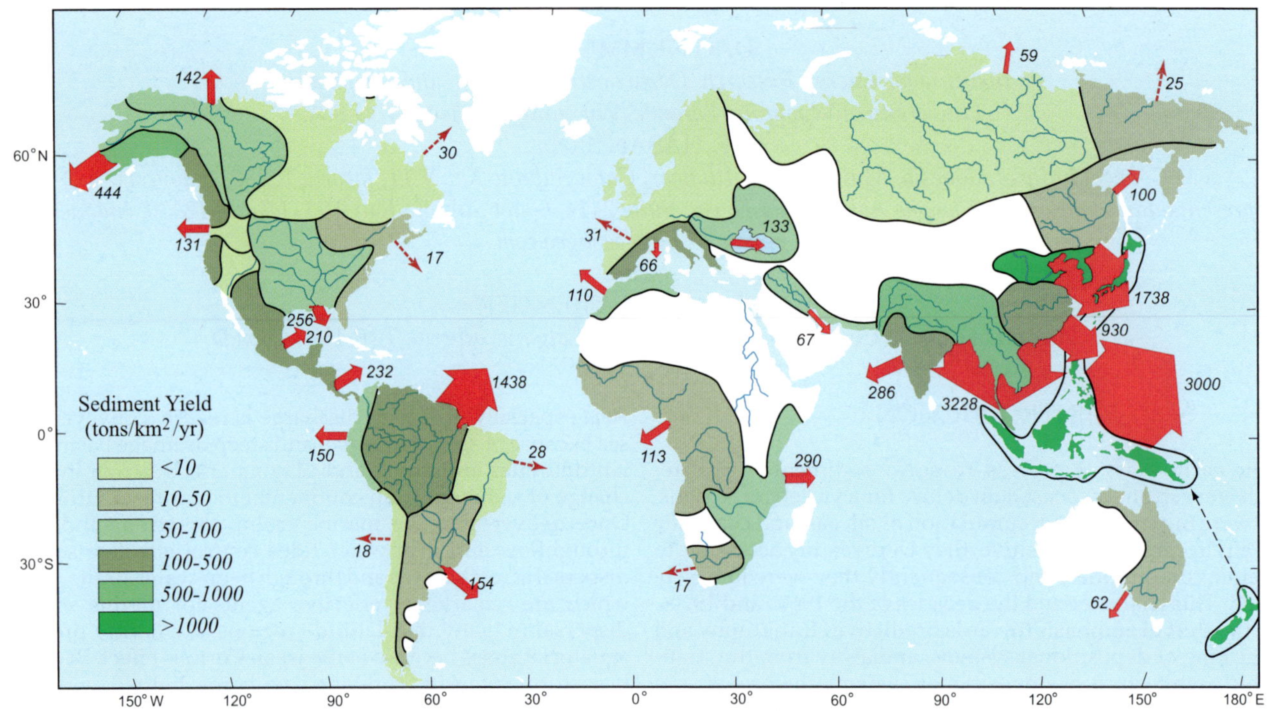

FIG. 1.—Annual discharge of suspended sediment from various drainage basins in the world; width of the arrow corresponds to relative discharge. Numbers refer to annual input in millions of tons. The sediment yields and major rivers of the various basins are also shown (from Milliman and Meade, 1983).

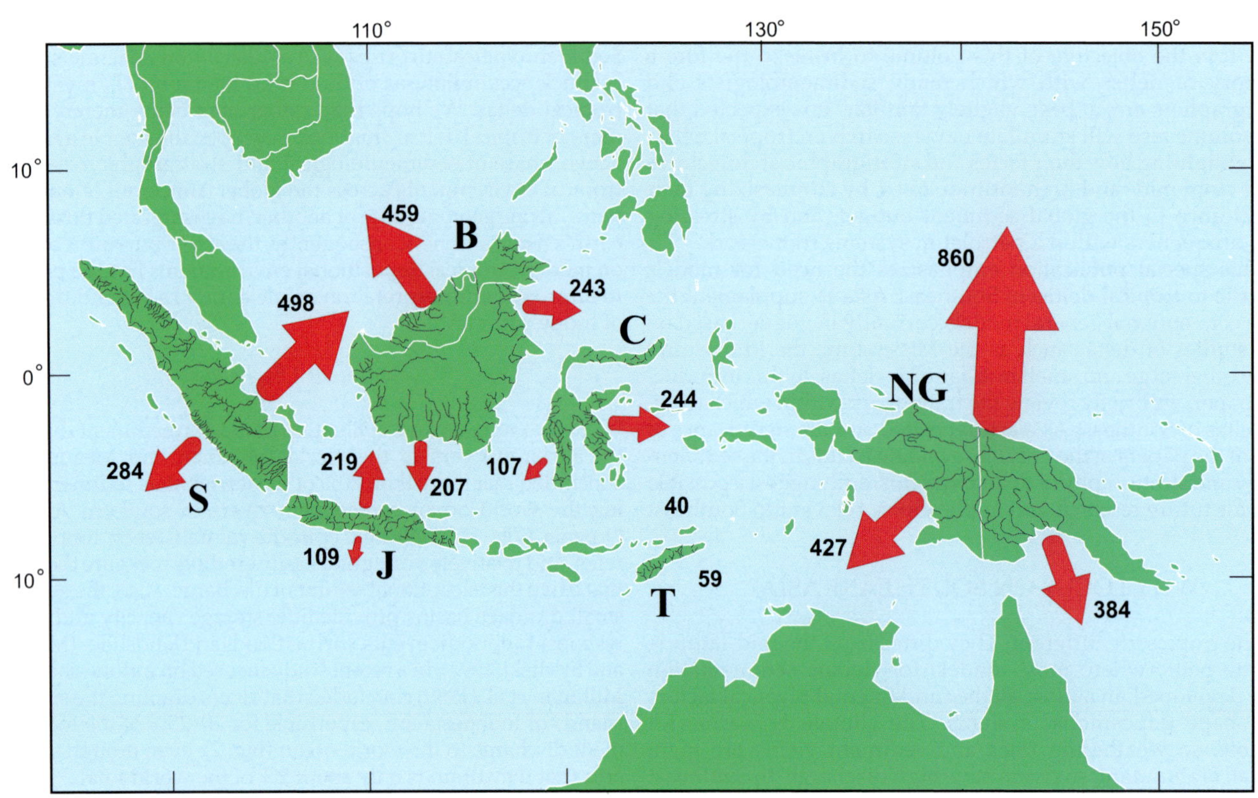

FIG. 2.—Sediment discharge (millions of tons per year) from six East Indies islands. Arrow width is proportional to the discharge. Letters are abbreviations for the names of the major islands (from Milliman et al., 1999).

The disproportionate sediment load of Southeast Asian streams affects all coastal depositional systems, deltas in particular. For example, this influence is evident in the large plume of prodelta mud off the Mahakam delta, deflected to the south by another unique regional process—the Indonesian throughflow (Roberts and Sydow, this volume). Another regionally unique aspect of the sediment load is evident in the thick bottom layers of fluid mud in the active distributaries of the Fly River delta (Dalrymple et al., this volume) and across the shelves of much of Indonesia (Milliman et al., 1999).

The present extreme sediment yield in the region, however, might not be representative of the geologic past. Anthropogenic effects, such as deforestation for agriculture and mining, the common occurrence of unchecked wildfires, and widespread poor land-use practices have driven current sediment yield far above the long-term average.

Rainfall

A quick glance at a map of global rainfall (Fig. 3) goes a long way towards explaining the sediment yield of Southeast Asia. Rainfall amount in the world's equatorial region in general, and in Southeast Asia in particular, is truly exceptional. Almost all of the region's high mountains, in Kalimantan, Papua New Guinea, and Sumatra, and many of the adjacent oceanic regions, have annual rainfall exceeding 3000 mm. It is sobering for those of us who live in the northern mid-latitudes to look at this map and remember that we tend to consider the US Gulf Coast as wet, yet rainfall is significantly less.

What is not evident in Figure 3 is the seasonality to the rainfall in the region, controlled primarily by the migration of the intertropical convergence zone (ITCZ). The position of the ITCZ follows the sun with a phase lag of about a month. This seasonality enhances the erosive potential of the rainfall. Intermittent dryness leads to locally exposed ground, and the rainfall, when it occurs, may be quite violent and highly erosive, prone to triggering flash flooding in constricted valleys and steep terrain. In Kalimantan rainfall is especially heavy in November during the early phase of the northwesterly (wet) monsoon. There is a secondary peak in April during the early phase of the southeasterly (dry) monsoon (MacKinnon et al., 1996).

In addition to the obvious effect of abundant rainfall on sediment supply, the high freshwater discharge into the sea has profound effects on oceanic circulation patterns of the adjacent shelf onto which it debouches. Being less dense than the ambient seawater, the freshwater often forms a "lid" on top of the saline waters (Murray et al., 1988). The resulting density gradient drives a baroclinic flow, which, in the case of the shelf off East Kalimantan, is superimposed on the shelf-parallel Indonesian throughflow. The resulting circulation consists of a weak seaward-directed flow component in the surface waters, compensated by a landward component in the bottom boundary layer. As a result, much of the mud delivered to these equatorial shelves can be locally

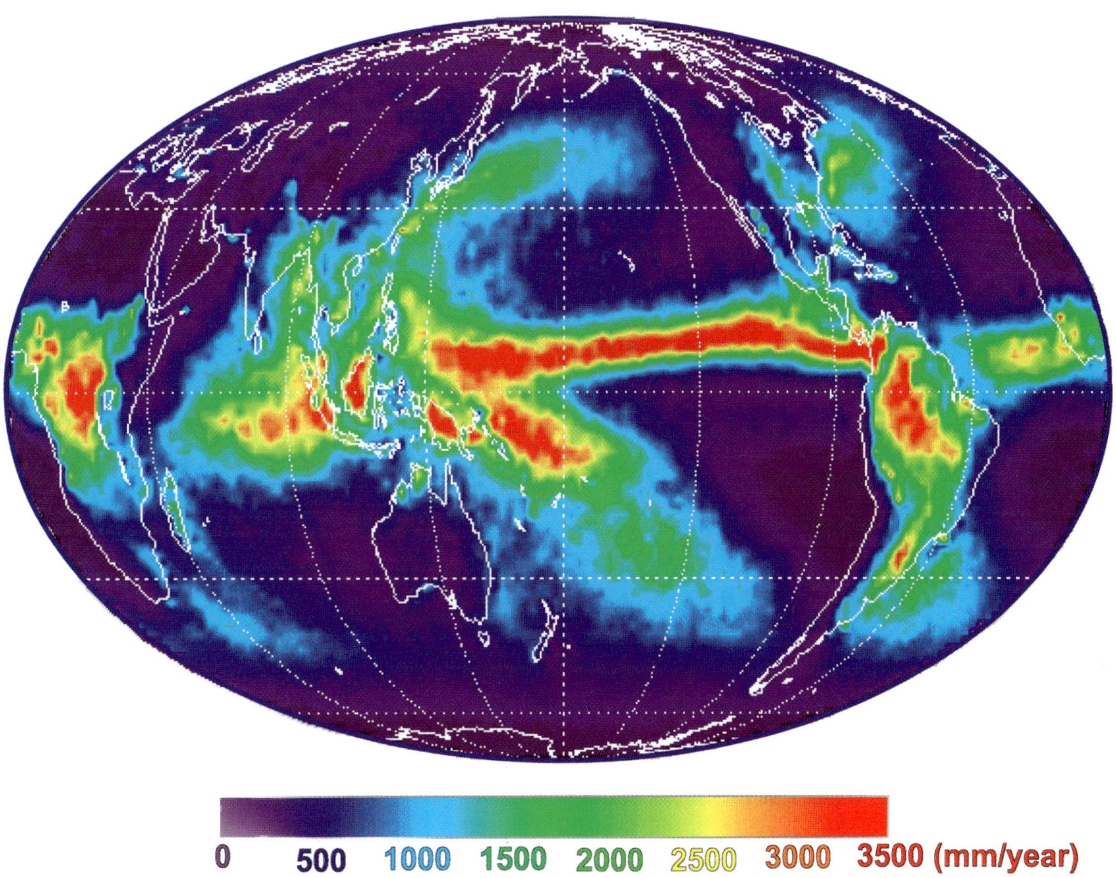

FIG. 3.—Annual mean rainfall across the globe. Data derived from Special Sensor Microwave Imager (SSM/I) flown aboard the Defense Meteorological Satellite Program (DMSP) series of satellites. Data collection began in 1987 (from website of Norman C. Grody, NOAA/NESDIS/Office of Research and Applications; http://www.ncdc.noaa.gov/ogp/papers/grody.html).

trapped in a band a few tens of kilometers wide near the river mouths. It is therefore quite common to encounter thriving carbonate environments offshore of, and contemporaneous with, prograding muddy deltas along shore. Several papers in this volume document such patterns.

Tides

Tropical tides are high around the globe (Fig. 4). It is therefore not the least surprising that most of the deltas discussed in this volume are either tide-dominated (Jones et al., this volume; Dalrymple et al., this volume) or tide-influenced mixed with fluvial (Roberts and Sydow, this volume). Tides in the tropics are high because the lunar tide-generating force produces maximum tidal elevation in the plane of the moon's orbit (Defant, 1958). With the lunar orbit inclined only at 28° relative to the earth's equator, the highest lunar tide generation is focused within the tropics. Although high tides occur outside of the tropics because of the complexity of the amphidromic system of the earth's tidal waves and the shoaling and growth of tidal amplitudes across wide continental shelves, what distinguishes the tropics is that *almost all coasts in this setting have high tidal ranges* (Fig. 4).

What makes a coastal depositional system "tide-dominated" is not simply the tidal range; rather it is the relative strength of the tidal range (tidal "energy") relative to wave energy and the magnitude of the fluvial input. Deltas have long been classified with reference to the interaction between tides, waves, and rivers (Galloway, 1975). Coastal environments other than deltas also are classified as wave-dominated, mixed, or tide-dominated (Nummedal and Fischer, 1979; Davis and Hayes, 1984; Boyd et al., 1992). Focusing on Southeast Asia, it seems that a major difference between tropical and extratropical deltas is the relative influence of waves; what can we infer about wave energies in the tropics? Is there a systematic difference between storms and wave energies in the tropics and those in the mid-latitudes, where so many earlier studies were conducted?

Storms and Waves

Again, the tropics—specifically the equatorial zone—are indeed different with regard to wave regimes. It is particularly noteworthy that tropical cyclones (i.e., hurricanes and typhoons) do not occur within the earth's equatorial belt (Fig. 5). In Southeast Asia and Oceania, in particular, data on storm tracks for a random year (1996) show a total absence of cyclones between latitude 5° north (northernmost Borneo) and latitude 10° south (southernmost tip of Timor and New Guinea). Tropical storms require several physical triggers. One is an atmospheric disturbance of some kind, another is warm sea-surface temperature (> 28°C), and a third is a Coriolis force strong enough to sustain the vorticity of a storm system. Close to equator, the Coriolis force is simply insufficient for the job. Extratropical cyclones, which derive their energy from conversion of potential energy across a frontal boundary in the mid-latitudes, do not even get close to

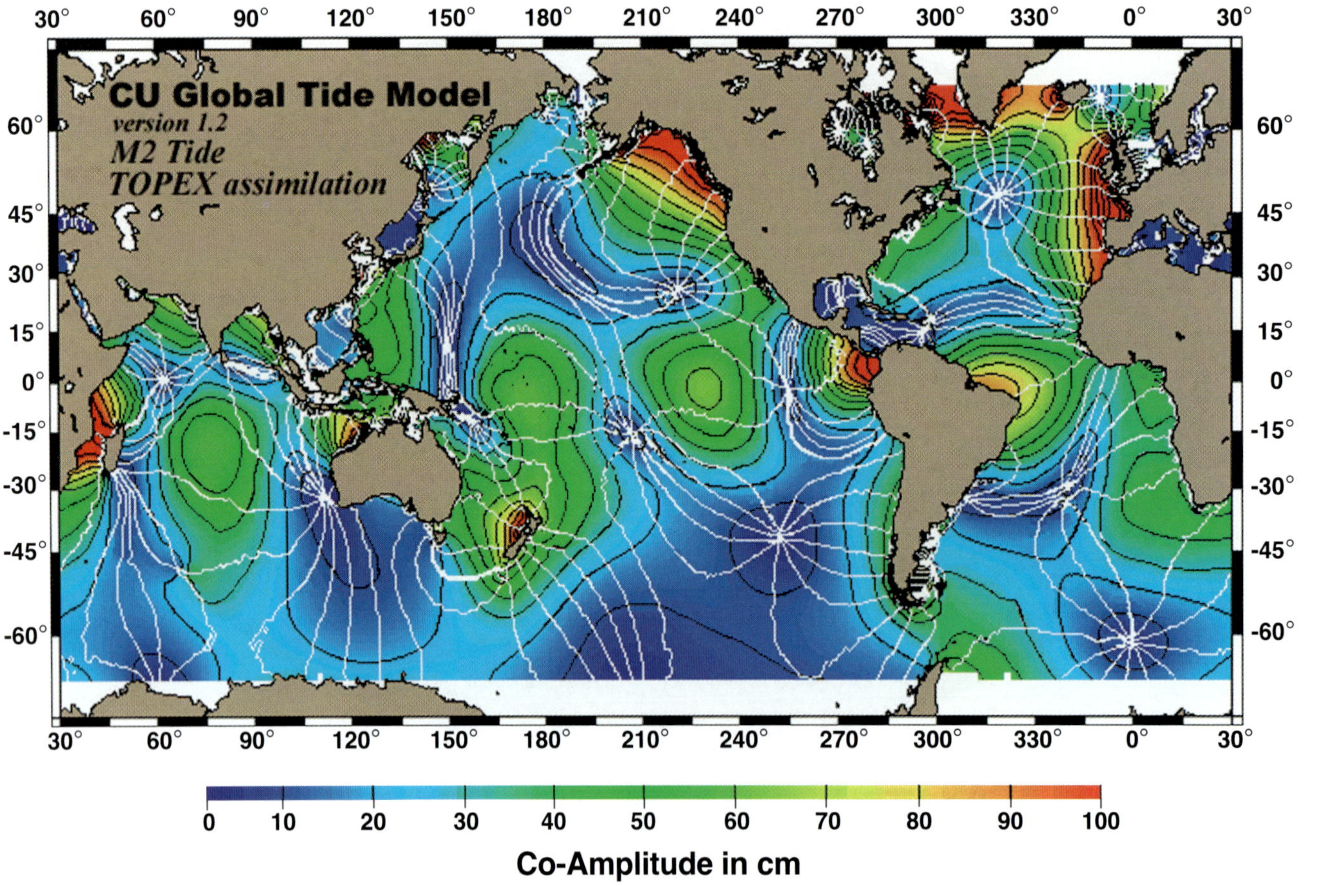

FIG. 4.—Global model of the lunar M2 tide. The chart illustrates well the equatorial band of high tides. Of particular relevance to this paper is the zone of high tides from Borneo south to Australia and east to New Guinea (from NOAA's website).

equatorial regions. Consequently, the wave climate of the Southeast Asian seas is mild, to which those who have stood on a tropical beach and watched the waves break—several inches high—can attest.

Long-term wave statistics for the globe are still incomplete, although real-time wave forecasts to mariners all over the world are now regularly accessible. One of the best global data sets was derived from radar altimeter measurements from the Geosat satellite in 1988 (Fig. 6). For the purposes of this discussion, we will assume that this was a typical year. We have chosen four seasons to cover the annual cycle of change from the northwesterly to the southeasterly monsoon. In March, near the end of the northwesterly monsoon, calm seas (blue in Fig. 6) prevail all through Southeast Asia, Oceania, and the northern Indian Ocean. In June, near the peak of the southeasterly monsoon (onshore flow to mainland South Asia), there is moderate to high wave energy in the northern Indian Ocean, yet Southeast Asia and Oceania remain calm. In September, towards the end of the southeasterly monsoon, wave heights in the Indian Ocean have decreased; Southeast Asia and Oceania remain calm. Finally, in December, near the peak of the northwesterly monsoon, Indian Ocean wave heights are reduced from earlier seasons, and most of Southeast Asia and Oceania remain calm. In fact, examining the entire map in Figure 6 reveals that Southeast Asia with Oceania is the only region in the world where seas are relatively calm all year. Only one area covered in this book, Taiwan, is affected by at least moderate waves during the year, and the "stormy" season is the Northern Hemisphere winter (December panel, Fig. 6). It is not coincidental, therefore, that the only example of a wave-dominated delta described in this book comes from Taiwan (Liu et al.), although wave influence is reported also in others.

The Indonesian Throughflow

Indonesia lies at the crossroads of the global ocean circulation system, accommodating all the "seepage" of warm tropical water from the western Pacific to the Indian Ocean (Murray and Arief, 1988; Gordon and Fine, 1996). The average discharge of this current is enormous, approximately 2 to 6 x 10^6 km^3/yr. For comparison, the mean annual discharge of the Amazon River is 0.006 x 10^6 km^3/yr (Milliman and Meade, 1983); the discharge of the Indonesian throughflow is a thousand times that of the Amazon. The current is always there, yet it varies in magnitude with the strength of the El Niño–Southern Oscillation cycle, being weaker during years of a strong El Niño. The strongest flow is focused at the level of the thermocline (about 50 to 200 m depth), where current velocity can be up to 25 cm/s. The current weakens significantly below that.

Most of the throughflow goes through Indonesia via the westernmost deepwater passage—the Makassar and Bali–Lombok straits (Fig. 7). Part of this flow is deflected eastward through the Flores and Banda Seas before it curls and exits past Timor into the Indian Ocean. Most of this water is derived from the North Equatorial Current via the Mindanao Current east of the Philippines. A secondary current, derived from the South Equatorial Current along the north shore of New Guinea, passes through the Molucca and Banda Seas east of Sulawesi. The surface currents are driven by the monsoon, and they change directions seasonally across the Java and Banda Seas (Fig. 7).

The Indonesian Throughflow clearly affects sedimentation throughout the archipelago, but it is most pronounced along the shores of the Strait of Makassar, where the asymmetry in the prodelta of the Mahakam (Allen et al., 1979; Allen and Chambers, 1998; Roberts and Sydow, this volume) is a direct consequence of this flow. The ultimate fate of much of the terrigenous clastics derived from the highlands of the Indonesian archipelago is greatly influenced by the outer-shelf and deep-water components of both the Indonesian Throughflow and strong deepwater tides, but these issues are not addressed in this book.

OUTLOOK OF DELTA RESEARCH IN SOUTHEAST ASIA

The Past

An understanding of the past sets the stage and creates a framework for understanding potential future trends in delta research. Geoscientists have been studying the sedimentology of coastal areas since ancient times. The first observations on deltas (by Herodotus) and on tidal currents in estuaries (by Ausone) were done by the ancient Romans. However, it was only in the

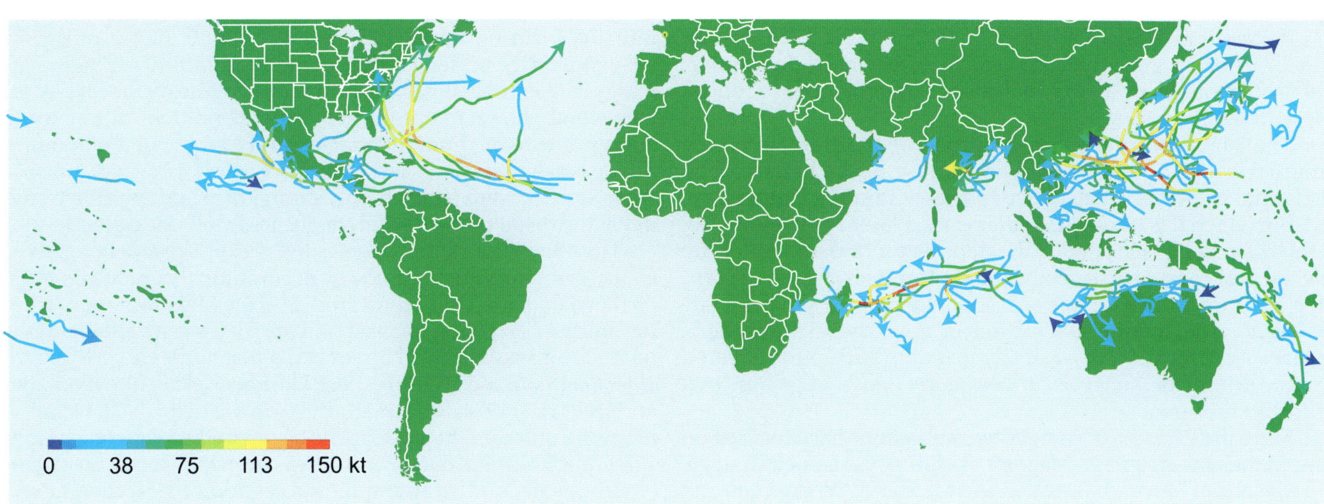

FIG. 5.—Worldwide hurricane and typhoon tracks for 1996. Note the total absence of typhoons between 6° N and 10° S in the Southeast Asia region. Maps for other years look about the same (from website http://www.solar.ifa.hawaii.edu/Tropical/GifArchitve/wld1996.gif).

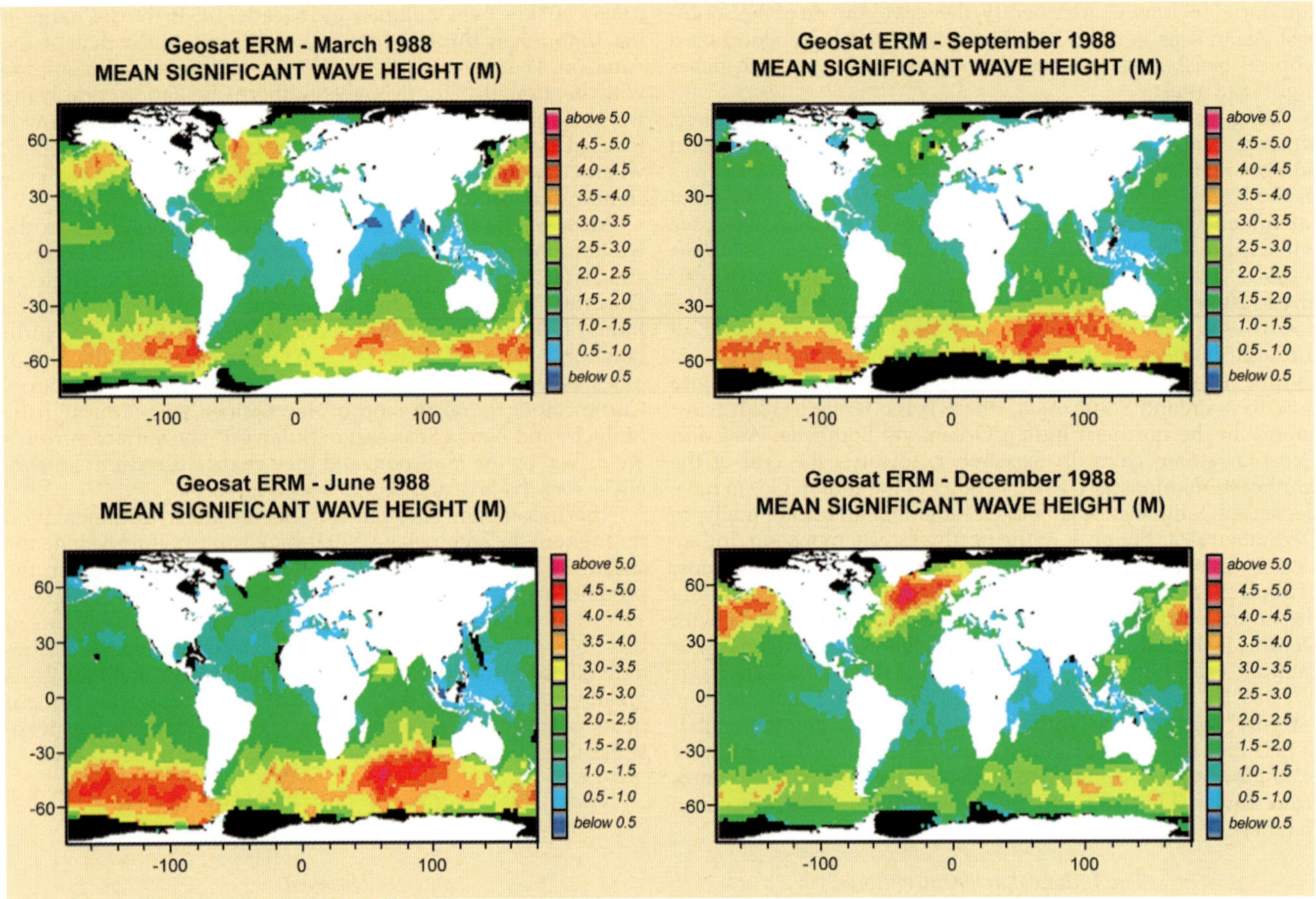

FIG. 6.—Global data on wave height from the Geosat altimeter. The figure includes four monthly average plots. Months were chosen on the basis of seasonal variations. Note that Southeast Asia is the only region on Earth where average wave heights are less than 0.5 m throughout the year. The Geosat satellite stopped transmitting data in 1989 (from website http://www.soc.soton.ac.uk/JRD/SAT/winds_waves/1989/wh_1989.html).

1960s and 1970s that geoscientists began to analyze sedimentary facies and identify dynamic sedimentary processes that focused on deltaic depositional environments (e.g., Fisk, 1961; Fisher, 1969). Prior to that, research tended to be more descriptive, focusing more on grain-size analysis and other static descriptions of sedimentary rocks.

The Mahakam delta has been one of the most intensely studied deltas in Southeast Asia because of the hydrocarbon potential of precursor deltaic deposits within the Kutai basin, within which it is located (Verdier et al., 1980). The present-day Mahakam delta is located at the depocenter of the Kutai Basin, on the east coast of Kalimantan, and comprises the most recent cycle within a thick deltaic succession deposited since the late Miocene. The unique association of an active delta system, nearby Tertiary outcrop equivalent, and prolific multi-layered reservoirs in the subsurface, has motivated a large research effort in this area.

Since the early 1970s numerous publications have focused on the Mahakam area (e.g., Magnier et al., 1975; Allen et al., 1979; Carbonel and Moyes, 1987, Allen and Mercier, 1988; Gastaldo and Huc, 1992; Debec et al., 1994; Gastaldo et al., 1993; 1995; Roberts and Sydow, 1996; Suiter, 1996; Allen and Chambers, 1998). In particular, the study of modern and Quaternary deltas in the Mahakam area benefited from the enthusiasm of the late Dr. George Allen (1942–1998), to whom this special publication is dedicated. His direct and indirect involvement in many research initiatives during the last 25 years of his life had a profound influence on the understanding of the stratigraphy and sedimentology of tropical deltas. These numerous studies of the Mahakam Delta have shown that it formed under conditions of low wave energy, low to medium tide ranges, and a large but nonflooding fluvial discharge (Allen and Chambers, 1998). However, most of these studies were focused only on the proximal portion of the delta; the distal portion still remains less well understood.

The other large delta system in the region is the Mekong River Delta, Vietnam, which was the object of study that predates work on the Mahakam. The earliest studies of the Mekong system were by Saurin (1962), who focused on Late Pleistocene alluvial sediments. This was followed by several other studies (e.g., Morgan, 1966; Gagliano and McIntire, 1968; Dornbush et al., 1969; Fontaine and Delibrias, 1974; Moorman, 1974) on the delta itself. Detailed mapping of sedimentary deposits began in the 1980s (e.g., Tran, 1980, unpublished data), with their principal focus on paleoenvironments and stratigraphic succession of Holocene mixed fluvial and tidal deltaic deposits. However, the sequence stratigraphic architecture corresponding to late Quaternary relative sea-level changes has not yet been systematically investigated. There are issues in addition to sequence stratigraphy, such as

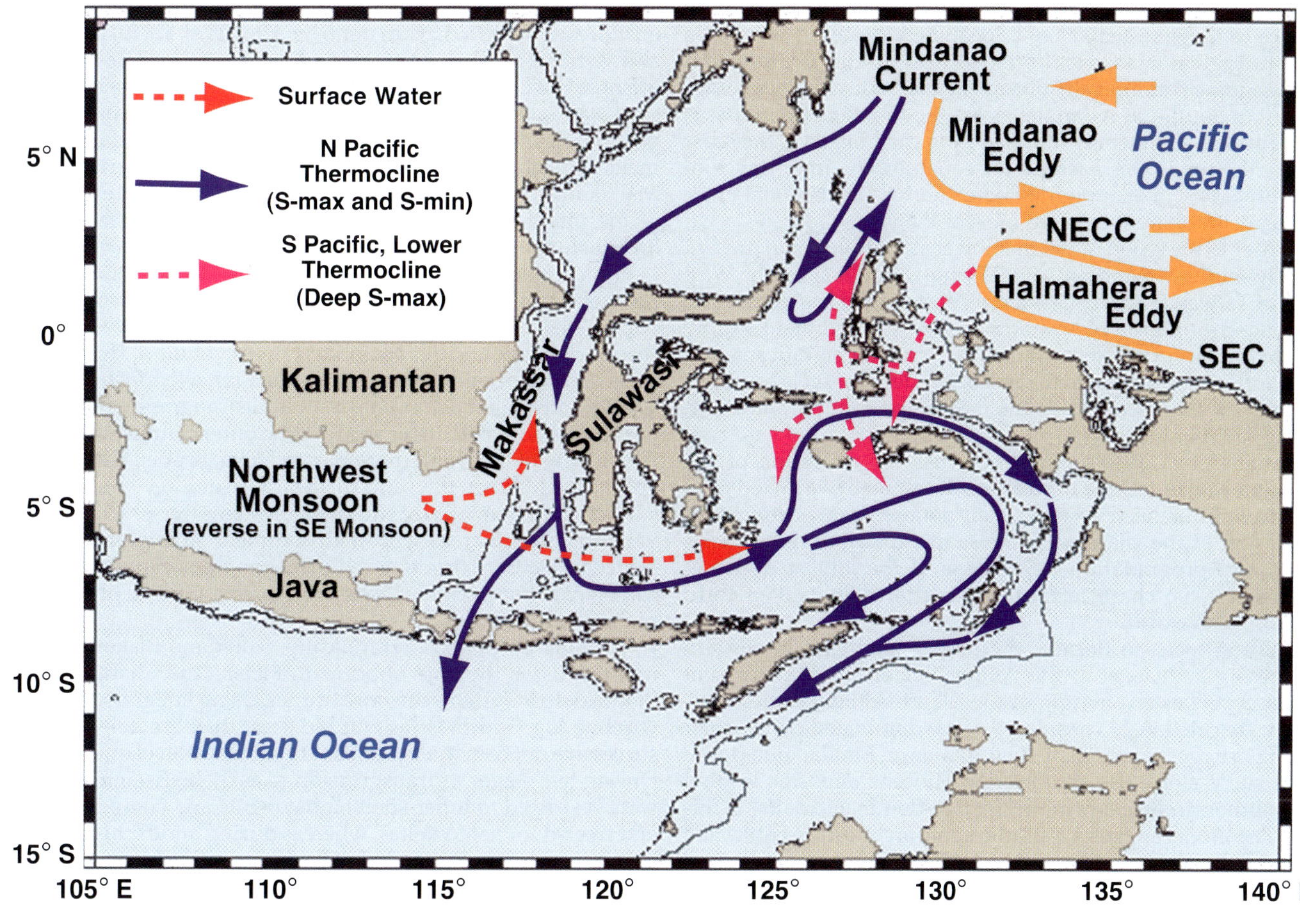

FIG. 7.—Schematic illustration of the thermohaline circulation patterns through the Indonesian seas. NECC is the North Equatorial Countercurrent, SEC is the South Equatorial Current. The Makassar Strait, which carries North Pacific water, is the dominant throughflow passway (from Gordon and Fine, 1996).

mechanisms of formation, delineation of geological boundaries, as well as the general evolution of the delta in response to sea-level changes, which remain largely unknown.

The Fly River delta in southern Papua New Guinea is the 17th largest river in the world in terms of sediment discharge, and its delta is a classic tide-dominated system (Fisher, 1969; Galloway 1975), with a funnel-shaped mouth and numerous distributaries separated by tidal bars. It lies in a foreland basin, in an area of long-term net deposition, and so is influenced minimally by preexisting bedrock topography. The low clay content in this delta is in marked contrast to the clay-rich sequences in many mesotidal systems in the Asia–Pacific region and some of the macrotidal deltas in Australia, where the rates of sediment supply and sediment reworking are much lower. From a research perspective, the Fly delta can be considered underexplored because relatively little work has been done on this system. Existing studies of the Fly have focused on such issues as rates of sedimentation and progradation (Harris et al., 1993) and dominant types of sedimentary processes (Baker et al., 1995).

The Present

More recently, there has been a trend toward a more unified vision, in which sediments are seen as an integral part of dynamic regional geological systems. This theme can be recognized in most papers presented in this volume; Figure 8 illustrates the locations of Southeast Asian deltas discussed here. The following section briefly summarizes the papers included in this volume.

Dalrymple et al. (this volume) present a comprehensive overview of the depositional processes, geomorphology, sedimentology, and stratigraphy of the Fly River delta, a tide-dominated delta on the southern margin of New Guinea. One of the regional insights they obtained is the predominance of mud-prone deposits caused by low-gradient river axis of the foreland basin. In this geological setting, the mouth-bar sands have the highest potential of forming a reservoir-prone facies, partly because of their relatively high thickness and large lateral extent (high width-to-thickness ratio) and partly because of the relative scarcity of mud baffles or barriers. However, more work is needed to understand the mouth-bar and delta-front region. The Quaternary stratigraphic architecture is characterized by delta-plain aggradation during late transgression, producing a complex stack of channel deposits, followed by mouth-bar progradation during subsequent regression.

Jones et al. (this volume) focus on deltas and estuaries in tropical northern Australia. These systems are linked to geologically mature landscapes and are characterized by complex morphological patterns that develop in response to a mixed tidal and fluvial influence. As a consequence of this complex origin, the essence of these systems is not adequately captured by simple morphological classifications, although there are some throughgoing common attributes. The major discrepancies with the rest of Southeast Asian deltas are the drainage basin that is drier and less sediment is supplied from catchments to the coast during the monsoon season. Consequently, the importance of tidal processes in shaping deltaic–estuarine depositional environments is apparent through most of the year.

Closer to the Asian mainland, Liu et al. (this volume) present a study on the wave-dominated Tseng-wen delta, on the west coast of Taiwan. They discuss a river system that has decreasing influence around the river mouth in terms of sediment supply and textural imprint. Reworking of sediment is significant, and wave influence—unlike in the rest of the deltas discussed in this volume—is considered to be the most important parameter affecting sediment dispersal.

Tanabe et al. (this volume) discuss the evolution of the Holocene Mekong Delta from a tide-dominated to a mixed tide- and wave-influenced system. Their detailed analysis includes discussion of the changes in the shape, distribution of facies tracts, and progradation rate of one of the largest deltas in Southeast Asia, characterized by a delta plain ranked third largest in the world.

Shifting focus to Borneo, the largest landmass in western Indonesia, Lambiase et al. (this volume) review the deltas present on the northwestern margin of the island. Although these systems were previously considered wave-dominated deltas, their findings suggest significant tidal influence. Similar tide-dominated successions also characterize Miocene deposits, leading these authors to suggest a model for tide-dominated deltas in this area. The model suggests a stratigraphic architecture controlled primarily by the interaction of tectonic and compaction-driven subsidence, a relatively high rate of sediment supply, and basin hydrodynamics. Their regional synthesis, based on observations from the modern as well as the ancient, resulted in a model characterized by a succession of predominantly aggradational, sandy intertidal deposits on the delta plain and a succession of predominantly aggradational delta lobes.

On the other side of the island, building upon studies of the onshore modern Mahakam delta by Allen et al. (1979) and Allen and Mercier (1994), Roberts and Sydow (this volume) focus on the offshore portion of the delta. Using high-resolution seismic and side-scan sonar, calibrated to bottom samples comprising 97 piston cores / vibracores, they developed the first detailed sedimentological and stratigraphic appraisal of the modern offshore Mahakam delta and shelf. They also present a detailed evaluation of the interaction between sea-level change and stratigraphic architecture.

To complement those studies that focus on stratigraphic architecture and sedimentary process, this volume also includes papers that specifically target the interactions between relative sea-level change and depositional system evolution. This emphasis is consistent with a trend towards a renewed focus on time stratigraphy and a de-emphasis on lithostratigraphy (Posamentier and Allen, 1999). In his study of the outer continental shelf in the vicinity of the Quaternary Baram Delta, Brunei, Hiscott (this volume) addresses the sand distribution and key stratigraphic surfaces that characterize successive systems tracts. A key observation is that in areas with high sediment supply and narrow shelves, extensive prodelta sediments and associated turbidites can blanket continental slopes even during periods of sea-level highstand.

Moving deeper stratigraphically, Saller and Blake (this volume) focus on the Late Miocene to Pleistocene Champion and Baram deltaic systems offshore Brunei. Using integrated seismic, wireline log, and biostratigraphic data, they are able to relate successive depositional systems to relative sea-level fluctuations. During late stages of transgression to early highstand, systems were restricted to inner-shelf embayments and comprised tide- and river-dominated deltas, whereas during middle to late highstand times wave-dominated deltas were far more common. The stratigraphic framework of these deltaic successions are also linked to syndepositional tectonism that affects the basin.

Staub and Gastaldo (this volume) address facies-tract distribution on the Rajan River delta, in southern Brunei. Building on earlier studies of peat development on this Quaternary system (e.g., Staub and Esterle, 1993), they demonstrate the spatial and

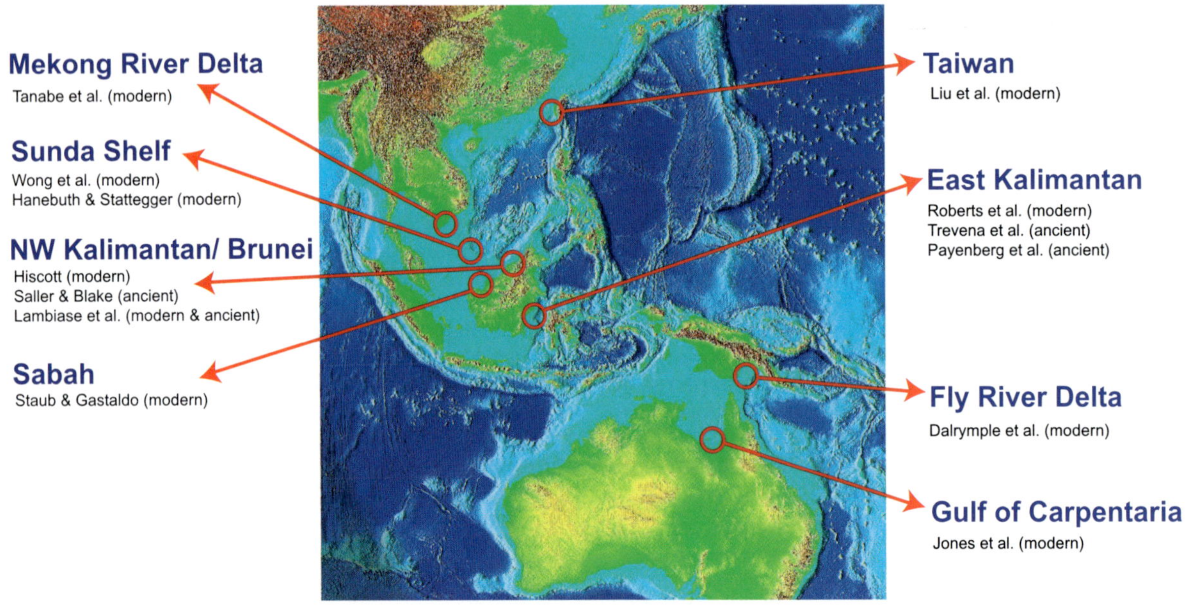

FIG. 8.—Graphical distribution of the papers presented in this volume. The source for the base map is www.fourmilab.com.

temporal relationship between thick peat deposits and siliciclastic sediments consisting of delta front/prodelta clays and silts, delta-front distributary-mouth sands, and shoreline sands.

Two papers from the Sunda shelf area focus on the stratigraphic complexity of Quaternary low-gradient shelfal successions. On the basis of 36 sediment cores Hanebuth and Stattegger (this volume) present a detailed reconstruction of the depositional succession in this area in response to sea-level changes over the past 50,000 yr. The broad Sunda shelf was not entirely emergent during the last high-glacial lowstand. No valley incision is evident within the study area, which is basically the outer shelf.

Wong et al. (this volume) complement the work of Hanebuth and Stattegger (this volume) with an analysis of high-resolution seismic profiles across the Sunda shelf. They conclude that sand ridges formed by transgressive reworking of Late Pleistocene material mark the modern water-depth range of 135–145 m, similar to what has been observed by Posamentier (2000, 2002) in the Miocene section offshore northwest Java.

In addition to studies focusing on the facies distribution and stratigraphic architecture of Quaternary and modern depositional systems, this volume also includes two studies that focus on detailed reservoir attributes. Trevena et al. (this volume) focus on detailed reservoir heterogeneity of Miocene deltaic reservoirs in offshore East Kalimantan. On the basis of observations from conventional cores and numerous wireline logs, they present a lithofacies classification for productive distributary channels and distributary-mouth bars. A key observation is that extensive burrowing within distributary-mouth-bar deposits resulting in a mixing of mud and sand may be the cause of suppressed sandstone resistivity and seemingly anomalous production characteristics. With regard to sequence stratigraphy, they suggest that the most porous and permeable hydrocarbon reservoirs accumulate in coarse-grained distributary-channel fills that formed during sea-level lowstands at a time when deltas prograded long distances across the shelves.

Onshore East Kalimantan, Payenberg et al. (this volume) present an analysis of sedimentary facies and paleocurrents based on conventional cores and electric logs (FMS). Their focus is on determining the distribution of dominant channel deposits within the Middle Miocene deltaic succession in this area. These reservoirs commonly comprise two stratigraphic units: single-story distributary channel sandstones and multistory alluvial-channel sandstones, each characterized by a distinctively different orientation. The orientation of the single-story distributary channels is determined predominantly by coeval tectonism, whereas the multistory systems likely are associated with incised valleys (Posamentier, 2001) and therefore belong to a totally different system.

In summary, these papers highlight the distinctive characteristics as well as the variety of deltaic systems that form in the tropical regions of Southeast Asia. The papers also highlight the value of an interdisciplinary approach essential for a robust understanding of such complex systems. The renewed focus on such deltas and subsequent new understandings with regard to facies models, stratigraphic architecture, sequence stratigraphy, and reservoir quality should have a profound influence on the exploration for and development of hydrocarbon reserves in this area.

The Future

In this region, the trends towards recognition of the unique attributes of tropical deltas will lead to continued advances in our understanding of how these systems work. The most effective approaches clearly will involve multidisciplinary teamwork, increased utilization of numerical models, and advanced imaging techniques. There will be a progressively greater need for more detailed understanding of stratigraphic architecture as more marginal fields are developed. The need for more accurate predictive models will be essential as exploration focus changes from frontier to mature basin.

In the petroleum business, the transition from frontier to mature basin status of many of the basins in Southeast Asia will be accompanied by the acquisition of progressively higher-resolution data as a more precise understanding of stratigraphic and lithologic relationships will be required to exploit the remaining hydrocarbon potential. 3D seismic and sophisticated borehole imagery will be vital in the development of more accurate sedimentary/stratigraphic models. In particular, recent studies have suggested that 3D seismic has the potential for leading to quantum shifts in our understanding of facies relationships. In particular, plan-view images extracted from 3D seismic volumes provide the basis for paleogeomorphic analyses (Posamentier, 2000, 2001, 2002), which, when integrated with seismic stratigraphic analyses, can yield geologic models with unparalleled detail (Fig. 9; Posamentier, 2001). Numerous examples of spectacular images derived from seismic geomorphological analyses of 3D seismic data illustrating landscapes and geometry of marginal marine deposits have been published (e.g., Brown, 1985; Hardage et al., 1996; Posamentier et al., 1998; Partyka et al., 1999; Posamentier, 2000, 2001, 2002) (Fig. 10).

The use of 3D clearly is moving beyond simple amplitude mapping towards progressively more sophisticated analyses of multiple seismic volumes and new approaches such as those based on neural network algorithms. Such seismic-based studies achieve their full potential when they are integrated with and calibrated to rock studies. It is essential to tie sedimentological insights (i.e., ground truth) to images of depositional elements extracted from 3D seismic volumes. Figure 11 is an example of a 3D amplitude extraction over a gas-bearing Miocene deltaic reservoir in West Natuna. The bright-amplitude distribution conforms to the inversion structure, but a detailed understanding of the stratigraphy and facies distribution of the upper delta plain would shed light on the potential stratigraphic compartmentalization of these deposits, which, in turn, would result in an improved reservoir model. Field development could be substantially optimized through this approach.

In our rapidly evolving understanding of tropical deltas, several specific areas for future research can be identified. Collaboration between academic and industry-based geoscientists will be essential in the future. 3D seismic data largely remains the proprietary domain of oil and gas exploration companies; advances will come from detailed analyses of such data. However, comparison with modern systems involving extensive detailed fieldwork as well as the analyses of wireline logs and core tend to lie more within the domain of academics and government geoscientists. Recently, there has been a trend for oil and gas companies to make available to academics the uppermost few hundred meters of 3D seismic volumes in light of the fact that these sections have no direct commercial value and hence are not considered proprietary. However, this trend is not yet apparent in Southeast Asia.

It is likely that new geologic concepts relative to delta development in tropical settings will emerge based on (1) the application of seismic geomorphology and stratigraphy using 3D seismic data, (2) comprehensive field work with a focus on modern and outcrop-based process sedimentology, (3) studies of temporal and spatial distribution of stratigraphic elements, and (4) studies of borehole-derived subsurface data. Clearly, the key to advancement will lie in the integration of these approaches. Sequence stratigraphic models will become more sophisticated, establishing robust relationships among the effects of tides, waves,

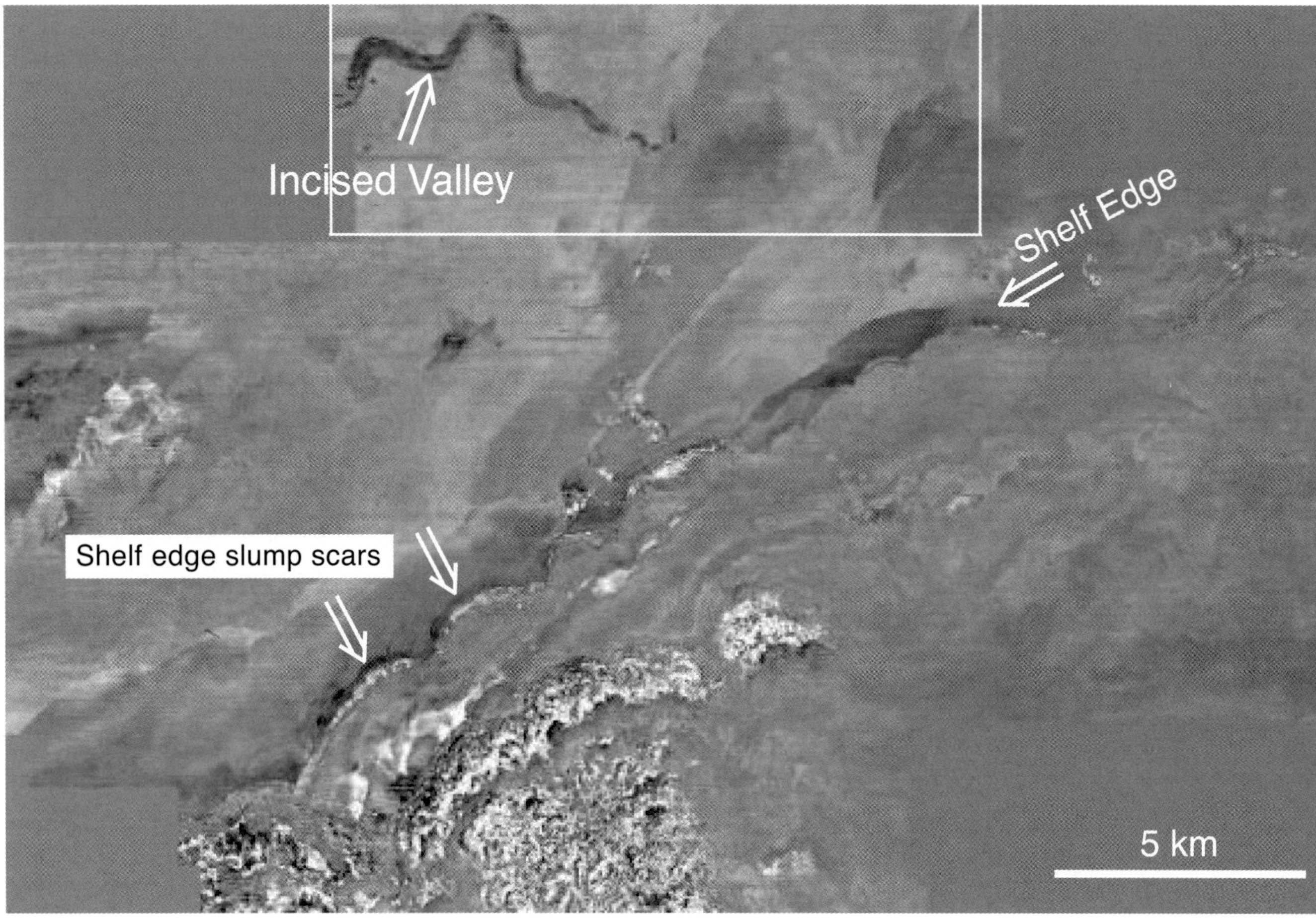

FIG. 9.—Sea floor at the Indonesian shelf edge east of Madura, Indonesia, derived from 3D seismic data. The shelf edge is characterized by numerous crescent-shaped slump scars. A small channel can be observed on the shelf near the shelf edge.

rivers, and relative sea-level change and a better understanding of the genesis of key bounding surfaces. Detailed sequence stratigraphic response models will be developed for different tectonic settings, for different types of sediment influx, and for mixed carbonate–siliciclastic systems. Reaching beyond the deltas, the dynamic linkage between shelf-edge deltas and deep-water depositional systems that lie outboard, on the slope and basin, will be better understood.

Numerical stratigraphic and sedimentological models have seen tremendous advances during the past decade. However, these models still fall short of formulating deterministic models at the scale of basins or even reservoirs. Whereas this is true of deltas in general, it is especially true of tropical deltas, which have largely been ignored in this respect. It is likely that, as both deterministic and stochastic models become more sophisticated, the modeling approach will increasingly become a standard tool for exploration and for reservoir development. Refinement and calibration of these models will require the accumulation of extensive data sets on stratigraphic geometry, dimensions of depositional element, dimensions of flow barriers, etc. These data will be derived from studies of outcrops as well as recent deposits.

Coastal and nearshore environmental and geological studies in tropical settings are still in their early development stage. Such issues as shoal management, preservation of sand dunes and reefs, and utilization of wave energy have been underappreciated and not studied adequately. Research in these areas, especially in the tropics, will go a long way towards a more comprehensive global understanding of these issues and will contribute directly to the development of coastal areas.

REFERENCES

ALLEN, G.P., AND CHAMBERS, J.L.C., 1998, Sedimentation in the modern and Miocene Mahakam delta: Indonesian Petroleum Association, Jakarta, Indonesia, 236 p.

ALLEN, G.P., LAURIER, D., AND THOUVENIN, J., 1979, Sediment distribution patterns in the modern Mahakam delta: Indonesian Petroleum Association, Jakarta, Fifth Annual Convention, Proceedings.

ALLEN G.P., LAURIER, D., AND THOUVENIN, J., 1979, Etude sédimentologique du delta de la Mahakam. Compagnie Française des Pétroles, TOTAL, Paris, Notes et Memoires, no. 15, 154 p.

ALLEN, G.P., AND MERCIER, F., 1988, Subsurface sedimentology of deltaic systems: Petroleum Exploration Society of Australia, Journal, v. 23, p. 30-44.

ALLEN, G.P., AND MERCIER, F., 1994, Reservoir facies and geometry in mixed tide and fluvial dominated delta mouth bars: Example from the modern Mahakam Delta (East Kalimantan): Indonesian Petroleum Association, 23rd Annual Convention, Proceedings, p. 261-273.

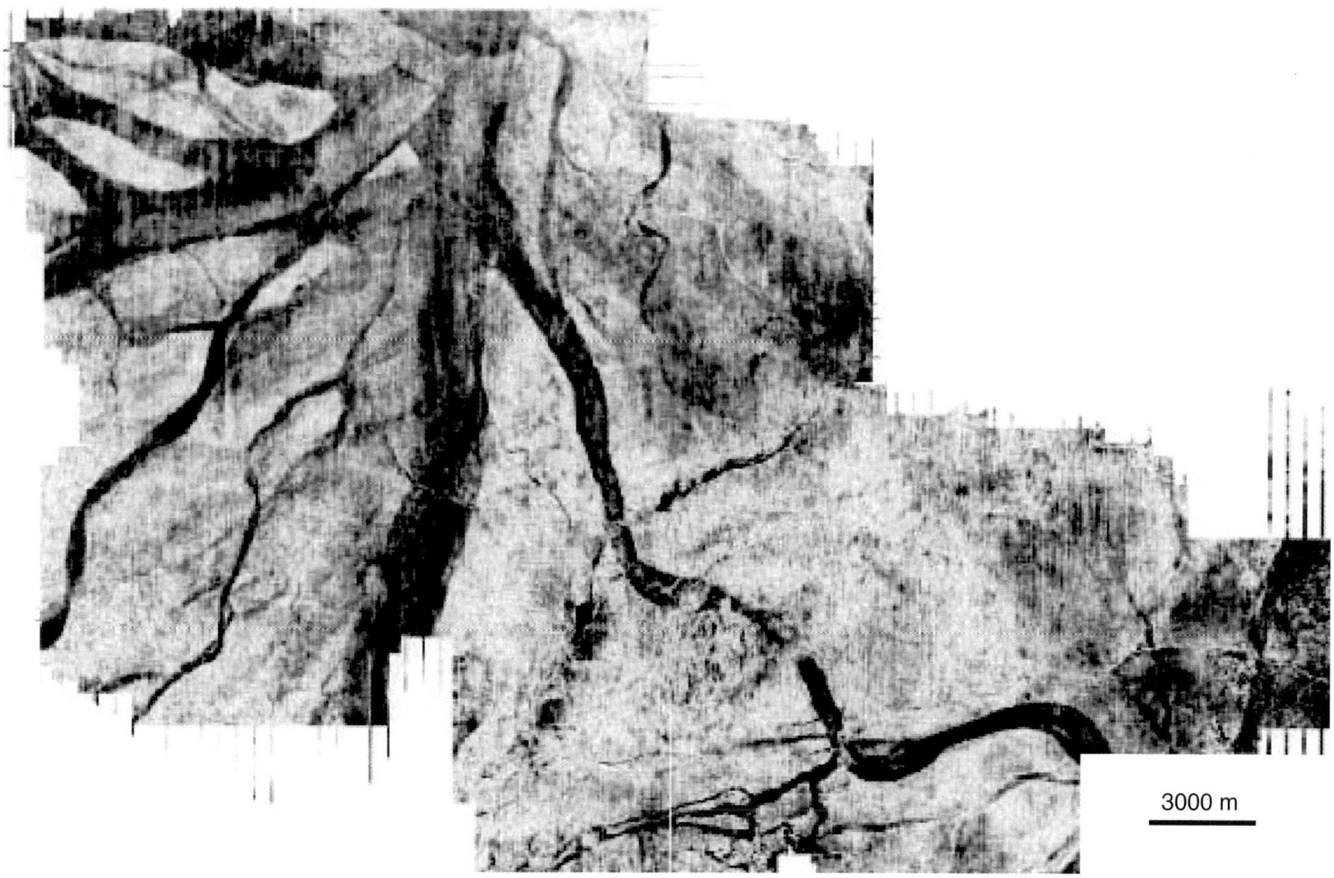

FIG. 10.—Distributary channels on a Pleistocene delta plain in the Gulf of Mexico imaged on 3D seismic data (from Partyka et al., 1999).

BAKER, E.K., HARRIS, P.T., SHORT, S.A., AND KEENE, J.B., 1995, Patterns of sedimentation in the Fly River Delta, *in* Flemming, B.W., and Bartholomä, A., eds., Tidal Signatures in Modern and Ancient Sediments: International Association of Sedimentologists, Special Publication 24, p. 193-211.

BOYD, R., DALRYMPLE, R.W., AND ZAITLIN, B.A., 1992, Classification of clastic coastal depositional environments: Sedimentary Geology, v. 80, p. 139-150.

BROWN, A.R., 1985, The role of horizontal seismic sections in stratigraphic interpretation, *in* Berg, O.R., and Woolverton, D.G., eds., Seismic Stratigraphy II: American Association of Petroleum Geologists, Memoir 39, p. 37-48.

CARBONEL, P., AND MOYES, J., 1987, Late Quaternary paleoenvironments of the Mahakam Delta (Kalimantan, Indonesia): Palaeogeography, Palaeoclimatology, Palaeoecology, v. 61, p. 265-284.

DAVIS, R.A., JR., AND HAYES, M.O., 1984, What is a wave-dominated coast?: Marine Geology, v. 60, p. 313-330.

DEBEC, P., BITOUN, G., AND MERCIER, F., 1994, The research of analogues in 3D seismic data: A tool for sedimentology and reservoir engineering: Indonesian Petroleum Association, 23rd Annual Convention, Proceedings, p. 275-288.

DEFANT, A., 1958, Ebb and Flow; The Tides of Earth, Air, and Water: The University of Michigan Press, Ann Arbor, 121 p.

DORNBUSH, W.K., MAY, J.R., AND COWEY, W.P., 1969, Distribution of coarse grained construction materials and potential construction sites in the Mekong delta, South Vietnam: US Corporation Engineering Technologic Report 69, Washington, D.C., p. 7.

FISHER, W.L., 1969, Facies characterisation of Gulf Coast Basin delta systems, with some Holocene analogs: Gulf Coast Association of Geological Societies, Transactions, v. 19, p. 239-261.

FISK, H.N., 1961, Bar-finger Sands of the Mississippi Delta, *in* Peterson, J.A., and Osmond, J.C., eds., Geometry of Sandstone Bodies: American Association of Petroleum Geologists, Tulsa, p. 29-52.

FONTAINE, H., AND DELIBRIAS, G., 1974. Anciens niveaux marins du Quaternaire au Vietnam: Archives Geologique Du Vietnam, Saigon, Vietnam, v. 17, p. 35-44.

GAGLIANO, S.M., AND MCINTIRE, M.G., 1968, Report on the Mekong Delta: Louisiana State University, Baton Rouge, Coastal Studies Institute, Technical Report 57, 143 p.

GALLOWAY, W.E., 1975, Process framework for describing the morphologic and stratigraphic evolution of deltaic depositional systems, *in* Broussard, M.L., ed., Deltas; Models for Exploration: Houston Geological Society, Houston, p. 87-98.

GASTALDO, R.A., ALLEN, G.P., AND HUC., A.Y., 1993, Detrital peat formation in the tropical Mahakam River delta, Kalimantan, eastern Borneo: Formation, plant composition, and geochemistry, *in* Cobb, J.C., and Cecil, C.B., eds., Modern and Ancient Coal-forming Environments: Geological Society of America, Special Paper 286, p. 107-118.

GASTALDO, R.A., ALLEN, G.P., AND HUC, A.Y., 1995, The tidal character of fluvial sediments of the modern Mahakam River Delta, Kalimantan, Indonesia, *in* Flemming, B.W., and Bartholomä, A., eds., Tidal Signatures in Modern and Ancient Sediments: International Association of Sedimentologists, Special Publication 24, p. 171-181.

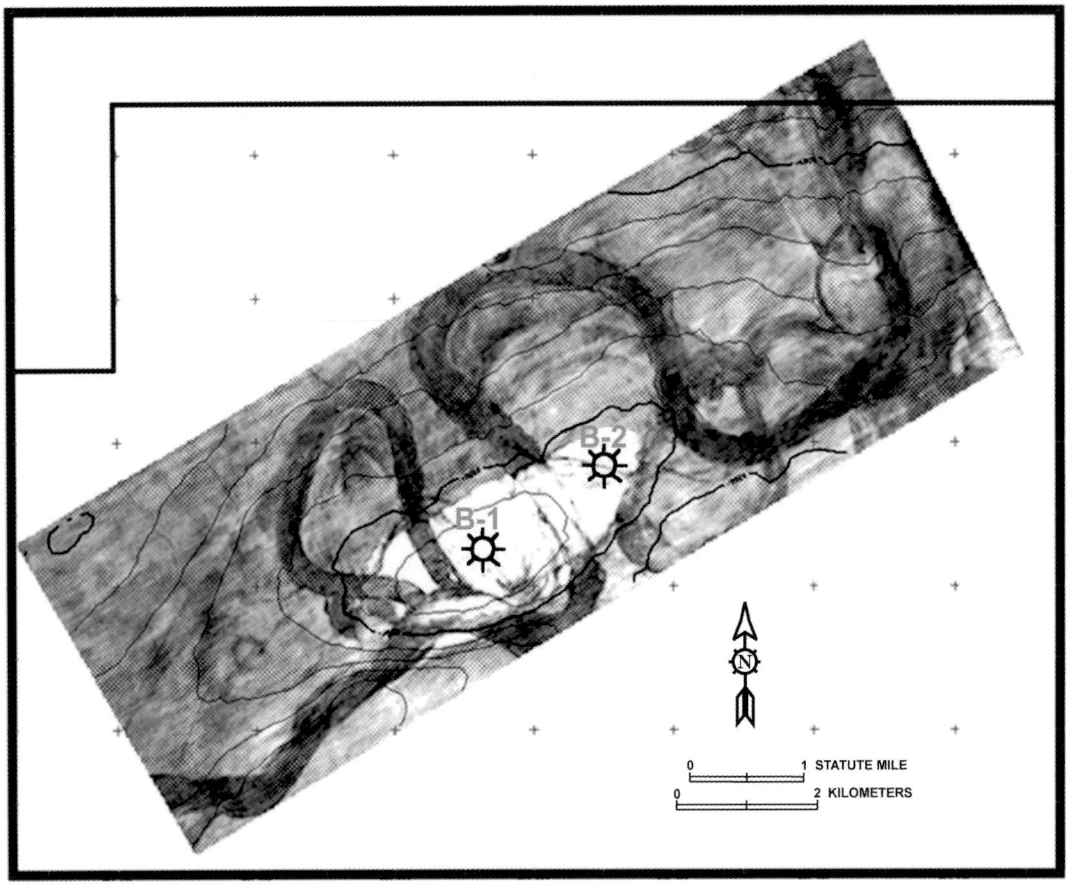

FIG. 11.—3D horizon slice over Miocene Arang reservoir in Bawal field, West Natuna overlaid with structural contours. In addition to the bright amplitude indicating the presence of shallow hydrocarbons, the images also provide a stunning view of the fluvial to upper delta-plain environment. Analyses of such detailed images provide great insights for the study of modern deltaic systems. Moreover, such studies provide templates applicable to the study of ancient deeply buried systems.

GASTALDO, R.A., AND HUC, A.Y., 1992, Sediment facies, depositional environments, and distribution of phytoclasts in the recent Mahakam River Delta, Kalimantan, Indonesia: Palaios, v. 7, p. 574-590.

GORDON, A., AND FINE, R.A., 1996, Pathways of water between the Pacific and Indian oceans in the Indonesian Seas: Nature, v. 379, p. 146-149.

HARDAGE, B.A., LEVEY, R.A., PENDLETON, V., SIMMONS, J., AND EDSON, R., 1996, 3-D Seismic imaging and interpretation of fluvially deposited thin-bed reservoirs, *in* Weimer, P., and Davis, T.L., eds., Applications of 3-D Seismic Data to Exploration and Production: American Association of Petroleum Geologists, Studies in Geology 42, p. 27-34.

HARRIS, P.T., BAKER, E.K., COLE, A.R., AND SHORT, S.A., 1993, Preliminary study of sedimentation in the tidally dominated Fly River Delta, Gulf of Papua: Continental Shelf Research, v. 13, p. 441-472.

MACKINNON, K., HATTA, G., HALIM, H., AND MANGALIK, A., 1996, The Ecology of Kalimantan: Periplus Editions (HK) Ltd., Dalhousie University, Halifax, Nova Scotia, 802 p.

MAGNIER, P., OKI, T., AND KARTAADIPUTRA, L., 1975, The Mahakam Delta, Kalimantan, Indonesia: 9th World Petroleum Congress, Proceedings, p. 239-250.

MILLIMAN, J.D., FARNSWORTH, K.L., AND ALBERTIN, C.S., 1999, Flux and fate of fluvial sediments leaving large islands in the East Indies: Journal of Sea Research, v. 41, p. 97-107.

MILLIMAN, J.D., AND MEADE, R.H., 1983, World-wide delivery of river sediments to the oceans: Journal of Geology, v. 91, p. 1-21

MILLIMAN, J.D., AND SYVITSKI, J.P.M., 1992, Geomorphic/tectonic control of sediment discharge to the ocean: The importance of small mountainous rivers: Journal of Geology, v. 100, p. 525-544.

MOORMAN, P., 1974, Characteristics of mangrove soils in relation to their agriculture land use and potential: International Symposium on Biology and Management of Mangrove, Honolulu, Proceedings (special issue), p. 529-547.

MORGAN, J.P., 1966, Mekong river delta: distribution of physical environments by aerial photographs: Louisiana State University, Baton Rouge, Coastal Studies Institute, 11 p.

MURRAY, S., AND ARIEF, D., 1988, Throughflow into the Indian Ocean through the Lombok Strait, January 1985–January 1986: Nature, v. 333, p. 444-447.

MURRAY, S.P., ROBERTS, H.H., AND YOUNG, M.H., 1988, Control of terrigenous–carbonate facies transitions by baroclinic coastal currents—Nicaragua, *in* Doyle, L.J., and Roberts, H.H., eds., Carbonate–Clastic Transitions: Amsterdam, Elsevier, Developments in Sedimentology, v. 42, p. 289-304.

NUMMEDAL, D., AND FISCHER, I.A., 1979, Process–response model for depositional shorelines: the German and the Georgia bights: New York, American Society of Civil Engineers, 16th Coastal Engineering Conference, Proceedings, p. 1215-1225.

PARTYKA, G, GRIDLEY, J., AND LOPEZ, J., 1999, Interpretational applications of spectral decomposition in reservoir characterization: The Leading Edge, v. 18, p. 353-360.

POSAMENTIER, H.W., 2000, Seismic and lithologic expression of Miocene shelf ridges on the South Java Sea Shelf (abstract): Proceedings, American Association of Petroleum Geologists, Annual Meeting, New Orleans, April 16–19, 2000, p. A118.

POSAMENTIER, H.W., 2001, Lowstand alluvial bypass systems: incised vs. unincised: American Association of Petroleum Geologists, Bulletin, v. 85, p. 1771-1793.

POSAMENTIER, H.W., 2002, Ancient shelf ridges—a potentially significant component of the transgressive systems tract: case study from offshore Northwest Java: American Association of Petroleum Geologists, Bulletin, v. 86, p. 75-106.

POSAMENTIER, H.W., AND ALLEN, G.P., 1999, Siliciclastic Sequence Stratigraphy: Concepts and Applications: SEPM, Concepts in Sedimentology and Paleontology, no. 9, 210 p.

POSAMENTIER, H.W., SUYENAGA, W., RUFAIDA, D., MEYRICK, R., AND PEMBERTON, S.G., 1998, Stratigraphic analysis of the Main member of the Upper Cibulakan Formation at E Field, offshore northwest Java, Indonesia: Indonesian Petroleum Association, 26th Annual Convention, Proceedings, p. 69-93.

ROBERTS, H.H., AND SYDOW, J. 1996. The offshore Mahakam delta: Stratigraphic response of Late Pleistocene to Modern sea level cycle: Indonesian Petroleum Association, 25th Annual Convention, Proceedings, p. 147-161.

SAURIN, E., 1962, Carte géologique Vietnam, Cambodge, Laos' au 500,000e, feuille de VinhLong. Dalat 20, p. 49.

STAUB, J.R., AND ESTERLE, J.S., 1993, Provenance and sediment dispersal in the Rajang River delta / coastal plain system, Sarawak, East Malaysia, *in* Fielding, C.R., ed., Current Research in Fluvial Sedimentology: Amsterdam, Elsevier, p. 191-201.

SUITER, J.S., 1996, Shallow 3-D seismic analysis of Late Pleistocene lowstand deltas (Mahakam, Indonesia): Indonesian Petroleum Association, 25th Annual Convention, Proceedings, p. 348-351.

Verdier, A.C., Oki, T., and Suardy, A., 1980, Geology of the Handil Field (East Kalimantan, Indonesia, *in* Halbouty, M.T., ed., Giant Oil and Gas Fields of the Decade 1968–1978: American Association of Petroleum Geologists, Tulsa, p. 399–422.

Quaternary Systems

DELTAS IN THE GULF OF CARPENTARIA, AUSTRALIA: FORMS, PROCESSES, AND PRODUCTS

BRIAN G. JONES AND COLIN D. WOODROFFE
School of Geosciences, University of Wollongong, Wollongong, New South Wales 2522, Australia
briangj@uow.edu.au; colin_woodroffe@uow.edu.au
AND
GLENN R. MARTIN
School of Geosciences, University of Wollongong, New South Wales 2522, Australia
Present address: 16/27-29 Dutton St., Coolongatta, Queensland 4225, Australia

ABSTRACT: Deltas in the shallow marine epicontinental Gulf of Carpentaria have developed in the slowly subsiding Karumba Basin and are influenced by monsoonal fluvial discharge and a diurnal tidal regime. The McArthur delta is protected from wave action by offshore islands, and the restricted width of the delta is a function of bedrock outcrops near the delta mouth. The upper delta plain is characterized by fluvial lateral-accretion point-bar deposits. In the lower delta plain, progradation has resulted in a buildup of shelly delta-front sands overlain by muddy intertidal and supratidal deposits. The latter dry mudflats are areas of wind deflation and may equate to emergent surfaces in older analogues. Fluvially active distributary channels have a relatively uniform width whereas abandoned channels adopt a tapering tide-dominated form. This delta shows progressive influence of riverine processes downstream onto the lower delta plain. Although the tide-dominated channels have a high sinuosity, their patterns reflect former fluvial channels.

The Gilbert River, by contrast, is not bedrock-controlled and contains a prograding coastal wedge of Holocene sediment extending laterally for a distance of 125 km along the coast. The subaerial portion of the delta has prograded 15–20 km during the past 6,500 years, and the subsurface facies show prodelta mud overlain successively by delta-front and subtidal sands, strandline beach and chenier ridges, and intertidal to supratidal mudflats. Thin floodplain deposits cover the inner portion of the delta. Changes in the locus of sedimentation result from fluvial avulsions and account for age variations in the local subtidal sand and beach-ridge accumulations.

Deltas in the northern Australian region are not adequately defined by simple morphological classifications. They drain from geologically mature landscapes and illustrate complex morphological patterns that develop in response to specific tidal and fluvial regimes. No major incision would have occurred around the Gulf of Carpentaria during the last low stand of sea level because the onshore and offshore gradients are equivalent, and the Holocene deposits have built out as temporary wedges of sediment (up to 30 km wide and 10 m thick) adjacent to the present shoreline.

INTRODUCTION

Deltas form where rivers discharge into the sea. Detailed studies of the Mississippi River and delta plain have served to demonstrate the important role that the river, and the sediment load that it carries, play in shaping delta morphology (Wright and Coleman, 1973; Coleman, 1988). The Mississippi is an example of a highly river-dominated delta, draining a large catchment into a small basin in which tides and wave energy are minor. While models based upon the Mississippi have advanced the interpretation of modern deltas and the search for hydrocarbons in ancient deltas, elsewhere in the world deltas are influenced by a series of additional factors. In particular, it has proved useful to view deltas within the framework of a ternary diagram in relation to the proportional role of river, tide, and wave processes (Wright et al., 1974; Galloway, 1975).

The importance of tidal processes in determining delta morphology has been demonstrated for a number of deltas and estuarine systems in the Asia–Pacific region. For instance, the Ord River, in Western Australia, has been cited as an example of a system that is highly tide-dominated (Wright et al., 1973; Coleman and Wright, 1978), and the Klang River in Malaysia is also considered a classic example of tide domination (Coleman et al., 1970). The Fly River delta in southern Papua New Guinea is another classic tide-dominated delta with wide tapering channels, rapid sedimentation, and rapid seaward progradation of about 6 m/yr (Harris et al., 1993; Baker et al., 1995). This macrotidal delta is characterized by high current velocities, abundant suspended sediment load, and low salinity, which result in a sand-dominated sedimentary succession with minor silt deposition and very little clay (Baker et al., 1995). The low clay content in this delta is in marked contrast to the clay-rich sequences in many mesotidal systems in the Asia–Pacific region and some of the macrotidal deltas in northwestern Australia, where the rates of terrigenous sediment supply and sediment reworking are much lower. The rapid tapering of width with distance upstream and linear sandy shoals within the channel are diagnostic features of tide-dominated estuaries and deltas (Bhattacharya and Walker, 1992; Walker, 1992; Woodroffe, 2003).

Deltas in the Asia–Pacific region occur in a center of diversity in shallow-water marine biota and coastal wetland vegetation. Shallow-water environments tend, especially where they are muddy, to be covered in seagrass. Intertidal zones support productive mangrove forests, and a range of supratidal wetland communities are found across emergent coastal and deltaic–estuarine plains. The intertidal mangrove communities show a broad pattern of zonation that is reproduced across the region (Macnae, 1968; Woodroffe, 2003). The vegetation to landward of the mangroves, however, varies, controlled primarily by a regional climatic trend. In semiarid areas of Western Australia (i.e., the Ord and Fitzroy Rivers) extensive saline flats bare of vegetation are characteristic, and the only vegetation consists of low shrubs and herbs, such as samphire (Jennings, 1975; Thom et al., 1975). Broad convex alluvial plains, covered by seasonally inundated grass and sedgelands, are common in the wet–dry tropics of northern Australia (Woodroffe et al., 1989; Woodroffe et al., 1993). In the perhumid tropics of much of Malaysia and Indonesia, similar plains support domed accumulations of woody peat beneath peat-swamp forest (Anderson, 1964; Gastaldo et al., 1993; Staub and Esterle, 1994).

Tropical Deltas of Southeast Asia—Sedimentology, Stratigraphy, and Petroleum Geology
SEPM Special Publication No. 76, Copyright © 2003
SEPM (Society for Sedimentary Geology), ISBN 1-56576-086-7, p. 21–43.

The Mahakam delta shows a vegetation zonation from intertidal mangrove forests, through brackish communities of *Nipa fruticans* and *Heritiera littoralis,* to inland peat-swamp forest, and demonstrates that both river and tidal processes can dominate a delta. It is a fan-shaped lobate delta with a series of distributaries, discharging into the Makassar Sea, which has a tidal range of up to 3 m and relatively low wave energy. Some of the distributaries are active, especially in the northeast and southeast portions of the delta, carrying much of the fluvial discharge through relatively straight but actively migrating channels with lateral-accretion sandy point bars (Allen et al., 1979). However, even in these regions the fluvial channels are tide-affected and are characterized by the presence of mud drapes and sand–mud couplets (Gastaldo et al., 1995). Other distributaries, in the fluvially abandoned central section of the delta, comprise intricately meandering tide-dominated channels, which taper rapidly with distance inland and are filled with plugs of fluid mud (Allen et al., 1979).

A still clearer example of the geographical separation of dominance by river, tide, and wave processes occurs in the Red River delta. A river-dominated sector occurs in the upper part of the delta; the northern sector is tide-dominated, and the southern sector is wave-dominated (Mathers and Zaasiewicz, 1999). In contrast, the lower Mekong delta is tide dominated with wave-generated beach ridges between each of the long tapering tidal channels (e.g., Gagliano and McIntire, 1968; Morgan, 1970; Nguyen et al., 2000). The Mekong delta sequence is relatively thin (generally 60–100 m thick) and overlies bedrock, and therefore it tends to prograde more rapidly than other subsiding delta regions where vertical stacking of deposits is the rule.

Numerous delta styles are present throughout the Asia–Pacific region, and the complexity of their form and function is not adequately accommodated in simple morphological classifications (Wright, 1989). River processes are important in many of the deltas in the Indo–West Pacific region. On the one hand, very large sediment loads are brought to the coast by those large rivers draining tectonically uplifted continental regions with monsoonal climates (Allison et al., 1998; Milliman and Meade, 1983). On the other hand, the substantial role of rivers draining young tectonically active high islands, along plate margins, has also recently been realized (Milliman and Syvitski, 1992; Chappell, 1993; Hoekstra, 1993; Baker et al., 1995).

In contrast, the river systems of northern Australia fall in the same biogeographic province as the Southeast Asian deltas, but they drain from a geologically mature landscape. The annual rainfall is generally lower, and is significantly monsoonal. In these areas the fluvial processes operate in conjunction with significant tidal processes. A range of environments occurs along the northern Australian coast. In this paper we examine deltas that drain into a predominantly diurnal epicontinental sea, the Gulf of Carpentaria (Fig. 1), and illustrate the complex morphological patterns that develop through time and in space in response to variable tidal and fluvial regimes.

This paper looks primarily at the subaerial delta, that part of the deltaic–estuarine system above sea level, with less consideration for the generally obscured subaqueous delta. A river channel that seasonally is an important source of water and sediment input traverses the subaerial delta. The inland limit of saline water intrusion or tidal fluctuations marks the division between the alluvial plain and the delta plain. The delta plain can generally be divided into an active delta plain containing the major distributaries, and an abandoned delta plain in which former distributaries, preserved as paleochannels, carry only a very small proportion of the flow. In addition an upper and lower delta plain can usually also be recognized. The upper delta plain is generally influenced primarily by the river but is subjected to tidal oscillations and contains a subsurface salt-water wedge. The sediments are predominantly alluvial, typically including channel, levee, and floodplain deposits of the river as it has migrated across or avulsed in its lower course. The lower delta plain is subject to influence from marine processes and brackish water, and mangroves are characteristic vegetation.

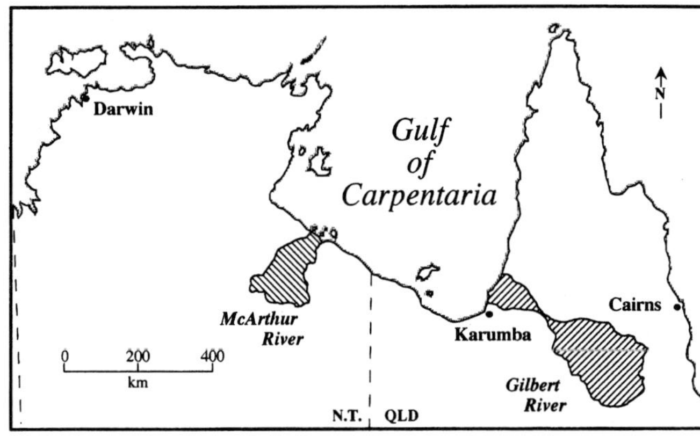

FIG. 1.—The Gulf of Carpentaria in northeastern Australia, showing catchments of the McArthur River (west) and Gilbert River (east).

It has become increasingly difficult to accommodate tidally dominated estuaries with tidally dominated deltas in any classification based simply on morphology, particularly where highly seasonal, monsoonal rivers may be river-dominated in terms of process for only a short period of the year. Deltas can be at different stages of development in time and across space. In many of these deltas, individual delta lobes go through an autocyclic change, with occupation, growth, diversion, and abandonment, marking cycles of river domination, and either wave or tide domination.

McARTHUR RIVER DELTA

The McArthur River (Fig. 1) rises on the Upper Proterozoic sandstone and siltstone of the Barkly Tableland, and is entrenched across the dissected mesa topography controlled by the flat-lying or gently warped plateau of Lower Cambrian Bukalara Sandstone. The lower part of the catchment is composed of moderately folded Lower Proterozoic sandstone and dolomite, with discontinuous cover of Cretaceous sandstone and Tertiary laterite. Extensive Quaternary alluvial sandy ridges flank the river channel in its lower course. It flows across a Holocene delta plain for the last 25 km of its course, and in these lower reaches the delta is topographically restricted to a zone approximately 25 km wide. This ensures frequent reworking of the sedimentary succession during delta channel migration. The delta is also protected from major wave activity by the presence of offshore islands.

The McArthur River has formed a delta, traversed by numerous meandering distributaries, in contrast to the single channel of adjacent rivers such as the Roper or Gregory Rivers. The complex delta may have formed as a result of the partial confluence of the McArthur River and smaller Batten Creek (Fig. 2). The greater part of the flow in these two channels continues seaward separately, Batten Creek as the relatively straight Carrington Channel (sinuosity 1.1) and the McArthur River as a meandering channel (sinuosity 1.3) discharging at Cora Point (Fig. 2). A narrow channel does join the two at the confluence, marking the apex of

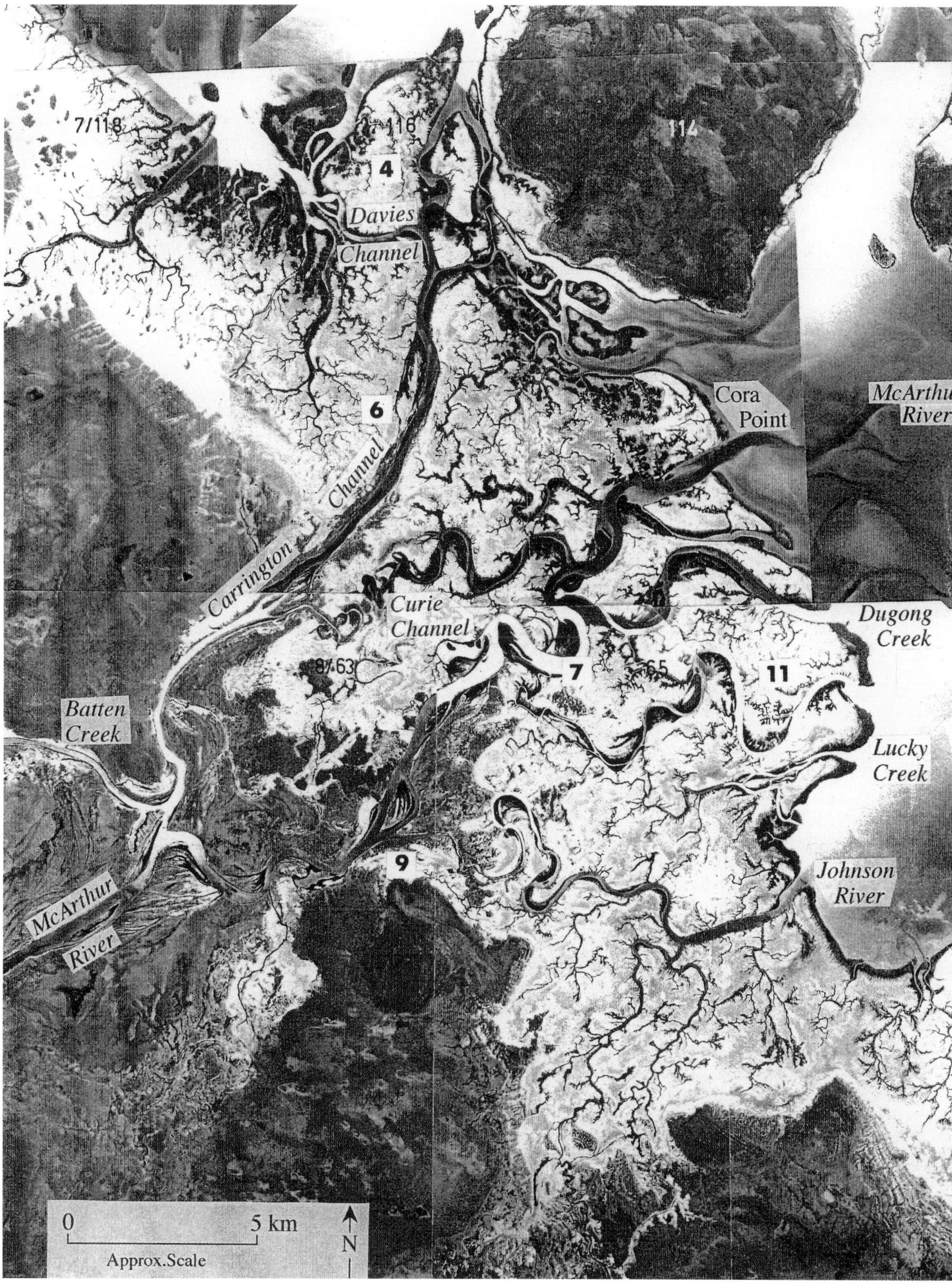

FIG. 2.—Aerial photomosaic of the McArthur River delta showing geographic localities referred to in the text (part of the Pellew [6226] 1:100,000 photomosaic series, copyright with the Commonwealth of Australia, and reproduced by permission of the General Manager, Australian Surveying and Land Information Group (AUSLIG), Department of Industry, Science and Resources, Canberra, ACT).

the delta. Smaller channels, such as Curie Channel, indicate past connections between the two distributaries.

In addition, the offshore Sir Edward Pellew group of islands, formed around Proterozoic rocks, has served as a control on the progradation of the coastal plain. These islands limit the wave energy that reaches the delta; they impede river discharge, accentuating the unusual development of this complex delta.

The McArthur River drains a catchment of around 18,400 km^2 (Fig. 1), much of which is dominated by open eucalypt woodland. It receives an annual rainfall of around 600 mm, with a pronounced wet season from December to March. The river drains into the Gulf of Carpentaria, a shallow epicontinental sea that is mainly between 20 and 60 m deep, and in which evaporation exceeds rainfall and runoff. Tides are complex in the Gulf of Carpentaria, with a pronounced seiche and a diurnal component seen in the tides. Tides penetrate up river to Borroloola. At Centre Island in the Pellew Group, there is a maximum tidal range of around 3 m, but a mean range of only 80 cm, and water levels can show a seasonal increase of elevation of up to 60 cm in the wet season (Forbes and Church, 1983).

Morphology of the McArthur Delta Plain

The subaerial delta plain can be divided into an upper delta plain, dominated by a sandy channel within a scroll plain of fluvial sandy ridges marking the positions of earlier point-bar deposits, and a lower delta plain with numerous distributaries incised through supratidal saline mudflats and mangroves. Seaward of the subaerial delta is a subaqueous delta comprising a low-gradient delta front of shelly sands, across which active distributaries continue as perceptible but shallow troughs flanked by subtidal or intertidal levees of well-sorted sand.

Observations of the delta were made during several fieldtrips and from aerial photographic coverage. Elevations across the plains were determined relative to a high-tide datum (HTL), which approximates the highest astronomical tide, inferred from the back of the mangroves. Along Carrington Channel it was possible to relate this to sparse benchmarks reduced to mean sea level (Australian Height Datum, AHD).

A detailed stratigraphy of the subaerial delta has been reconstructed on the basis of riverbank exposures, augering with a portable Jacro 100 drill rig, shallow pits, and tube-in-tube cores (Woodroffe and Chappell, 1993). Radiocarbon dating, particularly of organic peat or root material attributed to mangroves and bivalves from shell beds, has enabled a reconstruction of the mid to late Holocene evolution of the delta. A transgressive phase, as sea level flooded the pre-Holocene topography, can be inferred. However, most of the delta is considered to have developed through a stillstand and regressive phase, with exposure of a former delta front resulting from a gradual fall of sea level from around 1.5 m above present to its present level (Woodroffe and Chappell, 1993). This account builds upon the earlier stratigraphic data to relate elements of the geomorphology of the delta to the processes that have been, or continue to be, operative.

Geological inheritance has exerted the primary control on the basin within which the delta has formed through the prior bedrock topography. Rock outcrops constrain both Carrington Channel and the southern distributary Johnson River, in terms of their lateral migration. Drilling has encountered highly oxidized orange to red mottled firm gray Pleistocene clayey sand beneath the wedge of Holocene sediments. This occurs as a nearly horizontal terrace around 2 m below HTL under much of the upper delta plain. It was also found at depths of 6–7 m in the upper part of the lower delta plain, but it was not encountered in the top 10 m in an auger hole at Cora Point, near the mouth of the active delta.

Morphology, Depositional Environments, and Sedimentary Units

Upper Delta Plain.—

The upper delta plain comprises a series of nested and coalescing point-bar ridges of pale brown, well-sorted medium sand. These flank the present channel and mark former channel positions, indicating fluvial dominance in this part of the delta. They are covered with grass and open eucalypt woodland, and termite mounds are a prominent feature on the crests of the ridges. Elsewhere along the riverbanks in the upper delta plain laminated overbank successions of peat and mud are found. The highly organic peat is composed of black and reddish compressed woody fragments, and it is interbedded with organic-rich mud. In many places the mud contains nodular, ferruginous or limonitic casts of rhizomorphs up to a centimeter in diameter, protruding erect several centimeters above the mud surface. These are interpreted as casts of the pneumatophores of *Avicennia*, which is widespread amongst the mangroves of the present McArthur River, and further substantiate the mangrove origin of the organic matter in these laminated successions during the higher stand of sea level between 6,500 and 3,000 yr BP.

The upper delta plain can be interpreted as having been frequently, though incompletely, reworked by the lateral migration of the meandering river channels and distributaries. The nested scroll bars contain within them interbedded peat and mud layers. Radiocarbon dating of material exposed along the banks of the McArthur River channel at two close sites just downstream of the confluence with Batten Creek yielded ages of 6,140 ± 100 and 1,720 ± 100 yr BP (all radiocarbon dates are reported as conventional ages). These dates suggest that this apex of the delta has been characterized by such lateral migration, and reworking of the plain, throughout the Holocene evolution of the rest of the delta.

Lower Delta Plain.—

The lower delta plain consists of distributaries, littoral mangrove forests, and extensive saline mudflats, which are generally bare of vegetation except for some grass and samphire adjacent to the mangrove fringe.

Distributaries.— Carrington Channel runs northeast and, except for Curie Channel, does not bifurcate into other distributaries until within a couple of kilometers of South West Island, where a maze of small mangrove-covered islands occurs. The McArthur River, on the other hand, although maintaining a meandering channel to its exit at Cora Point, also bifurcates into a number of east-trending distributaries. Johnson River, the southernmost, is a relatively deep channel and carries a substantial discharge, whereas other distributaries have only narrow and shallow junctions with the main channel and no longer actively carry much of the fluvial discharge.

Littoral mangrove forests.— The mangrove fringe consists of prominently zoned forests, several hundred meters wide, along the delta's seaward margin, and more opportunistic, less clearly zoned littoral mangrove along creeks and riverbanks. The seaward mangrove fringe comprises a seaward zone dominated by *Avicennia marina*, a central zone of *Rhizophora stylosa* in which the dwarf shrub *Aegialitis annulata* forms an understory, and a landward zone of *Avicennia marina* and *Ceriops tagal* (Fig, 3). Much of the mangrove was devastated by tropical cyclone Kathy in 1984; *Rhizophora* was defoliated but left standing,

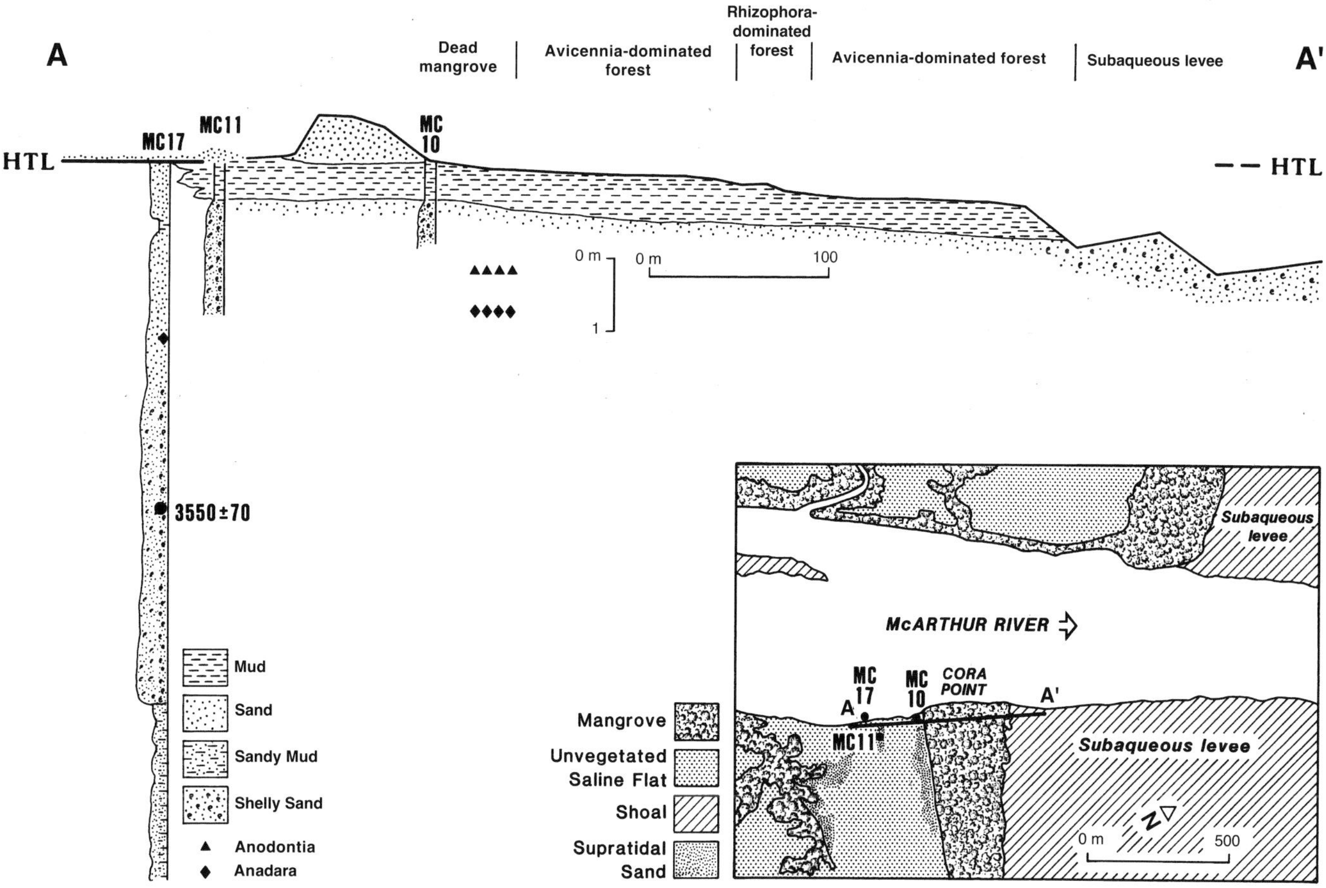

FIG. 3.—Stratigraphy and vegetation zonation at Cora Point, mouth of the McArthur River. The channel cuts through earlier seaward-dipping subtidal and intertidal deposits.

whereas other mangroves were uprooted and swept landward in a tangle of tossed debris (Bardsley, 1985). While a similar zonation might also be typical of gently sloping riverbanks, or the laterally accreting point bars of gradually migrating tidal creeks, the dynamics of creek and river bank sedimentation means that the generally complex patterning of mangroves is associated with the recent geomorphological history of particular sections of bank. Thus some areas of the bank are eroding and may have had the lower intertidal habitats truncated, such that *Rhizophora* is exposed on the riverbank. Elsewhere, still further truncation has left an erosional scarp above which *Ceriops* is exposed. At other sites, where there has been a complex history of episodic point-bar accretion, as at Goondi Point, the patterning of mangroves reflects this geomorphic history, with *Rhizophora* growing in those areas which are subject to flooding at each tide, and *Ceriops tagal* and *Bruguiera exaristata* on the areas less frequently flooded. *Avicennia marina, Lumnitzera racemosa,* and *Excoecaria agallocha* are found to the rear of these mangrove stands.

Bluish-gray mud and sandy mud containing woody fragments (Fig. 3) underlie mangrove forests. These muds are massive, bioturbated, and mottled. Similar sediments are found beneath the saline mudflat immediately adjacent to the mangrove fringe. These intertidal muds tend to be 50–100 cm thick at the seaward margin of the delta, and they generally grade downwards into gray medium sand with abundant shell fragments that represent a high-stand ravinement lag deposit.

At Cora Point, beneath the mangroves, an *in situ* shell bed of articulate bivalves of *Anodontia edentula* occurs at 1.5 m below high-tide datum (HTL) within the subtidal sand sheet. A parallel shell bed of *Anadara granosa* occurs at 2 m below HTL (Fig. 3). The shell beds continue laterally and are inferred to extend beneath the mangrove forest. Despite proliferation of these and other shellfish (particularly *Geloina* sp.) in midden deposits dating modern to only a few hundred years old throughout the surface of the McArthur delta, it is not clear where these shells are living today. *Geloina* sp. appears to have been extensively available for Aboriginal exploitation in the past but is rarely seen today.

Saline mudflats.— Most of the lower delta plain comprises supratidal saline mudflat, which is inundated during the wet season, and by storm surges, such as that which accompanied cyclone Kathy. Numerous mangrove-lined minor tidal creeks dissect, and form intricate dendritic patterns across, the saline mudflats. The flats dry out during the dry season and form intricate patterns of desiccation cracks, occasionally with associated gypsum. The flats are composed of sandy mud and muddy sand, in many places with abundant shell fragments. In some areas small accumulations of supratidal sand are present (described below); elsewhere prominent shell beds provide an important clue to the origin of these plains. Comparison of the stratigraphy and molluscan content of sediments exposed across the delta surface and those beneath the mangrove fringe, as at Cora Point (Fig. 3), indicate that much of the saline mudflat developed as a former delta front. The saline mudflat appears to be largely a deflational surface, formed within a subaqueous environment, rather than a further environment of deposition. This feature is distinctly different from the inundated mudflats typical of most tropical delta successions in the rest of Southeast Asia. It could be equivalent to emergent surfaces in older deltaic successions.

Occurrences of articulate bivalves across the surface of much of the saline mudflats are directly comparable with shell beds at the modern coast, and in riverbank exposures throughout much of the lower delta plain. Exposed in the bank of Davies Channel are shellbeds with *Anodontia* and *Anadara* in apparent growth position, together with *Gafrarium tumidum, Quadrans capsoides, Terebralia sulcata, Neritina ovalaniensis, Saccostrea* sp., *Thais* sp., *Cerithidea cingulata,* and *Clementia papyracea.* The assemblage is characteristic of an intertidal to shallow subtidal environment, and in this case a radiocarbon date of shell material yielded an age of 2,930 ± 100 yr BP (Woodroffe and Chappell, 1993). It probably represents a lag deposit produced by winnowing in a shoreface environment as sea level fell about 1.5 m since ~ 3,000 yr BP.

The shell beds are considered to indicate a former delta-front environment. At an exposure in Curie Channel, the basal bed of *Anadara* (about 0.5 m below HTL) gave a radiocarbon age of 4,660 ± 160 yr BP, while the overlying *Anodontia,* 70 cm higher, gave an age of 3,150 ± 160 yr BP. These ages would appear to support a gradual vertical rate of accretion of 0.5 mm / yr during a period of higher sea level. The majority of dates across the delta fall into an age range of 3,900–2,900 yr BP. Landward of the seaward margin of the delta at Cora Point ages of 2,400 ± 100 and 1,970 ± 70 yr BP were obtained, indicating that this part of the delta has built out more recently.

Radiocarbon ages of 2,950 ± 120 yr BP at the mouth of Johnson River, 2,880 ± 70 yr BP at the mouth of Lucky Creek, and 2,400 ± 110 yr BP in Davies Channel (Fig. 4), and those near Cora Point just referred to, appear to indicate that there has been relatively little build-out of the delta over the last 2,000 years. The only exception seems to be some active accretion around the small islands to the northwest of the delta, on one of which shell hash from about 60 cm depth (core MC4) has been dated 670 ± 150 yr BP (Fig. 4).

Supratidal sand sheets.— The upper surface of much of the lower delta plain appears to be deflational, exposing the shellbeds formed within a prior delta front, through the winnowing of sediment from the surface. A series of supratidal sand deposits occur intermittently across the surface. These form low grass-covered mounds or crescentic ridges, composed of well-sorted fine white sand. They form up to 30–80 cm above the general level of the saline mudflat, and in many cases appear to be eolian, perhaps aggrading as a result of deflation of the general surface. Grass and a weak caliche cementation give the surface of these ridges some stability.

Evolution of the McArthur Delta

Coring has revealed that shelly sand forms most of the wedge of Holocene sediments beneath the McArthur delta. These unconformably overlie a highly oxidized pre-Holocene surface (Fig. 5A), which presumably equates with the regional Holroyd surface, as defined by Doutch (1976) and Smart et al. (1978), formed by subaerial exposure of Pleistocene sediments during the last glacial maximum. There is little if any indication of a basal transgressive intertidal sedimentary unit equivalent to transgressive mangrove muds as found in estuaries in northern Australia (Woodroffe et al., 1989). Radiocarbon dates of 5,650 ± 310 yr BP at the base of a core near the junction between the upper and lower delta plain, and 6,140 ± 100 yr BP near the apex of the upper delta plain, may reflect the highstand period of sea level following the postglacial rise that drowned the prior surface. Sea level is likely to have been slightly higher than its present level 6,000 years ago (Fig. 5B). Much of the delta built up as a result of gradual vertical sediment accumulation over a mid-Holocene delta front. The shell beds are considered to form at depths typically 1.5–2.0 m below HTL (as they are found at present at Cora Point); their exposure at a level close to HTL over much of the plains is considered to result from a relative fall of sea level of about 1.5 m since 3,000 yr BP (Woodroffe and Chappell, 1993).

Figure 5 shows a schematic interpretation of the evolution of the McArthur delta based upon the stratigraphic and chronological details given by Woodroffe and Chappell (1993). Initial reworking of the subaerial landscape must have occurred as the sea rose 7,500–7,000 years ago (Fig. 5A), until the sea reached slightly above present level 6,000 years ago (Fig. 5B). Vertical and lateral accretion occurred across the broad area of the present delta, and the delta-front environments shallowed and were suitable for the

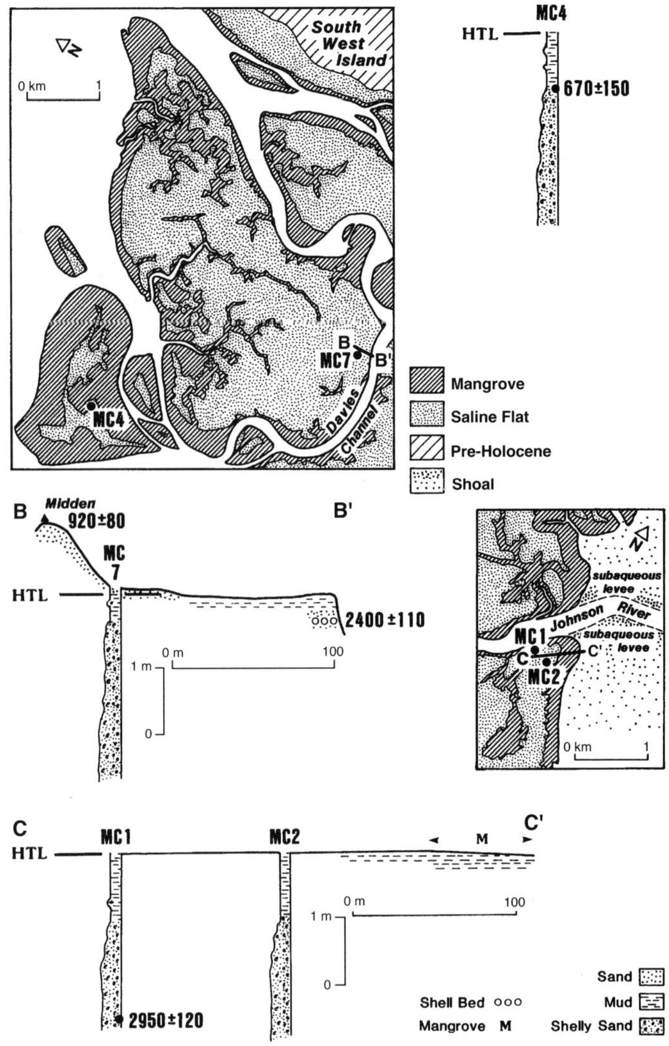

FIG. 4.—Stratigraphy and zonation of recently accreted islands on the northwestern part of the delta, and at the mouth of Johnson River in the south. In both cases the subaerial delta extends out over subtidal sand sheets.

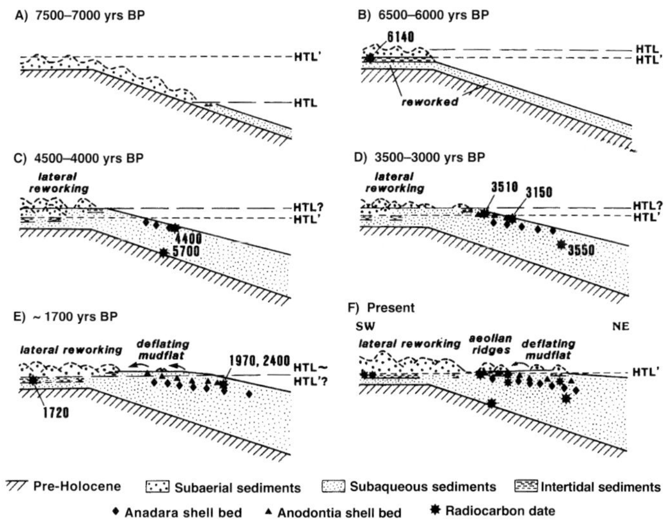

FIG. 5.—Schematic evolution of the McArthur River delta. Shellbeds and radiocarbon dates are drawn as determined from the stratigraphy of drill holes and riverbank exposures described in this paper and by Woodroffe and Chappell (1993). HTL' = current high tide level.

development of shellbeds of *Anadara* by at least 4,000 years ago (Fig. 5C). Subsequently much of the present area of the delta formed as a delta front, with extensive shellbeds, especially of *Anodontia*, under a slightly higher sea level (Fig. 5D). Sea level fell about 3,000 years ago, exposing this former delta front, and the sediments that composed it have been eroded through deflation (Fig. 5E). Throughout this time, and also currently, extensive lateral migration of the river-dominated channel has occurred, reworking much of the upper delta plain, and some of these fluvial deposits have extended farther downstream across the delta, with deposition of levees and a new subtidal delta (Fig. 5F).

Fluvial Versus Tidal Processes in the McArthur Delta.—

The influence of lateral migration by the river is clearly imprinted upon the landscape of the upper delta plain through the nested patterns of scroll bars, and the active levees of both McArthur River and Carrington Channel. These sandy levees appear to form in the wet season as point bars on the insides of river meanders. There is a gradual decline in the elevation of their upper surfaces, determined on the basis of surveys for Carrington Channel, from 2.0–2.5 m above HTL at the delta apex to around 1 m above HTL in mid delta (Fig. 6). Fluvial processes extend downstream into the lower delta plain, both through lateral migration and point-bar accretion, and also by infilling of former distributaries with fluvial sand.

The formation of active point bars can be seen clearly in the region around Goondi Point, where the McArthur River makes a particularly sharp turn (Fig. 7). Mangroves cover Goondi Point itself, and point bars are actively extending downstream from the inside of the meander. The river is especially narrow at this point and is actively carving into the bank on the northern side of the river, resulting in erosion and exposure of a sequence of shellbeds. Rhizomorphs occur discontinuously along the exposed and eroded bank, but they are less obvious where the sediment is sandier. The *in situ* shell beds, on which a radiocarbon date of 3,510 ± 90 yr BP has been determined (Fig. 7), indicate that this bank is eroding into the former delta front. A similar pattern of development is occurring at the next bend. Again mangroves cover the point-bar sequence, which is less prominent but can be seen to be extending from the next bend also. Shellbeds of similar age occur to the east on the opposite, eroding bank (Fig. 7).

In several instances the process of point-bar accretion has left an entire broad, shallow muddy reentrant within a chute bar abandoned by channel lateral migration. In the case of the McArthur River this is illustrated by a feature termed Mugg's Mistake, which traps unwary fishermen returning upstream (Fig. 8). The reentrant lies within a sequence of point-bar ridges; a particularly prominent levee forms the riverbank, rising to an elevation of more than 2 m above HTL, on the higher parts of which termite mounds occur. Older ridges are extremely subdued but can be discerned from the aerial photographs. This reentrant, and a similar one farther downstream on the left bank of the river (opposite Goondi Point), are presumably slowly filling. Muds, typical of mangrove environments, occur in the lowest depressions between ridges and are exposed in the lower bank of the levee.

Elsewhere, the river, or a chute channel thereof, has reinvaded the fluvial ridge sequence to leave an arcuate island within the channel. The clearest example of this in the McArthur delta is Kangaroo Island, which occurs at the confluence with Johnson River. Kangaroo Island is an asymmetrical island with grass-covered ridges running its length (Fig. 9). The ridges on either side of the island, flanked by the river, are the highest, and both support termite mounds and shell middens. The mangroves *Avicennia marina* and *Lumnitzera racemosa* are typical of the low-lying depressions between ridges.

Exposures in the riverbank reveal mangrove mud underlying the sandy ridges. The form of the ridges indicates that they have been constructed as the result of downstream extension of point bars. Kangaroo Island would appear to have formed as a result of the lateral accretion of the river with episodic formation of point bars, and then avulsion and reoccupation of a course through the sequence of ridges. A similar island occurs in the Carrington Channel at the point where Curie Channel bifurcates (see Fig. 2).

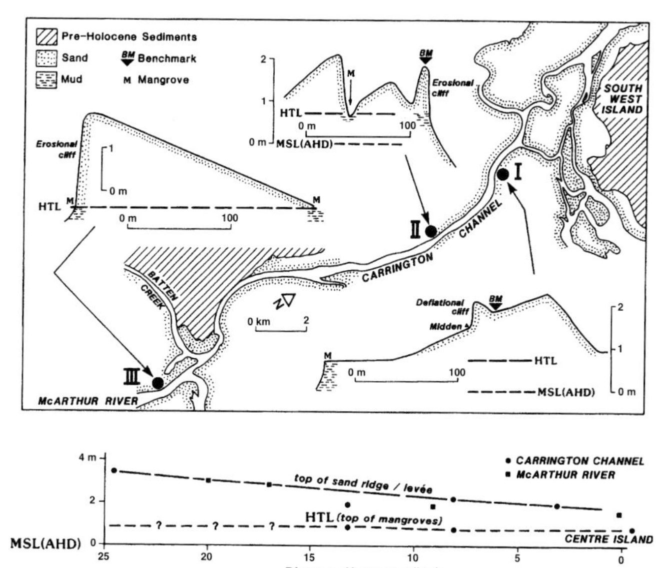

FIG. 6.—Elevation and morphology of river banks along Carrington Channel (with comparative data for McArthur River channel).

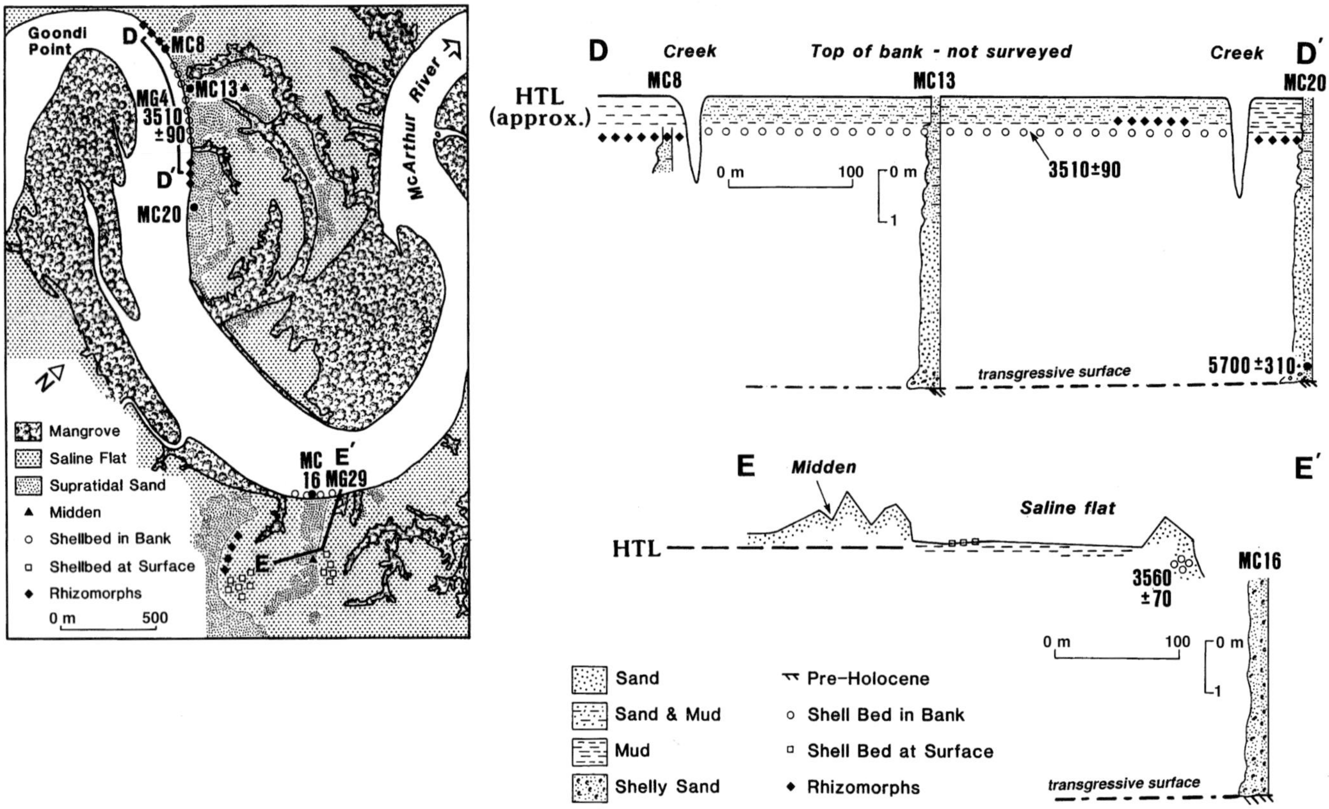

FIG. 7.—Stratigraphy and vegetation zonation at Goondi Point on the McArthur River.

On the lower delta plain there is less obvious influence by fluvial deposition. Sands that reach the lower delta plain tend to be deposited below HTL and are therefore less obvious. The numerous distributaries that cross the lower delta plain can be divided into those that are active and those that are abandoned. McArthur River and Carrington Channel are active, whereas many of the shallow, rapidly tapering distributaries on the delta margin are abandoned. In these the principal flows are tidal. Johnson River (sinuosity 1.7) is an anomaly in that it shows evidence of both river and tide domination.

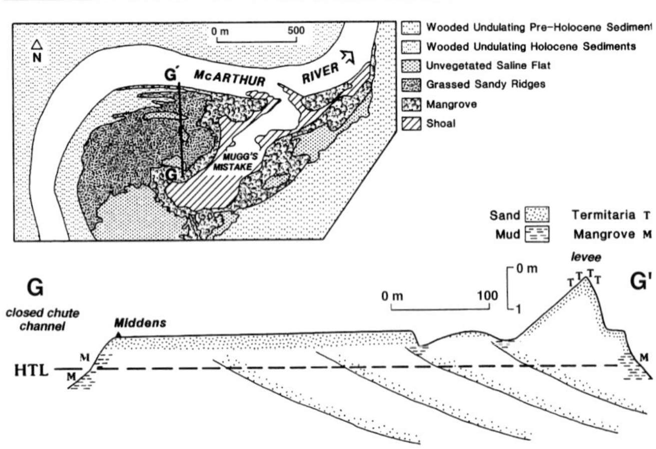

FIG. 8.—Morphology and vegetation of a reentrant (Mugg's Mistake) on the McArthur River.

Mangroves generally flank tide-dominated channels, particularly on the convex depositional bends. In contrast, on the concave erosional banks the combined tidal and fluvial flows in the distributaries frequently cut into former delta-front sediments, deposited during the higher stand of sea level, as the channel migrates, and older shellbeds are exposed as in the main river.

The morphology of distributaries provides a clue to the processes that shape them. Fluvial distributaries maintain relatively constant width with distance (Fig. 10), presumably reflecting peak discharge during the wet-season floods. On the other hand, those channels that are flooded primarily by tides show an exponential decrease in width with distance from the coast as tidal volume decreases exponentially upstream. They also become shallower inland, and several of the tidal distributaries cannot be followed through to the main channel in a boat at low tide.

The courses of former river-dominated distributaries can be discerned from aerial photographs. This is especially the case if the area of mangroves is considered in combination with the presently tide-dominated distributary. Thus both the McArthur River and Carrington Channel vary little in width from the apex of the delta to the delta mouth (Fig. 10). Lucky Creek (sinuosity 1.6) and Dugong Creek (sinuosity 1.2), on the other hand, are rapidly tapering, and scarcely connect through their shallow headwaters to meet the modern McArthur River. However, the width of the present channel, together with the littoral mangrove, shows a much more constant width with distance and implies that each of these distributaries was once the active mouth of the McArthur River through which most of the flow was carried (Fig. 10).

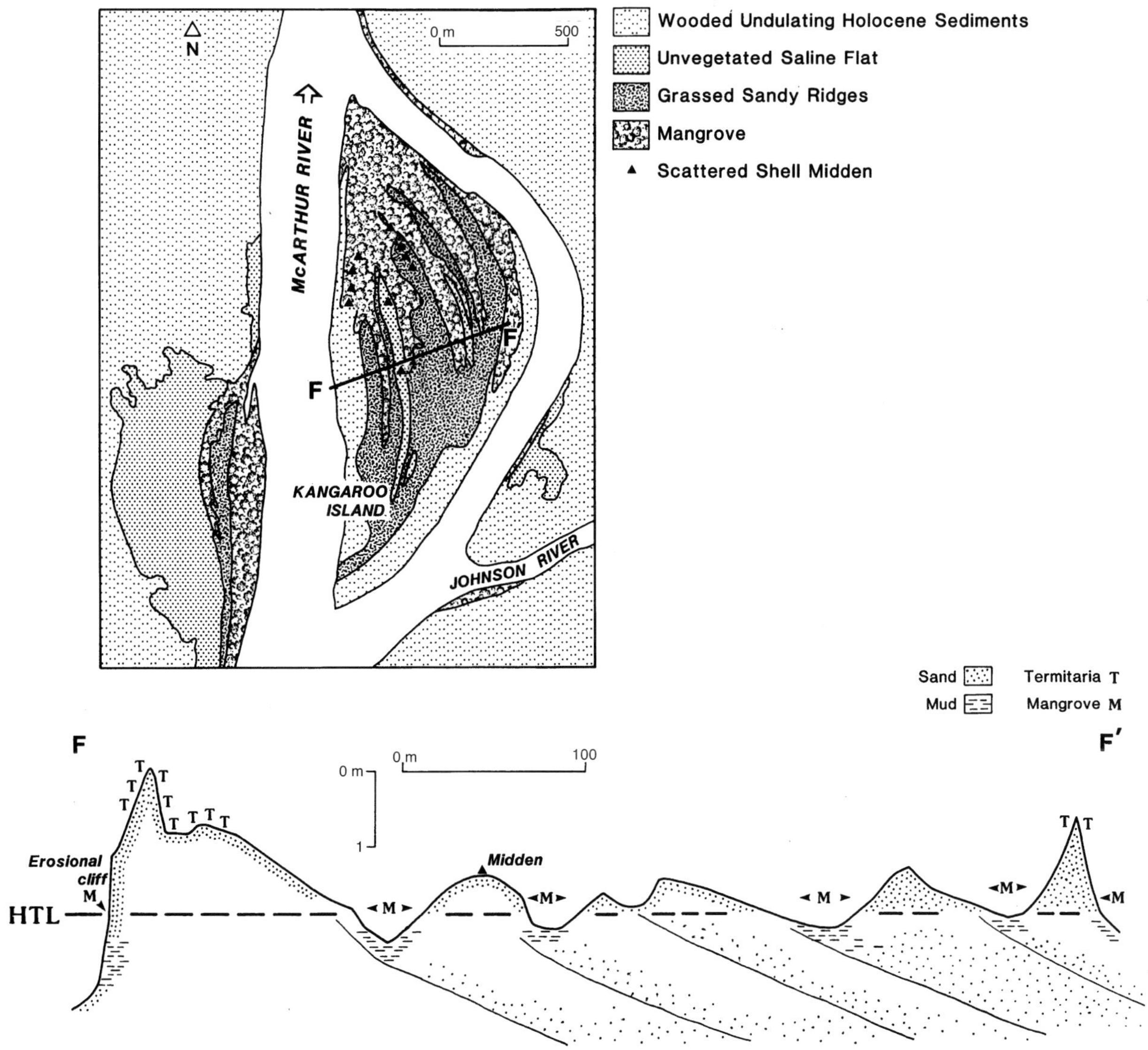

FIG. 9.—Morphology and vegetation of the arcuate Kangaroo Island point bar at the junction between the McArthur and Johnson Rivers. The present McArthur River occupies an enlarged and deepened prior chute channel.

Augering was undertaken to investigate the sediments that underlie the mangroves flanking such tide-dominated channels. Auger holes MC19 and MC21 penetrated through the sandy mud and muddy sand underlying the mangrove forests into clean medium sand, and then at depths of several meters into shelly sand characteristic of the delta front. The clean sands are interpreted as fluvial sands that have infilled the former fluvial distributaries after their abandonment, i.e., channel-fill deposits. Radiocarbon dates on shells in the underlying shelly sand indicate that the original depositional delta-front sediments were deposited around 3,000 years ago. The absence of material suitable for dating in the fluvial sands means that the time of abandonment and infilling of the distributaries cannot be established. In Figure 11 the form of earlier distributaries, as interpreted from the width of the channel and the fringing mangroves, is indicated. The transition into shelly sand in auger holes MC19 and MC21 is taken to mark the former distributary floor. Each of the now tide-dominated distributaries, such as Dugong and Lucky Creeks, appears to have been the course of a former distributary of the main river. Johnson River does not fit into the tide-dominated category, and it appears to have always maintained a strong fluvial influence.

It seems most likely that these various distributary channels have been occupied sequentially. Channel avulsion is indicated because dating of the shellbeds at exposures and in auger holes between the distributaries has indicated ages coeval with the delta front, so the intervening sediments have not been reworked by migration of distributaries.

Sand-Body Geometry.—

The lateral migration of river channels in the upper delta plain is constricted by pre-Holocene outcrop (Fig. 2), and they occur in extensively reworked channel belts 2–4 km wide. Channel widths are up to 400 m, and channel depths are generally about 5 m with scours up to 10 m deep. The channel belt shows evidence of numerous point-bar deposits, chute channels, and occasional abandoned channels (Fig. 9). These channel-belt deposits are dominated by fluvial sand and generally show an upward-fining grain-size signature. In some areas the sand bodies are multistory. Mud occurs in the upper part of the point-bar deposits, in inactive chute channels, and forms extensive plugs in abandoned channels. The channel-belt sand bodies are moderately interconnected but, in places, are separated by areas of floodplain mud accumulation.

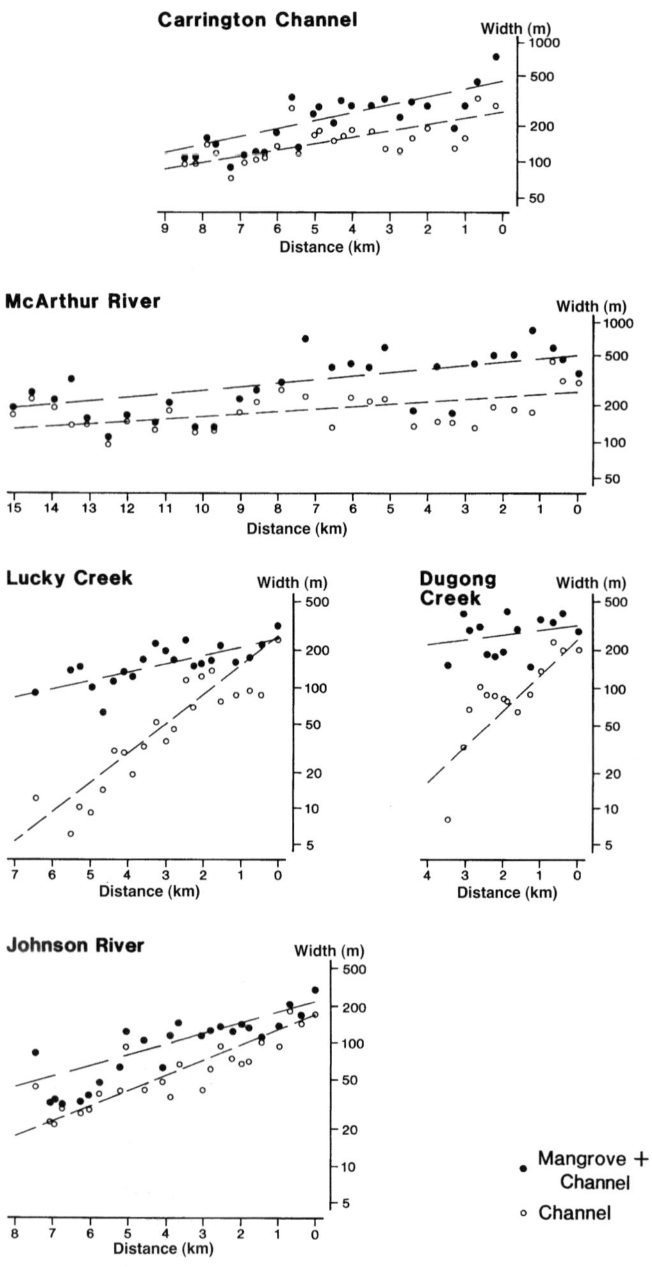

FIG. 10.—Width–distance relationships (measured from the channel mouth) for distributaries of the McArthur delta (see Fig. 2 for location). River-dominated distributaries (Carrington Channel and McArthur River) show a gradual decrease in width upstream, whereas tidal-dominated distributaries (Lucky and Dugong Creeks) are rapidly tapering. Johnson River appears intermediate, though with river domination. Note that when flanking mangrove is considered together with channel width, the river-dominated form of the paleochannels now occupied by Lucky and Dugong Creeks is apparent (see Fig. 11).

The distributary-channel facies in the lower delta plain occurs as a series of channel belts ranging from 1 to 3 km wide and up to 10 km long. Most of the channels show a meandering form with lateral accretion in point bars as the dominant depositional mechanism (Fig. 8). The channel deposits are up to 6 m thick and show upward fining, but they contain a higher proportion of mud than in the equivalent upper-delta-plain fluvial deposits.

Mangrove mud occurs along the channel margins whereas saline mudflats cover much of the lower delta-plain area between the tidal channels and distributaries. Over most of this area the mud forms a blanket, up to 1.5 m thick and 5 km laterally, over older channel facies and especially over older subtidal sand sheets and mouthbar deposits. Most of the smaller tidal channels are dominated by mud and tend to become filled with organic-rich mud when abandoned. Deflation of parts of the mudflats during the dry season has released sand and produced some shell lag deposits.

Supratidal sand sheets occur as a series of low sand ridges consisting of fine-grained sand reworked from the dry mudflats or exposed point-bar successions (e.g., Fig. 7). They are formed by eolian action and tend to occur as discontinuous and disconnected ridges and patches tens to a few hundred meters in lateral extent and 30–80 cm thick, but ranging up to 3 m thick in places.

Offshore from the delta the sea is very shallow and the delta mouthbars are reworked to form extensive subtidal sand sheets dissected by subtidal channels (Fig. 2). Although these modern subtidal delta-front sand sheets have not been studied in detail they extend around most of the 25 km margin of the delta, and they also extend 2–4 km seaward. Equivalent sand sheets formed during the mid-Holocene highstand of sea level have been intersected in many of the drill holes on the delta and are typically 4–8 m thick (Figs. 3, 4, 7, 11). They generally consist of poorly sorted slightly muddy and shelly sand, and they may contain distinct shellbeds, especially in the upper part of the unit. The subtidal sand sheet typically overlies sandy prodelta mud either gradationally, forming an upward-coarsening package, or abruptly above a minor highstand ravinement surface.

GILBERT RIVER DELTA

The Gilbert River system rises in the Great Dividing Range and drains westward into the southeastern Gulf of Carpentaria (Fig. 1). The river system drains Proterozoic and Paleozoic sequences in its headwaters, passes through the Mesozoic Carpentaria Basin sequence, and spreads out to form a broad unconfined low-angle arcuate alluvial plain and delta approximately 175 km long and up to 125 km wide at the coast (Fig. 12). In the past the Gilbert and Einasleigh Rivers have also contributed detritus to the Staaten River delta farther north.

In contrast to the relatively confined McArthur delta, the Gilbert delta consists of a broad but thin expanse of Upper Tertiary and Quaternary deposits that have accumulated in this slowly subsiding region (Smart et al., 1978). No bedrock controls affect the delta form in this area, but the tropical monsoonal climate, together with possible drier phases operating during the Quaternary glacial periods, has produced well-cemented calcrete and ferricrete soil profiles that provide erosion-resistant barriers to free lateral migration of the river channels (Nanson et al., 1991). As a result, discharge switching on the delta margin is

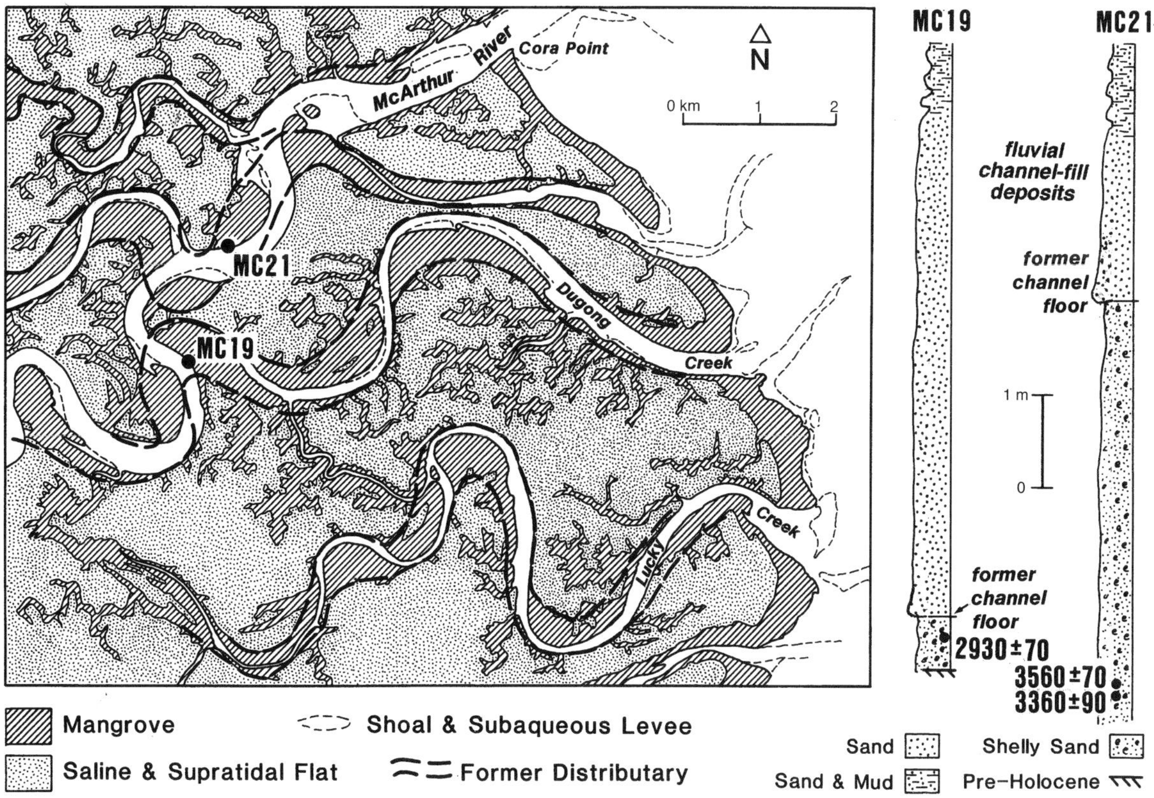

FIG. 11.—Form and stratigraphy of active and former distributaries of McArthur River, showing courses of paleochannels, and stratigraphy of cores that penetrate to former channel floor (after Woodroffe and Chappell, 1993).

largely controlled by upstream avulsion of the river, with seven major distributary channels identified as operating during the Holocene. Late Quaternary and Holocene sediments comprise a veneer across the alluvial and delta plains, being thickest (10–15 m) along the fluvial axes but forming only a thin floodplain sheet (< 2 m) and coastal wedge (1–10 m) over the southern third of the delta (Nanson et al., 1991; Jones et al., 1993).

The coastal portion of the delta is influenced by mesotidal conditions and is exposed to wave action from the Gulf of Carpentaria. However, this southeastern portion of the gulf is very shallow, with maximum depths of about 15 m occurring 80 km west of the delta. Wave energy is kept low by frictional resistance across this very gently shelving offshore region except during major cyclones. However, the wave energy is sufficient to have built up a series of chenier and beach ridges marking successive shorelines along the delta front.

Morphology of the Gilbert Delta Plain

The upper part of the delta plain is characterized by broad low-angle muddy floodplains interspersed with a number of major sand-dominated distributary channel axes. The Holocene fluvial channels are concentrated mainly in the northern half of the delta, with Walker Creek providing the only active Holocene channel on the southern margin of the Gilbert delta (Fig. 12). Thermoluminescence dating has shown that individual channel axes were accumulating sediment during the early Holocene (Nanson et al., 1991), and several of the channels have been occupied two or more times during the Holocene as a result of avulsion events.

The Holocene delta overlies a Late Pleistocene paleosol, equivalent to the Holroyd surface, that generally consists of strongly mottled, highly compacted, fine mud, which may contain carbonate nodules, pisolitic iron nodules, or, rarely, coarse quartz sand.

The lower delta plain is tide-dominated for most of the year, but the active distributaries become fluvially influenced during the monsoon period from December to March (Fig. 13). The abandoned distributaries show a strong tidal influence and have a distributary network of smaller incised channels extending into the adjacent supratidal mudflats. A fringe of mangroves flanks most of the distributaries. A series of stranded beach and chenier ridges forms prominent features in the lower delta plain (Jones et al., 1993) and provides an added constraint on the location of fluvial distributaries.

Wave action and longshore tidal currents have modified the subaqueous extension of the delta. This has produced a series of shallow subtidal sandy bars, with the main tidal channel being partly blocked by mouthbar deposits (Fig. 13). The seaward margins of the subtidal sand bodies show a rapid increase in water depth to about 3 m, followed by a very gently shelving offshore profile. A considerable quantity of mud is moved out onto the subtidal delta during the monsoon. The mud is reworked by wave and tidal action during the rest of the year and moved into the prodelta area or back shorewards to form the intertidal and supratidal deposits.

Observations of the delta have been made during numerous fieldtrips combined with aerial photographs and satellite images plus radiocarbon and thermoluminescence dating (Table 1). Elevations on the delta-plain successions have been related to a high-tide datum (HTL) at the back of the mangrove deposits.

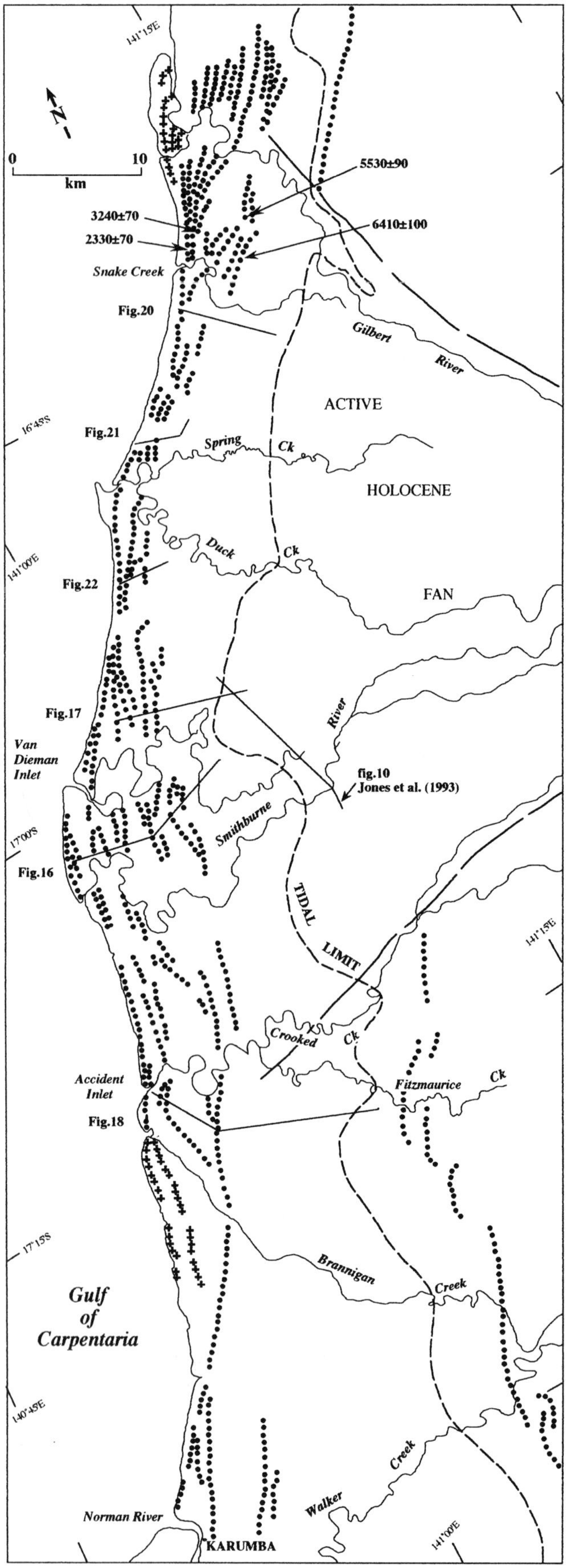

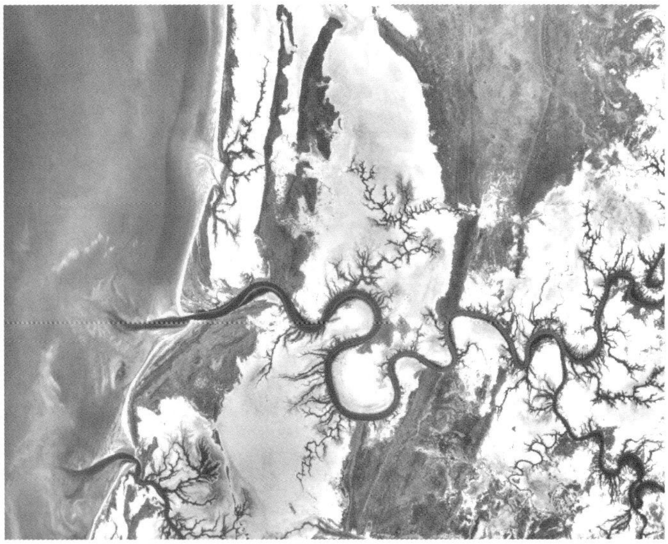

FIG. 13.—Beach ridges and supratidal facies (white salt flats) in the Accident Inlet area. The main channel is strongly meandering and is flanked by numerous small tidal distributary channels. The subtidal delta sandbars and the shallow muddy intertidal zone along the coast are clearly visible. Field of view ~ 16.5 km. (Part of photograph Normanton Run 2, CAB321, 1967, copyright with the Commonwealth of Australia, reproduced by permission of the General Manager, AUSLIG, Department of Industry, Science and Resources, Canberra.)

Where practical, the upper-delta-plain deposits have been corrected to Australian Height Datum from benchmarks. Stratigraphic data have been obtained mainly by a series of auger hole traverses and associated topographic surveys. In places shallow pits and riverbank exposures have augmented these data. The Holocene evolution of the delta can be related to sediment accumulation during the latter stages of sea-level rise, but the extensive lower-delta-plain sequence has been exposed during a gradual regressive phase that has operated since about 3,000 yr BP and helps account for the stranded beach and chenier ridge systems.

Morphology, Depositional Environments, and Sedimentary Units

Upper Delta Plain.—

The upper delta plain extends landwards from the upper limit of tide-dominated channels to approximately 5 km landward of the current tidal limit (see figs. 1 and 3 in Jones et al., 1993), which represents the maximum landward extent of mangrove muds and intertidal estuarine deposits. The latter deposits have a maximum elevation of 3 m AHD and may be overlain by 2–4 m of fluvial and floodplain deposits. The width of the upper delta plain thus ranges from about 7 to 15 km (Fig. 12). It

←

FIG. 12.—Beach (large dots) and chenier (crosses) strandline ridges in the tidally influenced portion of the Gilbert River lower delta plain in the southeastern portion of the Gulf of Carpentaria. Locations of the drill-hole sections (Figs. 16–18 and 20–22) are indicated. The delta extends from Gilbert River to Walker Creek, with the northern half of the region (down to Crooked Creek) being active during the Holocene.

TABLE 1.—Radiocarbon and thermoluminescent dates from the Gilbert River Delta area.

Laboratory Code	Grid Reference	Analyzed material	Hole	Depth (m)	Location	Conventional ^{14}C Age
ANU-7709	M134428	shell	90/19	7.2	Kelso Pocket	6,430 ± 120 yr BP
ANU-7710	M081346	shell	90/54	2.7	15 km NE Van Dieman Inlet	1,990 ± 90 yr BP
ANU-7711	M226498	shell	89/7	2.5	Snake Creek	5,920 ± 100 yr BP
ANU-7712	M029266	shell	90/68	1.2	6 km NE Van Dieman Inlet	4,000 ± 70 yr BP
ANU-7713	M093241	shell	90/65	4.9	Kelso Pocket	4,520 ± 70 yr BP
ANU-7714	K940989	shell	89/26	1.5	1 km S Accident Inlet	2,100 ± 70 yr BP
ANU-7715	K935997	shell	89/29	3.2	Accident Inlet	2,670 ± 60 yr BP
ANU-7716	K973947	shell	89/31	2.7	5 km SE Accident Inlet	3,330 ± 80 yr BP
ANU-7717	K979941	shell	89/33	1.6	6 km SE Accident Inlet	6,050 ± 80 yr BP
ANU-7718	M047243	shell	90/57	5.4	5 km W Van Dieman Inlet	5,450 ± 60 yr BP
Wk7776	M231561	shell	99/6	0.55	2 km E Snake Creek	2330 ± 70 yr BP
Wk7777	M232557	shell	99/5	2.5	2 km E Snake Creek	3240 ± 70 yr BP
Wk7778	M272556	shell	99/3	0.85	5 km E Snake Creek	5,530 ± 90 yr BP
Wk7796	M269543	shell	99/2	1.55	5 km E Snake Creek	6,410 ± 100 yr BP
UW-W1196	M161431	fluvial sand	90/26	3.0	3 km E Kelso Pocket	15,400 ± 2,500 yr BP
UW-W1197	D125200	fluvial sand	90/47	1.2	18 km E Smithburn River	14,400 ± 2,500 yr BP
UW-W1198	M091342	strandline	90/50	4.0	15 km NE Van Dieman Inlet	2,500 ± 300 yr BP
UW-W1199	M062282	strandline	90/67	1.0	9 km NE Van Dieman Inlet	3,400 ± 500 yr BP

ANU = Australian National University Radiocarbon-Dating Laboratory; Wk – University of Waikato Radiocarbon-Dating Laboratory; UW = University of Wollongong Thermoluminescence Laboratory; Grid references refer to the Australian 1:100,000 Topographic Survey sheets 7063 Karumba (K), 7163 Double Lagoon (D) and 7164 Macaroni (M).

has a very low gradient in the upper fluvial portion and is almost flat in the lower portion, where it has been ponded behind the innermost sand ridge (e.g., Fig. 13). Most of this region consists of open grassland with bands of riverine eucalypt woodland along the channel banks, giving way downstream to mangrove forest (dominated by *Avicennia marina*). The lower part of the upper delta plain gives way to samphire marsh and saline scalds and limited areas of bare supratidal mudflat. Most of the floodplain soils are oxidized, contain little organic matter, and may contain very small calcareous or ferruginous nodules. In places they show evidence of abandoned channel-fill deposits (Fig. 14). In contrast, the marsh and mudflat areas tend to have reduced saline soils with progressively more preserved organic matter. Near the upper limits of present tidal activity along most distributaries many of the soils are partly oxidized, with oxidation extending down along rotted mangrove pneumatophores.

Along the channel axes reworking and partial erosion of the tidal sequence has resulted from phases of active fluvial transport along each of the major distributaries when they became active as a result of avulsion processes operating up to 150 km inland in the upper fluvial portion of the delta. These reworked channel axes range from about 1 to 5 km in width, with individual sand bodies generally 4–6 m thick. Reoccupation of the main channel axes by successive active distributaries has produced some multistory

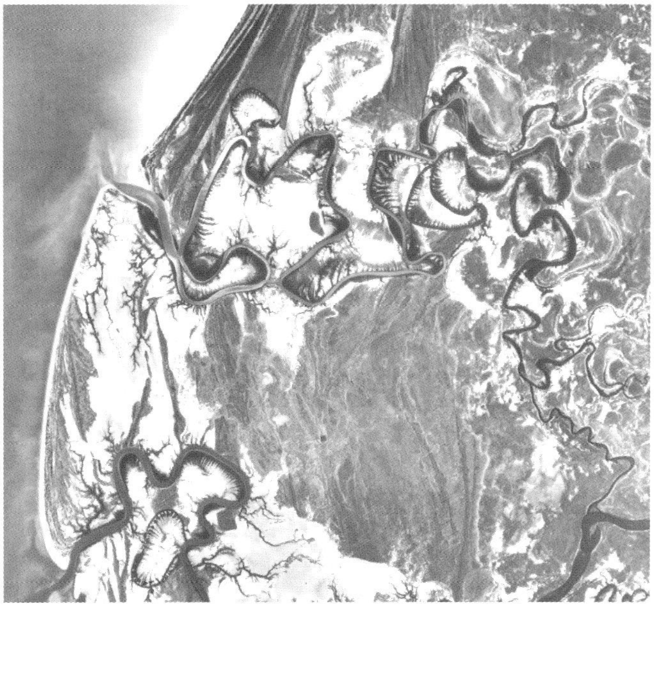

→

FIG. 14.—Area between the Smithburne River (south) and Van Diemen Inlet with the latter showing evidence of abandoned channel-fill deposits and meander loops both in the alluvial plain area (east) and in the supratidal mudflat area. Between the two rivers, the central part of the photograph shows evidence of older Pleistocene beach ridges. A chenier ridge is present on the Holocene supratidal surface whereas younger beach ridges occur adjacent to the modern coast. Field of view ~ 16.5 km. (Part of photograph Burketown Run CT1, 3319, 1966, copyright with the Commonwealth of Australia, reproduced by permission of the General Manager, AUSLIG, Department of Industry, Science and Resources, Canberra.)

sand bodies up to 15 m thick. The intervening floodplain areas between distributary channels range in width from about 7 to 15 km, with the average floodplain area increasing in the southern half of the delta plain, where the distributary channels were active mainly during the Late Pleistocene. Over most of these floodplain areas Holocene deposition has been much slower, with typically only 1–2 m of floodplain mud being deposited during the infrequent major flood events.

Lower Delta Plain.—

The lower part of the Gilbert delta plain is characterized by the presence of a series of low beach and chenier ridges (Fig. 15) with extensive saline mudflats between the older stranded sand ridges (Fig. 13). The intricate dendritic networks of distributary and larger tidal channels are lined with mangroves. The mangrove forest has an outer zone characterized by *Avicennia marina*, whereas the inner zone contains variable proportions of *Aegialitis annulata*, *Aegiceras corniculatum*, and *Bruguiera exaristata* and more restricted occurrences of *Rhizophora stylosa*. Shoreline mudflat areas have a littoral mangrove fringe, 50–200 m wide, dominated by *Avicennia marina*. The lower parts of the supratidal mudflats adjacent to the main tidal channels or shoreline mudflats are generally covered with algal mats (mainly the filamentous blue-green algae *Lyngbya aestuarri*), which tend to become desiccated and removed by wind action from the upper, drier portion of the mudflat. Minor samphire and grasses are present around the mudflat areas and the lower parts of the sand ridges.

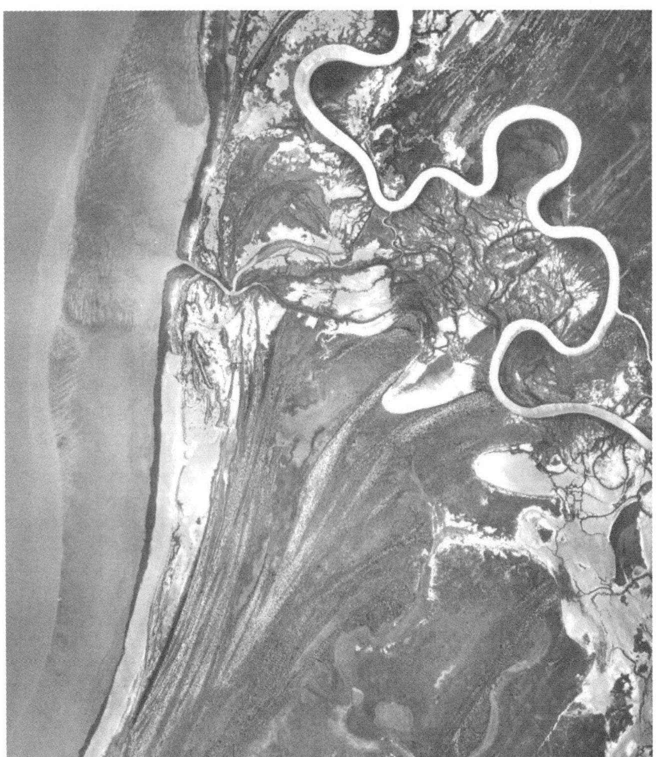

FIG. 15.—A broad Holocene beach-ridge system is present south of the Gilbert River. An area of alluvial plain that overlies earlier supratidal deposits separates the oldest (eastern) beach ridge from the younger beach ridges. The coastal fringe consists of intertidal and supratidal mudflats with a low chenier ridge. Field of view ~ 13 km. (Part of photograph Galbraith Run 9, CAB116, 1958, copyright with the Commonwealth of Australia, reproduced by permission of the General Manager, AUSLIG, Department of Industry, Science and Resources, Canberra.)

Strandline sand ridges.—Strandline sandy beach ridges and cheniers, elongated parallel to the coastline, are a prominent component in the lower delta plain in the northern part of the Gilbert delta (Fig. 15). The beach ridges represent successive strandlines deposited seaward of their predecessors on the prograding shoreline. They are contemporaneous with local delta-front development, and successive beach ridges reflect pulses of rapid sediment input (as noted by Rhodes, 1982). In contrast, chenier sand ridges build up on a mudflat (Fig. 14) as sediment is moved towards the shore from longshore-migrating shoals in the shallow marine zone. They are therefore younger than the sediment on which they rest and probably formed during periods of low sediment flux, when sorting and winnowing can occur.

Commonly six to eight sand ridges are present as isolated or amalgamated features covering up to half of the lowest 10 km of the Gilbert delta plain (e.g., Fig. 16). Most of the ridges have a relief of 2–5 m and may either have a single crest or consist of a series of ridges and swales indicating successive accretion of the ridge in a seaward direction. The lower portions of the ridges commonly consist of medium- to coarse-grained calcareous sand with numerous broken shell fragments and scattered whole shells (Fig. 17). In a few areas these lower sands have been cemented to form beachrock, which may show evidence of low-angle stratification representing beach-face deposits. This occurs mainly in the older strandline deposits, whereas scattered irregular calcareous concretions and nodules are present in some of the younger beach ridges. The innermost sand ridges are the oldest, with a radiocarbon age of 6,050 yr BP (Fig. 18). The strandline sand ridges become progressively younger seawards, with a 1,990 yr BP date from the rear of a series of three subrecent coastal beach ridges (Fig. 16). Most the drill holes show that the sand ridges gradationally overlie delta-front or intertidal sand bodies (Fig. 17), not mud, and are therefore beach ridges.

In the southern half of the Gilbert delta, south of the Smithburne River, strandline sand ridges are sparse. The innermost ridge is up to about 3 m high and is relatively continuous, representing sand reworked shorewards forming a beach ridge at the point of maximum Holocene transgression. Seaward from this line chenier sand ridges are rare, low, and discontinuous. Very extensive saline mudflats dominate the lower delta plain (Fig. 18). This reflects the very low sand supply to this part of the coast, because sand brought down by the Norman River is mainly moved to the west and southwest by longshore currents.

In the northern half of the delta a discontinuous coastal chenier ridge is present north of Snake Creek and breaks up to form four low ridges south of the current Gilbert River mouth. These chenier ridges have also formed in areas of slow sedimentation after most sediment supply to Snake Creek had been cut off by avulsion.

The beach ridge or strandline sands can be distinguished from eolian reworked sands by the presence of scattered marine and intertidal shells. The eolian portions of the beach ridge and dune barriers are characterized by moderately well sorted fine- to medium-grained sand but generally show no internal sedimentary stratification, which has probably been destroyed by roots and burrowing activity. Thermoluminescence dating has suggested that eolian activity may have continued for some time after the formation of the beach ridge (Fig. 17). Apart from the modern back-beach area, where small wind-blown dunes are present,

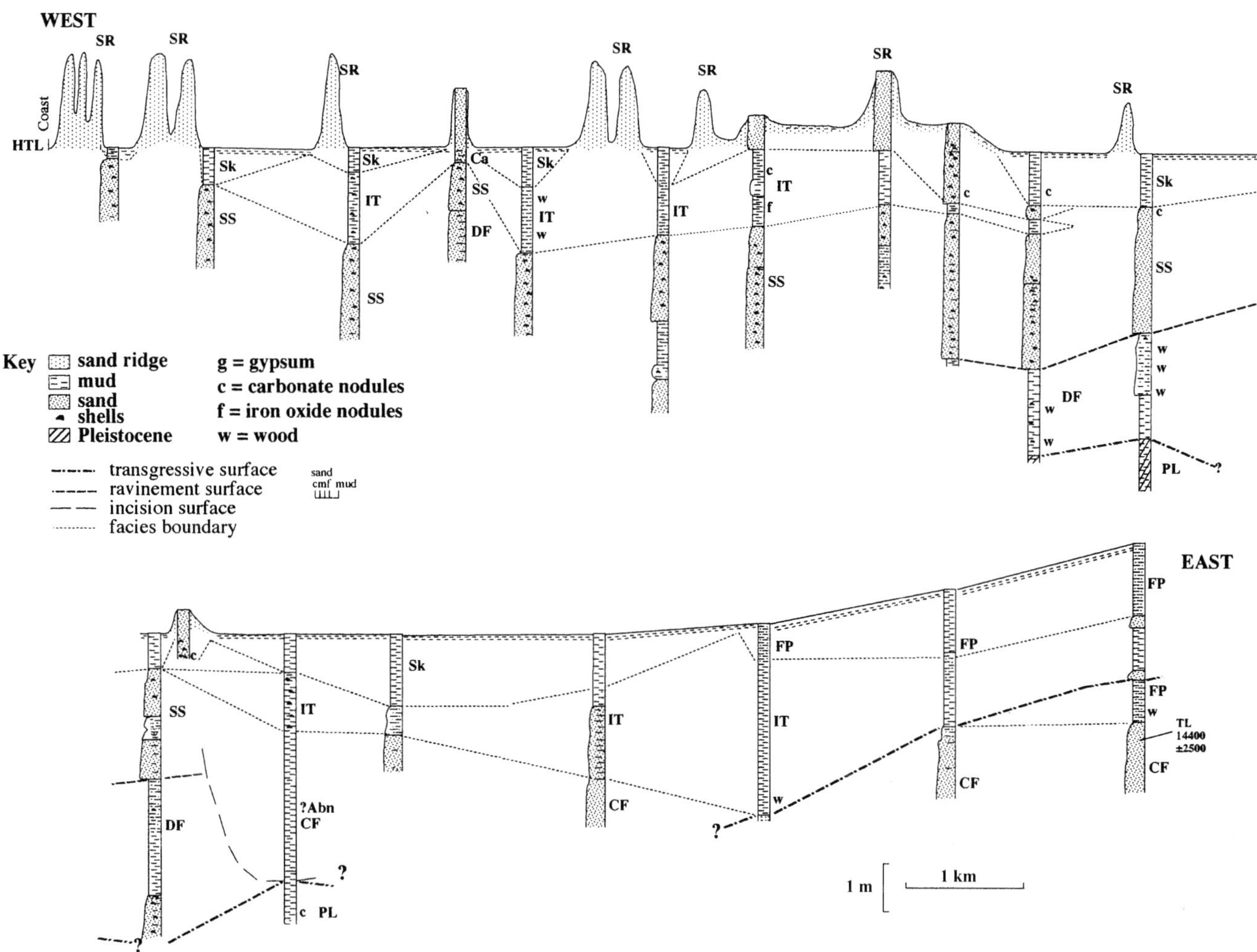

FIG. 16.—West–east drill-hole cross section between the Smithburne River and Van Diemen Inlet showing the beach ridges overlying thick subtidal delta deposits. This reflects periodic delta growth in the region. The deltaic deposits pass landwards into fluvial-channel and floodplain deposits. The presence of Late Pleistocene fluvial deposits in this area indicates that no major incision occurred during the low stand of sea level. CF = channel fill; DF = delta front; FP = flood plain; IT = intertidal; PL = Pleistocene; PM = prodelta mud; Sk = supratidal (sabkha), So = soil; SR = strandline sand ridge; SS = subtidal sand; and TC = tidal channel.

most of the older beach ridges are partially to completely vegetated, with very little remaining mobile sand. Open eucalypt woodland is present on some of the older sand ridges, but most of the younger ridges are covered in vines. Large termite mounds occur on many of the older landward sand ridges and are useful for identifying partly buried ridges with a subdued elevation.

Distributaries.—Seven major distributaries and / or tidal channels emerge onto the coast in the Gilbert delta area (Fig. 12). Very few minor channels breach the beach ridges to reach the coast in the northern half of the delta, whereas small channels become more numerous southwards from the Smithburne River to Karumba, where there is essentially no modern beach ridge and the supratidal mudflats merge into the intertidal and subtidal sandy mud.

The seaward ends of the major channels are either straight or gently curved over distances of 2–4 km, and they typically show a landward-tapering channel cross section with the width decreasing to approximately half the mouth width (see fig. 10 in Jones et al., 1993). Above this short straight reach the channels become highly sinuous (Figs. 13, 14), with active lateral migration leading to the formation of point-bar sequences, neck cut-offs, and oxbow lakes partially filled with peat and organic-rich mud. The channels are tide-dominated for up to 15 km from the coast, where they become river-dominated for the remaining tidally influenced section (Fig. 12), as well as farther inland. Jones et al. (1993) have provided details of the tidal and fluvial channel facies.

The tide-dominated portions of the channel are characterized by mud deposition, with sand being confined to the lower part of the 5–6 m deep channel (e.g., Fig. 17). The extent and thickness of the tidal-channel sand bodies could not be assessed accurately because auger-drilling recovery failed in the cleaner unconsolidated sand. The exposed channel banks are all covered with poorly consolidated mud. Tidal-channel facies can commonly be recognized by the presence of woody debris, especially in abandoned channel-fill deposits. The mud in the tidal-channel facies is brought in from the subtidal delta and offshore mud facies during the dry season, when river flows are minimal. The mud settles onto the channel banks and mudflat surface during the

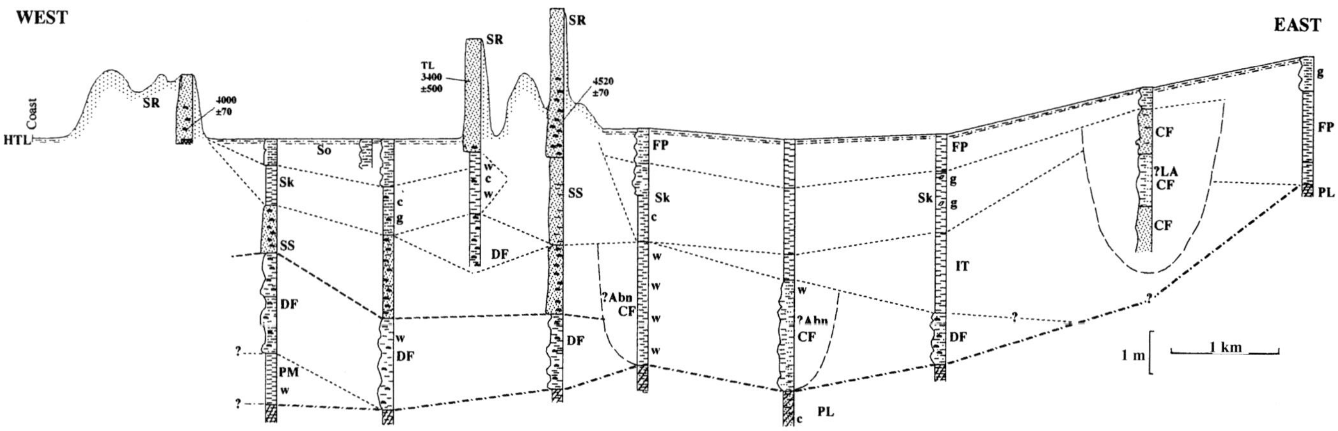

FIG. 17.—Cross section north of Van Diemen Inlet showing two major episodes of delta growth with associated beach sand-ridge development. The landward portion of this section is dominated by channel-fill deposits overlain by either supratidal or floodplain muddy successions. Details of channel-fill deposits in the area are given in Fig. 3 in Jones et al. (1993). For symbols see Fig. 16.

slack water period at high tide, and the ebb tide is generally much clearer than the flood tide.

In contrast, the river-dominated portions of the upper delta plain contain a greater proportion of sand, and lateral-accretion deposits on the point bars show alternating sand and mud layers, with the proportion of mud increasing both vertically and downstream (Figs. 14, 16). Tidal activity in the upper delta plain area, near the junction with the alluvial plain, produces a loss of competence within the river flow, and large quantities of sand are deposited during the waning stages of monsoonal flows. This sand tends to fill the river channels, further reducing their transport capacity, and results in numerous thin splay sand sheets extending away from the channel for up to a few hundred meters. The current Gilbert River channel is too small and clogged with sand to be able to carry the whole monsoonal flood (up to 30,000 m^3/s in January 1974). This leads to extensive overbank flow in the lower alluvial and delta-plain areas during floods, thus partly reactivating many of the older distributary channels, especially Snake Creek. However, apart from the distributaries feeding into Snake Creek and the localized splay sand deposits, the overbank flow carries very little sand. The older main distributary channels branching off higher in the alluvial plain typically carry only fresh water and mud to their respective delta plains.

The channel banks in the tide-dominated areas support a luxuriant fringe of mangroves, 5–10 m wide, whereas the scroll bars and point bars may support more extensive mangrove communities (Fig. 19). Farther upstream the mangroves give way to eucalypts in the alluvial-plain area.

Intertidal facies.—Intertidal facies occur in sheltered areas along the coastal fringe, along tidal-channel margins and in cut-off channels. This facies consists mainly of homogeneous blue-gray clay containing woody material, rhizomorphs, or strong vertical mottling attributed to root traces. It displays the characteristics of sediment accumulation in the zone of tidal movement that are suitable for development of mangrove vegetation. It may interdigitate with delta-front or tidal-channel facies as a result of migration of coarser material within these facies. The size and thickness of the intertidal deposits (Fig. 20) is directly related to the tidal range, and this facies is less prominent in the rivers entering the Gulf of Carpentaria than in more tidally active areas (see Thom et al., 1975; Woodroffe et al., 1989).

Supratidal mudflats.—Supratidal mudflats extend for up to 20 km landwards from the coast and are present adjacent to all the major and minor distributary and tidal channels (Fig. 13). The mudflats are very prominent south of the Smithburne River, where they occupy most of the lower delta plain, which wedges out southwards between Accident Inlet and Karumba. All the mudflats have a very low relief, with maximum elevations reaching about 2 m AHD. During the monsoon period the combination of tidal and fluvial flow causes frequent inundation of the mudflats and allows algal growth and deposition of mud. For much of the year the mudflat surface is exposed, because spring high tides inundate only about half the area for a few days each month. During the dry season the margins of the mudflats are subject to wind deflation and reworking of pelletized mud, whereas the area covered with precipitated salt suffers very little deflation. Some of the windblown mud accumulates around vegetation, especially along the network of distributary channels, and as low clay dunes in the lee of some sand ridges.

The supratidal mudflats occupy much of the area between the strandline sand ridges as well as covering, or partly covering (e.g., Fig. 14), former channel and point-bar successions. In most areas the mudflats gradationally overlie shelly intertidal estuarine deposits consisting of variable quantities of fine sand and mud (Fig. 20). The contact between mudflat and tidal point-bar deposits is also usually gradational, with the upper muddy point-bar deposits giving way to an increasing quantity of sand downwards towards the channel base. Where a mudflat covers an abandoned channel succession, the mudflat grades down into an organic-rich peaty mud filling the old channel, whereas when a mudflat facies overlies channel-fill sand deposits the contact is commonly sharp.

The mudflat facies is generally only 1–2.5 m thick and consists of massive khaki-brown clay that typically shows an upward-fining sequence in vertical sections. The lower part of the mudflat sequence may contain occasional thin interbeds of sand. This is overlain by gray bioturbated silty clay that may show minor oxidation mottles, especially along mangrove root traces. Rare shells may be present in this facies, particularly adjacent to shelly beach ridges. The upper part of the mudflat facies commonly shows no internal structures and consists of homogenized variably mottled clay with some algal detritus (generally $<3\%$). Some areas display a pelletal appearance indicating bioturbation. Small

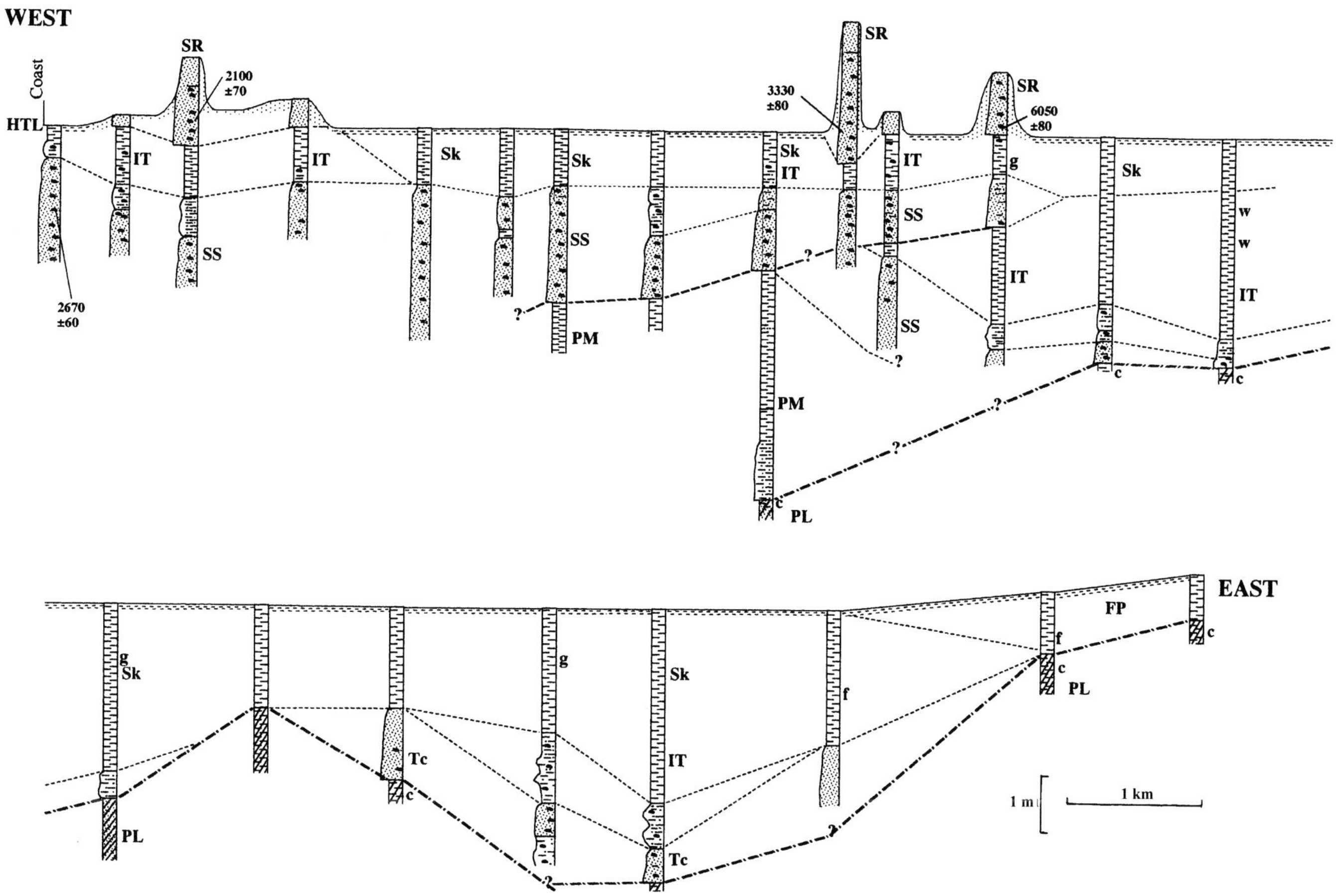

FIG. 18.—Low rates of sand supply are a characteristic feature in the area south of Accident Inlet. Low beach ridges occur above the restricted subtidal and intertidal sandy succession in this area. Thick prodelta, intertidal, and supratidal clays are the dominant lithology in this area, overlying the somewhat irregular topography on the shallow Pleistocene surface. For symbols see Fig. 16.

discoidal and rosette gypsum crystals may be present in the lower part of this unit and the upper part of the underlying unit, particularly around the margins of the mudflats where desiccation is more pronounced and flooding is less frequent (Figs. 17, 20–22). The surface of the mudflat is typically covered with desiccation cracks and a halite efflorescence in the drier portions of the mudflat, whereas pustular and laminated algal mats are common in the lower parts of the mudflats, especially adjacent to the channels. During the dry periods the algal mats are black and usually 2–5 mm thick, with high temperatures (40–50°C) being recorded in the sediments immediately below the mat. These elevated temperatures may account for the development of early diagenetic dolomite and gypsum. The supratidal mudflat facies is equivalent to a paralic sabkha (Sonnefeld and Perthuisot, 1989) or a marginal marine sabkha (Moore, 1989).

Subtidal delta.—The subtidal deltas generally occur as broad arcuate sand bodies 2–4 km wide that extend seaward from each of the major distributary and tidal channels. These subtidal sand bodies merge both north and south into narrower intertidal sand flats flanking the modern beach ridges.

The arcuate subtidal sand bodies have gently sloping surfaces that become extensively exposed, especially during spring low tide. Their surfaces are dissected by a number of tidal channels, commonly up to 3 m deep and 25 m wide, that serve to flood and drain the extensive intertidal flats. One or more deeper tidal channels usually meander across each subtidal delta to link with the main tidal channel extending inland. The main ebb tide–dominated channels are lined with medium-grained sand and shell detritus, and constitute the coarsest facies in the subtidal delta. However, even these channels are not always navigable at low tide. The subtidal sand deposits are subject to wave action, and much of the muddy detritus is winnowed out of the system but accumulates as clay drapes in the offshore portion of the delta and delta slope, as well as clay deposits along the tidal channels landwards from the delta mouth. The delta-top sands tend to be poorly to moderately sorted and very fine- to medium-grained, and they commonly show ripple bedding or have been homogenized by bioturbation. Lenses of shell detritus represent lag layers formed during periods of stronger wave and tidal action. The tidal channels show evidence of migration across the subtidal delta, with lateral accretion deposits defined by the presence of mud drapes. A low island is present on the Smithburne subdelta, where wave action has produced a raised beach and allowed accumulation of low back-beach dunes. North of the Smithburne subdelta the arcuate shape of the subtidal sand bodies becomes elongated northwards as a result of longshore tidal currents impinging on the coast. Although subdeltas are present at Van Diemen Inlet and at Duck Creek, an asymmetrical form is most apparent at the current mouth of the Gilbert River, where a

FIG. 19.—The meandering Gilbert River channel in the tide-dominated area is fringed by mangroves, whereas the scroll and point bars support more extensive mangrove communities.

seaward sand barrier has built up on the southern margin and the subdelta extends about 6 km north of the present river mouth.

The seaward edges of the subtidal deltas typically show an abrupt increase in depth from about 1 m to about 3 m over a lateral distance of a few tens of meters. This increase in depth also marks an abrupt change from sand to prodelta mud facies. The deeper-water facies consists of blue-gray silt and clay with a few thin sandy interbeds and shells and rare wood, especially in the proximal areas adjacent to prograding deltaic sand bodies. These coarser beds may represent storm-redeposited sand or proximity to ebb-tidal-channel sand lobes. The prodelta mud facies forms a very extensive shallow marine unit along the eastern and southern shores of the Gulf of Carpentaria. It becomes thinner in a seaward direction, where it forms only a thin veneer over Last Interglacial lowstand sand deposits.

Between the subtidal-delta sand bodies the intertidal sand flats are generally 50–100 m wide and pass seawards into subtidal sand and mud with no abrupt changes in gradient. The intertidal flats consist of ripple-bedded sand with occasional thin mud drapes. Bivalve shells are scattered across the intertidal flats and tend to become concentrated as thin lag deposits, especially along the base of the beach slope and near the top of the swash zone. Gastropod shells are less abundant but still form an important component. During periods of strong onshore winds the upper part of the intertidal sand flats dries out and sand is blown up onto the beach ridges, forming small low dunes.

Evolution of the Gilbert Delta

The Gilbert delta occupies an area of the Karumba Basin that has been slowly subsiding since the Late Tertiary (Doutch, 1976), with at least episodic deposition of fluvial deposits and development of soil profiles during this period. Dating of the older fluvial channel and floodplain successions in this basin is largely conjectural and is based mainly on correlated soil profiles. Most of the sand bodies contain some muddy matrix or broken down lithic grains that provide a high annual radiation dose and limit thermoluminescence (TL) dating to a maximum of about 80 ka (Nanson et al., 1991).

Thin fluvial-channel and floodplain deposits of oxygen isotope stage 3 age are present along the margins of a number of channel axes, where they have generally been strongly indurated by pedogenic processes. Still older sands beneath an extensive mud layer are beyond the range of TL dating but, based on the evidence from U/Th dating of pedogenic material, they may represent Last Interglacial deposits (Nanson et al., 1991). Preserved fluvial deposits are sparse during the period of the last glacial maximum but have been recorded near the Smithburne River and Kelso Pocket (Figs. 16 and 21). The lack of river-channel incision during the glacial lowstand of sea level (as noted by Jones et al., 1993, fig. 6, in cross-channel drillhole profiles) and the continued deposition of fluvial sand can be attributed to the essentially constant gradient of the onshore and offshore portions of the delta, i.e., no gradient advantage was generated by the last glacial fall in sea level. The early and late parts of this glacial period are also typified by the development of both calcrete and ferricrete in the soil profiles (Nanson et al., 1991). These soil profiles contain yellow-brown mottled clays in the upper parts of the profiles that are readily distinguishable from the unaltered Holocene deposits. This pre-Holocene oxidized surface represents the regional Holroyd surface defined by Doutch (1976).

The Holocene succession on the Gilbert delta can be divided into a number of zones. In the river-dominated alluvial and upper delta plain the Holocene sequence consists of sand-filled or partially filled channels representing the main drainage axes on the fan surface. Most individual channel axes have been occupied more than once during the Holocene as the river has avulsed across the fan surface. The abundance of preserved channel sand

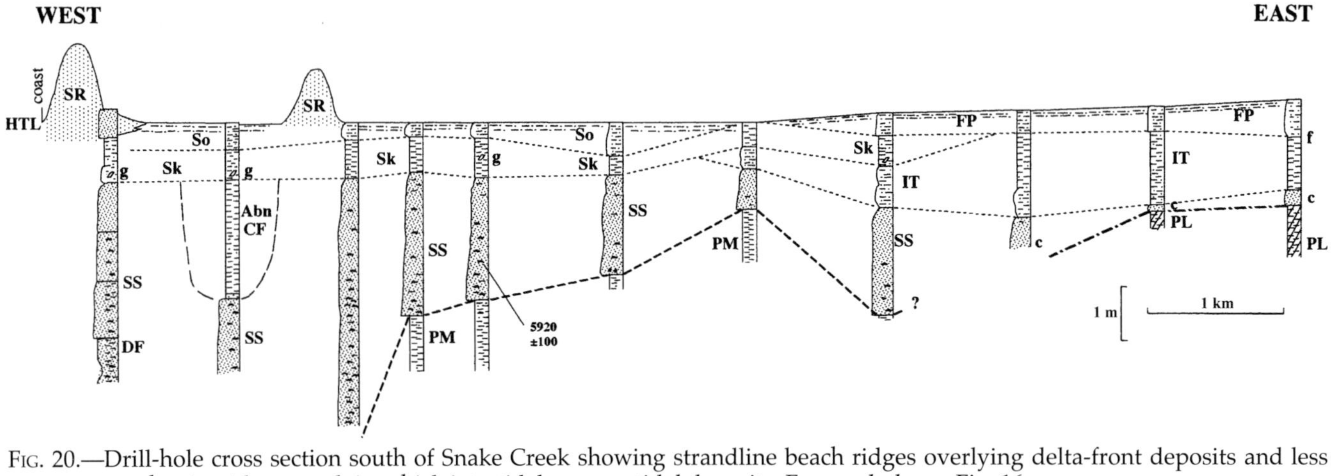

FIG. 20.—Drill-hole cross section south of Snake Creek showing strandline beach ridges overlying delta-front deposits and less common chenier ridges overlying thick intertidal or supratidal deposits. For symbols see Fig. 16.

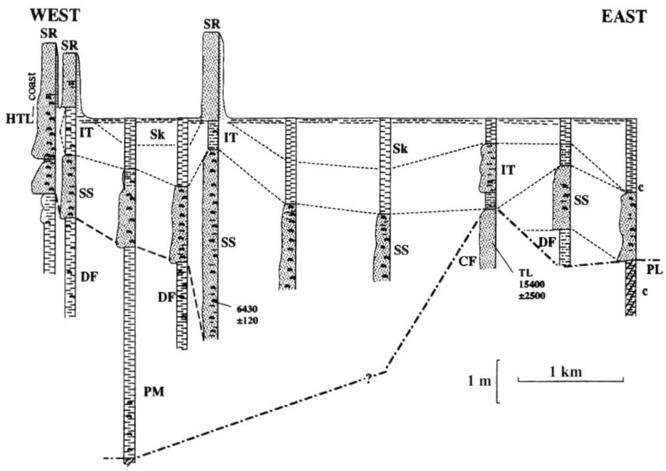

FIG. 21.—Drill-hole cross section at Kelso Pocket showing strandline beach ridges overlying delta-front deposits. For symbols see Fig. 16.

in the lower reaches of this zone would be a function of rising base level during the postglacial transgression and increasing tidal influence in the distal portions of the channels. As base level rose, competence of the channel to transport sand would decrease, especially during the waning stages of the monsoonal flow period. This would cause the channel to aggrade rapidly and become clogged with sand (as in the present Gilbert River near the limits of tidal activity). Succeeding floods would start with a large quantity of overbank flow that led to frequent avulsion events. Thus most of the major channel axes in the northern half of the delta were active at some period during and following the transgression, whereas intervening floodplain areas become more prominent in the southern part of the delta (Fig. 12).

The lower delta plain forms a wedge of sediment that has accumulated during and since the maximum Holocene transgression. Shell material in the landward portion of the easternmost strandline sand ridges gave radiocarbon dates of 6,410 ± 100 and 6,050 ± 80 yr BP, and the subtidal sands range up to at least 6,430 ± 120 yr BP. These deposits, therefore, represent the approximate age of maximum transgression in the Gilbert delta. Since this time the delta has extended 10–15 km seawards, with successive strandline deposits showing younger dates down to 1,990 ± 90 yr BP at the rear of a subrecent strandline ridge. The rate of accretion has not necessarily been constant. For example, near Snake Creek a phase of beach-ridge building (6,410–5,530 yr BP) was followed by a wide mudflat development during a period of low sand supply and then another period of beach-ridge formation (3,240–2330 yr BP). The mudflat and lower-delta-plain deposits between the sand ridges consist of a thin vertically accreted mud blanket overlying intertidal and / or estuarine muddy sands—the latter probably accumulating by longshore drift from the subtidal delta by longshore drift. Within the major channel areas the intertidal deltaic and estuarine sand deposits have been at least partly reworked by the meandering tidal channels forming a complex of sand and mud deposits. However, most of the lateral build-out of the lower delta plain is not attributed to extensive sedimentation but rather has been caused by slight hydroisostatic uplift along the gulf margin, which has resulted in a sea-level fall of about 1.5 m since 3,000 yr BP (Nakada and Lambeck, 1989; Woodroffe and Chappell, 1993).

The offshore intertidal and subtidal sand bodies adjacent to each distributary channel mouth represent contemporary accumulations of sand that are slowly being distributed along the shore to form new strandline ridges. Farther offshore the Holocene mud blanket is thin, leading in some cases to obtaining both Holocene and Pleistocene material in a single grab sample (Yassini et al., 1993). A possible drowned sandy strandline deposit was recorded some 4–5 km offshore in water depths of 5–7 m off Van Diemen Inlet by bathymetric profiling and sampling.

Wave, Tidal, and Fluvial Processes in the Gilbert Delta.—

Fluvial processes are dominant in the alluvial and upper-delta-plain areas. These fluvial processes are active during and immediately after the summer monsoon from December to March, with very low flows or no surface flows for the rest of the year. During the wet season the lower delta plain of the Gilbert River system is also river-dominated, with very large fluvial drainage rates and rapid dumping of material onto subtidal mouthbars. In contrast, during the dry season the subtidal mouthbar deposits dominantly experience wave energy conditions, with moderate amounts of tidal energy transporting material within channels in the lower delta plain. Jones et al. (1993) have previously discussed these depositional processes and have shown that the tidal-channel deposits are very heterogeneous. The prominent point-bar deposits in the meandering tidal channel reaches are characterized by alternating sand layers, deposited during the monsoon-generated fluvial discharge, and thicker mud layers laid down by tidal action during the dry season. During the dry season wave energy sorts the rapidly dumped material on the subtidal mouthbar, resuspends the mud fraction, and moves it landward to coat the tidal channels and cause vertical accretion on the adjacent tidal flats.

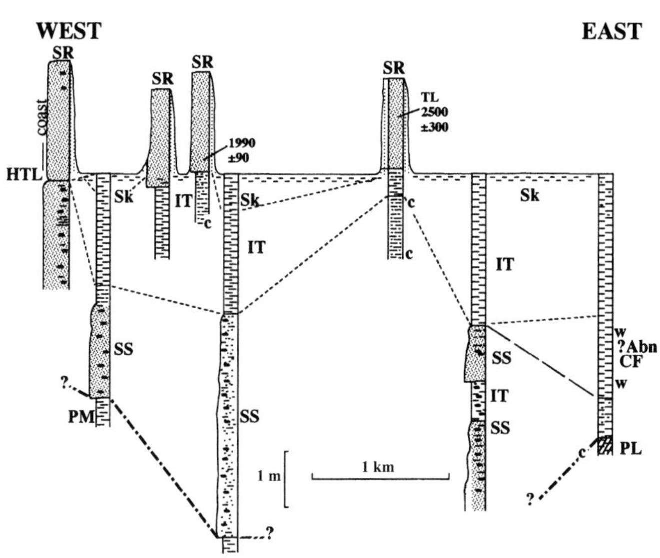

FIG. 22.—Drill-hole cross section south of Duck Creek showing strandline beach ridges overlying delta-front deposits. For symbols see Fig. 16.

Lateral and Vertical Distribution of Facies in the Gilbert Delta.—

Progradation of the delta has resulted from the high rate of sediment supply and the low regional gradient. Avulsion processes operating within the river system farther inland have controlled lateral variation in sediment supply rate and the local facies stacking pattern in the delta region.

In the vertical profiles, prodelta and delta-front facies form the lowermost units overlying a transgressive erosion surface cut into Pleistocene deposits. This surface exerts a local control on the lateral thicknesses of these lowest facies in the paleochannel areas (e.g., Figs. 16 and 17), but they are thin and widespread in the offshore region. The buildup of beach ridges during periods of active sand supply to the delta provides the protected environments necessary for accumulation of vertical-accretion deposits in the intertidal and supratidal facies. The intercalation of intertidal and delta-front facies on the inner portion of the subtidal delta is caused by periodic shoreward migration of coarse terrigenous and bioclastic material over the intertidal mudflats.

Most of the depositional facies boundaries are gradational, apart from highstand ravinement surfaces beneath the subtidal sands and the incision surfaces caused by channel migration. Thus the prodelta mud facies coarsens upwards into the delta-front deposits but the junction with the subtidal sand facies is commonly sharp. The latter facies normally grades upwards or laterally into the intertidal muddy sand to sandy mud succession (Fig. 21). The passage from both these latter facies into the supratidal mudflat facies is a gradational upward fining. The transition from the mudflat to alluvial deposits is also gradational, being marked by the development of soil profiles (Figs. 17 and 20). In other areas the subtidal sand facies coarsens up into the beach-ridge facies, which may then fine upwards into eolian ridge-top and back-ridge fine sands (e.g., Figs. 16, 17, 20). In contrast, the chenier sand ridges generally have sharp lower and upper or lateral contacts with intertidal or supratidal muddy facies (Fig. 18). The tidal and fluvial channel facies show abrupt facies changes along the erosional bank and thalweg (Fig. 18). However, the point-bar and abandoned-channel-fill deposits may grade upwards into finer overbank and supratidal mudflat facies (Fig. 17).

From Figure 12 it can be seen that the strandline beach ridges fan out northwards from each major river mouth and would have grown during periods of active sediment supply and longshore transport. Because upstream avulsion processes control river-mouth sedimentation, the variation in age of beach-ridge successions associated with each river mouth is not surprising. However, the dating program so far has been inadequate to clearly document the times of active delta growth at each river mouth and the consequent northward progradation of strandline beach ridges from the river mouth. During periods of low sediment supply, intertidal and supratidal mudflats may form on the seaward margins of the beach ridges and may form sites of development of chenier ridges, for example along the current coast north of Snake Creek (Fig. 15).

Sand-Body Geometry.—

Fluvial channel belts in the lower alluvial plain and upper delta plain are typically 3–5 km wide and contain up to 12 m of multistory sand bodies. Individual sand bodies are generally < 6 m thick, and lateral-accretion point-bar deposits become more prominent downstream, where a higher proportion of mud is preserved in the upper delta plain. These channel successions are separated by floodplain areas 10–20 km wide, especially in the southern half of the delta, that commonly have only a thin veneer (~ 1 m) of Holocene floodplain mud with occasional incursions of splay sand sheets. The latter are commonly elongate (0.5–1 km long and a few hundred meters wide) rather than lobate and typically have a maximum thickness of < 50 cm.

In the lower delta plain the tide-dominated channels are very similar in character to the fluvial channels, but they tend to be heterolithic with a higher proportion of mud. The mouth of the channel through the beach-ridge system is typically narrower (< 1 km), straight, and tapering inland to where the strongly meandering channels produce a meander belt up to 5 km wide. In addition to mud on the lateral-accretion surfaces, cut-off meanders and abandoned loops tend to be filled with a plug of mud up to 300 m wide and 6 m thick.

The beach and chenier ridges are typically 2–5 m high, and up to 500 m wide, and extend parallel to the shoreline for 5+ km. Some of the beach ridges coalesce to form a wider sand sheet, and most of the ridges terminate adjacent to tidal channels. Some of the older ridges have been partially buried by supratidal or alluvial-plain muds.

The supratidal mudflats form laterally extensive (up to 10 km x 20 km) thin (< 2 m) blanket deposits usually overlying intertidal mud or subtidal sand deposits but also occurring above tidal-channel facies. The intertidal sandy mud facies is also laterally extensive, generally being thin in interchannel regions, but it may form thicker lenses near the channels. In the interchannel areas this facies forms part of an upward-fining unit at the top of the subtidal sand sheet.

The subtidal sand sheet is produced as a lobate sand body, or subtidal delta, extending offshore for up to 5 km adjacent to the active channel mouth. These sand bodies are reworked by tidal and wave activity to produce a shore-parallel sand sheet, espe-

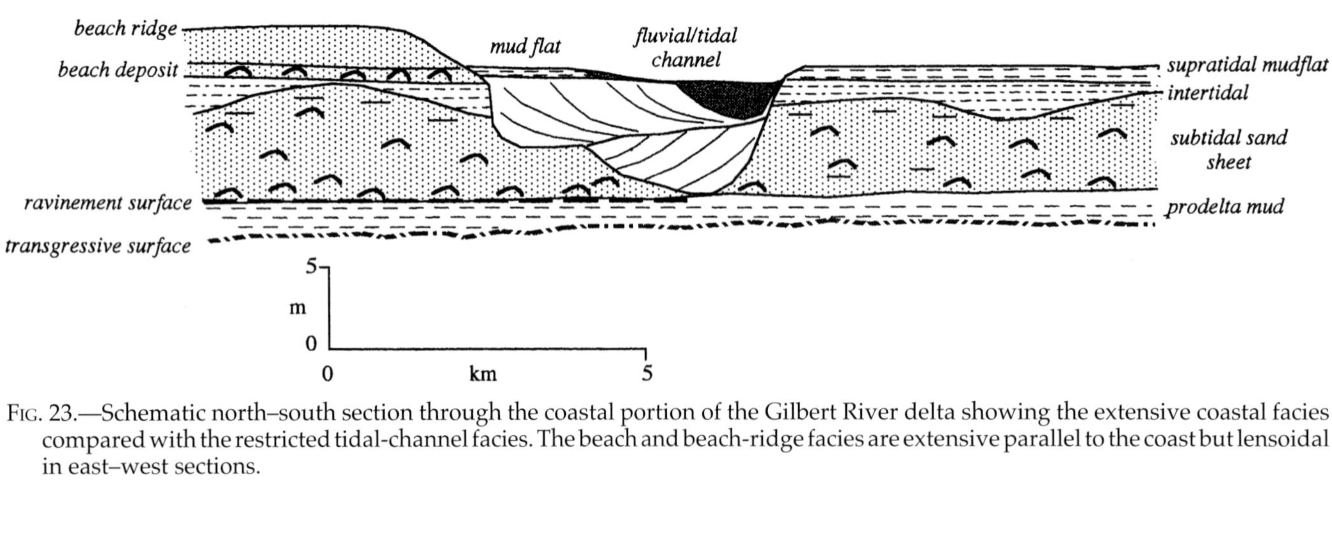

FIG. 23.—Schematic north–south section through the coastal portion of the Gilbert River delta showing the extensive coastal facies compared with the restricted tidal-channel facies. The beach and beach-ridge facies are extensive parallel to the coast but lensoidal in east–west sections.

cially in the northern half of the delta area. This sand sheet appears to extend beneath most of the subaerial portion of the delta and is up to about 8 m thick. Much of the sand is silty and shelly, and it commonly shows an upward coarsening to about low-tide level. Where it is directly overlain by beach facies it continues to coarsen upward. The sand sheet either gradationally overlies the prodelta facies or it shows an abrupt transition at a highstand ravinement surface. The prodelta silty mud facies occurs as an extensive thin deposit overlying the transgressive surface cut into the Pleistocene succession.

Figure 23 shows a schematic north–south section through a single distributary in the Gilbert delta system showing the vertical and lateral relationships between the coastal and tidal-channel facies.

DISCUSSION

The delta of the McArthur River occurs in a sheltered part of the Gulf of Carpentaria and is influenced by a highly monsoonal fluvial discharge and a predominantly diurnal tidal regime. The delta reflects fluvial dominance in the scroll plain of the upper delta plain, within which lateral migration of the channel occurs with lateral-accretion deposits (primarily point bars). In the lower delta plain, the extensive saline mudflats are the result of emergence of shelly sands deposited in a delta-front environment. Active distributary channels are characterized by nearly constant width. Abandoned distributaries adopt a rapidly tapering form, but the lower-lying areas adjacent to them, which are now covered by mangrove forests, indicate the former extent of river-dominated distributaries (marking earlier mouths of the McArthur River). These tide-dominated channels meander, but they have apparently adopted their present course through avulsion, in that the saline mudflats between them (and into which they are partially incised) have not been reworked but contain shell beds indicative of the original delta-front deposition. Although these tide-dominated channels are characterized by high sinuosity, their pattern is in fact inherited, and partly constrained, from former fluvial channels.

Within the McArthur delta, features such as reentrants and the arcuate mid-channel islands reflect progressive influence of riverine processes downstream onto the lower delta plain. River–tidal interactions not only are apparent in the modern delta morphology but evidently have been through a complex pattern of evolution through time. The Gilbert River, by contrast, shows a gradually prograding sequence of beach and chenier strandline ridges separating areas of supratidal mudflats and floodplain deposits. These latter vertical-accretion deposits partly bury the older strandline ridges, which generally have a subdued relief. The progradation of the Gilbert delta is evident from the subsurface facies distribution, whereby prodelta mud is overlain successively by subtidal delta sands, strandline beach ridges, and supratidal mudflats. This progradation has occurred over the past 6,500 years and has built up a coastal wedge of Holocene sediment up to 10 m thick and extending laterally for a distance of 15–20 km seaward and 125 km along the coast.

The deltas and estuaries of northern Australia show many of the broad characteristics typical of those of the Asia–Pacific region (Woodroffe, 2003). In particular they occupy the same Indo–West Pacific biogeographical region and are dominated by similar vegetation, with mangrove forests in intertidal areas and grass-sedgelands or tropical forest in supratidal areas. For much of the year the northern Australian continent is dry, and relatively little sediment is supplied from catchments to the coast during the monsoon season. As a consequence, the importance of tidal processes in shaping deltaic–estuarine depositional environments is apparent through most of the year. The balance between tidal and fluvial processes forms part of a continuum of monsoonal systems ranging from the microtidal, predominantly diurnal Gulf of Carpentaria to the macrotidal semidiurnal tidal rivers of northern and northwestern Australia. The morphology of deltaic–estuarine landforms, together with seasonal observations of the systems and stratigraphic and radiometric-dating control, provide insights into the dynamics of these systems.

In the Gulf of Carpentaria both the Gilbert and McArthur River systems have developed delta forms, as have other rivers around the margin of this low-gradient epicontinental sea. The Gilbert River has formed a broad (120 km), gently arcuate delta front with beach ridges separating a succession of tidal channels. The latter represent previous river mouths during the last 6,500 years, reflecting the avulsion processes operating in the sand-dominated fluvial portion of the delta. Sand is carried to the delta mouth during the monsoon period, forming an extensive thin subtidal sand sheet whose distribution is controlled by the northward circulatory current in the gulf. Tidal processes are dominant in the seaward portions of the active and abandoned distributary channels for about nine months of the year. During these periods wave-suspended mud is transported landwards by tidal action and has produced extensive supratidal mudflats landwards of, and interspersed with, the beach ridges. Under the microtidal conditions in this area, the channel assumes a meandering form close to the channel mouth, producing extensive estuarine meander-belt deposits.

In the case of the McArthur River, a delta has developed downstream of the incomplete confluence of two river channels. Radiocarbon dating indicates gradual progradation of a now emergent Holocene delta front across which the river channel has episodically avulsed. Upon abandonment, prior distributaries become tide-dominated and are infilled to adopt a tapering and shoaling morphology.

In contrast, the seasonal rivers of the Top End of Australia adopt an estuarine form crossing broad horizontal Holocene coastal plains, usually in one discrete, well-mixed tidal channel. Whereas sinuous meandering, mangrove-lined channels dominate many rivers (e.g., the Adelaide River), others (e.g., the South Alligator River) contain reaches with cuspate estuarine channels (Woodroffe, 2003). The channel migrates across the plains, though at different rates, reworking different proportions of the plain (Woodroffe et al., 1989; Woodroffe et al., 1993). The relative balance of tide versus river is evidently a factor. The efficacy of tidal processes alone is clearly demonstrated by the Mary River, which until the 1950s appeared not to reach the sea as a single channel. Over the last 50 years tidal channels have incised into the plains, exploiting paleochannel sand bodies and leading to the rapid salinization of wetlands and an expanding tapering tidal network (Mulrennan and Woodroffe, 1998).

In contrast, estuaries in northwestern Australia that are affected by macrotidal conditions show a distinctly different character with a very extensive tidal prism. Typical examples are the Ord River, in Cambridge Gulf, and the Fitzroy River, in King Sound (Wright et al., 1973; Jennings, 1975; Thom et al., 1975). These estuaries exhibit a broad low-sinuosity funnel shape with a high width-to-depth ratio. The large tidal flux produces efficient sand transportation and results in the development of large bedforms and longitudinal bars in the lower two-thirds of the estuary, with an abrupt change to meandering-style estuarine point-bar accretion in the upper third. The tidal channel is flanked by up to 100 m of mangrove swamp, followed landward by broad supratidal mudflats. The efficiency of the tidal currents means that the tidal channels act as transport corridors, and apart from minor vertical accretion on the supratidal mudflats there is no net delta growth. The intensity of tidal processes thus controls the delta–estuary form and produces characteristic depositional sequences.

CONCLUSIONS

McArthur delta has its shape and extent controlled by bedrock. It is protected from significant wave action by the presence of offshore islands and hence contains no major beach-ridge systems. This allows easy migration of the delta mouths and a closer spacing of both large and small tidal channels across the seaward margin of the delta. The offshore area is very shallow, and the delta mouthbars are reworked to form extensive subtidal sand sheets dissected by subtidal channels. The subtidal sand sheets extend 2–4 km beyond the margin of the delta.

The Gilbert delta is more exposed to wave action and has built up elongate beach ridges and cheniers that restrict fluvial and tidal access to the coast. Channels become strongly meandering landward of the beach-ridge constriction, but on the alluvial plain lateral channel movement is restricted by the development of soil profiles. Along the front of the Gilbert delta, mouthbar deposits are restricted laterally to areas around the mouths of the active distributaries, but longshore movement of sand does occur under the influence of stronger tidal circulation and waves.

Thus even though the McArthur and Gilbert deltas occur in a similar mesotidal setting, their morphology is controlled by local factors such as the presence of bedrock and a higher wave energy regime along the eastern coast of the gulf. Tide domination characterizes those channels that are not presently receiving substantial fluvial flows, and active distributaries become progressively shaped by tidal processes, despite the relatively limited tides in the Gulf of Carpentaria, once they are abandoned.

ACKNOWLEDGMENTS

This research was funded by the Australian Research Council, the University of Wollongong, and the Australian National University. We thank Comalco Aluminium Ltd. for logistical support and Morr Morr Pastoral Company Pty. Ltd. for access in the Gilbert River delta area. John Chappell, Eugene Wallensky, Kristin Bardsley, Damien Kelleher, Colin Holbert, Peter Ward, Wang Pixian, and Russell Hanley provided valuable field assistance in the McArthur River delta area. The Australian National University Radiocarbon-Dating Laboratory, University of Waikato Radiocarbon Dating Laboratory, and David Price from the University of Wollongong Thermoluminescence Laboratory provided the new dates recorded in this paper. D.M. Martin and R. Miller drafted the figures. The aerial photographs (Figs. 2, 13–15) are Crown Copyright and have been reproduced by permission of the General Manager, Australian Surveying and Land Information Group, Department of Administrative Services, Canberra.

REFERENCES

ALLEN, G.P., LAURIER, D., and THOUVENIN, J., 1979, Étude sédimentologique du delta de la Mahakam: Compagnie Français des Pétroles, Notes et Mémoires, v. 15, p. 13–156.

ALLISON, M.A., KUEHL, S.A., MARTIN, T.C., and HASSAN, A., 1998, Importance of floodplain sedimentation for river sediment budgets and terrigenous input to the oceans: insights from the Brahmaputra–Jamuna River: Geology, v. 26, p. 175–178.

ANDERSON, J.A.R., 1964, The structure and development of the peat swamps of Sarawak and Brunei: Journal of Tropical Geography, v. 18, p. 7–16.

BAKER, E.K., HARRIS, P.T., KEENE, J.B., AND SHORT, S.A., 1995, Patterns of sedimentation in the Fly River Delta, *in* Flemming, B.W., and Bartholomä, A., eds., Tidal Signatures in Modern and Ancient Sediments: International Association of Sedimentologists, Special Publication 24, p. 193–211.

BARDSLEY, K.N., 1985, The effects of Cyclone Kathy on mangrove vegetation, *in* Bardsley, K.N., Davie, J.D.S., and Woodroffe, C.D., eds., Coasts and Tidal Wetlands of the Australian Monsoon Region: Canberra, Australian National University, North Australia Research Unit, p. 167–185.

BHATTACHARYA, J.P. and WALKER, R.G., 1992, Deltas, *in* Walker, R.G., and James N.P., eds., Facies Models: Response to Sea Level Change: Toronto, Geological Association of Canada, p. 157–177.

CHAPPELL, J., 1993, Contrasting Holocene sedimentary geologies of lower Daly River, northern Australia, and lower Sepik-Ramu, Papua New Guinea: Sedimentary Geology, v. 83, p. 339–358.

COLEMAN, J.M., 1988, Dynamic changes and processes in the Mississippi River delta: Geological Society of America, Bulletin, v. 100, p. 999–1015.

COLEMAN, J.M., GAGLIANO, S.M., and SMITH, W.G., 1970, Sedimentation in a Malaysian high tide tropical delta, *in* Morgan, J.P., ed., Deltaic Sedimentation; Modern and Ancient: Society of Economic Paleontologists and Mineralogists, Special Publication 15, p. 185–197.

COLEMAN, J.M. and WRIGHT, L.D., 1978, Sedimentation in an arid macrotidal alluvial river system: Ord River, Western Australia: Journal of Geology, v. 86, p. 621–642.

DOUTCH, H.F., 1976, The Karumba Basin, northeastern Australia and southern New Guinea: BMR Journal of Australian Geology and Geophysics, v. 1, p. 131–140.

FORBES, A.M.G., and CHURCH, J.A., 1983, Circulation in the Gulf of Carpentaria. II Residual currents and mean sea level: Australian Journal of Marine and Freshwater Research, v. 34, p. 11–22.

GALLOWAY, W.E., 1975, Process framework for describing the morphologic and stratigraphic evolution of deltaic depositional systems, *in* Broussard, M.L., ed., Deltas: Models for Exploration: Houston, Houston Geological Society, p. 87–98.

GAGLIANO, S.M., and MCINTIRE, W.G., 1968, Reports on the Mekong River delta: Baton Rouge, Louisiana, Louisiana State University, Coastal Studies Unit, Technical Report 57, 78 p.

GASTALDO, R.A., ALLEN, G.P., AND HUC, A.Y., 1993, Detrital peat formation in the ropical Mahakam River Delta, Kalimantan, eastern Borneo: Sedimentation, plant composition, and geochemistry, *in* Cobb, J.A., and Cecil, C.B., eds., Modern and Ancient Coal-Forming Environments: Geological Society of America, Special Paper 286. p. 107–118.

GASTALDO, R.A., ALLEN, G.P., AND HUC, A.Y., 1995, The tidal character of fluvial sediments of the modern Mahakam River delta, Kalimantan, Indonesia, *in* Flemming, B.W., and Bartholomä, A., eds., Tidal Signatures in Modern and Ancient Sediments: International Association of Sedimentologists, Special Publication, 24, p. 171–181.

HARRIS, P.T., BAKER, E.K., COLE, A.R., and SHORT, S.A., 1993, Preliminary study of sedimentation in the tidally dominated Fly River delta, Gulf of Papua: Continental Shelf Research, v. 13, p. 441–472.

HOEKSTRA, P., 1993, Late Holocene development of a tide-induced elongate delta, the Solo delta, East Java: Sedimentary Geology, v. 83, p. 211–233.

JENNINGS, J.N., 1975, Desert dunes and estuarine fill in the Fitzroy estuary, northwestern Australia: Catena, v. 2, p. 215–262.

JONES, B.G., MARTIN, G.R., and SENAPATI, N., 1993, Riverine–tidal interactions in the monsoonal Gilbert River fandelta, northern Australia: Sedimentary Geology, v. 83, p. 319–337.

MACNAE, W., 1968, A general account of the fauna and flora of mangrove swamps and forests in the Indo–West-Pacific region: Advances in Marine Biology, v. 6, p. 73–270.

MATHERS, S., and ZAASIEWICZ, J., 1999, Holocene sedimentary architecture of the Red River delta, Vietnam: Journal of Coastal Research, v. 15, p. 314–325.

MILLIMAN, J.D., and MEADE, R.H., 1983, World-wide delivery of river sediment to the oceans: Journal of Geology, v. 91, p. 1–21.
MILLIMAN, J.D., and SYVITSKI, J.P.M., 1992, Geomorphic / tectonic control of sediment discharge to the ocean: the importance of small mountainous rivers: Journal of Geology, v. 100, p. 525–544.
MOORE, C.H., 1989, Carbonate Diagenesis and Porosity: Amsterdam, Elsevier, Developments in Sedimentology, no. 46, 338 p.
MORGAN, J.P., 1970, Depositional processes and products in the deltaic environment, *in* Morgan, J.P., ed., Deltaic Sedimentation; Modern and Ancient: Society of Economic Paleontologists and Mineralogists, Special Publication 15, p. 31–47.
MULRENNAN, M.E., and WOODROFFE, C.D., 1998, Holocene development of the lower Mary River plains, Northern Territory, Australia: The Holocene, v. 8, p. 565–579.
NAKADA, M., AND LAMBECK, K., 1989, Late Pleistocene and Holocene sea-level changes in the Australian region and mantle rheology: Geophysical Journal v. 96, p. 497–517.
NANSON, G.C., PRICE, D.M., SHORT, S.A., YOUNG, R.W., and JONES, B.G., 1991, Comparative uranium–thorium and thermoluminescence dating of weathered Quaternary alluvium in the tropics of northern Australia: Quaternary Research, v. 35, p. 347–366.
NGUYEN, V.L., TA, T.K.O., and TATEISHI, M., 2000, Late Holocene depositional environments and coastal evolution of the Mekong River Delta, southern Vietnam: Journal of Asian Earth Sciences, v. 18, p. 427–439.
RHODES, E.G., 1982, Depositional model for a chenier plain, Gulf of Carpentaria, Australia: Sedimentology, v. 29, p. 201–221.
STAUB, J.R., and ESTERLE, J.S., 1994, Peat-accumulating depositional systems of Sarawak, east Malaysia: Sedimentary Geology, v. 89, p. 91–106.
SMART, J., GRIMES, K.G., DOUTCH, H.F., and PINCHIN, J., 1978, The Mesozoic Carpentaria and Cainozoic Karumba Basins, Australia, Queensland: Australia Bureau of Mineral Resources, Geology and Geophysics, Bulletin, v. 202.
SONNEFELD, P.S., and PERTHUISOT, J.P., 1989, Brines and Evaporites: Washington, D.C., American Geophysical Union, Short Course in Geology, v. 3, 126 p.
THOM, B.G., WRIGHT, L.D., and COLEMAN, J.M., 1975, Mangrove ecology and deltaic–estuarine geomorphology, Cambridge Gulf–Ord River, Western Australia: Journal of Ecology, v. 63, p. 203–222.
WALKER, R.G., 1992, Facies, facies models and modern stratigraphic concepts, *in* Walker, R.G., and James, N.P., eds., Facies Models: Response to Sea Level Change: Toronto, Geological Association of Canada, p. 1–14.
WOODROFFE, C.D., 2000, Deltaic and estuarine environments and their Late Quaternary dynamics on the Sunda and Sahul Shelves: Journal of Asian Earth Sciences, v. 18, p. 393–413.
WOODROFFE, C.D., 2003, Coasts; Form, Process and Evolution: Cambridge, U.K., Cambridge University Press, 623 p.
WOODROFFE, C.D., and CHAPPELL, J., 1993, Holocene emergence and evolution of the McArthur River delta, southwestern Gulf of Carpentaria, Australia: Sedimentary Geology, v. 83, p. 303–317.
WOODROFFE, C.D., CHAPPELL, J., THOM, B.G., and WALLENSKY, E., 1989, Depositional model of a macrotidal estuary and floodplain, South Alligator River, Northern Australia: Sedimentology, v. 36, p. 737–756.
WOODROFFE, C.D., MULRENNAN, M.E., and CHAPPELL, J., 1993, Estuarine infill and coastal progradation, southern van Diemen Gulf, northern Australia: Sedimentary Geology, v. 83, p. 257–275.
WRIGHT, L.D., 1989. Dispersal and deposition of river sediments in coastal seas: models from Asia and the tropics: Netherlands Journal of Sea Research, v. 23, p. 493–500.
WRIGHT, L.D., and COLEMAN, J.M., 1973, Variations in morphology of major river deltas as functions of ocean wave and river discharge regimes: American Association of Petroleum Geologists, Bulletin, v. 57, p. 370–398.
WRIGHT, L.D., COLEMAN, J.M., and ERICKSON, M.W., 1974, Analysis of major river systems and their deltas: morphologic and process comparisons: Baton Rouge, Louisiana, Louisiana State University, Coastal Studies Institute, Technical Report 156, 114 p.
WRIGHT, L.D., COLEMAN, J.M., and THOM, B.G., 1973, Processes of channel development of a high-tide range environment: Cambridge Gulf–Ord River delta, Western Australia: Journal of Geology, v. 81, p. 15–41.
YASSINI, I., JONES, B.G., and JONES, M.R., 1993, Ostracods from the Gulf of Carpentaria, northeastern Australia: Senckenbergiana Lethaea, v. 73, p. 375–406.

SEDIMENT DISPERSAL PATTERN OFF AN ERODING DELTA ON THE WEST COAST OF TAIWAN

JAMES T. LIU, RAY T. HSU, AND JEONG-SHANG HUANG,
Institute of Marine Geology and Chemistry, National Sun Yat-sen University, Kaohsiung, Taiwan 804-24, Republic of China
e-mail: james@mail.nsysu.edu.tw
AND
SHENN-YU CHAO
Horn Point Environmental Laboratory, University of Maryland Center for Environmental Science, Cambridge, Maryland 21613-0775, U.S.A.

ABSTRACT: The Tseng-wen River is a small monsoon-regulated mountainous river on the wave-dominated west coast of Taiwan. Historical maps reveal a trend of shoreline accretion through delta progradation during the past two and a half centuries until the mid 1970s. Two decades after the completion of a reservoir in the upper reaches of the river in 1974, the depositional setting at the river mouth gradually changed from a progradational deltaic system to an estuarine system. Presently, during the flood season, most of the river-borne sediment is in the mud-size fraction. On the time scale of a tidal cycle, the area of initial river sediment discharge is restricted to the nearshore region in the immediate vicinity of the river mouth. Because of tidal movements in the Taiwan Strait, dispersal patterns of river-borne sediment are dominated by longshore tidal currents that display bilateral swings in the course of a tidal cycle. Yet, on a long-term basis the river discharge is the most important sediment source for the coast and nearshore region. One other important sediment source for the area is the reworked offshore palimpsest sediments. After initial deposition, wave agitation is the most important mechanism affecting sediment entrainment. Several lines of evidence, including grain-size distributions on the sea floor (on the scales of years), measurements of in situ flow and turbidity (on the scale of days), and numerical model simulations of the plume (on the scale of tidal cycles) indicate net northward (flood-oriented) sediment dispersal. The Tseng-wen River represents the type of river that has dwindling influence on the surrounding coast in terms of sediment supply and textural imprint.

INTRODUCTION

Where a river discharges into a coastal sea, the quantity and pattern of sediment supply and how the available sediments are dispersed in a coastal depositional system determine not only the configuration and geomorphology of the system but also the long-term behavior of the system, such as accretional or erosional (Galloway, 1975; Orton and Reading, 1993; Wright and Coleman, 1972). Wright and Nittrouer (1995) proposed a four-stage conceptual model for the dispersal of river sediments in the coastal ocean. These stages comprise: I, supply of sediments via river plumes; II, initial deposition; III, resuspension and transport; and IV, long-term net accumulation. They point out that these stages might be overlapping in space and time, and thus, not restricted to the two-dimensional and sequential relationships presented in the illustration (fig. 2 in Wright and Nittrouer, 1995). Stages I and II include primary processes by which the sediment-laden river effluent (plume) spreads immediately seaward of the river mouth. The dispersal and deposition of the river sediments are to a large extent determined by the type of the plume, whether it is hypopycnal or hyperpycnal.

River-derived sediments are the major sediment source for most coastal depositional systems in the world. Yet, sediment loads for most rivers in the world do not reflect their natural state, but rather represent the results of human activities (Milliman and Syvitski, 1992). Activities such as river damming and diversion have dramatically reduced the amount of sediments discharged by rivers. The Nile is one of the better-known examples. Erosion on the Nile Delta shorelines has increased greatly since the construction of the Aswan High Dam in 1964 (Fanos, 1995). Damming not only reduced the sediment load of the river; it also altered the content of the load (Guillén and Jimenéz, 1995; Palanques et al., 1990). In the case of the Ebro River in Spain, damming and building reservoirs not only greatly reduced the sediment load, it also reduced the sand fraction in the sediments discharged by the river, causing intense reshaping of the nearshore deltaic area (Guillén and Jiménez, 1995) and the coarsening trend of the beach sediments in the Ebro Delta (Guillén and Palanques, 1996). On the west coast of Taiwan, similar effects also took place on the Tseng-wen River (Liu et al., 1998). Consequently, the morphodynamic characteristics of the river mouth have shifted from a wave-dominated delta to a wave-dominated estuary (Liu et al., 1998).

In a recent flume study on a microdelta system, Endo et al. (1996) have shown that grain-size variations on an experimental microdelta are strongly governed by the textural composition of the source sand, which is analogous to the river sediment load in a natural system. Their results support the hypothesis that the textural composition of a depositional system is largely determined by the textural composition of its sediment input (Swift et al., 1971). The observed grain-size patterns result from the interaction between the transport and deposition of sediments being delivered to the system, and the reworking of in situ sediments. The nature and amount of sediment input can determine the overall textural characteristics on the receiving end of the depositional system, as illustrated by the coupling between the Amazon River and shelf (Nittrouer et al., 1995).

Therefore, a correlation exists between the textural characteristics of a marine depositional system and the processes by which sediments are delivered and deposited. Liu and Zarillo (1990) found that the spatial distribution patterns of individual grain sizes on a wave-dominated shoreface reveal the influence not only of their sources but also of the hydrodynamics that have reworked and redistributed the sediments during the transport process. The unique grain-size distribution patterns suggest that different grain sizes have different responses to various hydrodynamic processes on the shoreface, and therefore, they can be used as natural tracers to indicate how the substrate responds to the time-averaged hydrodynamic processes (Liu and Zarillo, 1989). The combination of using individual grain sizes as natural tracers

Tropical Deltas of Southeast Asia—Sedimentology, Stratigraphy, and Petroleum Geology
SEPM Special Publication No. 76, Copyright © 2003
SEPM (Society for Sedimentary Geology), ISBN 1-56576-086-7, p. 45–70.

to indicate sediment sources and transport process, and the sea floor topography, which also reflects the long-term interplay between substrate and process, has proven to be effective in defining the shoreface boundaries, identifying the sources and sinks, and determining transport pathways in a coastal depositional system (Liu and Zarillo, 1990).

Recently Liu and Hou (1997) studied the distribution of grain-size classes in the sand fraction and the distribution of mud in the surficial sediments around the entrance of a harbor. Their findings show close morphotextural relationship among sediment patterns, shoreline morphology, and bathymetry.

Sediments discharged into coastal areas by rivers and streams originating in the Central Mountain Range of Taiwan are the most important sediment source for the coastal plains along the west coast of Taiwan. Therefore, the west coast of Taiwan is often classified as alluvial (Wang and Aubrey, 1987). In this study, we focus on the coastal depositional system of one of these rivers. Our goal is to determine the major sources and to identify processes that affect the dispersal of river sediments in the study area. Our approach is threefold. Firstly, distribution patterns of individual grain sizes on the surface of the sea floor were used as natural tracers to indicate long-term interplay among sediment source, dispersal process, and bathymetry. Secondly, short-term process–response observations at the mouth of the river were made to quantify directly the processes that might be related to the grain movements. Lastly, the Princeton Ocean Model (POM) was used to simulate the behavior of the river plume in order to further understand its effect on the dispersal of river sediments. Results from these three components of the study are subsequently integrated to render a comprehensive presentation for the dispersal of river sediments.

STUDY AREA

The Tseng-wen River is the second-largest river in southern Taiwan in terms of the drainage area (1176 km^2), estimated mean annual runoff (2.36 x 10^9 m^3) and estimated sediment discharge (31.13 x 10^6 metric tons) (Water Resources Planning Commission, 1992). Owing to the influence of monsoon climate, the river experiences distinctive dry and flood seasons. Approximately 84% of the annual river discharge takes place between June and September. On average, the discharge in August is two orders of magnitude greater than that of each month in the dry season between November and February. This great disparity between dry and flood seasons implies that the sediment outflux from Tseng-wen River is also concentrated during the flood season.

The cuspate shoreline geomorphology near the mouth of the river (Fig. 1) puts its delta near the wave-dominated end member in the triangular delta-classification scheme originally proposed by Galloway (1975) and on another, similar, classification scheme proposed by Davis (1985). Historically (in the last 300 years), sediments discharged from Tseng-wen River have been the major source for the coastal progradation in the area (Liu et al., 1998).

The wave field offshore of Tseng-wen River is closely coupled with the seasonal wind patterns. In the summer, waves are influenced by southwesterly winds and typhoons, which are characterized as having longer and steeper waves. In winter, the wave field is dominated by northeasterly winds. During the southwest monsoon in summer, the mean significant wave period is 4.7 s and the mean wave height is 0.49 m. During the northeast monsoon in winter, the mean significant wave period is 6.1 s and the mean wave height is 0.52 m (Liu et al., 1999). The tides in the study area are mixed but predominantly semidiurnal, having a tidal range of less than 2 m (microtidal). Yet, the tidal range does increase northward in the study area, mainly because of the increase of the amplitude of the M2 component. Longshore currents in this area are dominated by tidal currents (Lin et al., 1995; Liu et al., 1998; Wen et al., 1995). Low-frequency nontidal currents also exist, but their orientations are probably a seasonal phenomenon, which is still a subject of future investigations (Wen et al., 1995).

The combined influence of the prevailing wave and tidal energy puts the morphodynamic distinction of the study area in the wave-dominated category according to the classification scheme proposed by Nummedal and Fischer (1978) and Hayes (1979) for mixed-energy coasts worldwide (Liu et al., 1998). There seems to be a demarcation point on the shoreline located immediately north of the river mouth. North of this point, the shoreline trends NNE–SSW, and its geomorphology is characterized by barrier islands and lagoons. South of this point, the coastline trends NW–SE, and is geomorphology is characterized by barrier spits and lagoons (Liu et al., 1998). This subtle morphological distinction suggests that wave-induced shore-parallel longshore sediment transport is more dominant in the south region (by SW summer waves), whereas tidal influence on the coastal sediment transport (shore-normal as indicated by greater number of tidal inlets) in the north is stronger.

Past Coastal Changes

A study of the shoreline change of this area between 1904 and 1983 indicates that north of the river mouth the barrier islands showed trends of landward migration, whereas accretion occurred on the inner shores of the lagoons (Liu et al., 1998). Accretion was especially significant in the river delta region, and consequently, during this period, the river mouth moved seaward and migrated to the south (Shih, 1979). During the same period, coastal areas south of the river mouth displayed growths of barrier spits, gradual infilling of several small coastal lagoons, and shoreline accretion immediately south of the river mouth. The geomorphic evidence suggests the pre-1983 coastal geomorphology to have been represented by a healthy strandplain deltaic system nourished by ample river-borne and in situ sediment.

In 1974, the dam of a reservoir in the upper reaches of the river was completed. As the reservoir began to retain river runoff, it also began to retain bedload sediments. Up until 1996, over 80,000 m^3 of sediment were deposited in the reservoir. Recently, barriers to the north of the Tseng-wen River mouth have migrated landward as much as 1 km since 1975 and in places have become welded to the mainland. To the south, the coastline has remained stable (Fig. 2; Liu et al., 1998). On the basis of changes in river-mouth geomorphology, spatial changes of mineral composition in river bed sediments of the entire river, and the grain-size trends near the mouth area of the river, Liu et al (1998) conclude that the mouth of Tseng-wen River has changed from a deltaic to an estuarine system.

With the decreasing river sediment discharge and the amount of bedload material, the delta region has already started the reshaping process. This is indicated not only by the shoreline retreat near the river mouth but also by the fact that the shoreline at the mouth is gradually losing its arcuate form and becoming straight, an indication of increased wave dominance.

Bathymetry

The major offshore bathymetric feature adjacent to the Tseng-wen River is a relatively gently sloping submarine platform

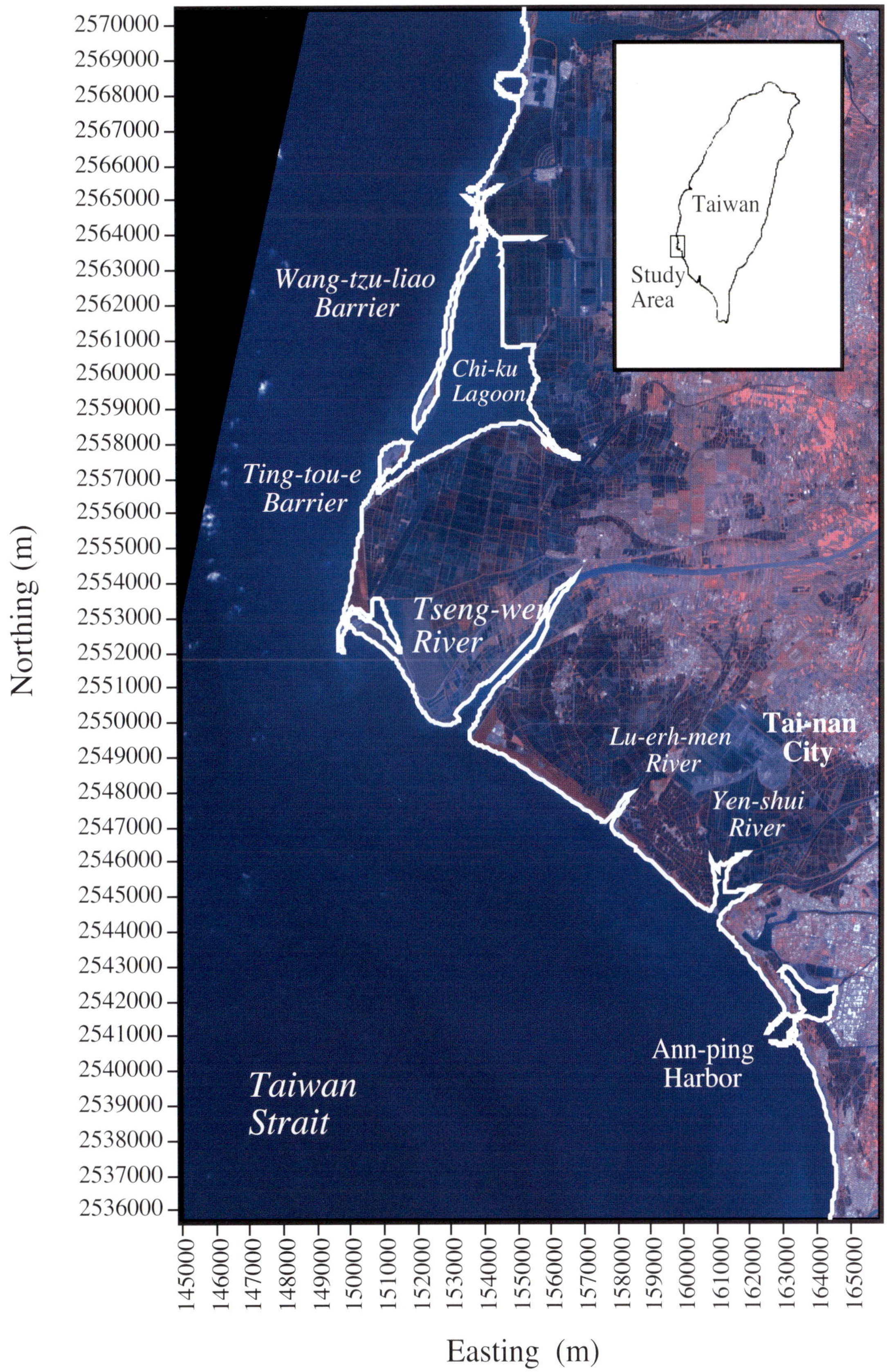

FIG. 1.—Composite map of the study area based on a geo-referenced SPOT satellite image taken on 13 August 1998. The shoreline was digitized from the image and then replotted. Inset shows the study area on the west coast of Taiwan.

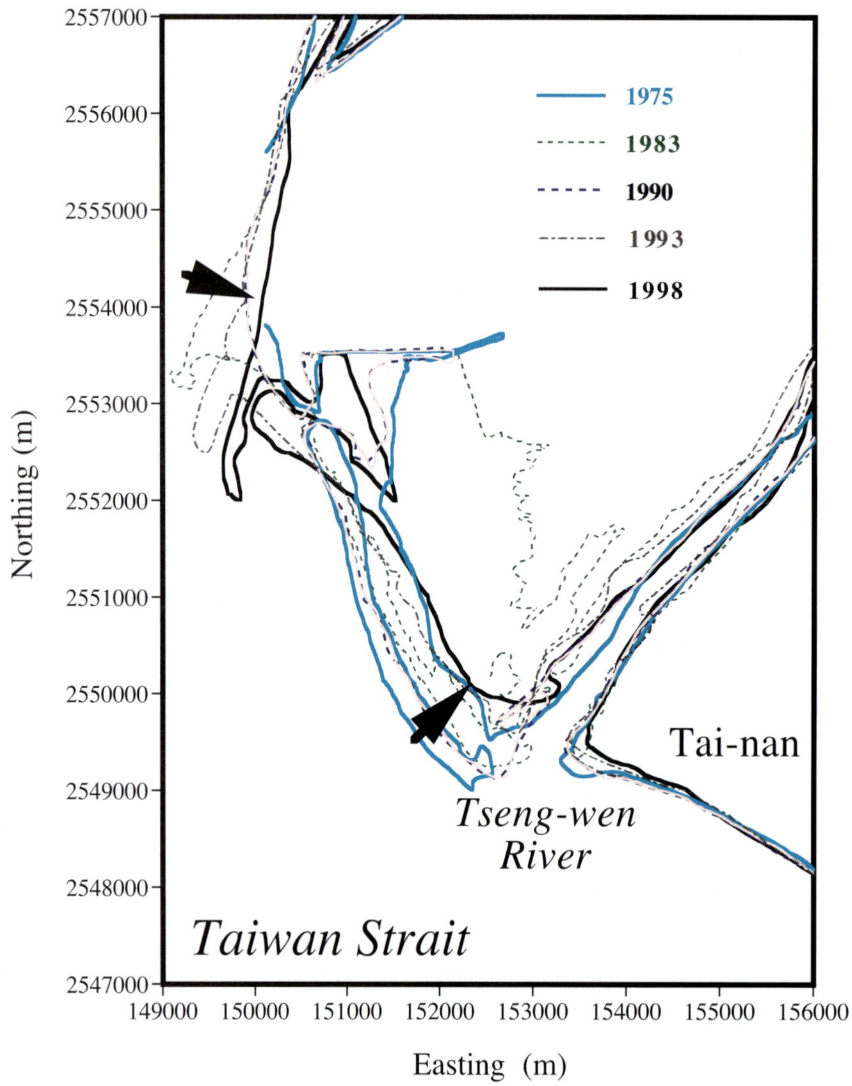

FIG. 2.—Map showing the shoreline positions in the vicinity of the river mouth. The arrows indicate the recessional trend of the shoreline movements.

whose seaward edge is defined by the 20 m isobath (Fig. 3). A gully in the narrowest part of the shelf divides the platform into a northern and a southern part, both having distinctive morphological features. North of the point of the shoreline deflection, the edge of the platform is oriented nearly north–south. Because of the NNE–SSW trending shoreline, the platform widens to 16 km at the northern boundary of the study area. The platform is narrowest (2 km) at the shoreline promontory. South of that point, the edge of the platform generally follows the trend of the shoreline, maintaining a constant width of about 12 km. Seaward of the edge of the platform, the depth of the sea floor drops abruptly to 50 m in approximately 2 km, forming a steep slope. A small and shallow shore-normal submarine gully is located at the narrowest point of the submarine platform, the head of which cuts into the platform at the 4 m isobath. Immediately north of the gully wall there is a shoal that slopes gently northward. The seafloor elevation immediately south of the gully is about 6 m lower than its northern counterpart.

North of the gully, a submarine ridge 12 km in length is located along the seaward edge of the submarine platform whose base is approximately delineated by the 16 m isobath. The height of the ridge varies between 4 to 6 m. The ridge is narrower (about 1.5 km wide) at the southern end and gradually widens to about 2.5 km near its northern end. Immediately landward of the ridge there is a wide, shallow depression whose deepest part is just over 18 m. The sea floor landward of the depression shows concave-upward geometry, typical of a wave-dominated shoreface (Niedoroda et al., 1985).

Immediately south of the shoreline deflection is a shoal delineated by the 8 m isobath. It is speculated that this shoal is the remnant part of the subaqueous delta deposited during the 1950s and the 1960s when the river mouth was located in the area (Liu et al., 1998). The platform south of the present river mouth is relatively featureless with steepening shoreface slope southward.

Wright and Coleman (1972) stated that a quasi-equilibrium exists between the subaqueous and subaerial morphology and the long-term wave climate. The Tseng-wen delta-front profile along a dashed line (Fig. 3) extending seaward from the present-day river mouth parallel to the axis of the river was plotted

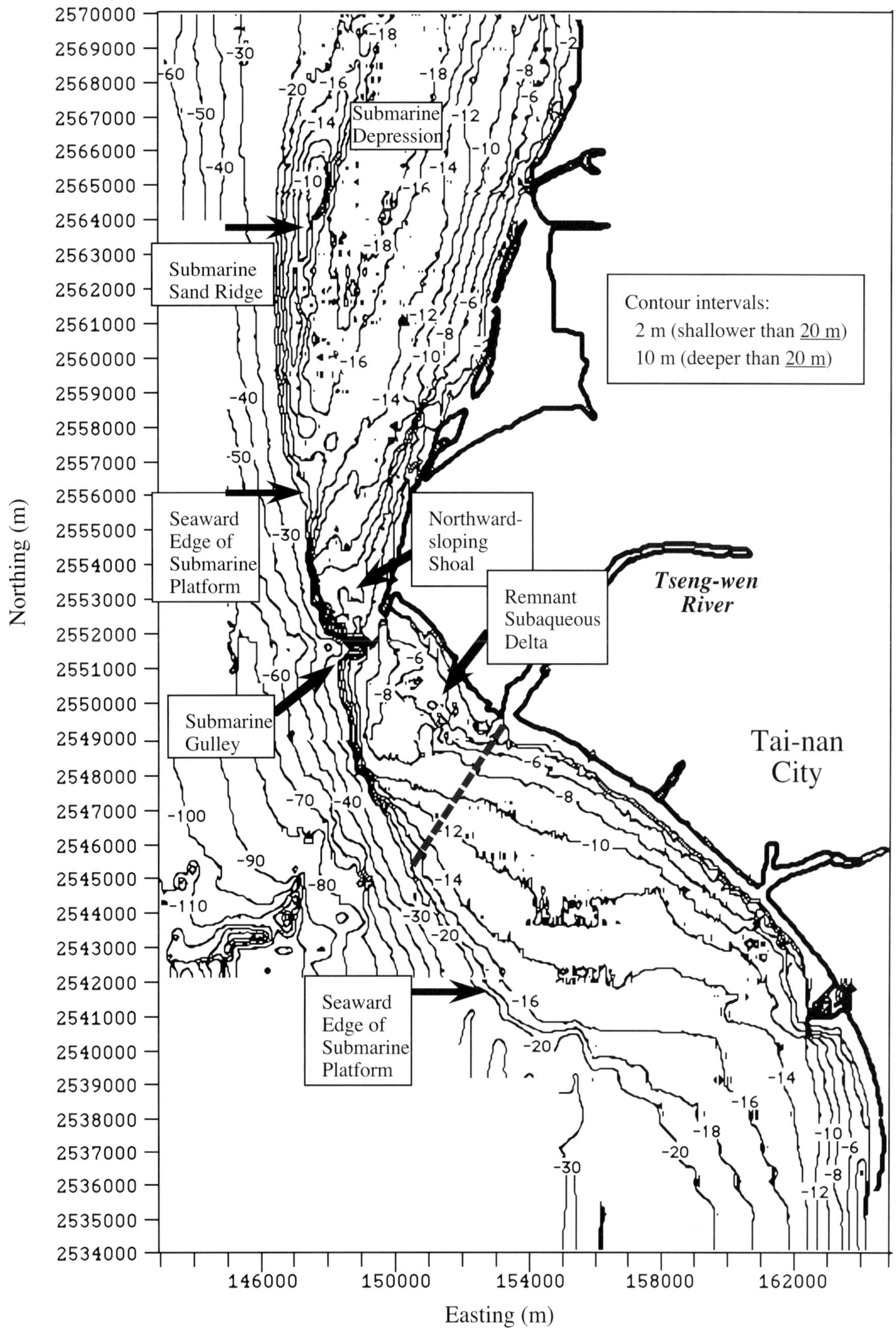

FIG. 3.—Bathymetric map of the study area. The thick dashed line extending from the mouth of the Tseng-wen River indicates the profile location shown in Figure 4.

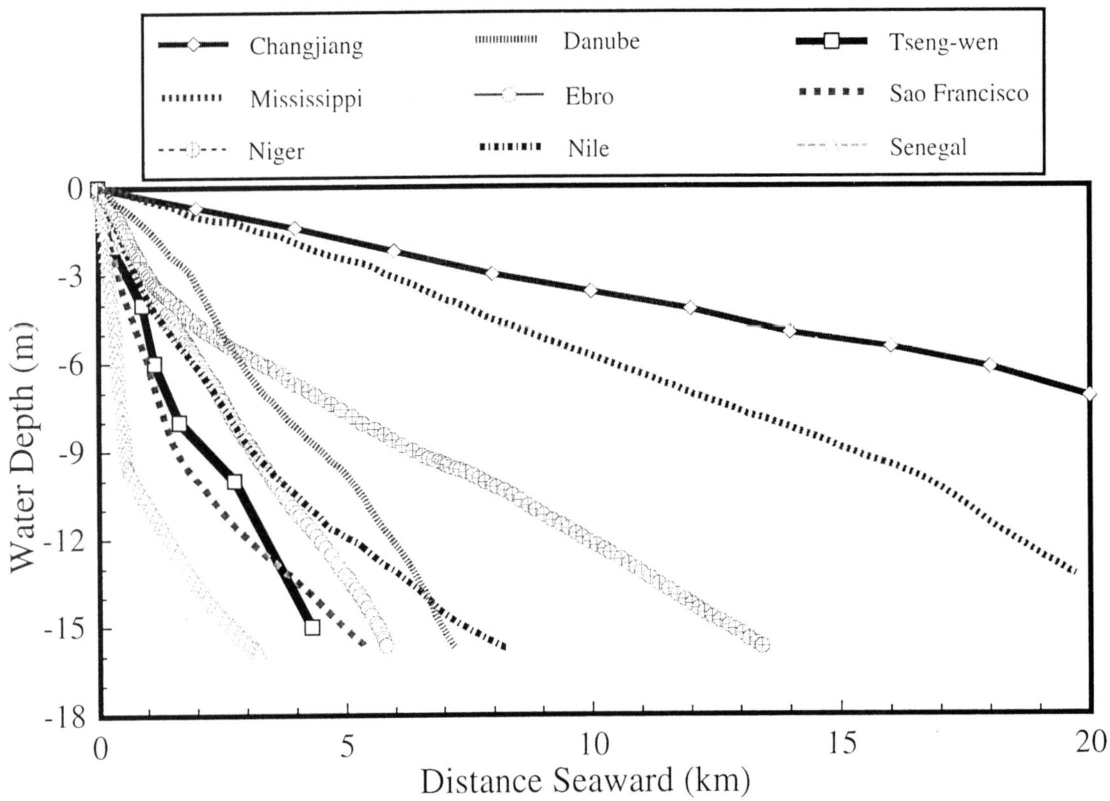

FIG. 4.—Comparison of subaqueous slopes of nine river deltas (modified from Wright and Coleman, 1972).

along with subaqueous profiles of various river deltas in the world (Fig. 4). These deltaic profiles reflect the relative power of river discharge and the reworking ability of waves (Fig. 4). For example, the Changjiang (Yangtze) represents the river-dominated end member, with high wave attenuation, whereas the Senegal River represents the wave-dominated end member, with low wave attenuation.

FIELD WORK

The objective for the field work of this study was twofold. The first objective was to augment the already existing 57 sampling stations of Industrial Technology Research Institute (ITRI, Fig. 5). In total, 194 surficial sediment samples were taken along 17 east–west oriented transects between Chiang-chung River and Er-hian River. On each transect, samples were collected from the surf zone seaward to depths beyond 60 m using a custom-made grab sampler. Sampling was densest near the mouth of Tseng-wen River. Ten sediment samples were also taken from inside the river mouth along the thalweg of the river. The sediment sampling spanned over two years during the flood season of 1994 between May and August, and the flood season of 1995 between May and July.

Additional comprehensive field work was carried out to monitor the sediment and hydrodynamic characteristics of the Tseng-wen River plume during the flood season between 9 and 15 June 1996. This period coincided with high river discharge (Liu et al., 1999). The field experiment included the deployment of an instrumented tripod about one kilometer seaward from the river mouth. Two instruments were mounted on the tripod, including a directional wave, current, and tide gauge with an additional OBS (optical backscatter sensor) and a conductivity, temperature, and depth (CTD) meter. The readers are referred to Liu et al. (1999) for detailed descriptions of the field work.

LABORATORY WORK: GRAIN-SIZE FREQUENCY ANALYSIS

All sediment samples were processed and analyzed following the procedure similar to that described by Liu and Hou (1997). Sediment samples were treated with 30% hydrogen peroxide solution to eliminate organic materials and then wet-sieved to exclude gravel (larger than 2 mm, or -1 ϕ) and silt and mud (finer than 0.063 mm, or 4 ϕ). The amount of the fines was recorded. The sand fraction was analyzed for grain-size frequency distribution using a custom-built rapid sediment analyzer (RSA). Grain size was recorded at quarter-phi intervals. Only 16 grain-size classes were registered in the sand fraction (between 0.84 mm and 0.063 mm). Consequently, the data set consisted of 17 grain-size classes (sand fraction plus the fines). In other words, the number of data points in the data set totals 3298 (17 x 194). Some samples also contain shell fragments. The algorithm used by the RSA converts the grain settling velocity to a hydraulic equivalent diameter of a quartz sphere. However, the inclusion of the carbonate fraction in the analysis did not adversely affect the RSA results, owing to the similarity of quartz and calcite densities. Furthermore, the hydraulic equivalent diameter produced by the RSA implicitly accounts for shape differences between the carbonate fraction and terrigenous clastics (Liu and Zarillo, 1990; Liu and Hou, 1997). Grain-size data were further manipulated and analyzed on a personal computer.

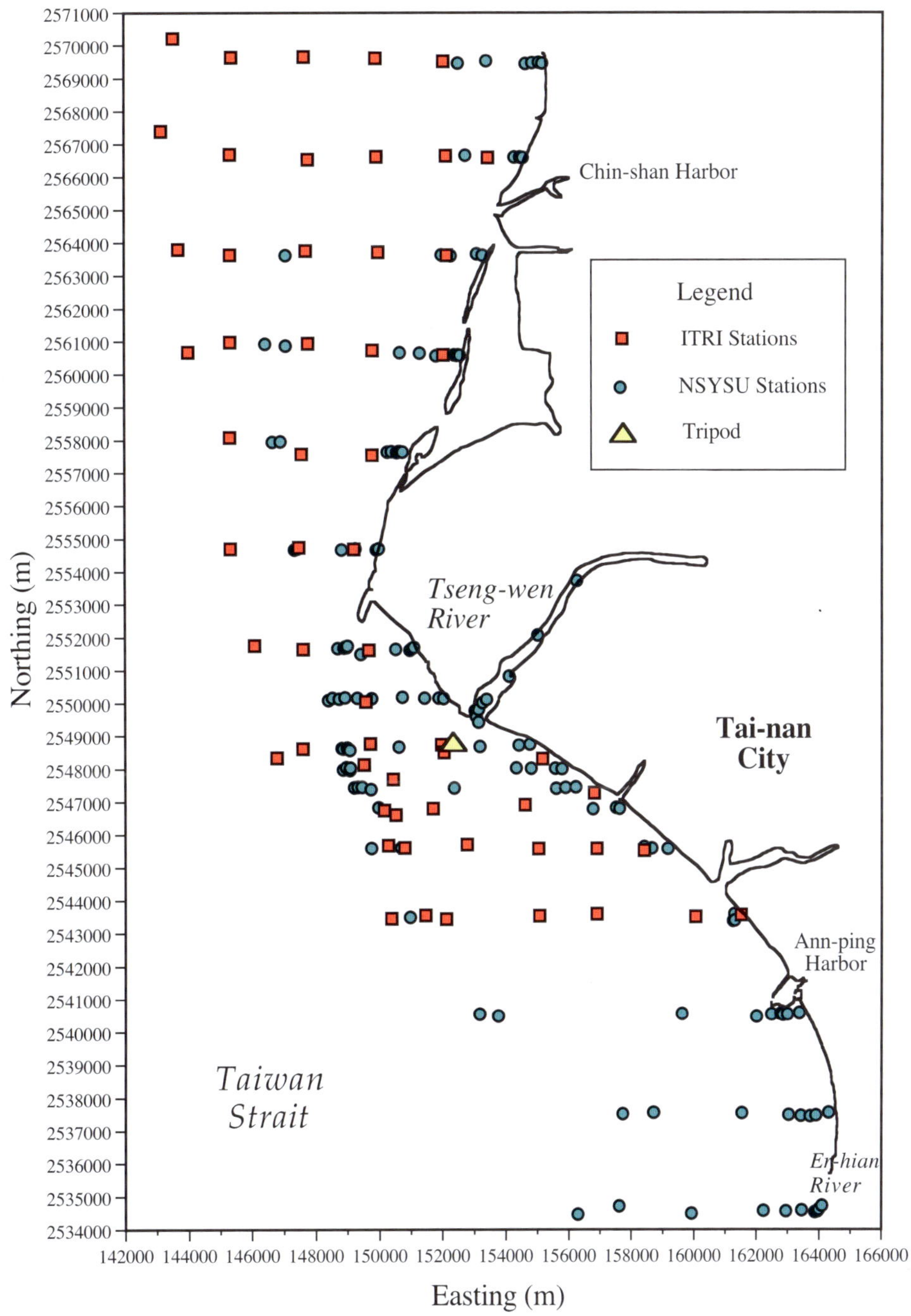

FIG. 5.—Locations for sediment sampling and the tripod deployment. ITRI stands for Industrial Technology Research Institute, NSYSU stands for National Sun Yat-sen University.

RESULTS OF FIELD OBSERVATIONS

Grain-Size Distributions

Because of geographical differences in the sediment sampling density and the shoreline orientation, for the distribution of each grain size the study area was also divided into five overlapping sections. In each section, the spatial distribution of the sediment abundance was gridded and contoured. Also, the bathymetric data from each corresponding section were recontoured at 5 m intervals. For each section, the bathymetric contours were then overlaid on top of the colored patterns of the grain abundance in percent weight, forming a morphotexture map. The five morphotexture maps of different sections were then pieced together electronically to form one composite map with the digitized shoreline superimposed. We need to point out that the mismatches of the colored patterns between two adjacent sections were caused by the contouring process due to different sampling density and different ranges of grain-size abundance in each section. Because the coarse size classes in the sand fraction are relatively scarce in the system, and their distribution patterns are limited to small scattered patches in the study area, only six representative grain sizes finer than 0.30 mm are presented (medium sand, Fig. 6).

Close morphotextural relationships have been revealed from these grain-size patterns. Different grain sizes tend to concentrate at different locations in the study area. For example, grain sizes in the fine-grained sand fraction (0.21 to 0.15 mm) are distributed mostly in a north–south orientation, being associated with the submarine sand ridge, the edge of the platform, and the seaward

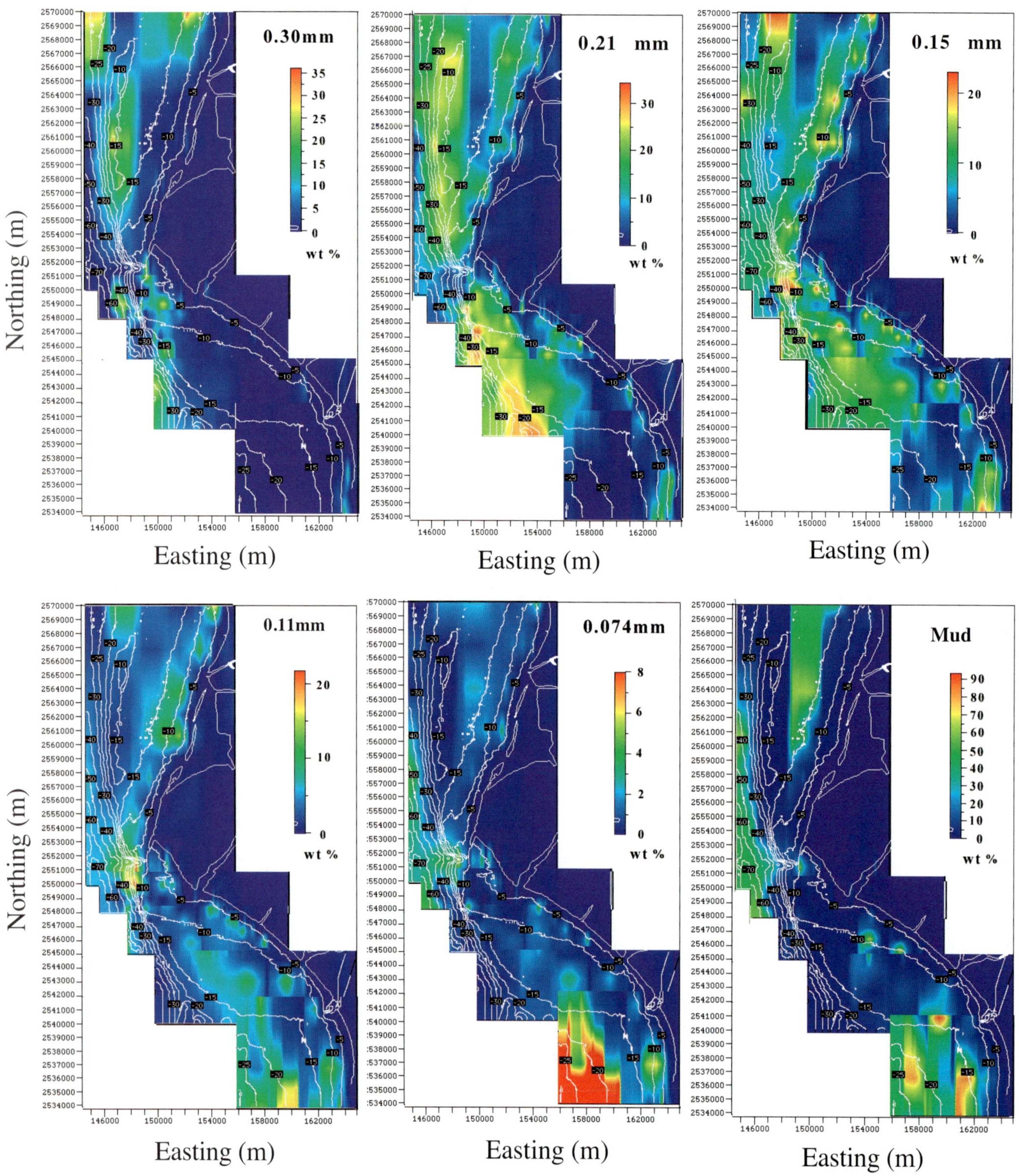

FIG. 6.—Spatial distribution patterns of selected grain sizes. The scale bar in the upper right indicates the range of scale in percent weight. The light-colored lines are bathymetric contours and the 1993 shoreline.

slope (Fig. 6). Some of the finer grain sizes in the fine-grained sand fraction are concentrated on the shoreface landward of the sand ridge. Upper fine-grained sand (0.21 mm) is the most abundant and most uniformly distributed group in the system. Lower fine-grained to upper very fine-grained sand (0.15 and 0.11 mm) is concentrated immediately south of the gully. In general, the finer the grain size, the more abundant it is on the platform near the south end of the study area. The mud distribution is particularly worth noting. It is concentrated only in three major areas: seaward of the platform in depths deeper than 30 m, in the depression landward of the submarine sand ridge, and on the platform near the south end of the study area.

Near-Bed Sediment Dynamics

The observed currents at about 1 m above the sea bed are essentially alongshore at predominantly semidiurnal frequencies, having a slightly onshore tendency during the flood (northward) and an offshore tendency during the ebb (southward, Fig. 7A). Flood currents dominate over ebb currents in terms of magnitude and duration. The wave field during the deployment is presented by the variations of the burst-averaged maximum orbital velocity (Fig. 7B). The influence of a weather front on the wave field is indicated by the abrupt increase of the maximum orbital velocity (follows almost the same pattern as the significant wave height) by threefold on 13 June after the passage of a low-pressure system (Fig. 7B).

The turbidity record, also measured at about 1 m above the sea bed, is presented in the manufacturer's calibration unit (Formazin Turbidity Unit, FTU), showing the influence of both the wave field and the tide (Fig. 7B). On a longer time scale for the entire sampling duration, the FTU fluctuations almost mimic those of the significant wave height. On shorter time scales, a closer examination of these peaks reveals that during the time of low wave energy prior to 13 June, double peaks often appear during periods of low water.

On a weekly time scale, the salinity decreased gradually while temperature rose gradually during the tripod deployment period (Fig. 7C). Below the weekly time scale, sudden salinity drops (double dips) were often synchronized with sudden temperature rises (double peaks). This negative relationship is apparently caused by the outflow of warm and fresher water from the river.

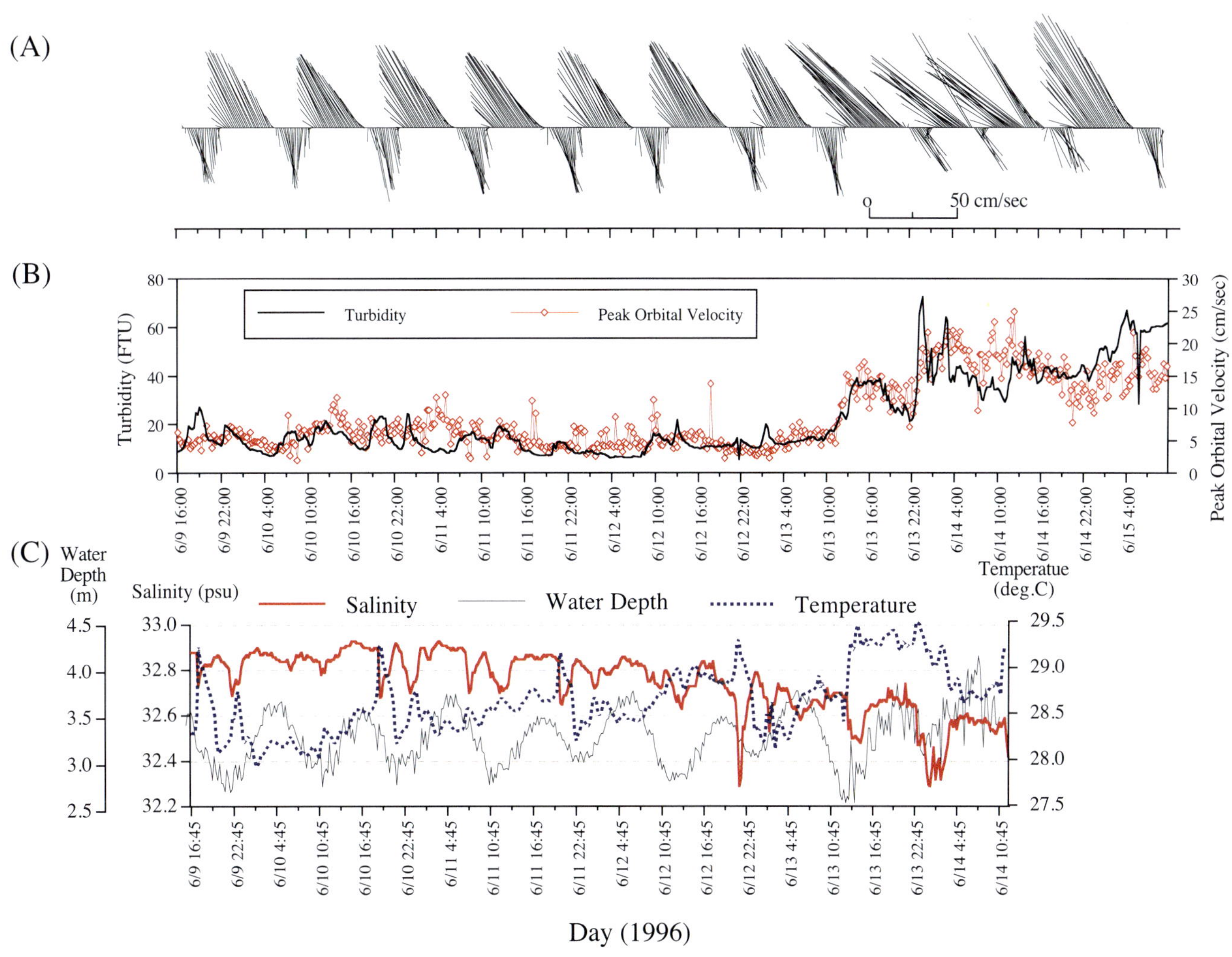

FIG. 7.—Data measured by SP2100 include: **A)** Stick diagram of the observed currents. The north is directed toward the top of the figure. **B)** Burst-averaged peak orbital velocity and turbidity (in Formazin Turbidity Unit, FTU). Data measured by TD-410 include **C)** salinity, temperature, and water depth. The TD-410 record is shorter than that of the SP2100. The starting points of both records are also different.

On visual inspection, all the double peaks of the turbidity and temperature, and double dips of the salinity, occur concurrently. The wiggles on the water-depth curve are caused by wave aliasing, which increased during the frontal passage on 13 and 14 June, when incident waves became greater.

In order to understand the causality among the time series of salinity (Sal.), water temperature (Temp.), turbidity (Turb.), significant wave height (H_s), maximum wave-orbital velocity (U_{orb}), and the longshore (U_x) and cross-shore (U_y) components of the current, cross-correlation analysis was performed and the cross-correlation coefficient ($\rho_{x,y}$) was calculated. In the analysis, for each pair of time series a time lag and lead were introduced by one observational time increment (15 min.) at a time up to five hours. This is a technique that was also used by Green at al. (1997) when examining synoptic records of suspended-sediment concentration and flow. However, because the in-phase and out-of-phase relationships among the analyzed parameters are apparent (Fig. 7), and because we are more interested in the extent to which two parameters are related, the absolute values of the cross-correlation coefficient ($|\rho_{x,y}|$) are plotted (Fig. 8). The minus time lag indicates that the first time series in the pair is leading the second one, and positive lag indicates that the first one is lagging behind the second one (Fig. 8).

The results show that the maximum wave-orbital velocity and the turbidity have the highest correlation, and the response between them is instantaneous. The significant wave height is the second-most-important factor influencing the turbidity, having a lead roughly about three hours. This is because the wave height itself is modulated by the tide (water depth and current velocity and direction). The turbidity and salinity also have an instantaneous correlation. Turbidity and temperature have high correlation, with the temperature leading by 15 min. The correlation between the turbidity and current components fall below 0.5. Yet, the cross-shore currents have higher correlation than the longshore currents, having a time lead of about three and a half hours.

These findings suggest that resuspension of sediments by waves is the most important factor influencing the behavior of suspended sediments. However, the instantaneous correlation between turbidity and salinity also suggests that the plume is the second important factor controlling the suspended-sediment concentrations. The low correlation between the tidal currents and turbidity is understandable. In the vicinity of the river mouth, the role of currents in resuspending bottom sediments in the water column is less than that of waves and the river plume. A closer examination of the turbidity values suggests the existence of background concentration or washload in the water column (Green et al., 1997; Wright et al., 1999). This fine-grained material is either kept in suspension by wave-orbital motions or transported laterally by the river plume. It would be beneficial if the background turbidity could be differentiated from high-energy resuspensions on the basis of grain-size compositions when sediment dispersal is concerned. Because no field data were available, we resorted to a crude method by comparing the observed wave-orbital velocity with the theoretical threshold of sediment movement under oscillatory wave motions proposed by Komar and Miller (1974) as an indication. The theoretical threshold (U_c) is expressed as (for grain diameter less than 0.5 mm)

$$U_c = \left[0.21\left(\frac{\rho_s - \rho}{\rho}\right)gD\left(\frac{d_o}{D}\right)^{1/2}\right]^{1/2} \quad (1)$$

in which r is the density of water, d_o is the observed orbital diameter of the wave motion, g is the acceleration of gravity, and D and ρ_s are the diameter and the density of the grains, respectively. Four grain diameters were used in the calculations to represent clay (1 μm), very fine-grained silt (5 μm), very fine-grained sand (63 μm), and fine sand (190 μm). Equation 1 does not consider the complex effect of wave–current interactions in the boundary layer, which is important (Li and Amos, 1998; Li

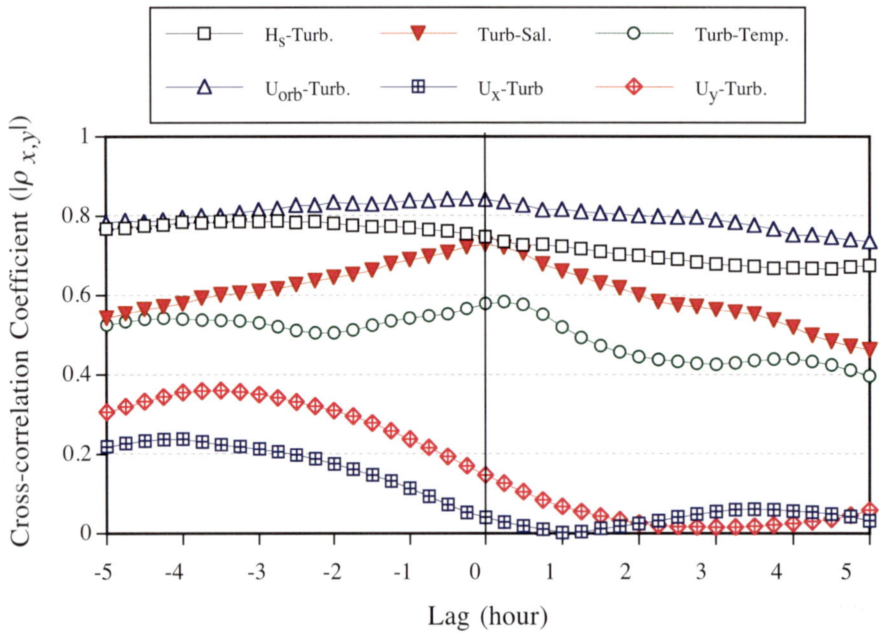

FIG. 8.—Cross-correlation analysis results among the time series of salinity (Sal.), water temperature (Temp.), turbidity (Turb.), significant wave height (H_s), maximum wave-orbital velocity (U_{orb}), and the longshore (U_x) and cross-shore (U_y) components of the current velocity. The minus lag indicates time lead, and the positive lag indicates time lag.

et al., 1997; Wright et al., 1999). However, because the observed d_o is already the product of the wave–current interactions, we believe that Equation 1 is sufficient to use for the illustrative purposes.

The results show that during the sampling period the threshold for clay was always exceeded and that for very fine-grained silt was partially exceeded (Fig. 9). The threshold for very fine-grained sand was exceeded during only part of the high-energy period, and that for fine sand was largely not exceeded during the sampling period. Therefore, one can deduce that the background turbidity is composed of grain sizes finer than very fine-grained silt. Only during high-energy events is very fine-grained sand mobilized and transported. Grain sizes coarser than fine-grained sand are rarely entrained. This can explain why the modal grain size found on the sea floor near the tripod site is fine-grained sand (Liu et al., 1998).

Because the OBS and EM current meter were co-located at 1 m above the sea bed, the instantaneous point sediment flux at that height was estimated by multiplying the observed suspended-sediment concentration and the current speed to render instantaneous flux (F_i):

$$F_i = U_i C_i \tag{2}$$

which displays strong tidal signatures (Fig. 10A). We assume that both U_i and C_i are composed of a steady (nontidal) and oscillatory (tidally related) components (Barua et al., 1994). In other words,

$$U_i = \overline{U} + u' \quad \text{and} \quad C_i = \overline{C} + c' \tag{3}$$

in which

$$\overline{U} = \frac{1}{T}\int_T U_i dt \quad \text{and} \quad \overline{C} = \frac{1}{T}\int_T C_i dt \tag{4}$$

and T is the period of the M2 tide. Therefore,

$$F_i = U_i C_i = (\overline{U} + u')(\overline{C} + c') = \overline{UC} + \overline{U}c' + \overline{C}u' + u'c' \tag{5}$$

In the last part of Equation 5 the first term represents the nontidal part of the suspended flux (Fig. 10B), whose frequency is lower than the semidiurnal tidal frequency. The rest of the terms represent the sum of all the tidally related fluxes (Fig. 10C). They include fluxes due to the co-oscillation between the nontidal U and the tidal part of C, the co-oscillation between the nontidal C and the tidal part of U, and U and C co-oscillation over the tidal cycle, respectively.

The nontidal flux indicates a steady northward transport of sediments throughout the sampling period. Furthermore, the northward flux was enhanced during the frontal passage primarily because of the increased suspended-sediment concentration and suppressed ebb flow (Figs. 7A, B). The tidal component of the flux was also enhanced during the frontal passage, and yet the flux caused by the ebbing tide was even more pronounced (Fig. 10C).

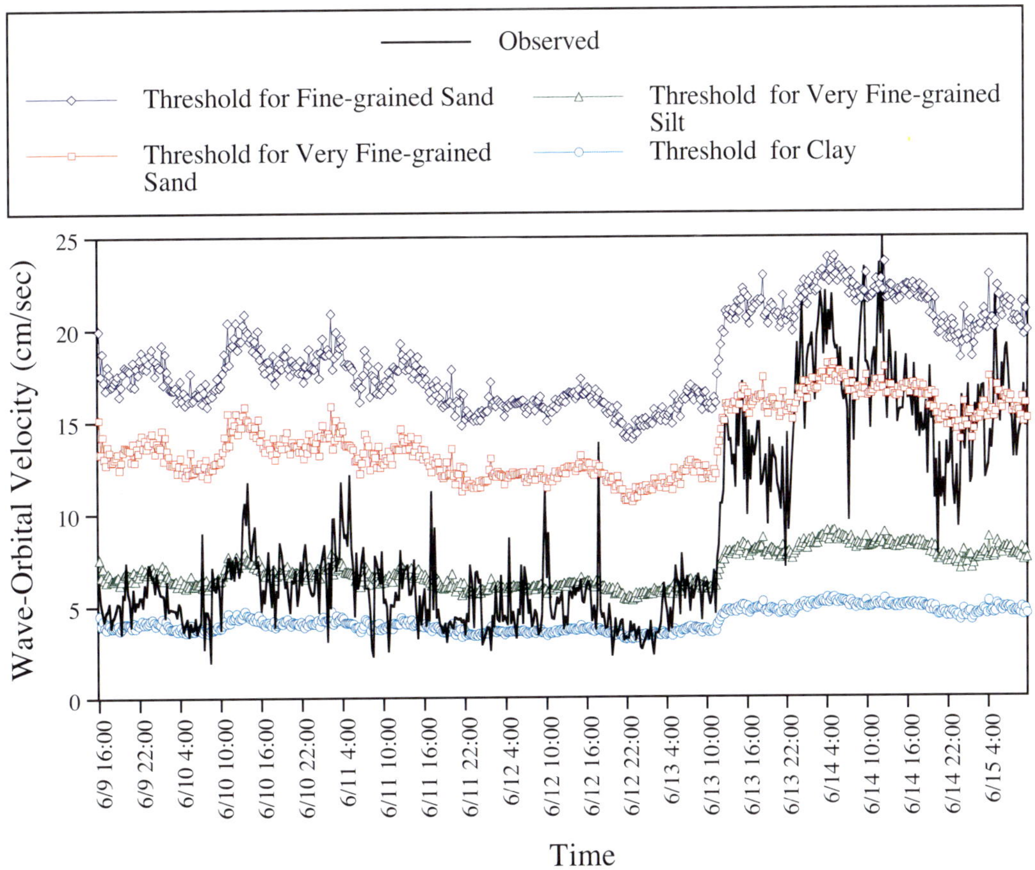

FIG. 9.—Observed and theoretical threshold wave-orbital velocity for fine-grained sand (190 μm), very fine-grained sand (63 μm), very fine-grained silt (5 μm), and clay (1 μm).

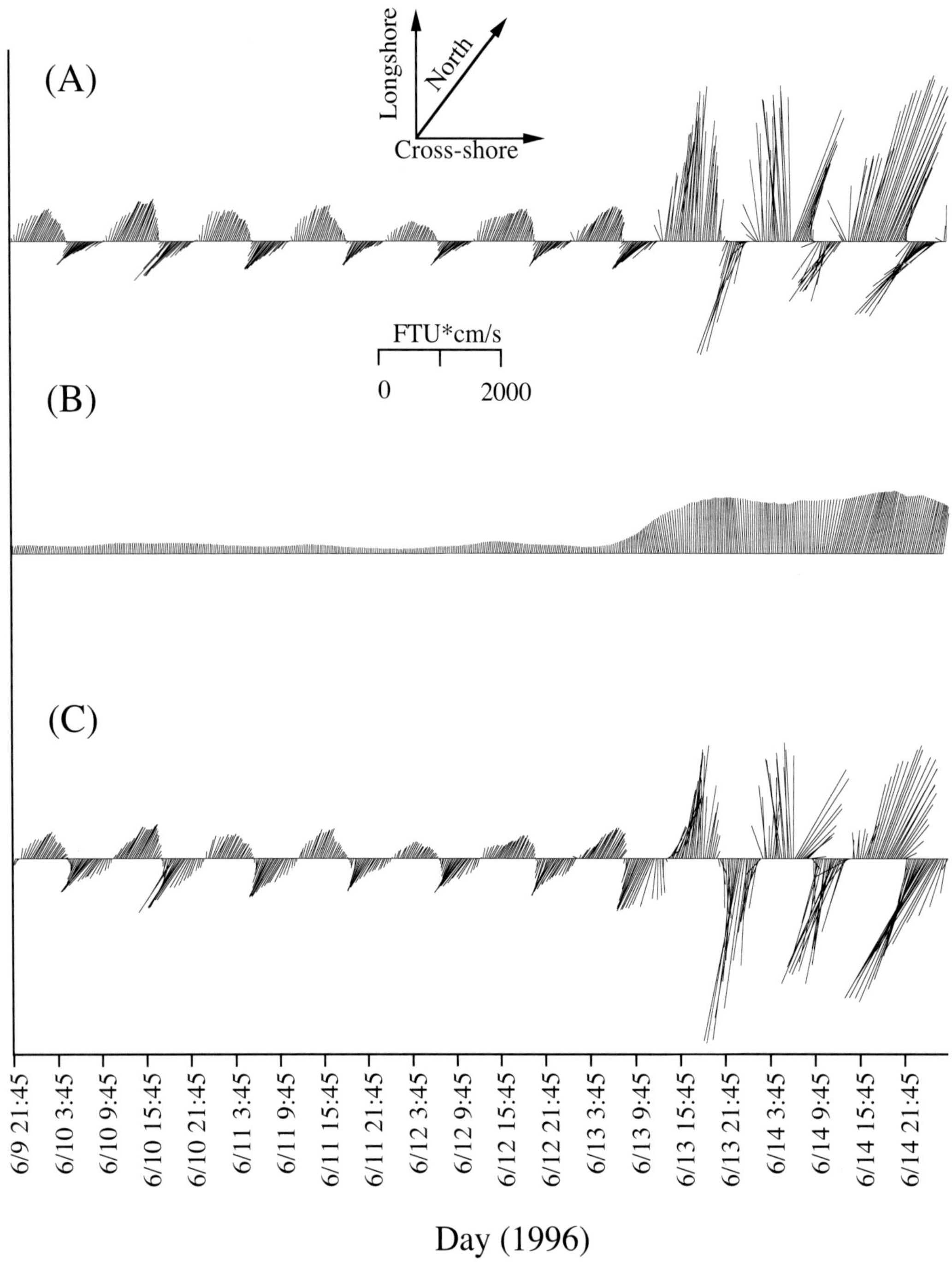

FIG. 10.—Stick diagrams of computed sediment fluxes at 1 m above sea bed, **A)** instantaneous (total) flux, **B)** the nontidal part of the flux (*U.C*), and **C)** the tidal part of the flux due to *U* and *C* co-oscillation over the tidal cycle ($u'c'$). Note: the positive longshore direction is toward the top of the page.

GRAIN-SIZE ANALYSIS

Hypotheses

Conceptually, factors influencing the patterns of grain-size distribution in the study area fall into three major categories: where the sediments come from (sediment sources); how they are moved about (sediment transport and deposition process); and how in situ older sediments are reworked. The observed morphotextural relationship between grain-size patterns and the bathymetry is the result of the interplay of the three categories through time. In a recent flume study imitating a river delta setting, Endo et al. (1996) demonstrate that both the suspended-load and bed-load sediments sampled along the deltaic depositional axis bear textural resemblance to the grain-size characteristics of the source. The challenge at hand is to determine what the major sediment sources for the study area are, on the basis of observed grain-size information. In the following we propose several hypothetical sediment sources and transport processes that might have influenced the observed grain-size patterns in the study area. The likely sediment sources include riverine sediments discharged by Tseng-wen River, sediments on the remnant river delta, the submarine sand ridge, and the palimpsest sediments offshore. Other possible influencing factors include sediments delivered to the study area at the northernmost and southernmost boundaries (via southward and northward sediment transports, respectively) and the effect of wave sorting and reworking. In order to minimize variability, samples of distinctive grain-size characteristics were carefully selected to represent each factor. We briefly discuss each of these factors.

Sediments Discharged by the Tseng-wen River.—

The grain-size composition of this sediment source is represented by the average grain-size frequency composition of the sediment samples taken inside the river mouth. It is apparent that the grain-size signal from the river is dominated by fine-grained sediments whose mud content exceeds 60% by weight (Fig. 11). The sand fraction of the river discharge is dominated by fine-grained sands.

Palimpsest Sediments.—

On the basis of relatively scarce offshore (depths deeper than 20 m) sediment samples taken in Taiwan Strait, Boggs et al. (1979) proposed a belt of muddy, very fine-grained sand palimpsest sediments that lies between the inshore modern fine-grained sand and the offshore relict fine- to medium-grained sand. After carefully examining the grain-size frequency distributions of all samples, we concur with Boggs et al. that sediment samples seaward of the submarine platform can be described as palimpsest sediments (Fig. 11), on the basis of the definition given by McManus (1975):

"Palimpsest sediment is a relict deposit that mainly contains particles supplied to the shelf before the present but also includes some particles now being supplied to the shelf."

In the sand fraction, these sediments generally have the coarsest modal grain size (0.30 mm) among all samples plus a small "tail" of mud (less than 5%). The sand portion fits the description of relict sediments (coarse-grained, well sorted) by Nino and Emery (1961) and Emery (1967) for this region. It also resembles the relict sediments Liu and Zarillo (1990) reported off the south shore of Long Island, New York. The mud "tail" is undoubtedly the modern contribution to the shelf. Therefore, we selected samples from locations seaward of the submarine platform that have not been extensively covered by the modern mud to represent the palimpsest sediments.

Remnant Delta Sediments.—

Grain-size frequency distributions on the remnant delta are fairly similar. The textural characteristics of the remnant delta are coarser than in the surrounding wave-dominated environments. This can be interpreted in two ways: (1) the remnant delta was deposited before the completion of Tseng-wen Dam, when a large amount of coarser-grained bed-load sediments were not trapped by the dam, and (2) because of the cessation of riverine sediment supply, finer grain sizes have gradually been winnowed out from the remnant delta sediments, turning them into lag deposits. Their elevated topography and the spatially coarsening nature of the remnant delta and the sand ridge suggests

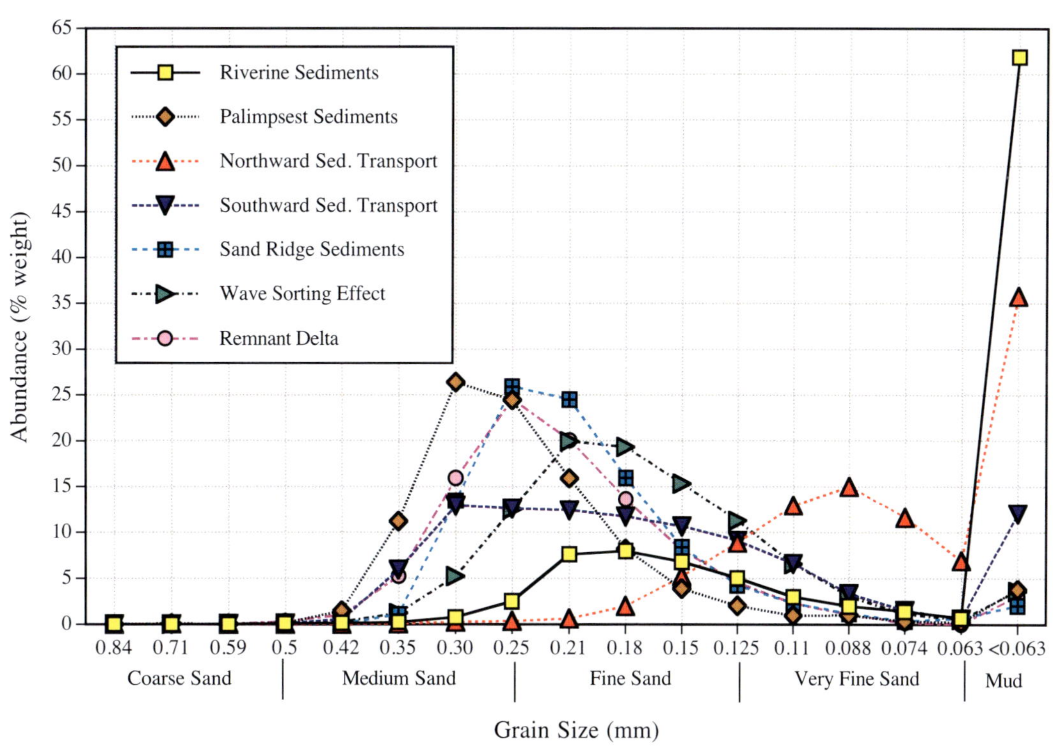

FIG. 11.—Grain-size distributions for all the possible influencing factors in the study area.

that they are probably being eroded (Guillén and Jiménez, 1995), suggesting that they are sediment sources in the system.

The Submarine Sand Ridge.—

Large-scale bedforms on continental shelves have been reported worldwide with different terminologies (Duane et al., 1972; Fenster et al., 1990; Liu and Zarillo, 1990; McBride and Moslow, 1991; McCave, 1979; Swift and Field, 1981) regarding their physical characteristics and origin. On the basis of seismic and stratigraphic information, this ridge is regarded as a drowned barrier island (ITRI, 1996). However, we are concerned only with this ridge as a sediment source. The textural signal of the ridge is represented by averaging the grain-size composition of six samples taken on the ridge. The representative grain-size distributions show that its modal size in the sand fraction is 0.25 mm, which is the second coarsest next to that of the palimpsest sediments (Fig. 11). The sand ridge contains the least amount of mud among all the possible sediment sources.

The Effect of Wave Sorting.—

Sorting by shoaling waves is the primary mechanism for sediment reworking on wave-dominated shorefaces (Liu and Zarillo, 1990). Wave sorting can be visualized by the progressive changes of individual grain-size abundance patterns from the beach seaward to the inner shelf (Liu and Zarillo, 1989). It can also be illustrated by the progressive change of grain-size frequency distribution along a typical transect in the study area (Fig. 12). As the water depth becomes deeper (Fig. 12A), the modal size in the sand fraction becomes progressively finer-grained from 0.25 mm in the surf at 1 m depth to 0.088 mm at 25 m depth (Fig. 12B). Concurrently, the mud content also increases from less than 5% to over 80%. Also, shallower than 11 m, the grain-size distribution

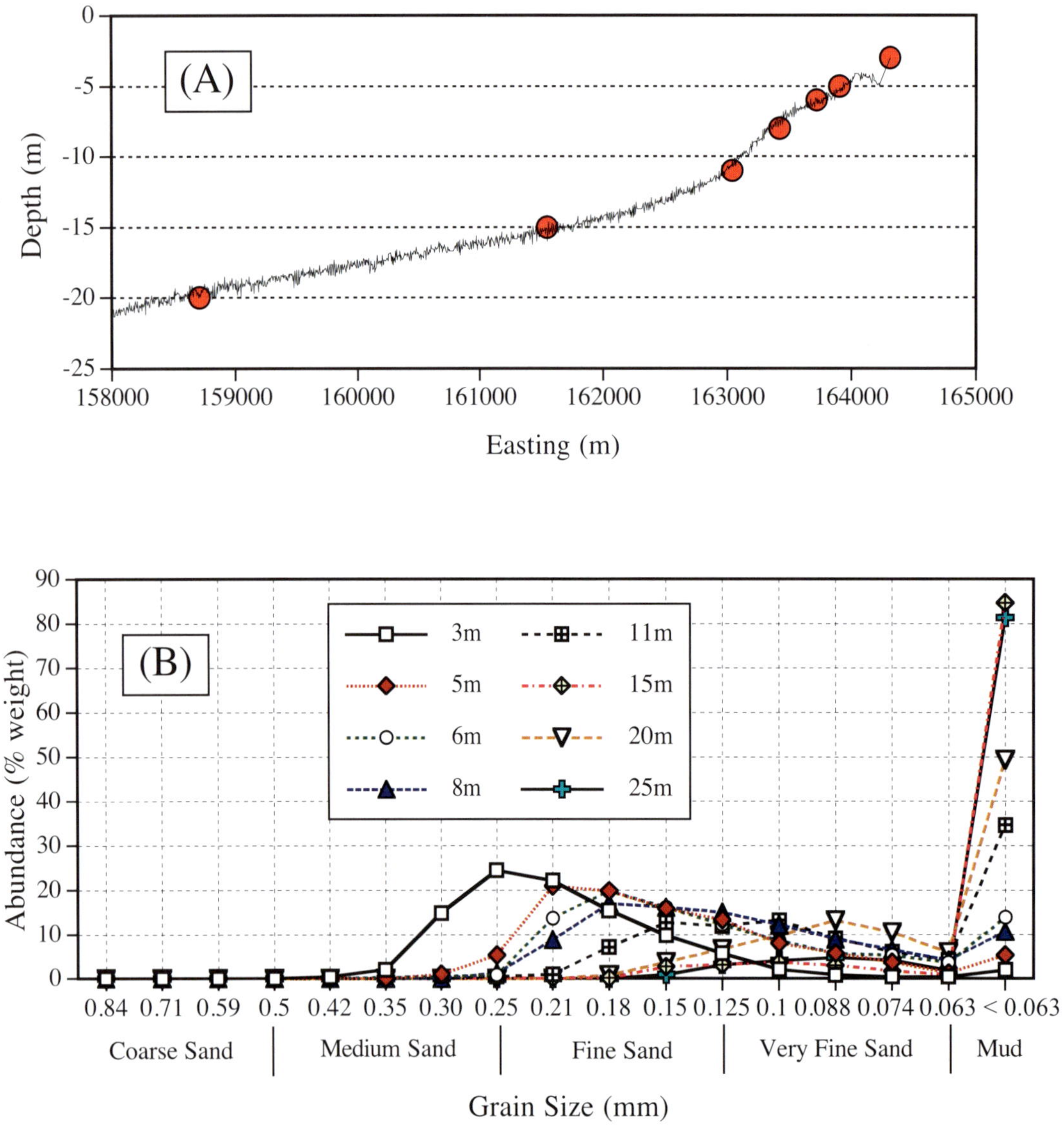

FIG. 12.—**A)** Measured bathymetric profile that corresponds to the second southernmost transect. The circles represent the depths along the transect where sediment samples were taken. **B)** Grain-size frequency distribution curves for samples taken along the above transect.

in the sand fraction is coarse-skewed, but becomes fine-skewed at deeper depths. Therefore, the 11 m contour seems to be a demarcation marking the average seaward limit of active wave agitation along this transect. Consequently, we took the average of grain-size frequency distributions of all samples that were shallower than 10 m geographically separated by the river mouth. After comparison, we determined that the representative grain-size curves of wave sorting for both regions north and south of the river mouth are indistinguishable. Subsequently, we used the average of all nearshore samples shallower than 10 m to represent to effect of wave sorting for the entire study area.

Northward Sediment Transport.—

As suggested by the estimate of short-term sediment flux presented above, northward sediment transport does exist. It is very likely that sediment patterns in the study area are influenced by transport of sediments that entered the study area at the southernmost boundary. This possible source is composed largely of fine-grained materials (Fig. 11).

Southward Sediment Transport.—

By the same token, because of the seasonality of the incident wave field, southward sediment transport in the study area is also likely. Therefore, sediments that enter the study area at the northernmost boundary are also a likely source. This hypothetical source is characterized by a platykurtic distribution in the sand fraction and a higher mud content, very different from the possible southern source (Fig. 11).

Data Filtering and Empirical Orthogonal (Eigen) Function (EOF) Analysis

In a coastal depositional system where there are multiple sources of sediment input, and where there are oscillatory wave motions, tidal currents, and low-frequency currents, the cause-and-effect relationships among the observed grain-size patterns and possible influencing factors are not straightforward. To test the above hypotheses and to decipher the sedimentary response to transport processes, an analytical methodology using the combination of filtering and the EOF analysis technique (Liu and Hou, 1997; Liu and Zarillo, 1990) was used. In this approach, the influence of each of the above hypothetical factors was filtered out by subtracting its representative signatures (grain-size distributions; Fig. 11) from the data set, thereby creating an associated deviation field (Fig. 13), in which the influence could be identified and quantified. Because the nature of the sediment data, including the deviation fields, is multivariate, showing correlations among the sampling locations (space correlated) and different grain-size classes (size correlated), the EOF analysis technique was able to separate orthogonal modes contained in the data sets. Eigenmodes were then ranked according to the amount of data (covariance) they explained. The better correlated the data, the more deviation can be accounted for by the first few eigenmodes. Each eigenmode also reveals a certain type of grain-size grouping and its corresponding spatial associations that might have source and/or process implications. Also, unlike the previous studies, the deviation fields in this study were not standardized. Because of the prevalent presence of mud and fine-grained sands in the system, it could serve as an important "tag" for identifying sediment sources and sinks. By not standardizing the data, the "signals" of mud and fine-grained sands would not be "smeared", which later turned out to be helpful in the interpretation of the EOF results.

NUMERICAL MODELING

Model Description and Sediment Formulation

Recently, numerical models have been used to study sediment dispersal by river plumes (Arnoux-Chiavassa et al., 1999; Chao, 1998; Xing and Davies, 1999). The widely used Princeton Coastal Ocean Model (POM) was chosen for this study to elucidate the influence of river plume, tidal currents, and sediment

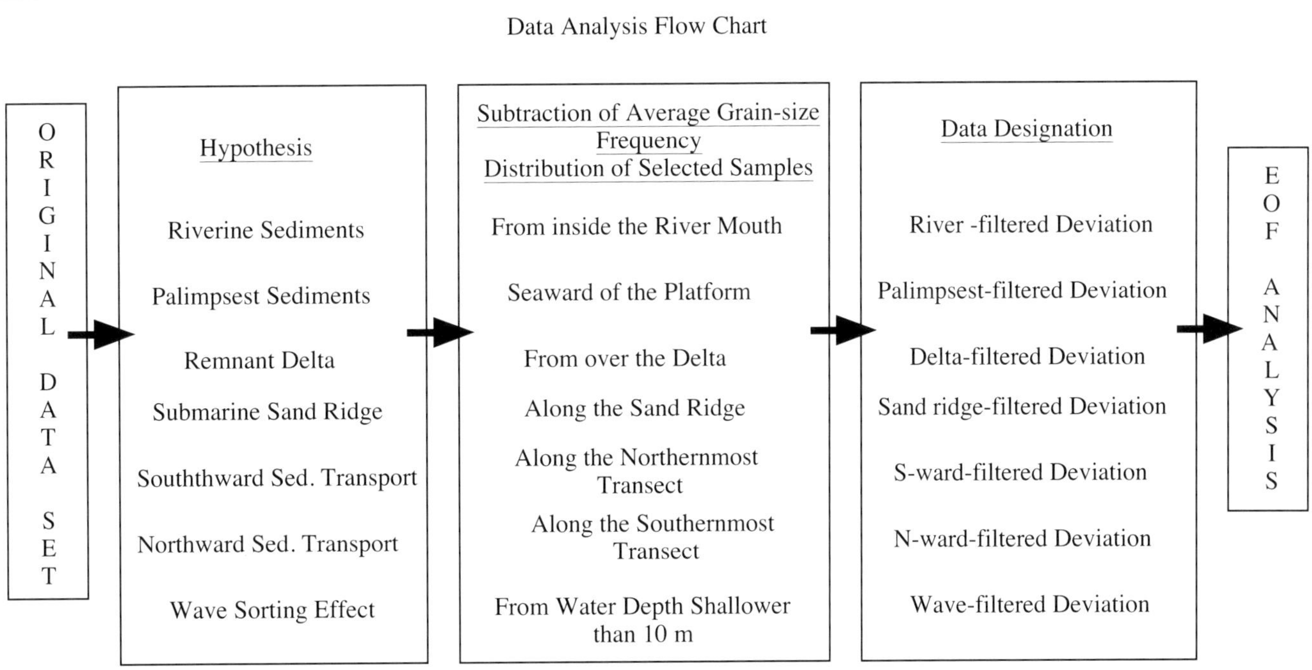

FIG. 13.—Data-analysis flow chart.

behavior on the net sediment transport in the study area. POM is a three-dimensional numerical model that applies finite-difference formulation and mode-splitting technique (Blumberg and Mellor, 1987; Mellor, 1998). The hydrostatic assumption and the Boussinesq approximation were used in the formulation of dynamic equations. For detailed description of the model, readers are referred to Mellor (1998).

Cartesian coordinate grids were set up in the horizontal over the modeling area. There are 70 grid points in the x (E–W) direction, and 80 grid points in the y (N–S) direction. The grid spacings are 228 m for dx and 262 m for dy. In the vertical, the sigma grid of twenty-one layers was used. The model water depth at each grid point was digitized from a surveyed bathymetric map. Although the contours of the resulting model bathymetry are jagged, they nevertheless reflect truthfully the shoreline and sea-floor geomorphology of the modeled area (Fig. 14). The sea-land interface (the shoreline) is represented by zero water depth.

Owing to the constraint of the grid size, the model is most stable if the river mouth is configured in the N–S orientation instead of being shore-normal (Fig. 14). Thus, this configuration introduces a deviation of about 45 degrees between the model river mouth and the true river mouth with respect to the shoreline orientation.

Sediment-transport module was added to the original hydrodynamic model in the same fashion as in Chao (1998). In the model, the density of sediment-water mixture (ρ_c) is related to the concentration of the suspended sediments (C_0):

$$\rho_c = \rho + (1 - \rho_o / \rho_s) C_0 \quad (6)$$

where the density of clear seawater (ρ) is a function of salinity and pressure, ρ_o is a reference density of clear seawater, and ρ_s is the density of the solids (Chao, 1998).

A sediment module is added to the POM in which the suspended-sediment concentration is subject to advection and diffusion in three dimensions by the seawater, while sinking with a characteristic settling velocity (w_s). The governing equation for C is

$$\frac{\partial C}{\partial t} + \frac{\partial uC}{\partial x} + \frac{\partial vC}{\partial y} + \frac{\partial (w - w_s)C}{\partial z} = A\left(\frac{\partial^2 C}{\partial x^2} + \frac{\partial^2 C}{\partial y^2}\right) + \frac{\partial}{\partial z}\left(k_v \frac{\partial C}{\partial z}\right) \quad (7)$$

where u, v, and w are velocities in x, y, and z directions, $A = 10^6$ cm^2/s is the horizontal eddy diffusivity, and K_v is the vertical eddy diffusivity, determined by the 2.5-level turbulent closure scheme (Chao, 1998). The settling velocity (w_s) for this grain size was calculated on the basis of an empirical formulation for spherical quartz particles given by Gibbs et al. (1971).

Boundary Conditions

On the north and south open boundaries of the modeling domain, sea-level fluctuations (η) about mean sea level are prescribed by simultaneously observed water levels at Chinshan Harbor and Ann-ping Harbor, respectively (Fig. 15A). On the west open boundary, η is linearly interpreted between the north and south boundaries. At the landward boundary of the river (head of the estuary), η is raised by 0.1 cm above the mean sea level in this study. This water head produces an average

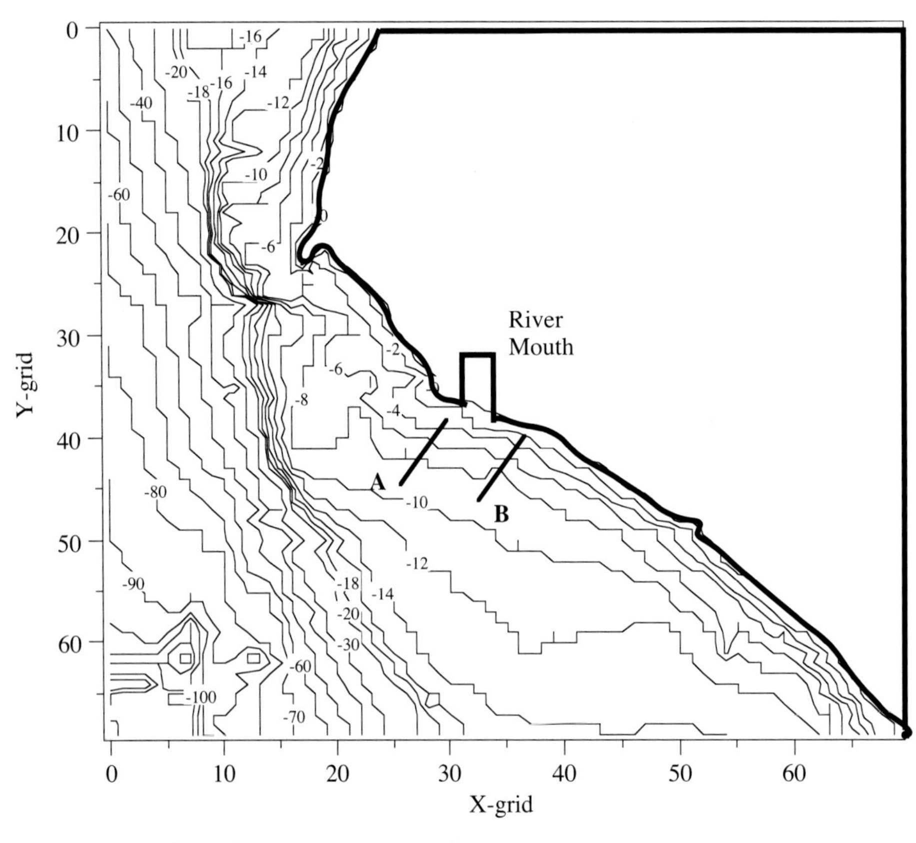

FIG. 14.—The modeling domain and the bathymetric contours of the model grid. Lines A and B indicate two 1-km-long shore-normal transects located 500 m from the sides of the model river mouth.

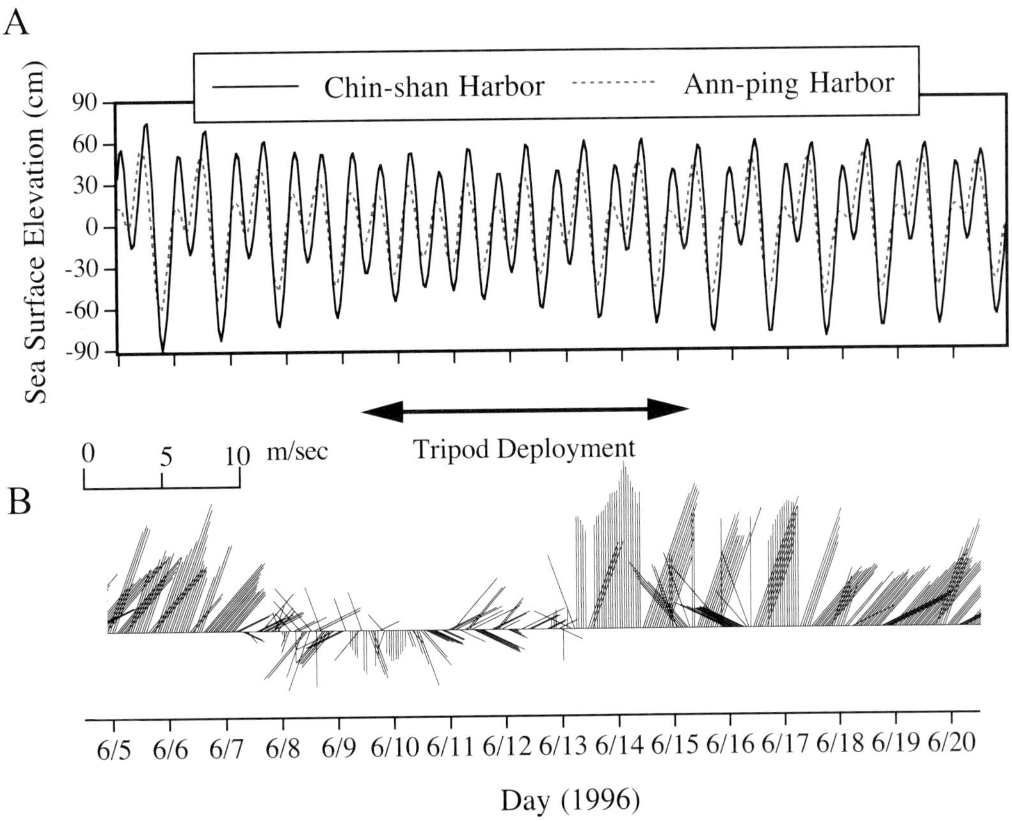

FIG. 15.—**A)** Observed sea-surface elevation at Chin-shan Harbor and Ann-ping Harbor. **B)** Stick diagram of observed wind speed and direction plotted according to the oceanographic convention.

river discharge of 31 cubic meters/second evaluated at the model river mouth, which coincides with the observed runoff range during the early part of the June 1996 field work.

Within the model domain, the salinity value was set at 33.5 psu. At the river boundary, a salinity difference (ΔS) of 10 psu was prescribed. In other words, a river plume was initiated by the introduction of a less saline river effluent.

Initial sediment concentration of 100 mg/l was set at the river open boundary to simulate river-sediment discharge. The wind stress was computed from the observed wind data (Fig. 15B).

The vertical boundary conditions include a free surface at which fluxes of momentum, mass, and salt are set to zero. No sediment flux is permitted either, so that

$$W_s C + K_v \frac{\delta C}{\delta z} = 0 \quad (8)$$

At the ocean bottom, salt and mass fluxes are also set to zero. The current boundary shear stress (τ_b) is calculated according to the quadratic stress law:

$$\tau_b = \rho C_D U^2 \quad (9)$$

where the dimensionless drag coefficient is defined as $C_D = \max\{k^2[\ln(z/z_o)]^2, 0.0025\}$, where $k = 0.04$ is the von Kármán constant, z is the vertical distance between the bottom and the first grid point nearest the bottom, and z_o is the bottom roughness height (Chao, 1998).

Sediment flux at the bottom is the difference between the deposition rate (D) and the erosion rate (E), so that

$$W_s C + k_v \frac{\delta C}{\delta Z} = D - E \quad (10)$$

Erosion takes place when t_b exceeds a critical value (t_c) and E becomes

$$E = M(\tau_b/\tau_c - 1)H(\tau_b - \tau_c) \quad (11)$$

where M is an empirical constant and H is the Heaviside step function. The τ_c was calculated on the basis of Equation 9 using the threshold velocity under conditions of unidirectional flow given by Miller et al. (1977) In our formulation, erosion and deposition occur independently of each other and also can occur simultaneously.

$$D = w_s C \quad (12)$$

where the value of w_s was based on an empirical relation between grain diameter and the terminal settling velocity for spherical quartz particles given by Gibbs et al. (1971). The model was run for eight days starting on 8 June, one day before the tripod deployment, to avoid any spin-up effects.

RESULTS

Grain-Size Analysis

The covariance explained by the first two eigenmodes of all deviation fields are listed in Table 1 in descending order. The

TABLE 1.—The first two eigenmodes of all deviation fields.

Deviation Field	Mode	Covariance Explained (%)	Accumulative Percentage
River-filtered	1	90.7	90.7
	2	6.0	96.7
Palimpsest-filtered	1	66.8	66.8
	2	25.6	92.4
Northward-filtered	1	82.6	82.6
	2	8.9	91.5
Delta-filtered	1	65.7	65.7
	2	25.5	91.2
Ridge-filtered	1	68.9	68.9
	2	22.2	91.1
Wave-filtered	1	65.1	65.1
	2	23.7	88.8
Southward-filtered	1	56.1	56.1
	2	26.0	82.1

river-filtered deviation field is the best correlated (over 96%), which means that the sediments discharged by Tseng-wen River exert the greatest influence on the observed grain-size patterns. The palimpsest-filtered deviation field has the second highest influence on the sediment data set (over 92%). This implies that sediments discharged by the river (modern) and the palimpsest sediments (relict) are the two major sources influencing the sediment texture in the study area. The northward-filtered deviation field outperforms the competing southward-filtered counterpart, suggesting that northward transport is the dominant direction of sediment transport in the study area. The remnant delta and the sand ridge are minor sediment sources for the study area. Although wave sorting in the cross-shore direction is the prevailing means of sediment reworking, its degree of influence is less than those of the sources and the dominant longshore sediment transport. We examine only the first modes of the deviation fields by plotting the eigenvectors (lower panels, Fig. 16) and corresponding eigenweightings (upper panels, Fig. 16), because all except one can explain over 60% of the data.

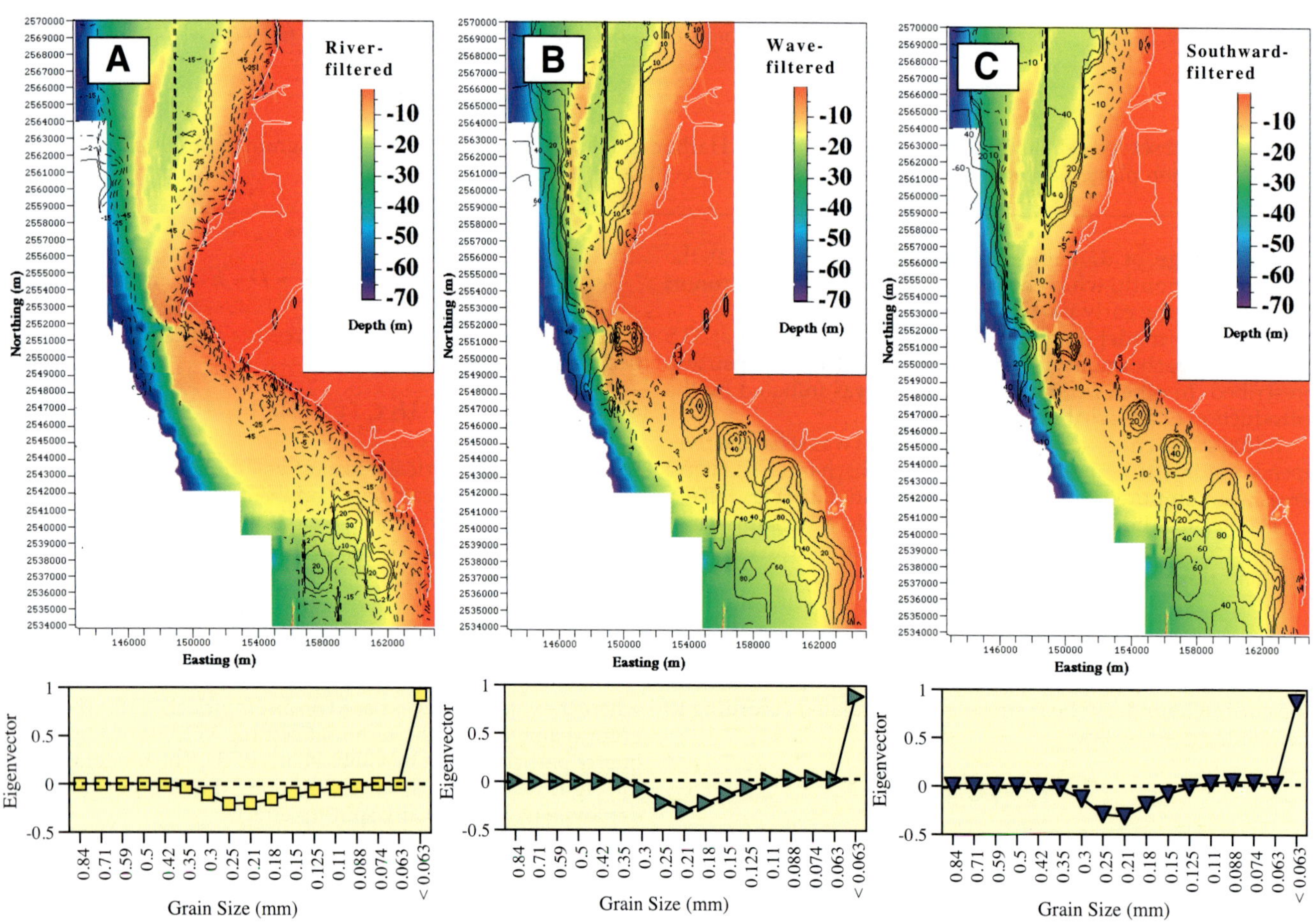

FIG. 16.—The first-mode eigenweightings (the upper panel) and the eigenvectors (the lower panel) of deviation fields of **A)** river-filtered, **B)** wave-filtered, **C)** southward-filtered, **D)** delta-filtered, **E)** palimpsest-filtered, and **F)** northward-filtered. (Negative eigenweightings are contoured with dashed lines.)

Sand (Negative Group) vs.
Mud (Positive Group) Assemblage.—

It appears that the seven deviation fields can be classified into two assemblages based on the grouping of grain sizes by the eigenvector. The first assemblage includes the river-filtered, wave-filtered, and southward-filtered deviation fields (Figs. 16A, B, C). In this assemblage, the 17 grain-size classes are divided into a positive group and a negative group according to the sign of their eigenvectors. Apparently the positive group is dominated by the mud alone, and the negative group is dominated largely by fine-grained sand, especially the 0.21 mm size, which is the most abundant size in the study area. The spatial associations of these two grain-size groups are indicated by the contours of the corresponding eigenweightings. Despite the identical nature of the grain-size groupings in this assemblage, there are some differences in the associated eigenweighting patterns.

River-filtered deviation field.—Because the river is the most important sediment source, the eigenweighting patterns of the river-filtered deviation field reveal locations where sediment transport processes and other sediment sources affect the substrate. These locations include the shoreface within 10 m of water, in the depression on the platform, and on the slope beyond the platform. It is barren elsewhere in the study area (Fig. 16A). Except for a small region on the southern part of the platform and at the foot of the platform, the spatial eigenpatterns are dominated by the sand group (dashed contours). The mud is weighted on the platform seaward of the shoreface in the southern region of the study area, and offshore on the seaward slope of the platform.

Wave-filtered deviation field.—The spatial eigenpatterns of the wave-filtered deviations indicate locations where there are deviations from the representative wave sorting grain-size curve (Fig. 11). In other words, these are places where the wave sorting is less effective or where there is excess input of grain sizes that are out of equilibrium with the hydrodynamic energy regime. When one inspects the eigenweighting contours of the wave-filtered deviation (Fig. 16B), one can see that (1) they are spatially dominated by the mud (positive contours) at locations where mud concentrates, and (2) they are absent on the shoreface. These results suggest that the sand group is generally in equilibrium with wave-related transport processes (agitation and transport) on the shoreface, and the accumulation of mud is likely due to excess input from land (on the platform) and reduced wave agitation (in the depression and on the lower slope of the platform). It is worth noting that the sand group shows strong correlation with the sand ridge, whose depths are shallow enough to be agitated by average waves. Also, this deviation field highlights the channeling effect of mud by the gully.

Southward-filtered deviation field.—The spatial eigenpatterns of the southward-filtered deviation field looks like the mixture of those of the previous two deviation fields. Because the southward sediment transport is the least viable factor influencing the grain-

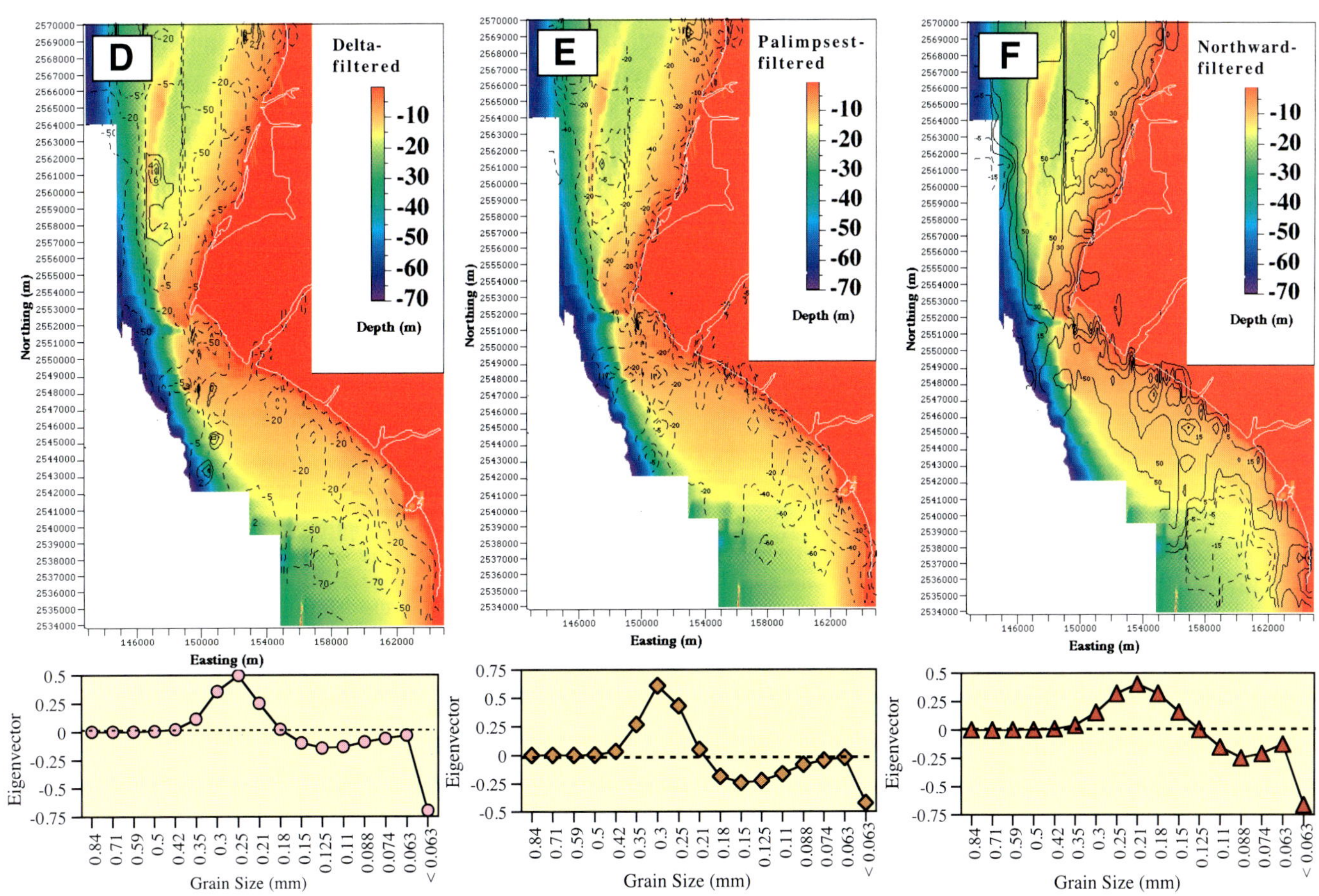

FIG. 16 (continued).—

size patterns in the study area (Table 1), its spatial eigen patterns are somewhat "fuzzy", showing combined signals of the previous two deviation fields (Fig. 16C).

Medium-Grained and Fine-Grained Sand (Positive Group) vs. Fine-Grained Sand to Mud (Negative Group) Assemblage.—

This assemblage includes the delta-filtered, palimpsest-filtered, ridge-filtered, and northward-filtered deviation fields (Figs. 16D, E, F). Because the eigen result of the ridge-filtered deviation field is identical to that of the delta-filtered deviation field in the amount of data explained (Table 1), and in the grain-size groupings and the eigenweighting patterns, it is not presented.

Both the spatial eigenpatterns of the delta-filtered and palimpsest-filtered deviation fields are dominated by the fine-grained group (indicated by negative contours). This is because the original hypothetical curves used in the analysis have coarse modal sizes and very little fines (Fig. 6). Thus the eigen results are more sensitive to the variations of the fines in the data. The contours of the spatial associations of these grain sizes do not indicate identifiable correlations to any morphological features in the study area, nor can any particular transport process be inferred from them (Figs. 16D, E). They are, however, more uniformly distributed compared to the patterns of the previous assemblage. On the other hand, the northward-filtered deviation field is dominated by the coarse-grained group (Fig. 16F), whose spatial eigen patterns are also "fuzzy".

Numerical Model Simulation

Model-simulated longshore and cross-shore currents, salinity, and suspended-sediment concentration (SSC) are compared with the tripod observations for verification (Fig. 17). In the model verification, the 63 mm size was chosen because it is fine enough to be carried by the plume and also to be resuspended by the currents, but is still heavy enough to display the settling effect. In general, the model reproduced the tidal aspects of the observed currents satisfactorily in terms of the phase, longshore amplitude, and the northward and offshore-transport dominated nature (Figs. 17A, B). Two inherent differences between the model and the reality might contribute to the discrepancy between the simulation and the observation. Firstly, the model river mouth is oriented N–S instead of being shore-normal, which might cause underprediction of the current speeds. Secondly, the river discharge and the initial strength of the river plume are constant in the model. This aspect precludes variabilities in the simulated flow caused by the change of river flow.

The simulated salinity also shows a good agreement with the observation except that the simulation shows stronger tidal fluctuations than those observed (Fig. 17C). This difference is largely an artificial effect determined by the ΔS between the river boundary and the model interior. The greater the ΔS value used, the greater the tidal amplitude in the simulated salinity. In order to simulate the observed surface plume in the salinity structure and the most optimal currents, the present ΔS value was chosen. A minor cause for the salinity discrepancy is also the invariant expression of river discharge in $\Delta\eta$ and ΔS at the river boundary.

The discrepancy between simulated and observed SSC is caused largely by the fact that the POM is unable to incorporate the effect of wave–current interactions in the simulation (Fig. 17D). Therefore, the effect of increased resuspension in the field due to wave action could not be simulated. Yet, the double peak feature is simulated. Nevertheless, for the purpose of investigating the primary processes by which the sediment-laden river effluent disperses immediately seaward of the river mouth (Stages I and II; Wright and Nittrouer, 1995), the POM is considered adequate.

The near-surface (1 m below the surface) features of the simulated plume, including current, salinity, and SSC, at various stages of a typical tidal cycle are plotted in Figure 18. Because of the weak river runoff, the dispersion of the plume is dictated primarily by the longshore tidal current, showing lateral swings as indicated by the salinity distribution. Consequently, the seaward extent of the plume is limited to be within about 2 km from the shoreline. Because of the distance between the river open boundary, where the sediments were introduced, and the river mouth, most grains settled out of the water column before they reach the river mouth. Throughout the tidal cycle, there is always a trace of SSC immediately seaward of the right-hand side of the river mouth. This plume is maintained by resuspension by both the flood and the ebb currents but is enhanced by the suspended sediment in the river effluent during the ebb.

In order to further examine the plume-related sediment transport, two one-kilometer-long shore-normal cross sections approximately 500 m from the center of the river mouth on either side were selected (lines A and B in Figure 14) to evaluate spatial sediment transport patterns. On each cross section, the sediment transport for each model grid ($q_{i,j}$) was integrated over approximately 10 tidal cycles (120 hourly simulations), so that

$$q_{i,j} = \sum_{t=1}^{120} C_t U_t \tag{13}$$

in which C_t, U_t represent simulated hourly values for suspended-sediment concentration and velocity component perpendicular to the cross section, respectively. Therefore, $q_{i,j}$ has the units of mass per unit area (g/m^2). The total transport across the cross section (q_A) is then obtained by integrating $q_{i,j}$ over the area of the cross section.

The values of $q_{i,j}$ are contoured for both cross sections (Fig. 19), which show stratified structures having large values near the surface and sea floor. This suggests that the net (time-integrated) unit sediment transport is contributed by the river plume due to the river sediment discharge, and by sediment settling and current resuspension. It is also noticeable that the net unit transport for the northern cross section is one order of magnitude greater than that of the southern counterpart. In fact, the total net transport q_A is 449 kg for the northern cross section (over 4927 m^2) and 84 kg for the southern cross section (over 12,317 m^2).

To further investigate the extent to which the river-borne background concentration can be carried by the river effluent, a clay size (1 μm) was substituted for the very fine-grained sand used in the next model run with the bottom resuspension turned off. The results show a continuous SSC plume leaving the river mouth, basically following the pattern of the salinity but with a smaller spatial range. This experiment suggests that under normal flood-season runoff conditions, the river, as a delivery mechanism for sediment, is rather weak. Its range of direct influence on the time scale of a tidal cycle is limited to the nearshore region in the vicinity of the river mouth.

DISCUSSION AND CONCLUSIONS

The short-term observations of near-bed sediment dynamics and transport during the flood season suggest that wave agitation is the most important mechanism for sediment suspension and resuspension. The river plume is the second most important factor influencing the near-bed sediment concentration, which indicates the importance of river-sediment discharge. Despite the role of waves, it is the tide-dominated longshore currents that determine the direction of nearshore sediment transport. The

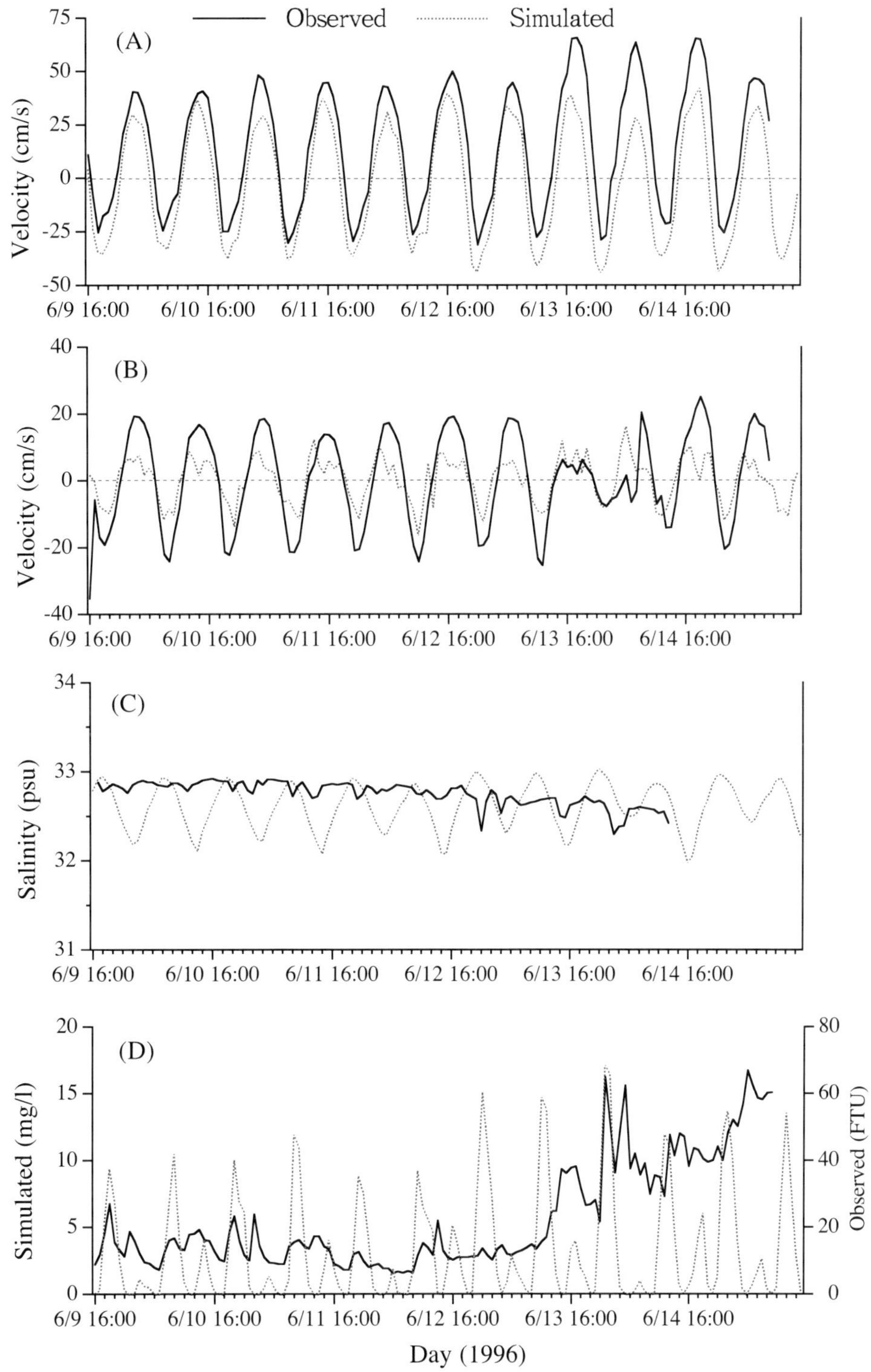

FIG. 17.—Comparisons between the model simulation and the observation on **A)** longshore current component, **B)** cross-shore current component, **C)** salinity, and **D)** suspended-sediment concentration (SSC).

sediment transport is determined by the combined effect of wave- and tide-dominated longshore currents. We make no attempt to identify the contribution of wave-induced currents on the sediment transport in this study. The direction of net (nontidal) sediment transport is northward.

The exceedingly high percentage of data explained by the first mode of the river-filtered and northward-filtered deviation fields suggests that the river is a major source of sediments in the study area. Northward longshore transport is the dominant direction of long-term sediment transport.

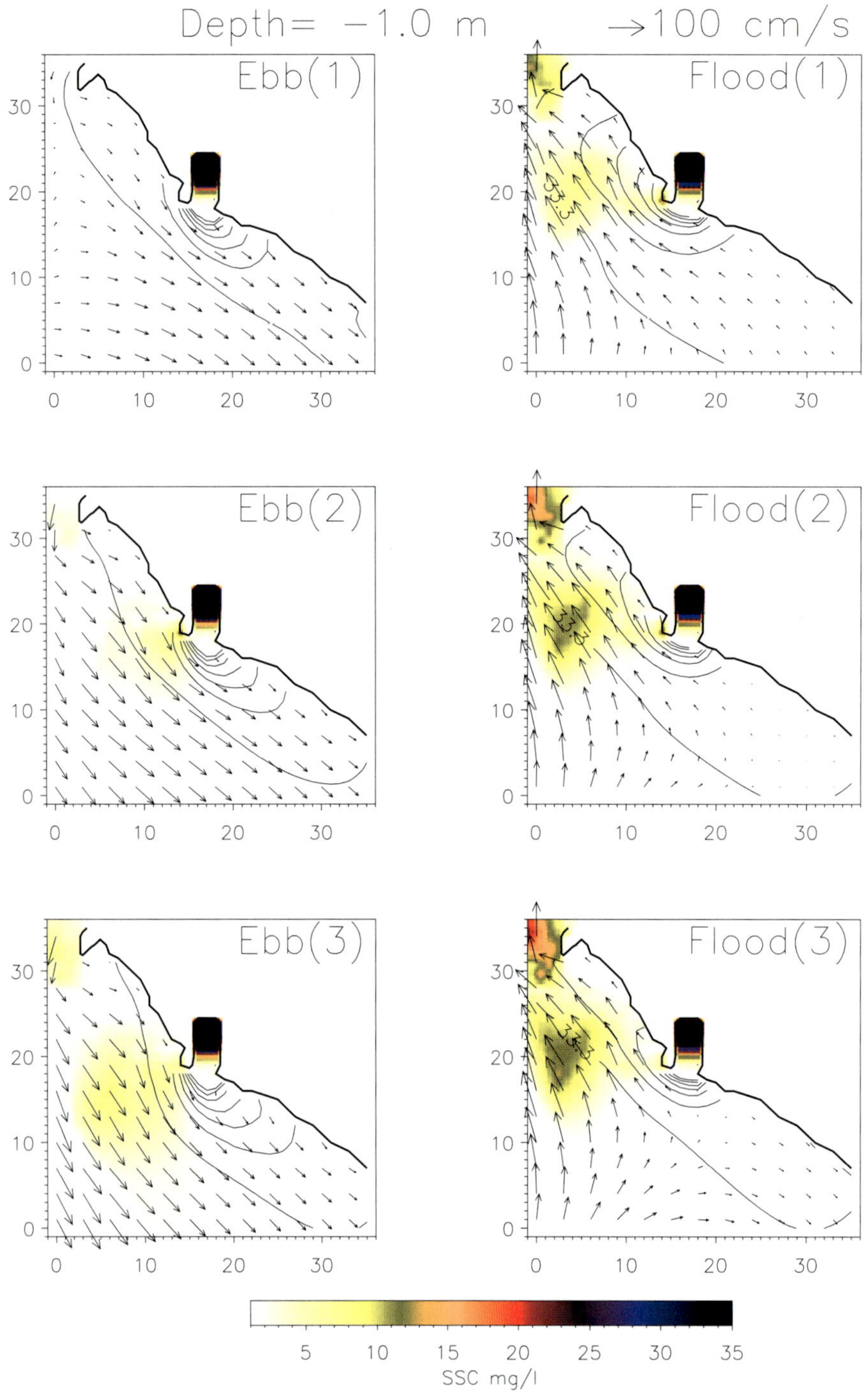

FIG. 18.—Model simulation of the plume at one meter below the surface in a typical tidal cycle, showing the spatial characteristics for each of the beginning, the middle, and the end of the ebb and flood half cycles. The arrows represent the current, the salinity variation is represented by the isohalines (at the interval of 0.3 psu with the outer boundary at 33.3 psu), and the range of suspended-sediment concentration (SSC) is expressed by the color bar. Very fine-grained sand was used in the river sediment discharge, and resuspension was allowed.

The modeling simulation of the sediment fluxes suggests the following: (1) the net nearshore sediment transport on either side of the river mouth is northward, (2) the surplus sediment flux on the north side of the river mouth is caused largely by the river sediment discharge, and (3) on the time scale of a tidal cycle, the spatial extent of river-sediment influence is limited to the nearshore region in the vicinity of the river mouth. These findings agree with the short-term direct sediment dynamics observed, and the indirect evidence from the grain-size analysis. Also, the dynamics of the river plume enhance the northward sediment transport on the right-hand side of the river mouth.

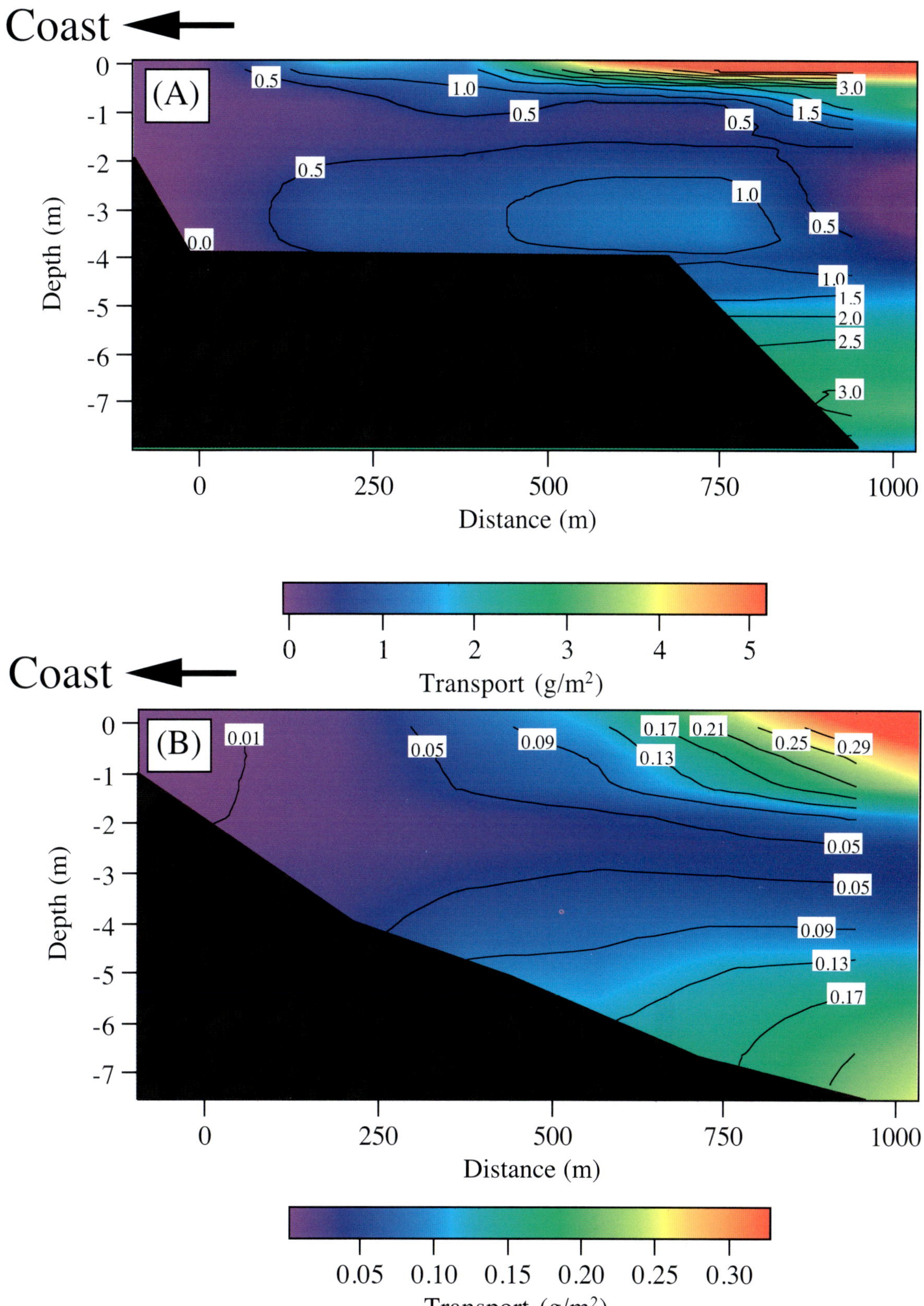

FIG. 19.—Contour plot of the net (time-integrated) sediment transport per unit area across two short shore-normal transects 500 m from the center of the model river mouth at **A)** the north side and **B)** the south side of the mouth. The positive value indicates northward direction.

On the basis of the first-mode grain-size groupings and their corresponding spatial patterns of all the deviation fields, the two dominant grain-size groups in the study area are mud and medium- to fine-grained sands (0.30–0.15 mm). The "behavior" of these two groups is distinctively different. The mud is distinctly river-borne, and thus modern in the study area. Although mud accounts for the major part of the sediment load of Tsengwen River, the river is not the only source for mud in the study area. Lu-erh-men River, Yen-shui River, and Er-hain River are small rivers with high sediment yield (load per unit of drainage

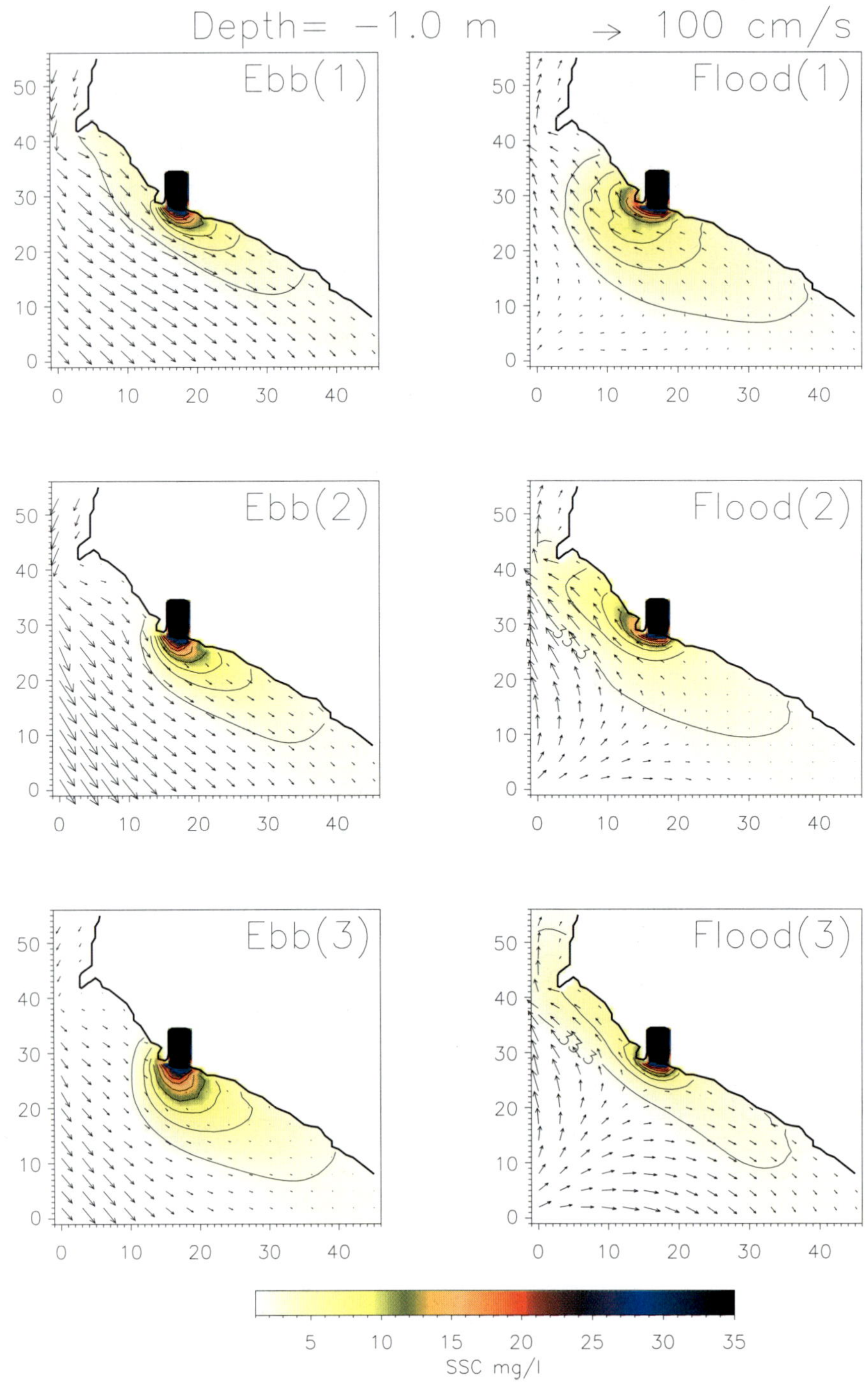

FIG. 20.—Model simulation of the plume at one meter below the surface in a typical tidal cycle, showing the spatial characteristics for each of the beginning, the middle, and the end of the ebb and flood half cycles. The arrows represent the current, the salinity variation is represented by the isohalines (at the interval of 0.3 psu with the outer boundary at 33.3 psu), and the range of suspended-sediment concentration (SSC) is expressed by the color bar. Clay was used in the river sediment discharge, no resuspension was allowed.

area). These rivers are likely to be important secondary sources for mud in the system. This is indicated by the patches immediately seaward of the mouths of Lu-erh-men River and Yen-shui River, and the large patch on the platform between Ann-ping Harbor and the Er-hain River, as highlighted by various deviation fields.

To put the findings of this study in a global perspective, two other deltaic systems are briefly compared: the Volta Delta on the western Bight of Benin, West Africa, and the well documented Ebro Delta, on the Spanish Mediterranean coast. Historically, the Volta River Delta was the major sediment source for the barrier coast immediately to the east via strong eastward longshore transport (Anthony and Blivi, 1999). Because of recent reduction of direct fluvial supply and the construction of a dam 60 km upstream from the sea in 1961, the downdrift delta-mouth barrier and shoreface began to erode. Consequently, an imbalance exists between the deltaic sand supply and the high demand of the littoral drift (Anthony and Blivi, 1999). As a result, the delta-mouth barrier and nearshore deltaic deposits began to erode. A similar scenario could develop at the Tseng-wen River mouth. The downdrift barriers of Ting-tou-e and Wang-tzu-liao (Fig. 1) are likely to experience erosion to satisfy the requirement of northward net longshore drift. The erosion probably will be exacerbated by the heavy coastal armoring of the updrift delta.

At the Ebro Delta, because of damming and building of reservoirs, not only is the sediment load of the Ebro River reduced, the sand fraction is also decreasing (Guillén and Jiménez, 1995; Jiménez and Sánchez-Arcilla, 1993; Palanques et al., 1990). Consequently, the progradational trend in the deltaic evolution after several centuries has changed from an intermediate river–wave dominated system to a wave-dominated system (Jiménez and Sánchez-Arcilla, 1993). Similarly, the depositional setting at the mouth of Tseng-wen River has changed from a deltaic system to a wave-dominated estuarine system (Liu et al., 1998). For both systems, the wash load (silt and clay) consists of the most important part of the river sediment supply but it does not contribute to the accretion of the adjacent coast.

A "mud belt" is reported in water depths of 20 m and 80–90 m surrounding the prodelta front at the Ebro Delta (Palanques and Drake, 1990). This prodelta mud has been linked to the mud discharged by the Ebro River (Diaz et al., 1996). Another offshore mud patch has been linked, through textural affinities, to the mud discharged from Gironde estuary (Weber et al., 1991). In our case, because the lower slope is below the depth of fair-weather wave base, and the adjacent mud-draped submarine depression is sheltered by the sand ridge from incident waves, this mud is the least likely to be resuspended.

In summary, without the rivers in the study area, the surficial textural characteristics of the submarine platform would have remained relict. It is the deposition of modern river-borne mud that essentially gave the otherwise coarse-grained and better-sorted relict sediments a "mud tail", creating the distinctive textural characteristics of the palimpsest sediments. Because of the weakness of the river runoff, direct river sediment discharge (initial deposition) exists only in the immediate vicinity of the river mouth. After the initial deposition by the river plume, waves and currents played an important role in the long-term sediment dispersal of river sediments. Wave agitation keeps the fine-grained sediments suspended and resuspended in the nearshore, leading to the eventual deposition of mud in the depression on the platform and at the foot of the platform. Wave action also maintains the relict appearance on the platform, especially on the submarine sand ridge and the remnant delta. The time-integrated effect of currents is responsible for the net northward longshore sediment transport.

ACKNOWLEDGMENTS

The funding for this study was provided by the R.O.C. National Science Council under the grant numbers NSC 84-2621-M-110-001, NSC 85-2621-P-110-006, NSC-86-2621-P-110-003, and NSC-87-2621-P-110-001. This study was also supported by the National Sun Yat-sen University and the Industrial Technology Research Institute through Mr. Jan-ming Chyan. Satellite images are provided by the Center for Space and Remote Sensing Research, National Central University. Captain Peter Tsao provided invaluable assistance at sea and in other areas of the field operation. Courtesies of the Tai-nan Fish Culture Branch of the Taiwan Fisheries Research Institute and Mr. Quan-da Wu to let us place the DGPS base station on their premises are acknowledged. Mr. Hwa-young Chang also assisted in the field work.

REFERENCES

Anthony, E.J., and Blivi, A.B., 1999, Morphosedimentary evolution of a delta-sourced, drift-aligned, sand barrier-lagoon complex, western Bight of Benin: Marine Geology, 158, p. 161–176.

Arnoux-Chiavassa, S., Rey, V., and Fraunie, F., 1999, Modeling of suspended sediments off the Phone River mouth: Journal of Coastal Research, v. 15, p. 61–73.

Barua, D.K., Kuehl. S.A., Miller, R.L., and Moore, W.S., 1994, Suspended sediment distribution and residual transport in the coastal ocean off the Ganges–Brahmaputra river mouth: Marine Geology, v. 120, p. 41–61.

Blumberg, A.F., and Mellor, G.L., 1987, A description of a three-dimensional coastal ocean circulation model, *in* Heaps, N.S., ed., Three Dimensional Coastal Ocean Models: American Geophysical Union, Coastal and Estuarine Sciences 4, p. 1–16.

Boggs, S., Wang, W.-c., Lewis, F.S., and Chen, J.-c., 1979, Sediment properties and water characteristics of the Taiwan Shelf and Slope: Acta Oceanographica Taiwanica, v. 10, p. 10–49.

Chao, S.-Y., 1998, Hyperpycnal and buoyant plumes from a sediment-laden river: Journal of Geophysical Research, v. 103, p. 3067–3081.

Davis, R.A., 1985, Coastal Sedimentary Environments, Second Edition: New York, Springer-Verlag, 716 p.

Diaz, J.I., Palanques, A., Nelson, C.H., and Guillén, J., 1996, Morphostructure and sedimentology of the Holocene Ebro prodelta mud belt (northwestern Mediterranean Sea): Continental Shelf Research, v. 16, p. 435–456.

Duane, D.B., Field, M.E., Meisturger, E.P., Swift, D.J.P., and Williams, S.J., 1972, Linear shoals on the Atlantic inner continental shelf, Florida to Long Island, *in* Swift, D.J.P., Duane, B. and Pilkey, O.H. eds., Shelf Sediment Transport; Process and Pattern: Stroudsburg, Pennsylvania, Dowden, Hutchinson & Ross, p. 447–499.

Emery, K.O., 1967, Relict sediments on continental shelves of world: American Association of Petroleum Geologists, Bulletin, v. 52, p. 45–464.

Endo, N., Masuda, F., and Yokokawa, M., 1996, Grain-size distributions of sediment carried by single transportation modes in an experimental microdelta system: Sedimentary Geology, v. 102, p. 297–304.

Fanos, A.M., 1995, The impact of human activities on the erosion and accretion of the Nile delta coast: Journal of Coastal Research, v. 11, p. 821–833.

Fenster, M.S., FitzGerald, D.M., Hohlen, W.F., Lewis, R.S., and Baldwin, C.T., 1990, Stability of giant sand waves in Eastern Long Island Sound, USA: Marine Geology, v. 91, p. 207–225.

Galloway, W.E., 1975, Process framework for describing the morphological and stratigraphic evolution of deltaic depositional systems, in Broussard, M.L., ed., Deltas, 2nd Edition: Houston, Texas, Houston Geological Society, p. 87–98.

GIBBS, R.J., MATTHEWS, M.D., AND LINK, D.A., 1971, The relationship between sphere size and settling velocity: Journal of Sedimentary Petrology, v. 41, p. 7–18.

GREEN, M.O., BLACK, K.P., AND AMOS, C.L., 1997, Control of estuarine sediment dynamics by interactions between currents and waves at several scales: Marine Geology, v. 144, p. 97–116.

GUILLÉN, J., AND JIMÉNEZ, J.A., 1995, Processes behind the longshore variation of the sediment grain size in the Ebro Delta coast: Journal of Coastal Research, v. 11, p. 205–218.

GUILLÉN, J., AND PALANQUES, A., 1996, Short- and medium-term grain size changes in deltaic beaches (Ebro Delta, NW Mediterranean): Sedimentary Geology, v. 101, p. 55–67.

HAYES, M.O., 1979, Barrier island morphology as a function of tidal and wave regime, *in* Leatherman, S.P., ed., Barrier Islands from the Gulf of St. Lawrence to the Gulf of Mexico: New York, Academic Press, p. 1–27.

ITRI (INDUSTRIAL TECHNOLOGY RESEARCH INSTITUTE), 1996, Assessment of Offshore Sand Resource in Tai-nan: Report 85-EC-2-A-17-0112, 88 p. (in Chinese).

JIMÉNEZ, J.A., AND SÁNCHEZ-ARCILLA, A., 1993, Medium-term coastal response at the Ebro delta, Spain: Marine Geology, v. 114, p. 105–118.

KOMAR, P.D., AND MILLER, M.C., 1974, Sediment threshold under oscillatory water waves: Journal of Sedimentary Petrology, v. 43, p. 756–775.

LI, M.Z., AND AMOS, C.L., 1998, Predicting ripple geometry and bed roughness under combined waves and currents in a continental shelf environment: Marine Geology, v. 18, p. 941–970.

LI, M.Z., AMOS, C.L., AND HEFFLER, D.E., 1997, Boundary layer dynamics and sediment transport under storm and non-storm conditions on the Scotian Shelf: Continental Shelf Research, v. 141, p. 157–181.

LIN, S.F., YANG, Y.J., AND TANG, T.Y., 1995, Three dimensional distributions of M2 tidal and net currents west of Tainan: 17th Ocean Engineering Conference, Proceedings, p. 1479–1488.

LIU, J.T., CHAO, S.-Y., AND HSU, R.T., 1999, The influence of suspended sediments on the plume of a small mountainous river: Journal of Coastal Research, v. 15, p. 1002–1010..

LIU, J.T., AND HOU, L.-H., 1997, Sediment bypassing and morphotextural characteristics around the entrance of Kaohsiung Harbor, Taiwan: Marine Geology, v. 140, p. 367–390.

LIU, J.T., YUAN, P.B., AND HUNG, J.-J., 1998, The coastal transition at the mouth of a small mountainous river in Taiwan: Sedimentology, v. 45, p. 803–816.

LIU, J.T., AND ZARILLO, G.A., 1989, Distribution of grain sizes across a transgressive shoreface: Marine Geology, v. 87, p. 121–136.

LIU, J.T., AND ZARILLO, G.A., 1990, Shoreface dynamics: evidence from sediment patterns and bathymetry: Marine Geology, v. 94, p. 37–53.

MCBRIDE, R.A., AND MOSLOW, T.F., 1991, Origin, evolution, and distribution of shoreface sand ridges, Atlantic inner shelf, U.S.A.: Marine Geology, v. 97, p. 57–85.

MCMANUS, D.A., 1975. Modern versus relict sediment on the continental shelf. Geological Society of America, Bulletin, v. 86, p. 1154–1160.

MCCAVE, I.N., 1979, Tidal currents at the north Hinder Lightship, southern North Sea: Flow directions and turbulence in relation to maintenance of sand bands: Marine Geology, v. 31, p. 101–114.

MELLOR, G.L., 1998, User's Guide for a Three-Dimensional, Primitive Equation, Numerical Ocean Model, Report: Princeton, N.J., Princeton University, Atmospheric and Oceanic Science Program, 41 p.

MILLER, M.C., MCCAVE, I.N., AND KOMAR, P.D., 1977, Threshold of sediment motion under unidirectional currents: Sedimentology, v. 24, p. 507–527.

MILLIMAN, J.D., AND SYVITSKI, P.M., 1992, Geomorphic/tectonic control of sediment discharge to the ocean: The importance of small mountainous rivers: Journal of Geology, v. 100, p. 525–544.

NIEDORODA, A.W., SWIFT, D.J.P., AND HOPKINS, T.S., 1985, The shoreface, *in* Davis, R.A., ed., Coastal Sedimentary Environments: New York, Springer-Verlag, p. 531–624.

NINO, H., AND EMERY, K.O., 1961, Sediments of shallow portions of East China Sea and South China Sea: Geological Society of America, Bulletin, v. 72, p. 731–762.

NITTROUER, C.A., KUEHL S.A., STERNBERG, R.W., FIGUEIREDO, A.G., AND FARIA, L.E.C., 1995, An introduction to the geological significance of sediment transport and accumulation on the Amazon continental shelf: Marine Geology, v. 125, p. 177–192.

NUMMEDAL, D., AND FISCHER, I.A., 1978. Process–response models for depositional shorelines: The German and the Georgia Bights: American Society of Civil Engineers, 16th Coastal Engineering Conference, Proceedings, p. 543–562.

ORTON, G.J., AND READING, H.G., 1993, Variability of deltaic processes in terms of sediment supply, with particular emphasis on grain sizes: Sedimentology, v. 40, p. 475–512.

PALANQUES, A., PLANA, F., AND MALDONADO, A., 1990, Recent influence of man on the Ebro margin sedimentation system, northwestern Mediterranean Sea: Marine Geology, v. 95, p. 247–273.

PALANQUES, A., AND DRAKE, D.E., 1990, Distribution and dispersal of suspended particulate matter on the Ebro continental shelf, northwestern Mediterranean Sea: Marine Geology, v. 95, p. 193–206.

SHIH, T.T., 1979, Landforms and their evolution of the Chinan Liman Coast in Southwestern Taiwan: Geographical Research, v. 5, p. 11–48.

SWIFT, D.J.P., AND FIELD, M.E., 1981, Evolution of a classic sand ridge field: Maryland sector, North American inner shelf: Sedimentology, v. 28, p. 461–482.

SWIFT, D.J.P., SANFORD, R.B., DILL, C.E., AND AVIGNONE, N.F., 1971, Textural differentiation on the shore face during erosional retreat of an unconsolidated coast, Cape Henry to Cape Hatteras, western North Atlantic Shelf: Sedimentology, v. 16, p. 221–250.

WANG, Y., AND AUBREY, D.G., 1987, The characteristics of China coastline: Continental Shelf Research, v. 7, p. 329–349.

WATER RESOURCES PLANNING COMMISSION, 1992, Hydrological Year Book of Taiwan: Ministry of Economic Affairs, R.O.C., 397 p.

WEBER, O., JOUANNEAU, J.M., RUCH, P., AND MIRMAND, M., 1991, Grain-size relationship between suspended matter originating in the Gironde estuary and shelf mud-patch deposits: Marine Geology, v. 96, p. 159–165.

WEN, J.T., LOU, J.S., SHEU, M.S., AND CHIANG W.S., 1995, Oceanic investigation of topography, wave, and current in the vicinity of Tainan City: 17th Ocean Engineering Conference, Proceedings, p. 1125–1138 (in Chinese with English abstract).

WRIGHT, L.D., AND COLEMAN, J.M., 1972, River delta morphology: Wave climate and the role of the subaqueous profile: Science, v. 176, p. 282–284.

WRIGHT, L.D., KIM, S.-C., AND FRIEDRICHS, C.T., 1999, Across-shelf variations in bed roughness, bed stress and sediment suspension on the northern California shelf: Marine Geology, v. 154, p. 99–115.

WRIGHT, L.D., AND NITTROUER, C.A., 1995, Dispersal of river sediments in coastal seas: six contrasting cases: Estuaries, v. 18, p. 494–508.

XING, J., AND DAVIES, A.M., 1999, The effect of wind direction and mixing upon the spreading of a buoyant plume in a non-tidal regime: Continental Shelf Research, v. 19, p. 1437–1483.

LATE QUATERNARY SEDIMENTATION AND PEAT DEVELOPMENT IN THE RAJANG RIVER DELTA, SARAWAK, EAST MALAYSIA

JAMES R. STAUB

Department of Geology, Southern Illinois University Carbondale, Carbondale, Illinois 62901-4324, U.S.A.

and

ROBERT A. GASTALDO

Department of Geology, Colby College, 5820 Mayflower Hill, Waterville, Maine 04901-8858, U.S.A.

ABSTRACT: Thick, domed peat deposits dominate most of the surface of the mesotidal to macrotidal Rajang River delta, tidally influenced alluvial valley, and adjacent coastal plain. Northeast-striking shoreline terrace sands that crop out along the landward margin of the delta and coastal plain and gravel outcrops in the alluvial valley are the surface expression of the VIIa highstand surface of 125 ka (oxygen isotope stage 5e). The upper few meters of the VIIa surface have undergone podzolization, are leached white, and are easily mapped. Near the present coast a peat/lignite bed, at a depth of 80 m, represents the IIIb highstand surface (oxygen isotope stage 3), indicating that 40 m of subsidence has occurred in the last 40 ka. In the alluvial valley, gravel dominates the base of an incised-valley fill 10 km wide and 45 m thick and is overlain by a fining-upward succession, the upper part of which is tidally influenced. Eroded Pleistocene terraces, mantled with thick peat, flank the Recent incised-valley fill. Within interfluve areas in the landward one-half of the northeast delta plain and adjacent coastal plain, thick Recent peat deposits (> 10 m) rest directly on or within a few meters of the buried remains of Pleistocene sediments. These peat deposits began accumulating between 7.3 and 5.8 ka as the rate of sea-level rise slowed. Recent siliciclastic sediments laterally adjacent to these peat deposits are composed of tidally influenced sands, silts, and clays. The seaward one-half of the northeast delta plain, delta front, and prodelta are composed of a seaward-thickening wedge of siliciclastic sediment up to 40 m thick that has accumulated in the last 5 ka. The base of the wedge is marked by a gravel lag that immediately overlies a rooted, yellow-brown alluvial soil. Siliciclastic sediments in this wedge consist of delta-front and prodelta clays and silts, delta-front distributary-mouth sands, and shoreline sands. Young (< 5 ka), reduced-thickness (< 10 m) peat deposits lie conformably on top of this wedge in this part of the delta plain. In contrast, the southwest part of the delta plain is not underlain by shallow-depth Pleistocene sediments, and it started to prograde into the South China Sea prior to 7 ka; its surface is dominated by beach ridges and gley soils mantled by mangrove–*Nipa* vegetation.

INTRODUCTION

In Southeast Asia and the Malay Archipelago many delta systems (e.g., Baram, Fly, Hari, Irrawaddy, Klang-Langat, Mahakam, Mekong, and Rajang) occur, but only a few of these are reported to contain extensive, thick (> 10 m), low-ash (< 5%) autochthonous peat deposits (Baram, Hari, Klang-Langat, Rajang). All of these peat-accumulating deltas are reported to be prograding into their depositional basins with shoreline accretion rates on the order of 10 m/yr. Many differences exist, however, between these peat-accumulating deltaic depositional systems. The most significant differences between these systems are peat thickness and age, indicating that system response to Late Pleistocene–Holocene sea-level rise has not been similar in all systems. In the Baram, Hari, and Klang-Langat the maximum reported peat thickness is on the order of 10 m and maximum peat age is 4.5 ka (Anderson, 1964; Coleman et al., 1970; Esterle, 1990). In the Rajang system maximum reported peat thickness is on the order of 20 m and maximum peat age is 7.3 ka (Anderson, 1964; Staub and Esterle, 1993, 1994).

What causative factors in the Rajang system resulted in peat deposits that are up to twice as thick and 3 ka older than other deltaic peat deposits in the region? This paper examines the Late Quaternary sediments in the Rajang in an effort to determine facies associations and developmental history and to ascertain what factors led to the early formation and spatial distribution of its peat deposits.

PHYSIOGRAPHIC SETTING, CLIMATE, AND PREVIOUS WORK

The Rajang River (Fig. 1) is the largest river in Sarawak and drains part of the Central Borneo Massif. Cretaceous to Eocene sediments dominate the massif; these rocks are accretionary complexes and forearc basin deposits that formed during the Tertiary opening and spreading of the South China Sea (Hamilton, 1979; Hutchison, 1989, 1996). In the Rajang River drainage basin, rock types are composed principally of folded and faulted siliciclastics with minor amounts of carbonates. Additionally, igneous intrusive and extrusive rocks are present, and many sedimentary rocks have been metamorphosed (Lam 1988).

The Rajang River drainage basin is about 50,000 km^2 in size, and elevations exceed 2,000 m. Hill slopes are steep and flood plains are of limited area. Soils are poorly developed and represent the mineral composition of the parent materials (Scott, 1985; McBeth, 1995). The Rajang River delta also receives discharge and sediment from the proximal hills region (Fig. 1) to the immediate south of the Rajang distributary. Four large rivers traverse the coastal plain to the east of the delta plain (Fig. 2). Their drainage basins average about 3,000 km^2 in size, and rock types and soils present are similar to those of the Rajang River drainage basin. The physiographic separation between the delta plain and the coastal plain occurs at the midpoint between the Igan distributary of the Rajang River and the Oya River of the adjacent coastal plain (Fig. 2). The coastal plain extends eastward from this point to the approximate position of Kuala Bintulu.

The Rajang River delta plain (Fig. 3) covers 6,500 km^2, and peat greater than 3 m thick (Fig. 2) covers about 50% of the delta-plain surface, with laterally extensive deposits common to the northeast of the town of Daro. The surfaces of most peat deposits are domed or raised (Fig. 3). The surfaces of the domed peat deposits in the delta plain are as much as 4 to 6 m higher than spring high-tide levels, with elevation differences increasing inland from the coast. Maximum peat thickness is 15 m. To the east the adjacent coastal plain is 4,500 km^2 in area (Fig. 2), and about 80% of the surface is mantled by peat greater than 3 m thick.

Tropical Deltas of Southeast Asia—Sedimentology, Stratigraphy, and Petroleum Geology
SEPM Special Publication No. 76, Copyright © 2003
SEPM (Society for Sedimentary Geology), ISBN 1-56576-086-7, p. 71–87.

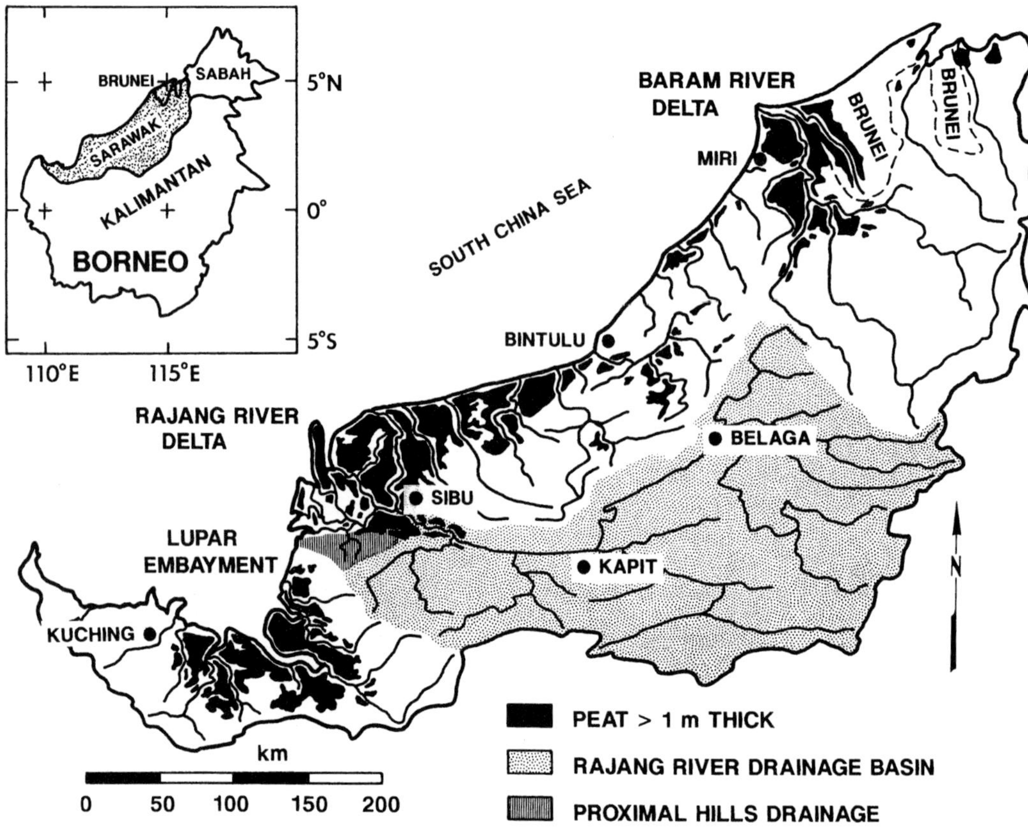

FIG. 1.—Map of Sarawak, East Malaysia, showing the locations of the major peat-forming regions and all peat deposits greater than 1 m thick (from Staub et al., 2000). The location of the Rajang River delta, the drainage basin of the Rajang River, and the proximal hills drainage that is adjacent to the delta are shown.

These coastal-plain peat deposits are also domed and are up to 15 m thick, and their domed surfaces are up to 7 m higher than the adjacent river channels during flood events. The alluvial valley is located between the towns of Sibu and Kanowit (Fig. 3). It covers 400 km^2, and maximum reported peat thickness is in excess of 20 m (Anderson, 1964; Scott, 1985). Peat greater than 3 m thick covers about 75% of the surface area (Fig. 2), and elevation differences between the surfaces of the raised peat deposits and the water level in the Rajang River are as much as 9 m at spring high tide.

Tides along the Sarawak coast (Fig. 1) increase from less than 1 m in the northeast (Baram River delta) to more than 9 m in the southwest (Lupar embayment). Tides are semidiurnal within the delta plain of the Rajang River; they range from mesotidal to macrotidal (Fig. 2, Table 1) and increase in range from northeast to southwest from 2.9 m at Kuala Igan to 5.8 m at Kuala Rajang. Tidal influence extends inland approximately 120 km to the town of Kanowit (Fig. 3). Tides are diurnal to semidiurnal in the coastal plain, with diurnal events occurring periodically during neap phases. In the coastal plain, tides range from microtidal to mesotidal and increase in range from northeast to southwest from 1.5 m at Kuala Bintulu to 2.2 m at Kuala Oya (Fig. 2). Winds and waves from the northeast monsoon (wet season) dominate from December to March and from the southwest monsoon (dry season) during the middle months of the year. Maximum wave heights are on the order of 2 m.

Rainfall averages in excess of 370 cm / yr, with the rainy or wet season coincident with the northeast monsoon. Temperature averages about 25° C, and humidity ranges from 55% in the day to almost 100% at night (Scott, 1985). Climate is classified as tropical ever-wet (Morley and Flenley, 1987). Typical single-month discharge for the Rajang River drainage basin range seasonally from about 1,000 m^3/s to 6,000 m^3/s, and the average monthly discharge is about 3,600 m^3/s (Fig. 4). This substantial variation in discharge causes seasonal changes (dry vs. wet) in estuarine circulation patterns in most of the distributary channel mouths (Table 1), with all distributaries functioning as tide-dominated estuaries for much or most of the year. Peak discharge rates during the northeast monsoon can exceed ≈ 25,000 m^3/s (Jeeps and Gates, 1963) and conversely can drop to less than 250 m^3/s during periods of extended drought (Staub et al., 2000). Typical discharge for the five coastal-plain rivers range from about 250 to 1000 m^3/s seasonally but can be substantially higher and lower during extreme weather events.

During the Quaternary, base-level lowering and/or epeirogenic uplift of the Central Borneo Massif is estimated at 0.19 + 0.03/-0.04 m/ky (Farrant et al., 1995) and, as a result, production of siliciclastic sediment from the upland drainage basin is substantial. Staub and Esterle (1994) estimated that before anthropogenic development the Rajang River drainage basin provided ≈ 30 million metric tons of sediment annually, of which about 24 million metric tons is deposited in the delta front and prodelta primarily during the wet season (Staub et al. 2000).

Inspection of aerial photographs covering the last 50 years reveals that the delta plain has been expanding at a rate of 1.0 to 1.5 km^2/yr (Scott 1985). Vertical changes in foraminiferal assemblages combined with ^{14}C dates in the Lassa distributary

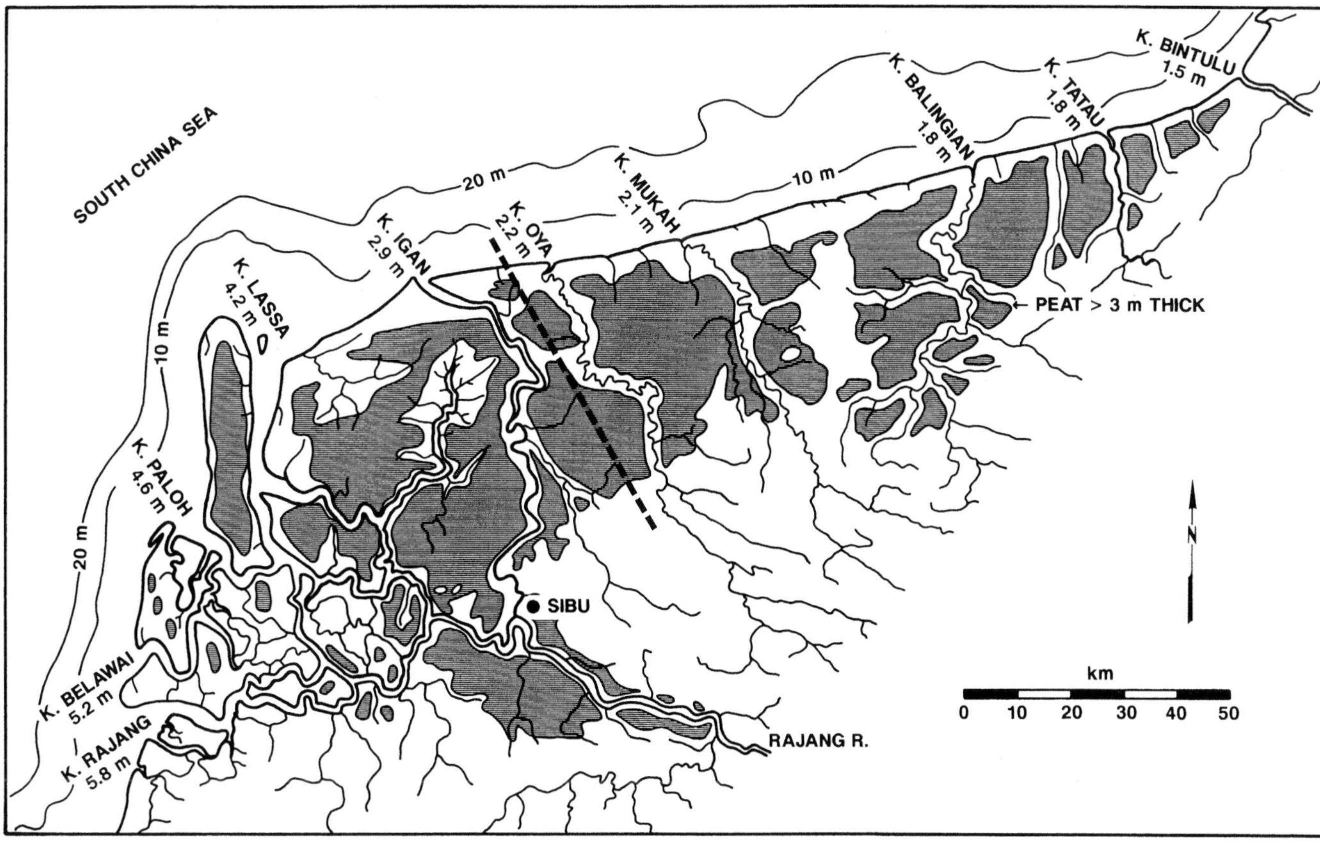

FIG. 2.—Map showing the areal distribution of thick (> 3 m) peat deposits in the Rajang River delta and adjacent coastal plain (modified from Staub and Esterle, 1994). The physiographic break (dashed line) between the delta plain (west) and the coastal plain (east) lies between the Igan distributary of the Rajang River and the Oya River. The alluvial valley is located upriver from the town of Sibu. Also listed is the maximum spring tidal range in meters at the mouth (kuala) of each of the delta distributaries and coastal-plain rivers.

demonstrate an ongoing regressive event for at least the last 2.2 ka in the northwestern part of the delta plain (Murphy, 1996). Long-term accretion rates, on the order of 8 m/yr, and aggradation rates of 1.5 mm/yr have existed in the delta plain for the last 7.5 to 8.5 ka (Staub and Esterle, 1993). The oldest dated peat deposits in the delta plain are in excess of 7.3 ka (Staub and Esterle, 1994).

DATA AND METHODS

Methods employed consisted of the compilation and analysis of published and unpublished data, as well as sampling and coring operations in the alluvial valley and delta plain of the Rajang River and the Oya River in the adjacent coastal plain. During field investigations, peat-swamp traverses were conducted and virtually all channels and major anthropogenic canals in the alluvial valley and delta plain of the Rajang River, as well as the Oya River drainage basin, were reconnoitered. Satellite images, low-altitude aerial photographs, and bathymetric data for coastal areas, distributary channels, and the alluvial valley channel were also analyzed.

Grab samples (Fig. 5) of channel-bottom, delta-front, midchannel-bar, point-bar, and beach sediments were obtained ($n = 173$). All samples were described in the field (grain size and shape, color and composition, and flora and fauna), and a cut of each sample was retained for laboratory analysis.

Subsurface data consists of vibracores, Eykelkamp peat auger cores, Edelman clay auger records, and deep bore-hole records. A total of 80 vibracores (Fig. 5) were recovered from delta-plain, distributary-channel, and delta-front and prodelta environments, and 30 Eykelkamp cores were obtained from peat-swamp environments. An additional 20 Eykelkamp core records with photographs and 120 Edelman clay-auger records (Fig. 5) were obtained from government reports. The records of 33 bore holes in the alluvial valley near Sibu and 10 bore holes in the vicinity of Daro in the delta plain were obtained from industry. Bore-hole records consisted of descriptions of materials encountered at depths of up to 100 m.

All vibracores and Eykelkamp cores were described in the field (peat type, grain size and shape, color and composition, sedimentary structures, and flora and fauna), and all vibracores and most Eykelkamp cores were photographed. Samples from each sediment type present in each vibracore and most Eylelkamp cores were retained for laboratory analysis. Most organic samples were analyzed for ash and sulfur content, and selected organic samples were sent to Krueger Enterprises, Inc., for ^{14}C age determinations.

Grain-size analysis for all siliciclastic-dominated sediment samples was conducted. Organic matter was destroyed with hydrogen peroxide. The > 63 μm part of each sample was analyzed according to Folk (1980), whereas the < 63 μm fraction of each sample was treated with dispersant and sized with a Spectrex

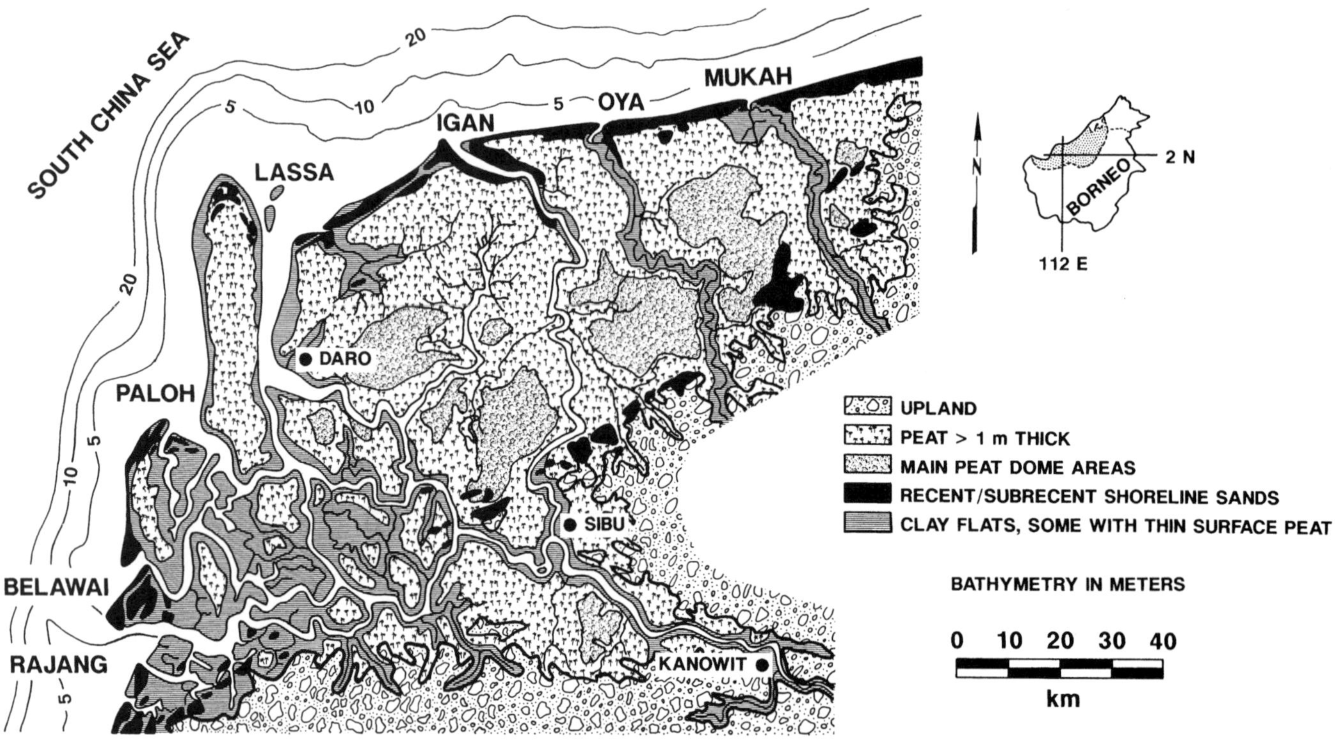

FIG. 3.—Physiography of the Rajang River delta and the part of the adjacent coastal plain (modified from Staub and Esterle, 1994). Peat domes dominate the northeastern part of the delta plain and adjacent coastal plain. Clay flats and shoreline sands dominate the southwestern delta plain.

laser particle counter. Mean ϕ and standard deviation were calculated following McBride (1971).

Sand obtained from grab samples ($n = 137$) and cores ($n = 160$) were subjected to petrographic analysis to determine mineral composition. Whole sand fractions (>63 μm) of sediment samples were used to generate epoxy grain-mount thin sections. Composition was determined by counting 300 grains per section. Angularity of grains was estimated following Powers (1953), and samples were classified according to Folk (1980).

Clays obtained from grab samples ($n = 56$) and cores ($n = 95$) were subjected to x-ray diffraction analysis to determine mineral composition. Sample preparation and identification procedures used were those detailed by Moore and Reynolds (1989).

Sediment types were grouped together into facies associations, and surficial and shallow core data were used to reconstruct the Recent progradation of the delta and coastal plains. Core and bore-hole data were used to construct cross sections. The resultant facies successions, as evidenced in plan maps and cross sections, were used to interpret the Late Quaternary history of the Rajang River alluvial valley, delta, and adjacent coastal plain. Ages of shoreline positions during the Holocene were estimated by using previously published (Staub and Esterle, 1994) long-term accretion rates (8 m/yr), combined with ^{14}C dated materials from 35 locations (Fig. 6) in the delta plain, alluvial valley, and coastal plain.

TABLE 1.—Tidal ranges and seasonal changes in estuary type.

Distributary Channel	Tidal Range[a] (m)	Estuary Type[b] Dry Season	Estuary Type[b] Wet Season
Igan	2.9	Salt Wedge	Partially Mixed
Lassa	4.2	Partially Mixed	Partially Mixed[c]
Paloh	4.6	Partially Mixed	Partially Mixed[c]
Belawai	5.2	Fully Mixed	Partially Mixed
Rajang	5.8	Fully Mixed	Fully Mixed

[a] Maximum spring tide
[b] Classification scheme after Postma (1980)
[c] Increased vertical stratification

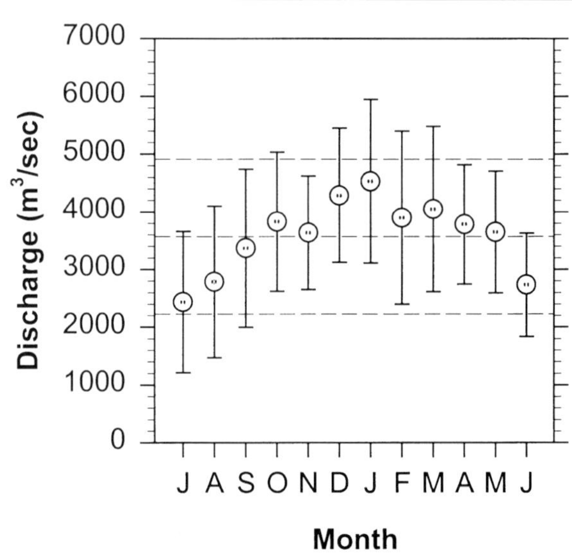

FIG. 4.—Hydrograph showing seasonal variation in discharge from the Rajang River drainage basin. Values are month-long averages based on 30 years of rainfall data. Error bars are one standard deviation (from Staub et al., 2000).

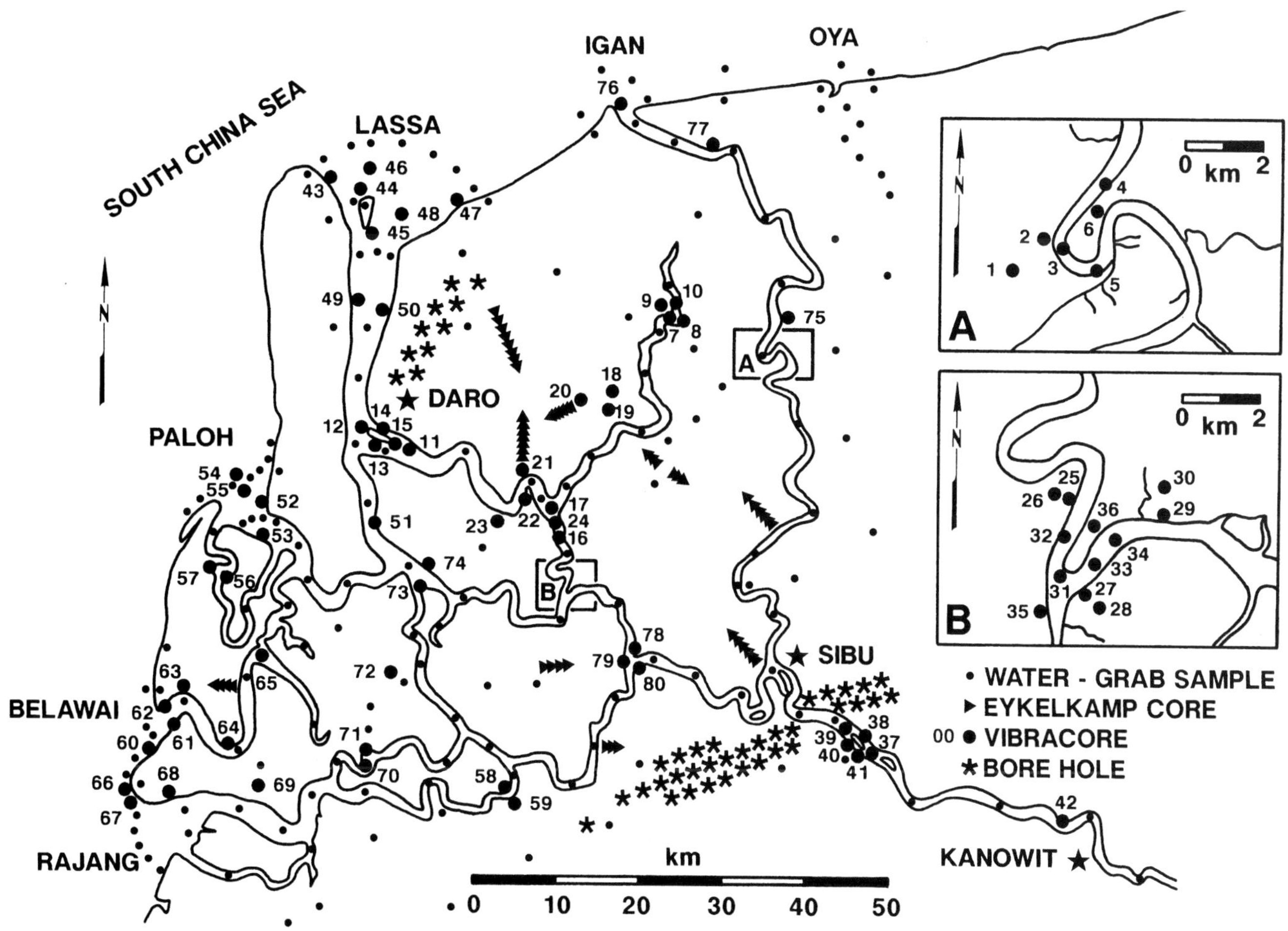

FIG. 5.—Map showing the location of grab samples, vibracores, Eykelkamp cores, and bore holes utilized in this study.

RESULTS

Vegetation Types and Peat Deposits

Mangroves dominate in brackish-water areas and show a general succession of *Avicennia–Rhizophora–Nipa–Oncosperma* that is related to decreasing salt tolerance (Scott, 1985). In coastal areas where shoreline sands are exposed, the freshwater tree *Casuarina* is dominant.

Mangroves, and particularly *Nipa* palms, dominate the surface of the southwestern delta plain. Various freshwater vegetation types dominate the northeastern delta-plain surface and adjacent coastal plain. Nutrient depletion caused by increasing peat thickness results in a general freshwater vegetation succession that culminates with *Shorea albida–Litsea–Parastemon* associations dominating (Anderson phasic community 4; Anderson, 1983) the main peat-dome areas (Fig. 3).

A peat deposit in the Rajang delta plain, where thick, is typically composed of high-ash, high-sulfur, degraded sapric peat at the base (< 50 cm) that is derived from pioneer mangrove–*Nipa* vegetation (Fig. 7). This basal layer is overlain by freshwater hemic peat that is low in ash and sulfur content. A thin sapric peat layer, derived from freshwater vegetation, caps the upper surface of most thick peat deposits. Ash content in this upper sapric layer is much lower than the basal sapric peat and sulfur content is low, unless brackish water siliciclastic sediments overlie it (Staub and Esterle, 1994). The peat deposit (as well as the water table) develops a convex or domed surface, the center of which is several meters higher than the adjacent channel systems.

Staub and Esterle (1994) noted that substrate composition is variable and that the pioneer vegetation in the succession is substrate dependent. If the pioneer vegetation is freshwater rather than brackish-water mangrove–*Nipa* (Fig. 7), then the basal sapric peat layer is reduced in thickness or absent, and ash and sulfur content are reduced.

Siliciclastic Sediments

Soils.—

Three basic siliciclastic soil types were encountered: marine-influenced gley soils, podzols, and freshwater alluvial soils. Gley soils are composed of dark gray clay or silty clay and show evidence of mangrove–*Nipa* root penetration. Clays are composed of mixed-layered illite–smectite, illite, and kaolinite with minor amounts of chlorite. Gley soils dominate much of the central and southwestern part of the delta plain.

Podzols are composed of gray-white to white clay with no or only slight yellow brown (iron) mottling. In some low-lying areas adjacent to uplands, and at locations in the landward part of the delta plain and alluvial valley, they occur as a residuum over noncalcareous sedimentary rocks or consolidated sediments. In

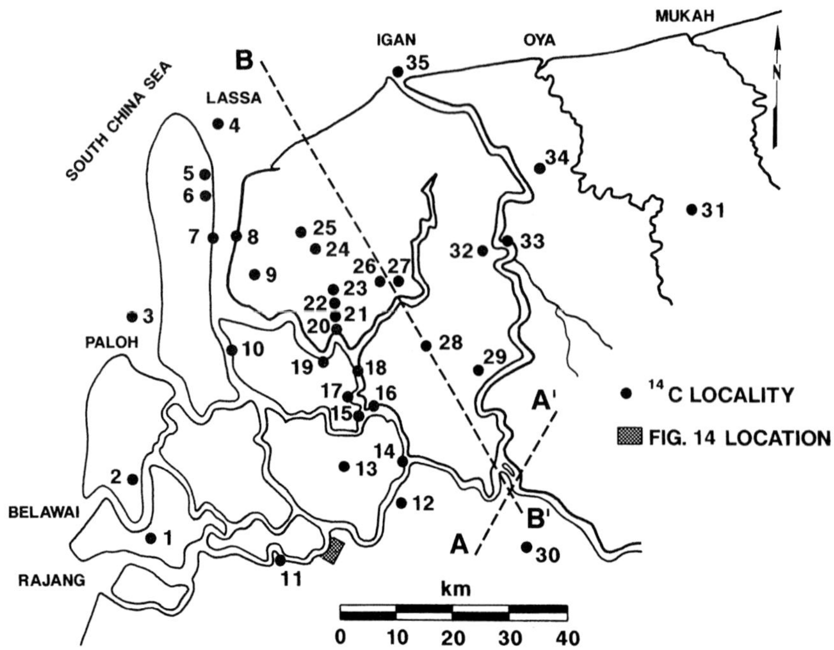

FIG. 6.—Map showing locations where ^{14}C dates were obtained (numbers are keyed to Table 2) and the location of Figure 11. All 120 Edelman auger records are from the Figure 14 location area. Also shown are the locations of cross sections A–A' and B–B' (Figs. 11 and 12, respectively).

the landward part of the delta plain they are found at various depths as a residuum over semiconsolidated intertidal to supratidal siliciclastic sediments in interfluve regions. Clays consist of kaolinite and illite with minor amounts of mixed-layer clays.

Alluvial soils are composed of mottled, yellow to brown clays and silty clays. They show evidence of root penetration of freshwater vegetation and are exposed in riverbanks and levees. Clays in alluvial soils are composed of mixed-layer illite–smectite and illite with kaolinite. Alluvial soils are found in the alluvial valley, in the landward part of the delta plain, and along channel margins in most of the coastal plain.

Alluvial-Valley and Delta-Plain Gravels.—

In the first 30 km upstream from Sibu, several longitudinal bars are present in the alluvial valley channel that contain subrounded to well-rounded pebbles and cobbles composed of siltstones, fine- to coarse-grained quartz arenites, medium- to coarse-grained sublitharenites and litharenites, and aphanitic to porphyritic volcanic rocks. Many sedimentary rock clasts show evidence of strain and/or low-grade metamorphism. Semiconsolidated outcrops (Fig. 8) in the alluvial valley, consisting of the aforementioned materials plus fine-grained sediments and wood fragments, attain heights of 5 to 7 m above MSL. Similar materials are reported at depth in bore-hole records in the alluvial valley.

The Rajang distributary channel at many locations in the southern part of the delta plain, adjacent to the proximal hills region, is floored by very angular to angular granules and pebbles. These granules and pebbles are composed mainly of vein quartz.

Alluvial-Channel and Distributary-Channel Intertidal Sediments.—

Sediment cores recovered from tidally influenced areas contain yellow-brown to dark olive-green laminated silts and sand–silt couplets (or tidal rhythmites) that are often wavy-bedded to lenticular-bedded. Rhythmites are best developed in distributary-channel bar forms in the middle part of the delta plain (Staub et al., 2000); these sediments contain foraminifera in brackish-water areas (Murphy, 1996).

Three other sediment types were observed in channel bar forms: yellow-brown to dark olive-green laminated silts, brown to gray cross-bedded sands, and brown to gray flaser-bedded sands. All contain layers of organic detritus, and many are burrowed when brackish water is present. Laminated silts are most common in the upper few meters of bar forms, whereas flaser-bedded and/or cross-bedded sands are found at greater depth. Sands in bar forms in the alluvial valley and the landward part of the delta plain are litharenites, whereas intertidal sands in the seaward part of the delta plain are sublitharenites.

Tidal-Creek and Tidal-Flat Sediments.—

Sediment cores recovered from brackish-water tidal creeks and the uppermost parts of tidal flats contain yellow-brown to dark olive-green to black laminated silts and clays or laminated clays. Burrowing often disrupts lamination. Evidence of slumping was observed in a few instances. Desiccation cracks develop on the upper surfaces of many tidal flats. Tidal-creek and tidal-flat sediments occur mainly in the southwestern part of the delta plain. Clays in these sediments are mixed-layer illite–smectite and illite with kaolinite incorporating minor amounts of chlorite.

Beach and Distributary-Mouth Sediments.—

The delta coastline is 160 km long and is composed of five distributary-channel mouths (Table 1) separated by sand beaches (Fig. 3). Widths of distributary mouths range from 2 to 12 km, and the intervening beaches are up to 35 km in length. Beach ridges

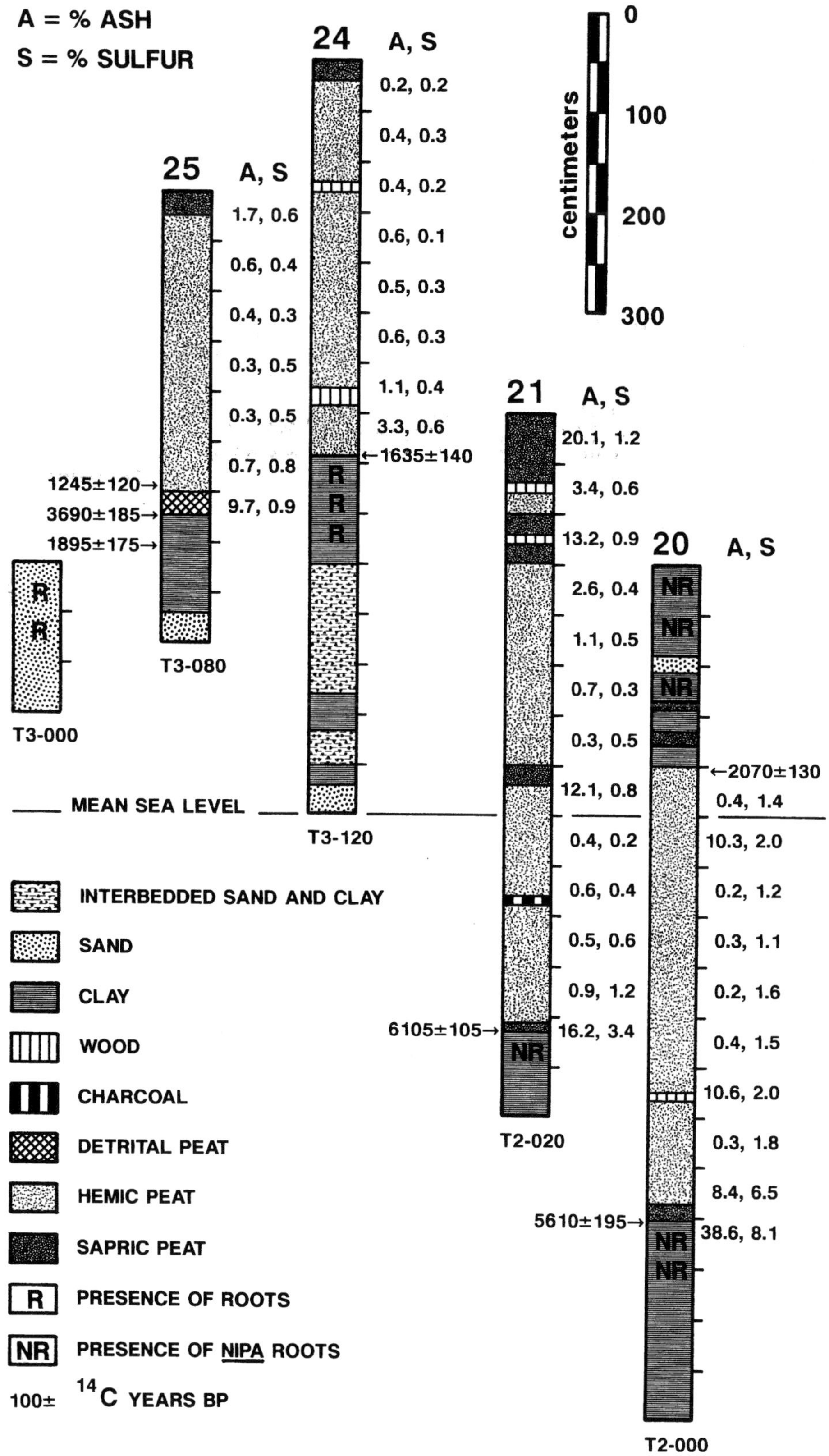

FIG. 7.—Core data showing vertical and lateral relationships of peat and siliciclastic sediment types, ash and sulfur content of the organic sediments, and ^{14}C dates. Hemic peats are better preserved, whereas sapric and detrital peats are more poorly preserved. Core locations (20, 21, 24, and 25) are the same as the ^{14}C locations shown in Fig. 6. Cores are also referenced to transect locations (e.g. T2-000) shown in Fig. 17 (modified from Staub and Esterle, 1994).

FIG. 8.—Pleistocene (VIIa) outcrop in the alluvial valley with gravel layers present. The fine-grained siliciclastic sediments in this outcrop are leached white. Lens cap is 52 mm in diameter.

Fig. 10. Recent shoreline accretion near the mouth of the Igan distributary channel in the northeast delta plain. Note how rapidly shoreline features are obscured by vegetation.

are readily apparent in aerial photographs up to 10 km inland from the coast in the southwestern part of the delta plain (Fig. 9), but ridges are rapidly obscured by vegetation or buried beneath a layer of peat in other areas (Fig. 10). Previous beach-ridge positions in peat-dominated areas can, in many areas, be observed along channel margins at low tide. Intertidal to supratidal beach sands are cross-bedded, brown to gray, and burrowed. Ridge-and-swale topographic relief is about 2 m, and beach sands are up to 5 m thick. Sands range from sublitharenites to quartz arenites, with quartz arenites more common.

Distributary-mouth sediments are composed of massive to cross-bedded, brown to gray, shallow-subtidal to intertidal sands. Layers of organic detritus also are present occasionally in these sands. Distributary-mouth sands are sublitharenites, up to 5 m thick, and sand-body geometry changes from one distributary mouth to the next in response to increasing tidal range. Sand-body geometry at the mesotidal Igan distributary mouth is lobate, at the macrotidal Lassa distributary mouth is elongate tidal-ridged, and at the macrotidal Paloh, Belawai, and Rajang distributary mouths is tidal-ridged (Pigott, 1995).

FIG. 9.—Recent beach-ridge deposits in the southwest delta plain near the mouth of the Rajang distributary channel.

Delta-Front and Prodelta Sediments.—

At depths of ≈ 3 m to 25 m below spring low tide, delta-front and prodelta sediments are encountered. The upper limit was determined by Staub et al. (2000); the lower limit is based on the analysis of bore-hole records and the work of Jackson (1962) and Pimm (1964). These sediments consist of gray to olive-black, burrowed, laminated silts and/or laminated silts and sands. Silt laminae and/or silt-and-sand interbeds are sometimes many centimeters thick, but average about 1 cm (Staub et al., 2000). Sands in these sediments range from sublitharenites to quartz arenites, with sublitharenites dominant. Layers of organic detritus and carbonate nodules also are present occasionally in these sediments. Delta-front and prodelta sediments contain a diverse foraminiferal assemblage that is the same as that found in the delta-front and prodelta sediments of the Mahakam River delta (Murphy, 1996).

Coastal-Plain River and Beach Sediments.—

The coastal-plain shoreline starts just to the west of the position of Kuala Bintulu (Fig. 2) and extends to the midpoint between the Oya River and the Igan distributary of the Rajang River. It is about 140 km in length and is composed of four large river mouths, which are separated by intervening areas consisting of sand beaches. Channel mouths are about 200 m wide and floored with quartz arenite sands. In their seaward two-thirds channel courses are floored with silts and clays. Clays are mixed-layer illite–smectite and illite with kaolinite. In the landward one-third, channel courses are floored by angular coarse-grained sand composed almost exclusively of quartz.

Beach ridges are apparent in shoreline areas; ridge-and-swale topographic relief is about 2 m. The prior positions of beach ridges can be observed along channel margins near the coast at low tide. Intertidal to supratidal beach sands are cross-bedded, brown to gray, burrowed, and composed of quartz arenites.

Terrace Deposits.—

A northeast-trending linear shoreline or terrace deposit, 5 to 7 m higher than MSL, is present adjacent to the upland (Fig. 3). It also crops out at several locations in the delta plain adjacent to the alluvial valley along its linear trend. The upper surface of this shoreline or terrace feature is flat and typically is constructed of compact, fine-grained quartz arenite sand.

Lag Deposits.—

Lag deposits are composed of subangular to rounded coarse sands, granules, and pebbles mixed with organic debris. These sediments are encountered at depths of 35 to 40 m below MSL in the vicinity of Daro. Typically, many meters of gray to black laminated silts and sand and silt interbeds immediately overlie them. Compact, rooted, yellow to brown clays and silts (alluvial soils?) immediately underlie lag deposits.

Age of Sediments

Table 2 lists the results of ^{14}C age determinations obtained on 46 samples from 35 different locations (Fig. 6). In general, all sediments in the delta become younger in a seaward direction. Freshwater root samples obtained from white podzols in the landward part of the delta plain average 17 ka, indicating a Pleistocene origin, whereas mangrove and *Nipa* roots in gley soils occurring in the same areas average only 7.4 ka. Basal peats in the landward part of the delta plain, irrespective of a *Nipa* or freshwater origin, average 6.3 ka in age. Overbank siliciclastics that overlie the margins of several peat deposits in the delta plain were deposited at about 2 ka. Mangrove and *Nipa* roots in gley soils at depth in the seaward one-half of the northeast delta plain average 3.4 ka. Basal peats in the seaward part of the northeast delta plain, all of which were of freshwater origin where sampled, average 1.6 ka in age and become progressively younger toward the coast.

Intertidal sediments and gley soils occurring between beach ridges in the southwest delta plain, near the mouth of the Rajang distributary channel, are 4.7 ka in age. Intertidal to subtidal sediments in the lower reaches of the Lassa and Paloh distributary channels are 2.3 ka in age, and sediments obtained at depth in the present delta-front areas average about 0.9 ka.

Samples obtained from within 4 m of the surface of peat domes in the alluvial valley and coastal plain were 6.3 and 5.2 ka in age, respectively. Organic samples obtained from active intertidal to subtidal bar forms in distributary channels ranged in age from 0.3 ka to the present. Finally, in the vicinity of Daro, at a depth of about 80 m, a peat/lignite sample was obtained from a deep well with an age slightly in excess of 40 ka.

Cross Sections

Two generalized cross sections were generated from core and bore-hole data. Datum for the cross sections is approximate mean high tide (MHT). Cross section locations are shown in Figure 6. The first cross section, A–A', traverses the alluvial valley at the position of the town of Sibu (Fig. 11). The valley is floored by weathered Tertiary bedrock, which in most areas is overlain by Pleistocene sediments of varying thickness. Pleistocene sediments range from clay to gravel and generally are well compacted. Sediment color ranges from gray to brown to yellow, and organic matter is common. The transition from Pleistocene to Recent sediments is usually marked by a gray-white to white podzol several meters thick.

Siliciclastic Recent valley-fill deposits are about 10 km wide and up to 45 m in thickness. Gravel, sand, and interbedded sand and silt dominate the lower part of the fill, whereas silt and interbedded silt and clay dominate the upper part of the fill. Sediments in the upper part of the fill succession are tidally influenced. Thick Recent peat deposits, resting mainly on Pleistocene sediments with their bases below MSL, flank the Recent valley-fill sediments. Occasionally Pleistocene and Tertiary outcrops extend upward to the surface through Recent fill materials.

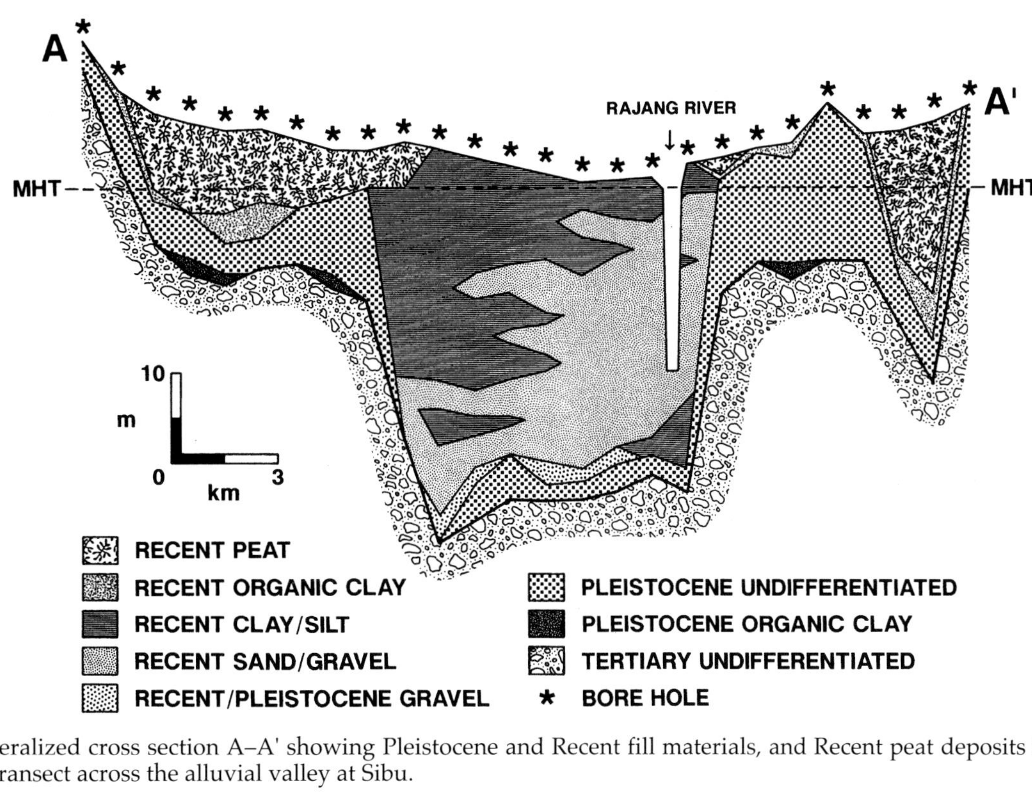

FIG. 11.—Generalized cross section A–A' showing Pleistocene and Recent fill materials, and Recent peat deposits in a northeast-trending transect across the alluvial valley at Sibu.

TABLE 2.—^{14}C dates obtained from all samples analyzed.

Map number	Field location	Age ^{14}C years BP	Elevation (MSL, m)	Material dated
01	Core V-69	4,690 ± 185	- 0.90	Intertidal organic material
02	Core RP-003	2,735 ± 165[a]	- 2.30	*Shorea* root in gley soil
02	Core RP-003	4,620 ± 245[b]	- 2.40	Organic material in gley soil
03	Core V-54	915 ± 90	- 6.80	Wood in delta front sediment
03	Core V-54	610 ± 80	- 4.50	Wood in delta front sediment
04	Core V-44	760 ± 105	- 6.40	Wood in delta front sediment
05	Grab 89-14	760 ± 110[a]	- 1.50	Mangrove root in gley soil
06	Grab 89-16	835 ± 330[b]	- 2.00	Organic material in gley soil
07	Core V-49	2,195 ± 165	- 3.30	Intertidal organic material
08	Core V-50	2,285 ± 160	- 2.80	Intertidal organic material
09	Daro 1F3	40,370 + 2,944 / -2,150	-78.00	Peat/lignite
10	Core V-51	1,955 ± 180	- 3.70	Intertidal organic material
10	Core V-51	2,690 ± 150	- 4.60	Intertidal organic material
11	Grab R-090-1	11,950 ± 250[b]	- 3.00	Freshwater root from white podzol
11	Grab R-090-2	12,480 ± 360[b]	- 3.00	Freshwater root from white podzol
11	Grab R-090-3	7,060 ± 280[b]	-2.70	Basal *Nipa* peat overlying white podzol
11	Grab R-090-4	6,820 ± 325[b]	- 2.70	Basal *Nipa* peat overlying white podzol
12	Core RB-01	7,340 ± 220[b]	- 0.60	Basal *Nipa* peat overlying gley soil
12	Core RB-01	1,895 ± 195[b]	+ 1.50	Upper contact, freshwater peat and siliciclastics
13	Core R-042	8,640 ± 195[a]	- 12.50	*Nipa* root in gley soil
14	Core V-79	26,760 ± 1,360	+ 1.50	Freshwater root from white podzol
15	Core V-35	5,485 ± 385	- 3.40	Freshwater peat matrix
16	Core V-29	230 ± 60	- 0.10	Plant fragments in intertidal sediments
17	Core V-26	7,415 ± 350	- 4.35	Wood fragment in gley soil
17	Core V-26	5,600 ± 60	- 1.35	Organic material in gley soil
18	Core V-16	170 ± 60	- 4.10	Organic material in midchannel bar
19	Core V-23	6,285 ± 270	- 4.40	Wood fragment in gley soil
20	Core T2-000	5,610 ± 195[b]	- 3.95	Basal *Nipa* peat overlying gley soil
20	Core T2-000	2,070 ± 130[a]	+ 0.50	Upper contact, freshwater peat and siliciclastics
21	Core T2-020	6,105 ± 105[a]	- 2.20	Basal *Nipa* peat overlying gley soil
21	Core T2-020	6,105 ± 105[a]	- 2.20	Basal *Nipa* peat overlying gley soil
22	Core T2-100	6,405 ± 120[b]	- 0.30	Basal *Nipa* peat overlying gley soil
23	Core T2-140	5,945 ± 205[b]	- 0.60	Basal *Nipa* peat overlying gley soil
24	Core T3-120	1,635 ± 140[b]	+ 3.60	Basal freshwater peat overlying alluvial soil
25	Core T3-080	1,245 ± 120[a]	+ 3.15	Basal freshwater peat overlying detrital peat
25	Core T3-080	3,690 ± 185[b]	+ 3.00	Detrital peat overlying alluvial soil
25	Core T3-080	1,985 ± 175[b]	+ 2.70	Organic material from alluvial soil
26	Core V-20	4,460 ± 110	- 2.10	Organic material in gley soil
27	Core V-19	2,345 ± 130	- 2.55	Organic material in gley soil
28	Core RL-03	6,055 ± 95	- 0.90	Basal *Nipa* peat overlying white podzol
29	Core RI-06	6,080 ± 100	- 1.60	Basal *Nipa* peat overlying white podzol
30	Grab R-006-1	6,290 ± 100	- 1.50	Freshwater peat from creek bottom in peat dome
31	Grab R-036-1	5,180 ± 220	+ 1.00	Freshwater peat from creek bottom in peat dome
32	Core V-01	4,090 ± 105	- 2.30	Intertidal organic material
33	Core V-03	modern	- 1.10	Organic material in point bar
34	Grab R-116-2	5,830 ± 100	- 1.00	Basal *Nipa* peat overlying white podzol
35	Core V-76	1,390 ± 140	- 6.60	Wood in delta front sediment

[a]from Staub and Esterle (1993). [b]from Staub and Esterle (1994)

The second cross section, B–B', traverses the Rajang valley at Sibu and continues to the northwest across the peat-dominated part of the delta plain (Fig. 12). Thick Recent peat deposits rest on Pleistocene sediments (gray-white podzols) with their bases below MSL in the landward one-half of the cross section (to about 5 km seaward of the abandoned arm of the Lassa). The margins of these thick peat deposits that flank distributary channels rest conformably on Recent sediments. In the seaward one-half of the cross section, thick peat deposits overlie Recent intertidal and shoreline sediments. These sediments, in turn, overlie Recent subtidal sands and deltaic deposits of the Rajang River. A yellow-brown mottled alluvial soil and gravel lag underlies the Holocene deltaic succession.

Channel and Coastal Bathymetry

Depths in the Rajang River channel in the lower part of the alluvial valley, and one-half the distance to the coast in the Lassa and Paloh distributaries, occasionally exceed 45 m (Fig. 13). The abandoned arm of the Lassa distributary, a large channel that drains to the southwest out of the peat-dominated part of the delta plain to join the main Lassa distributary, exceeds 45 m depth for 75% of its length. In contrast, the Igan, Belawai, and Rajang distributaries rarely exceed 20 m in depth. All distributary channels shallow toward the coast. With the exception of the Rajang distributary, distributary-channel mouths are less than 5 m in depth at spring low tide. The coastal-plain rivers have maximum depths of 25 m about half the distance across the coastal plain, and shallow toward both the coast and uplands. Coastal-plain river-mouth depths are less than 1 m at spring low tide.

The delta-front slope averages about 0.15% to depths up to 25 m. Slopes average about 0.10% to depths of 20 m along the coastal-plain shoreface. At depths of < 3 m below spring low-tide surface, sediments typically are sands, whereas at depths of > 3 m surface sediments are clays and silts. Shelf slope averages 0.03%, and shelf sediments, starting at distances of 30 km offshore, are composed of sand (Jackson 1962).

Bathymetric surveys do not reveal the presence of remnant incised valleys on the shelf in the vicinity of the Rajang River delta and adjacent coastal plain (Royal Malaysian Navy, 1990). However, 150 km to the northwest of the delta front, multiple incised valleys about 5 to 7 km wide and up to 60 m deep, trending north-northwest, are present. These incised valleys are attributed to the Lupar River (Haile et al., 1963). In addition, between Kuala Bintulu and Kuala Tatau (Fig. 2) a north-oriented incised valley can be observed traversing the shelf at water depths greater than 30 m. This valley, which is derived from the Tatau and Kemena rivers, is up to 40 m deeper than the flanking shelf.

Lateral Relationship of Pleistocene and Holocene Sediments

Figure 13 is a generalized map showing lateral relationships among Late Quaternary sediments. The distribution of the Pleistocene terrace deposits is shown, and areas where Pleistocene deposits are found in the shallow subsurface are delineated. The upper surface of these deposits consists a of gray-white to white podzol. The positions of the main peat domes also are shown. The bases of all thick peat deposits that overlie the Pleistocene surface are located below MSL. Basal peat ages in these deposits range from 5.8 to 7.3 ka, with an average age of 6.3 ka. Seaward of the area of thick peat and podzol shown in Figure 13 are a series of accretionary shorelines starting at about 7 ka and ending at the present. The first shoreline that can be mapped on a system-wide basis formed at about 5 ka.

Edelman auger borings (Fig. 14) delimit the amount of topographic relief on the podzol surface in a small area along the Rajang distributary channel in the southwestern part of the delta plain. Datum is MHT, and topographic relief on the surface ranges from less than -2 m to more than -6 m. In most areas where this surface is present, former topographically negative areas are filled with Holocene tidal-creek and tidal-flat sediments and capped by gley soils. Figure 15 shows an outcrop where the podzol surface is mantled in peat along the Rajang distributary channel.

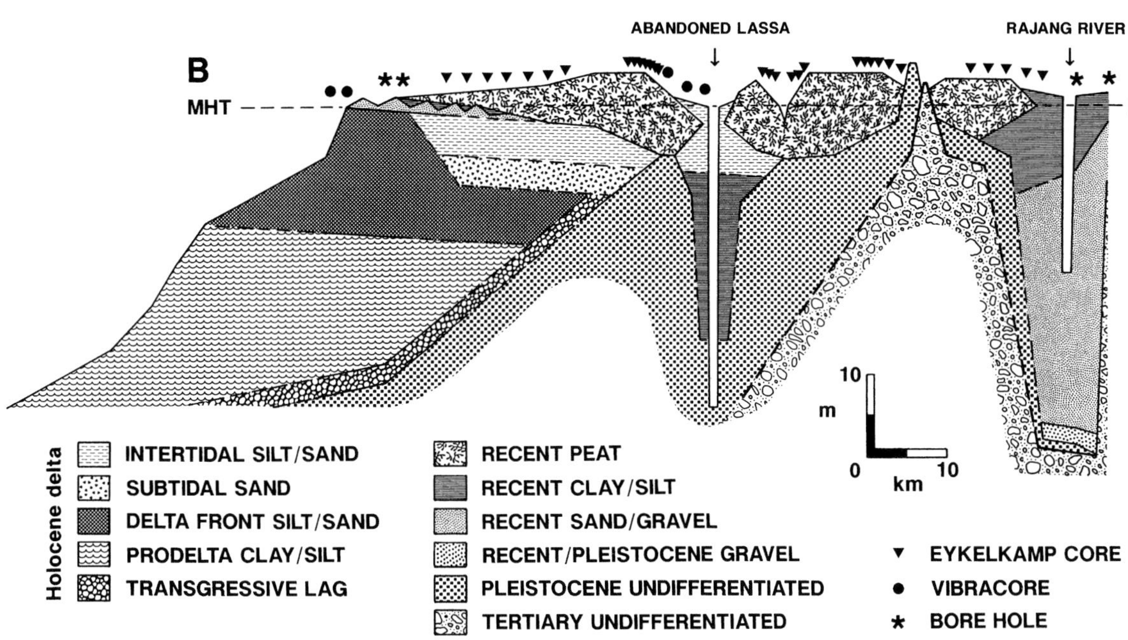

FIG. 12.—Generalized cross section B–B' showing Pleistocene and Recent fill materials, Recent peat deposits, and the Holocene deltaic sediment wedge in a southeast-trending transect from offshore to Sibu.

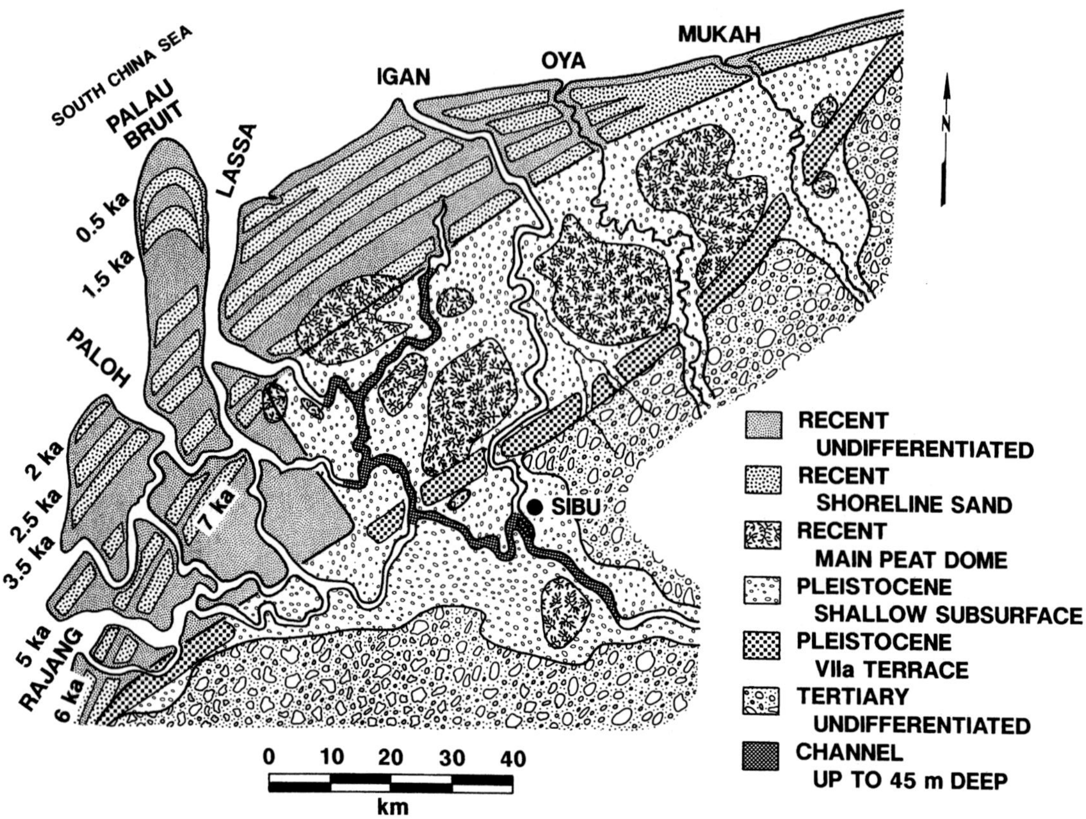

FIG. 13.—Generalized map showing the position of the Pleistocene (VIIa) terrace, areas where Pleistocene deposits are in the shallow subsurface, main peat dome areas, and the positions of accretionary shorelines starting at 7 ka and ending at the present. Also shown are areas where channel depths commonly are 45 m in the alluvial valley and delta plain.

INTERPRETATION AND DISCUSSION

Pleistocene Sediments

Beach or terrace sediments adjacent to upland areas are 5 to 7 m higher than MSL (Fig. 3). In addition, the maximum elevations of gravel outcrops in the alluvial valley are 5 to 7 m higher than MSL. The upper surface of these terrace and outcrop sediments can be correlated laterally in the alluvial valley, delta plain, and coastal plain with the gray-white podzol surface, which at every location dated returned results of more than 10 ka.

Previous studies (Anderson, 1964; Scott, 1985) have suggested that the terrace deposits formed about 5.0 to 5.5 ka during a Holocene elevated stand of sea level. However, studies on the Huon Peninsula in New Guinea (Chappell and Shackleton, 1986) and elsewhere in the region (Fleming et al., 1998) indicate that current sea level is the highest level of the Holocene.

On the Huon Peninsula at least seven reef terrace/highstand events (Fig. 16) have been identified that occurred in the last 125 ka (oxygen isotope stages 1 to 5). ^{14}C age dates plus elevation data indicate that a more accurate interpretation of the Rajang terrace deposits is that they are the local remains of the VIIa highstand event (oxygen isotope stage 5e) of 125 ka. This highstand surface formed at ≈ 6 m above present MSL (Bloom et al. 1974; Chappell and Shackleton 1986; Esat et al. 1999).

In the seaward part of the delta plain near the town of Daro, organic materials (peat/lignite) from a depth of 78 m below MSL returned an age of 40 ka. If peat development in coastal deltaic settings is associated with regressive to highstand system tracts, as proposed by some workers (Aitken and Flint, 1995; Martino, 1996; Bohacs and Suter, 1997; Staub, 2002), then this peat/lignite probably formed as part of the IIIb highstand event of 40 ka. The IIIb highstand event (oxygen isotope stage 3; Fig. 16) occurred at 41 ± 4 m below present MSL (Bloom et al., 1974; Chappell and Shackleton, 1986), indicating that about 40 m of subsidence has occurred. These results demonstrate that substantial subsidence and sediment accumulation has occurred during the Quaternary in coastal areas of the Rajang delta. Published Neogene sediment isopachs of the Rajang delta area indicate a total thickness of 1 to 2 km onshore and more than 4 km offshore (Hutchison 1989).

Although no bathymetric expression of incised valleys can be observed on the shelf in front of the Rajang River delta, an idea of scale can be garnered from other areas. Incised valleys from other rivers are present on the shelf to the east and west of the delta. These incised valleys are on the order of 5 to 10 km wide and 40 to 60 m in depth.

In cross section (Fig. 11), the Pleistocene incised valley in the vicinity of Sibu is up to 45 m deep and 10 km wide. In the alluvial valley and the landward one-half of the delta plain, with the exception of the Igan distributary channel, all major channels are up to 45 m deep (Fig. 13). Eroded remains of the VIIa highstand surface flank all these areas. This commonality in channel depth is thought to represent the maximum amount of incision of the Rajang River in these areas.

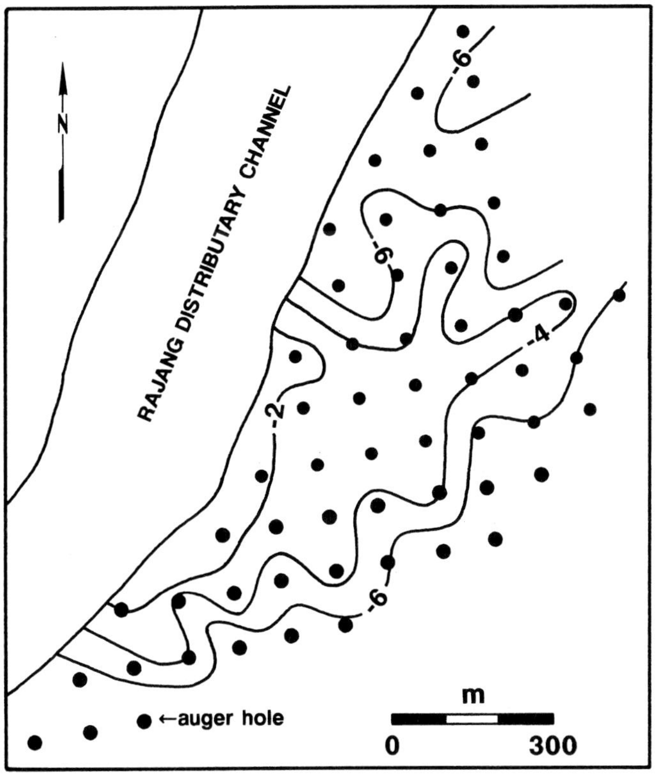

FIG. 14.—Map showing topographic relief on the Pleistocene (VIIa) surface in a small area adjacent to the Rajang distributary channel (see Fig. 6 for location; modified from Kuek, 1991).

FIG. 15.—Outcrop of the Pleistocene (VIIa) gray-white to white podzol surface capped by a Holocene *Nipa* peat along the margin of the Rajang distributary channel at mid-tide. The photo is from ^{14}C location 11 (see Fig. 6).

Recent Progradation

Regional.—

The incised valley in the Rajang alluvial valley (Fig. 11) is floored with gravel. This material is overlain by a fining-upward succession, the upper part of which is tidally influenced. The depth to which tidal influence extends in this 45-m-thick sediment fill is not known. Most of the surface of the valley floor is mantled in thick peat.

Most thick peat and all main peat dome areas are found in the alluvial valley, the northeastern part of the delta plain, and the adjacent coastal plain. In these areas, thick peat is lying on or in close proximity to the underlying gray-white to white Pleistocene podzol surface (Fig. 13). There is substantial relief on this surface (Fig. 14), and in the delta plain this relief has been filled mainly by intertidal sediments and gley soils. Peat deposition started in most of these areas concurrently at about 6.3 ka, when sea level was several meters below its present level. Accumulation of sediment, both organic and siliciclastic, in these areas has been aggradational since the initiation of peat deposition.

Seaward of the thick peat and podzol area are a series of system-wide accretionary shorelines. The southwestern part of the delta plain represents a reentrant in the general system-wide shoreline trend, and the remains of two older, shorelines of limited extent are present in this reentrant, with ages of 7 and 6 ka. The existence of these prograding shoreline deposits in the southwest delta plain demonstrate that the reentrant was flooded prior to 7 ka, and that the initial development of the Rajang River delta coincided with the Early Holocene deceleration in sea-level rise, as did many other deltas (Stanley and Warne, 1994).

The oldest identified system-wide shoreline formed at about 5 ka and coincides with the seaward margin of the area of thick peat and podzol in the northeast delta plain and adjacent coastal plain. Sands from this shoreline are found in the southwest delta plain, under thick peat deposits in the northeast delta plain, in the upper reaches of the abandoned arm of the Lassa distributary channel, and along the banks of the Igan distributary channel. Age dates from intertidal sediments and gley soils in this shoreline deposit in the southwest delta plain are 4.7 ka, which corroborates its estimated age. Next, a series of shorelines are identified that all weld into a single shoreline feature to the northeast. Estimated shoreline age ranges from about 3.5 to 2.0 ka and age dates from gley soils and peat deposits corroborate the reasonableness of the estimates. More recently, coastline erosion has truncated all of these former shorelines in the southwestern part of the delta plain.

A series of smaller prograding shoreline features can be observed in Palau Bruit (Fig. 13), starting at about 1.5 ka and continuing to the present. At present, the delta shoreline is

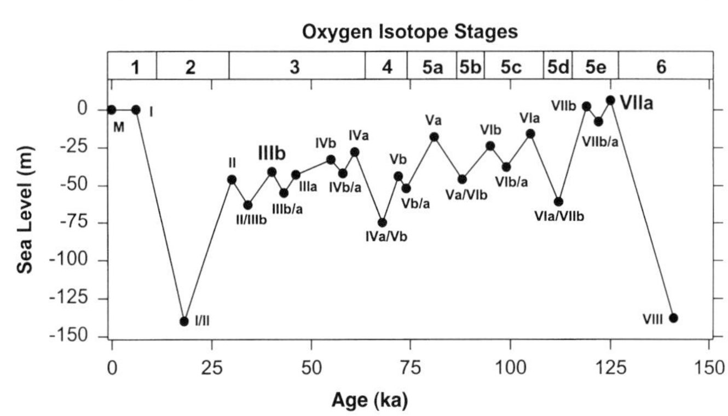

FIG. 16.—Oxygen isotope stages and sea-level changes for the past 140 ka based on correlation of deep-sea core data with ages of uplifted and drowned coral reefs of the Huon Peninsula, New Guinea (modified from Chappell and Shackleton, 1986).

actively prograding (Fig. 10) on the north coast, while some areas of the southwest coast appear to be undergoing erosion and retreat (Fig. 3). This erosion and retreat in the southwest is, in all likelihood, related to the fact that the Rajang distributary channel currently transports the smallest amount of the sediment to the coast on an annual basis. Staub et al. (2000) demonstrated that about 24 million metric tons of sediment is supplied directly to the delta front and prodelta annually, but less than 10% of this total is currently transported via the Rajang distributary.

The Pleistocene–Holocene transition occurs at a depth of about 40 m below MSL at the coast in the northwest part of the delta plain and is identified by a yellow-brown rooted alluvial soil overlain by a gravel lag. A coarsening-upward succession, capped by distributary-mouth sands and/or beach deposits, overlies the lag deposit. This Holocene–Recent deltaic succession, or wedge, thins and shallows in a landward direction to the position of the 5 ka shoreline.

Abandonment of the Lassa.—

Within the northeast delta plain, data indicate that the unusual southwest-flowing channel that presently drains much of the peat-dominated area of the delta plain (Fig. 17) was formerly an active distributary channel flowing north-northeast. It is about 45 deep for most of its length until it intersects the position of the 5 ka shoreline (Fig. 13), at which point it rapidly shallows to less than 5 m depth. The maximum extent of drainage directed to this channel is coincident with the position of the 3.5 ka shoreline. Sand composition in the area between these two shoreline trends is litharenite.

The 5 ka shoreline trend is roughly coincident with the T1 transect shown in Figure 17. At most core locations on this transect the peat is immediately underlain by quartz arenite beach sand. The maximum elevation of this sand is MSL. Mangrove and *Nipa* vegetation dominate the upper end of the drainage network, even though freshwater vegetation associations dominate the rest of the area and water in the channel averages less than 1 ppt salinity. The distribution of the mangrove–*Nipa* vegetation is not dissimilar to the shape of a distributary-channel mouth (e.g., Muara Lassa; Fig. 17). The existing mangrove–*Nipa* distribution in the upper reach of the channel is most likely a relict of the topographically negative area left when abandonment occurred at 3.5 ka.

Recent Peat Development

In the Malay Archipelago a generalized shoreline progradation model has been proposed by many investigators (e.g., Anderson, 1961; Coleman et al., 1970; Anderson and Muller, 1975; Scott, 1985; Esterle, 1990; Staub and Esterle, 1994) for the development of the deltaic peat swamps. In the Baram, Hari, and Klang-Langat river delta systems the maximum reported peat age is on the order of 4.5 ka. The maximum reported peat thicknesses are on the order of 10 m, and the bases of the peat deposits are very near, at or above, MSL (Anderson, 1964; Coleman et al., 1970; Esterle, 1990). Although these delta systems vary greatly in terms of physical processes (e.g., microtidal versus macrotidal; wave dominated versus tide dominated) they all occur in the same climatic setting (ever-wet), and judging from available information the general shoreline progradation model proposed is probably correct.

The Rajang system occurs in the same climatic setting (ever-wet), but unlike these other peat dominated delta systems most thick peat deposits in the Rajang River delta and adjacent coastal plain are found in areas underlain by near-surface Pleistocene erosional remnants. In addition, the Rajang peat deposits are larger and older. They cover two to five times more area than the deposits present in the Baram, Hari, or Klang–Langat systems. They are up to twice as thick (15–20 m), their bases occur well below MSL, and their average age is 6.3 ka, with a limited range of 5.8 to 7.3 ka. This narrow time window for peat development over such an extensive area (≈ 7,500 km^2) and the aggradational nature of the sediments indicates that the general progradation model as proposed in the case of the Rajang is not correct.

In order for peat to accumulate in any depositional system many criteria must be met, but probably the two most important are the presence of a substrate in which the plants can root and a rising water table or base level (McCabe and Parrish, 1992). In addition, if the water table or base level is rising, an equilibrium rate must be attained. If it is too rapid, the surface floods and peat

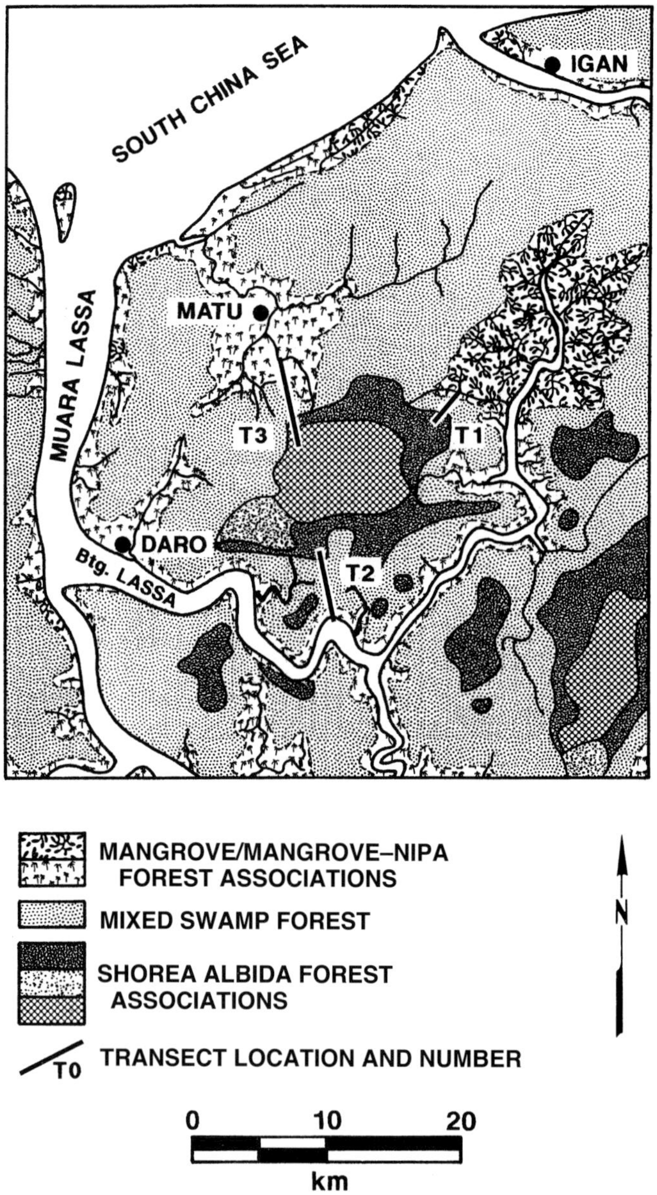

FIG. 17.—Map showing the abandoned arm of the Lassa distributary channel and the active distributary channel (Muara Lassa). Also shown is the surficial distribution of brackish-water and freshwater vegetation types (modified from Anderson, 1961).

accumulation ceases; if it is too slow, the organic material is oxidized and peat does not accumulate.

In the Rajang system, as sea level rose during the Early Holocene the existing topographic relief (minor incised drainage in interfluve areas) in the coastal plain and the northeastern delta plain filled with tidal creek–tidal flat sediments. This created an initial siliciclastic platform, and this platform was rapidly colonized by pioneer mangrove–*Nipa* vegetation. As the water table continued to rise, peat accumulation commenced. It may be coincidental, but the initial peat accumulation in the Rajang system coincides with the end of the catastrophic sea-level rise event 3 (CRE 3) of Blanchon and Shaw (1995). In this case the end of this event reflects the substantial deceleration in the rate of water-table and sea-level rise that was necessary for platform and peat development. Subsequently, as the shoreline has advanced, the peat domes have increased in thickness and expanded seaward. Winston (1994) developed theoretical numerical models that explain the types of domed peat accumulation that have occurred.

In the southwestern part of the Rajang delta plain, little significant peat accumulation has occurred. The surface of the area is dominated by beach ridges, saline gley soils, and *Nipa* palms. Pleistocene sediments are not found at shallow depths, except immediately adjacent to the VIIa terrace (Fig. 13). Data indicate that there was little or no prior existing substrate in this area circa 7 ka. This is also the part of the delta plain that experiences the highest tides. These factors have hindered peat accumulation, but they have not prevented it (see Fig. 2).

As stated earlier, peat accumulation started much later (≈ 4.5 ka) in the Baram, Hari, and Klang-Langat delta systems, and as a consequence the peat deposits are not as thick. Climate and rates of base-level rise have been similar for all four systems (Baram, Hari, Klang-Langat, and Rajang) during the Holocene, so why is there a substantial difference in the age of the peat deposits? Variation in physical processes (e.g., variation in tidal range, wave energy) appears not to be a significant factor. The only viable explanation for the later start in the Baram, Hari, and Klang-Langat delta systems is the absence of a substrate or platform in which plants could take root. The drainage-basin area of each of these systems is much smaller than that of the Rajang (Coleman et al., 1970; Esterle, 1990; Staub and Esterle, 1994), and as a consequence the rate of sediment delivery to the coast is substantially less. This reduced rate of sediment supply slowed the platform development rate, and as a result peat accumulation started about 2 to 3 ka later in these systems.

Deltaic Style

The Rajang River delta is a complex interplay between seasonal variation in discharge, tidal currents, and surface waves. Distributary channels dominated by tidal and estuarine processes (Table 1) and intervening beach-ridge plains characterize much of the Holocene delta. Harris et al. (1993) discussed the importance of tidal currents and surface waves for sediment dispersal and deposition. Galloway (1975) gave strong emphasis to the estuarine component in tide-dominated deltas. Another factor to be considered, according to Staub et al. (2000), is seasonal variation in discharge. All these factors are important.

Seasonal drainage-basin discharge varies by one order of magnitude during average years and two orders of magnitude in extreme years. During the dry season (April to November) much of the sediment supplied by the drainage basin to the delta plain is sequestered in the distributary channels by tidal and estuarine processes. During the wet season (December to March) this sediment is transported to delta-front and prodelta areas from the distributary channels and results in annual sedimentation events. Once sediment is in the delta front and prodelta, some of it is reworked during subsequent dry seasons into beach ridges (Staub et al. 2000).

The Rajang River delta is similar in many respects, in particular physical processes and sediment grain size, to other siliciclastic-sediment-dominated tropical river deltas such as the Fly, Irrawaddy, and Mekong (Harris et al., 1993; Orton and Reading, 1993). Even though the terms wet season and dry season are used to describe monsoon events in Sarawak, the Rajang River system is located in an ever-wet climatic region. The terms wet versus dry locally denote the relative amount of rainfall. In coastal lowland areas this ever-wet climatic regime enables the water table to remain constantly at or within a few centimeters of the surface, even during the dry season, and has enabled the accumulation of thick, extensive peat deposits. In Sarawak the presence of this climatic regime extends well back into the Pleistocene (Morley and Flenley, 1987).

Conversely, the Fly, Irrawaddy, and Mekong deltas are in areas of Southeast Asia and the Malay Archipelago that are truly seasonal in the sense that there is a distinct dry season during which there is little or no rainfall (Whitmore, 1984) and a wet season during which there is substantial rainfall. This climatic difference, in terms of rainfall frequency, means that there is substantial seasonal variation in the position of the water table relative to the ground surface, and this variation has hindered and / or prevented peat development in these deltas by allowing the organic materials to be oxidized.

SUMMARY AND CONCLUSIONS

The oldest identified Quaternary surface in the Rajang system is the VIIa highstand surface of 125 ka. Maximum elevations are 5 to 7 m above MSL, and maximum incision into this surface is on the order of 45 m. The upper part of much of this surface is capped by a gray-white to white podzol several meters thick. In the Rajang system, the alluvial incised valley is floored with gravel. This material is, in turn, overlain by a fining-upward succession, the upper part of which shows evidence of tidal influence.

The next younger Quaternary surface identified is the IIIb highstand surface of 41 ka. This surface is found at a depth of approximately 80 m below MSL in the seaward part of the delta plain, demonstrating that 40 m of subsidence has occurred in this area during the Late Quaternary. In this same region, the Pleistocene–Holocene transition occurs at a depth of about 40 m below MSL and is identified by a yellow-brown rooted alluvial soil overlain by a gravel lag. A Holocene coarsening-upward succession capped by distributary-mouth sands and / or beach deposits overlies lag deposits.

Peat formation started about 7.3 ka. Initial formation was limited to interfluve areas in the alluvial valley, the northwestern part of the delta plain, and the adjacent coastal plain. Development is primarily aggradational in this part of the system, and thick peat may rest in direct contact with Pleistocene podzols or Recent gley soils.

Shoreline progradation started at about 7 ka in the southwestern part of the delta plain. The oldest system-wide shoreline identified is 5 ka in age. Since then, shoreline progradation has been primarily to the north and northwest, and peat deposits have extended seaward with the shoreline.

The southwestern part of the delta plain is dominated by beach ridges, saline gley soils, and *Nipa* palms. This area was flooded by rising sea level prior to 7 ka and, at present, experiences the highest tides. These factors are thought to contribute to the limited peat development in this part of the delta.

ACKNOWLEDGMENTS

Permission to work in the Rajang River delta was granted by the State Secretary, Sarawak. The personnel of the Geological Survey of Malaysia, Sarawak, and the Soils Division, Department of Agriculture, Sarawak, made the research possible. Without their support, this research could not have been conducted. Financial support was provided by grants from the National Science Foundation (EAR-9104945 to JRS and EAR-9111842 to RAG), the Illinois Mining Institute, and Carbon Consultants International, Inc. We also would like to thank Katherine Bartram, Joan Esterle, Richard Esterle, Anne Raymond, and Jerome Ward for their assistance in Sarawak. Finally, this paper was improved by comments from the reviews of Joan Esterle, John Ferm, John Suter, Denis Tan, and Christopher Williams.

REFERENCES

ANDERSON, J.A.R., 1961, The ecology and forest types of the peat swamp forests of Sarawak and Brunei in relation to their silviculture: unpublished Ph.D. Dissertation, University of Edinburgh, Scotland, 117 p.

ANDERSON, J.A.R., 1964, Structure and development of the peat swamps of Sarawak and Brunei: Journal of Tropical Geography, v. 18, p. 7–16.

ANDERSON, J.A.R., 1983, The tropical swamps of western Malesia, *in* Gore, A.P.J., ed., Ecosystems of the World, Regional Studies 4B. Mires: Swamp, Bog, Fen, and Moor: Amsterdam, Elsevier, p. 181–200.

ANDERSON, J.A.R., AND MULLER, J., 1975, Palynological study of a Holocene peat deposit and a Miocene coal deposit from NW Borneo: Review of Palaeobotany and Palynology, v. 19, p. 291–351.

AITKEN, J.F., AND FLINT, S.S., 1995, The application of high-resolution sequence stratigraphy to fluvial systems: A case study from the Upper Carboniferous Breathitt Group, eastern Kentucky, USA: Sedimentology, v. 42, p. 3–30.

BLANCHON, P., AND SHAW, J., 1995, Reef drowning during the last deglaciation: Evidence for catastrophic sea-level rise and ice-sheet collapse: Geology, v. 23, p. 4–8.

BLOOM, A.L., BROECKER, S.W., CHAPPELL, J.M.A., MATTHEWS, R.K., AND MESOLELLA, K.J., 1974, Quaternary sea level fluctuations in a tectonic coast: new $^{230}Th/^{234}U$ from the Huon Peninsula, New Guinea: Quaternary Research, v. 4, p. 185–205.

BOHACS, K., AND SUTER, J., 1997, Sequence stratigraphic distribution of coaly rocks: Fundamental controls and paralic examples: American Association of Petroleum Geologists, Bulletin, v. 81, p. 1612–1639.

CHAPPELL, J., AND SHACKLETON, N.J., 1986, Oxygen isotopes and sea level: Nature, v. 324, p. 137–140.

COLEMAN, J.M., GAGLIANO, S.M., AND SMITH, W.G., 1970, Sedimentation in a Malaysian high tide delta, *in* Morgan, J.P., ed., Deltaic Sedimentation, Modern and Ancient: Society of Economic Paleontologists and Mineralogists, Special Publication 15, p. 185–197.

ESAT, T.M., MCCULLOCH, M.T., CHAPPELL, J., PILLANS, B., AND OMURA, A., 1999, Rapid fluctuations in sea level recorded at Huon Peninsula during the penultimate deglaciation: Science, v. 283, p. 197–201.

ESTERLE, J.S., 1990, Trends in petrographic and chemical characteristics of tropical domed peats in Indonesia and Malaysia as analogues for coal: unpublished Ph.D. dissertation, University of Kentucky, Lexington, 270 p.

FARRANT, A.R., SMART, P.L., WHITAKER, F.F., AND TARLING, D.H., 1995, Long-term Quaternary uplift rates inferred from limestone caves in Sarawak, Malaysia: Geology, v. 23, p. 357–360.

FLEMING, K., JOHNSON, P., ZWARTZ, D., YOKOYAMA, Y., LAMBECK, K., AND CHAPPELL, J., 1998, Refining the eustatic sea-level curve since the last glacial maximum using far- and intermediate-field sites: Earth and Planetary Science Letters, v. 163, p. 327–342.

FOLK, R.L., 1980, Petrology of Sedimentary Rocks: Austin, Texas, Hemphill, 184 p.

GALLOWAY, W.E., 1975, Process framework for describing the morphologic and stratigraphic evolution of deltaic depositional systems, *in* Broussard, M.L., ed., Deltas; Models for Exploration: Houston Geological Society, p. 87–98.

HAILE, N.S., KEIJ, A.J., AND PIMM, A.C., 1963, Preliminary report on the oceanographic cruise of HMS 'Dampier' in the South China Sea, 1963: Borneo Region, Malaysian Geological Survey, Annual Report, p. 119–145.

HARRIS, P.T., BAKER, E.K., COLE, A.R., AND SHORT, S.A., 1993, A preliminary study of sedimentation in the tidally dominated Fly River delta, Gulf of Papua: Continental Shelf Research, v. 13, p. 441–472.

HAMILTON, W., 1979, Tectonics of the Indonesian Region: U.S. Geological Survey, Professional Paper 1078, 345 p.

HUTCHISON, C.S., 1989, Geological Evolution of South-East Asia: Oxford, U.K., Oxford University Press, Oxford Monographs on Geology and Geophysics no. 13, 368 p.

HUTCHISON, C.S., 1996, The 'Rajang accretionary prism' and 'Lupar Line' problem of Borneo, *in* Hall, R. and Blundell, D., eds., Tectonic Evolution of Southeast Asia: Geological Society of London, Special Publication 106, p. 247–261.

JACKSON, A., 1962, Sea-bed sampling off northwest Borneo: British Borneo Geological Survey, Annual Report, p. 51–55.

JEEPS, M.D., AND GATES, R.I., 1963, Physical aspects of the January–February, 1963, floods in Sarawak: Sarawak Hydrologic Yearbook for the Water-Year 1962–1963, p. 40–95.

KUEK, K.S., 1991, Ball clay deposit in the Rumah Empaga area, Sarikei, Sarawak: Geological Survey of Malaysia, 22nd Geological Conference, Proceedings, no. 3, p. 58–74.

LAM, K.S., 1988, Sibu Area, Central Sarawak Malaysia, Explanation Sheet 2/111/12: Geological Survey of Malaysia, Geological Survey Report, Sarawak, 151 p.

MARTINO, R.L., 1996, Stratigraphy and depositional environments of the Kanawha Formation (Middle Pennsylvanian), southern West Virginia, U.S.A.: International Journal of Coal Geology, v. 31, p. 217–248.

MCBETH, J.L., 1995, Composition of coarse-grained siliciclastic sediment from a tropical peat-accumulating delta: Rajang River delta, Sarawak, East Malaysia: unpublished M.S. thesis, Southern Illinois University Carbondale, Carbondale, 104 p.

MCBRIDE, E.F., 1971, Mathematical treatment of size distribution data, *in* Carver, R.E., ed., Procedures in Sedimentary Petrology: New York, Wiley, p. 109–127.

MCCABE, P.J., AND PARRISH, J.T., 1992, Tectonic and climatic controls on the distribution and quality of Cretaceous coals, *in* McCabe, P.J., and Parrish, J.T., eds., Controls on the Distribution and Quality of Cretaceous Coals: Geological Society of America, Special Paper 267, p. 1–16.

MOORE, D.M., AND REYNOLDS, R.C., JR., 1989, X-ray Diffraction and the Identification and Analysis of Clay Minerals: Oxford, U.K., Oxford University Press, 332 p.

MORLEY, R.J., AND FLENLEY, J.R., 1987, Late Cainozoic vegetation and environmental changes in the Malay Archipelago, *in* Whitmore, T.C., ed., Biogeographical Evolution of the Malay Archipelago: Oxford, U.K., Oxford University Press, Oxford Monographs on Biogeography no. 4, p. 50–59.

MURPHY, S., 1996, Distribution and ecology of Holocene foraminifera within the Lassa distributary of the Rajang River delta, Sarawak, East Malaysia: unpublished M.S. Thesis, Southern Illinois University Carbondale, Carbondale, 121 p.

ORTON, G.J., AND READING, H.G., 1993, Variability of deltaic processes in terms of sediment supply, with particular emphasis on grain size: Sedimentology, v. 40, p. 475–512.

PIGOTT, J.D., 1995, A seismic classification scheme for clastic shelf wedges (deltas), *in* Oti, N.M., and Postma, G., eds., Geology of Deltas: Rotterdam, A.A. Balkema, p. 17–29.

PIMM, A.C., 1964, Seabed sediments of the South China Sea: Geological Survey, Borneo Region, Malaysia, Annual Report, p. 122–146.

POSTMA, H., 1980, Sediment transport and sedimentation, *in* Olausson, E., and Cato, I., eds., Chemistry and Biochemistry of Estuaries: Chichester, U.K., Wiley, p. 153–186.

POWERS, M.C., 1953, A new roundness scale for sedimentary particles: Journal of Sedimentary Petrology, v. 23, p. 117–119.

ROYAL MALAYSIAN NAVY, 1990, Navigation Chart SAR1. Director of Marine, Kuching, Sarawak, East Malaysia.

SCOTT, I.M., 1985, Soils of the Central Sarawak Lowlands, East Malaysia: Kuching, Sarawak, East Malaysia, Department of Agriculture, Soils Division, Soil Memoir 2, 302 p.

STANLEY, J.S., AND WARNE, A.G., 1994, Worldwide initiation of Holocene marine deltas by deceleration of sea-level rise: Science, v. 265, p. 228–231.

STAUB, J.R., AND ESTERLE, J.S., 1993, Provenance and sediment dispersal in the Rajang River delta / coastal plain system, Sarawak, East Malaysia: Sedimentary Geology, v. 85, p. 191–201.

STAUB, J.R., AND ESTERLE, J.S., 1994, Peat-accumulating depositional systems of Sarawak, East Malaysia: Sedimentary Geology, v. 89, p. 91–106.

STAUB, J.R., AMONG, H.L., AND GASTALDO, R.A., 2000, Seasonal sediment transport and deposition in the Rajang River delta, Sarawak, East Malaysia: Sedimentary Geology, v. 133, p. 249–264.

Staub, J.R., 2002, Marine flooding events and coal bed sequence architecture in southern West Virginia: International Journal of Coal Geology, v. 49, p. 123–145.

WHITMORE, T.C., 1984, Tropical Rain Forests of the Far East: Oxford, U.K., Oxford University Press, 352 p.

WINSTON, R.B., 1994, Models of the geomorphology, hydrology, and development of domed peat bodies: Geological Society of America, Bulletin, v. 106, p. 1594–1604.

LATEST QUATERNARY BARAM PRODELTA, NORTHWESTERN BORNEO

RICHARD N. HISCOTT

Department of Petroleum Geoscience, Universiti Brunei Darussalam, Jalan Tungku Link BE1410, Brunei Darussalam
Present address: Department of Earth Sciences, Memorial University of Newfoundland, St. John's, Newfoundland, A1B 3X5, Canada
e-mail: rhiscott@sparky2.esd.mun.ca

ABSTRACT: The Quaternary Baram Delta is more than 1 km thick on the outer continental shelf of Brunei, and includes mud-prone highstand delta lobes, sand-prone lowstand shelf-edge deltas, incised-valley fills, and transgressive sheet-like deposits lying on wave-cut ravinement surfaces. The shelf break is defined by a prominent fault scarp ~ 130 m below sea level. Beyond, the seabed descends at an average of 2–3° to the Borneo Trough at depths exceeding 2750 m. Uppermost Quaternary units thicken by a factor of 2–5 across *en echelon* shelf-edge growth faults, then progressively thin seaward as an extensive prodelta wedge that drapes irregular slope morphology. The slope owes its rugged relief to gravity-driven growth faulting, mud diapirism, cutting of submarine canyons, sediment sliding, and construction of levees alongside sporadically active turbidity-current channels that head in the region of shelf-edge deltas. During highstands, muddy lobes of the Baram Delta repeatedly extended to the upper slope as a consequence of high sediment yields and narrow shelf width. During the last lowstand, the Baram River cut an incised valley across the narrow shelf so that its muddy sediment load largely bypassed the continental margin through a major canyon. Delta deposits in the canyon head collapsed to generate a debris-flow complex that buries much of the canyon floor and that extends over large areas of the lower slope and Borneo Trough. This paper demonstrates that thick, extensive prodelta muds and associated turbidites can blanket basin slopes even during highstands in areas with high sediment supply and narrow shelves. Conversely, during lowstands, the suspended load of deltas can largely bypass the continental margin through incised valleys and canyons, resulting in reduced sedimentation on basin slopes.

INTRODUCTION

The continental shelf of Brunei and adjacent parts of Sarawak and Sabah, northwestern Borneo, is underlain by 8–9 km of post-Eocene siliciclastic sediments (Fig. 1) derived from the Rajang–Crocker ranges (Hamilton, 1979; Sandal, 1996). The shelf break occurs 50–70 km seaward of the Brunei coast in water depths of ~ 130 m, where there is a prominent scarp, ~ 20 m high, produced by active growth faulting (Figs. 2 and 4). The seabed then descends steeply to the floor of the Borneo Trough at depths exceeding 2750 m. This trough is interpreted as an abandoned subduction zone, inactive since the late Miocene (Hamilton, 1979; Hutchison, 1996). The large-scale structural features of the Baram Delta province closely resemble those of the larger Niger Delta (Damuth, 1994), in that there is an outer-shelf zone of extensional growth faults, an upper-slope zone of counter-regional faults and mud diapirs, and thrust faults in the deep subsurface at the toe of the slope (Fig. 1B).

The coastline of Brunei consists of extensive exposed sand beaches sculpted by waves from the South China Sea. Wave heights during the northeast monsoon can exceed 3 m. Significant wave heights of 1–1.5 m are exceeded > 20% to > 30% of the time most months of the year, whereas larger significant wave heights are exceeded only ~ 10% of the time and only during the peak of the northeast monsoon (Sandal, 1996, p. 33). Offshore, beneath the modern shelf, Hiscott (2001) described four highstand, mud-prone delta lobes separated from one another by a set of flat, shelf-crossing seismic reflectors (named, from youngest to oldest, α, β, δ, γ, and ε). These lobes are located directly seaward of the modern Baram Delta (Fig. 2), and their seismic and sedimentary facies, the latter from geotechnical borings, are identical to the modern delta. The shelf-crossing reflectors between the delta lobes truncate the tops of deltaic clinoforms and are interpreted as wave ravinement surfaces produced when successive lobes were abandoned, subsided, and transgressed. Only on the outer shelf was the rate of subsidence high enough (> 6 cm / 100 yr) to allow preservation of the topset-to-foreset transitions of these deltas during transgressions (Fig. 3).

On the basis of chronology from exploration wells and lack of fluvial channel incision at the flat shelf-crossing reflectors, Hiscott (2001) inferred that the four mud-prone delta lobes accumulated between the beginning of the last interglacial (oxygen isotope stage 5, ~ 125 ka) and the major sea-level fall leading to the last glacial maximum (beginning of isotope stage 3, ~ 24 ka). Hence, only the youngest of the four delta tops was ever subaerially exposed and dissected by fluvial channels. These four lobes are part of a highstand systems tract (HST),

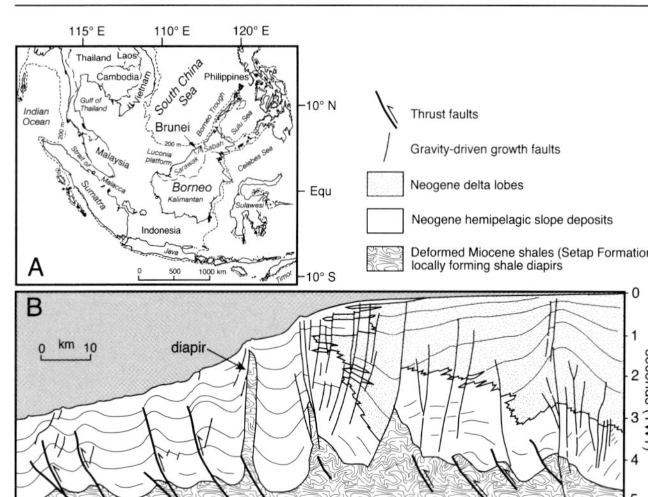

FIG. 1.—**A)** Regional geography of southeast Asia, with country names set in normal font, and italics used for names of islands, states, or provinces. **B)** Shelf-to-basin cross section of Cenozoic deposits offshore Brunei interpreted from a regional industry seismic profile (Sandal, 1996, p. 71). The more sand-prone deltaic successions (stippled) pass by facies transition into slope prodelta deposits. Thrust faults characterize the deep subsurface whereas gravity-driven growth faults and counter-regional faults dominate the deltaic successions of the shelf and upper slope. Loaded and deformed shales at the base of the pile locally rise to form diapirs (see also Figs. 9–11).

Tropical Deltas of Southeast Asia—Sedimentology, Stratigraphy, and Petroleum Geology
SEPM Special Publication No. 76, Copyright © 2003
SEPM (Society for Sedimentary Geology), ISBN 1-56576-086-7, p. 89–107.

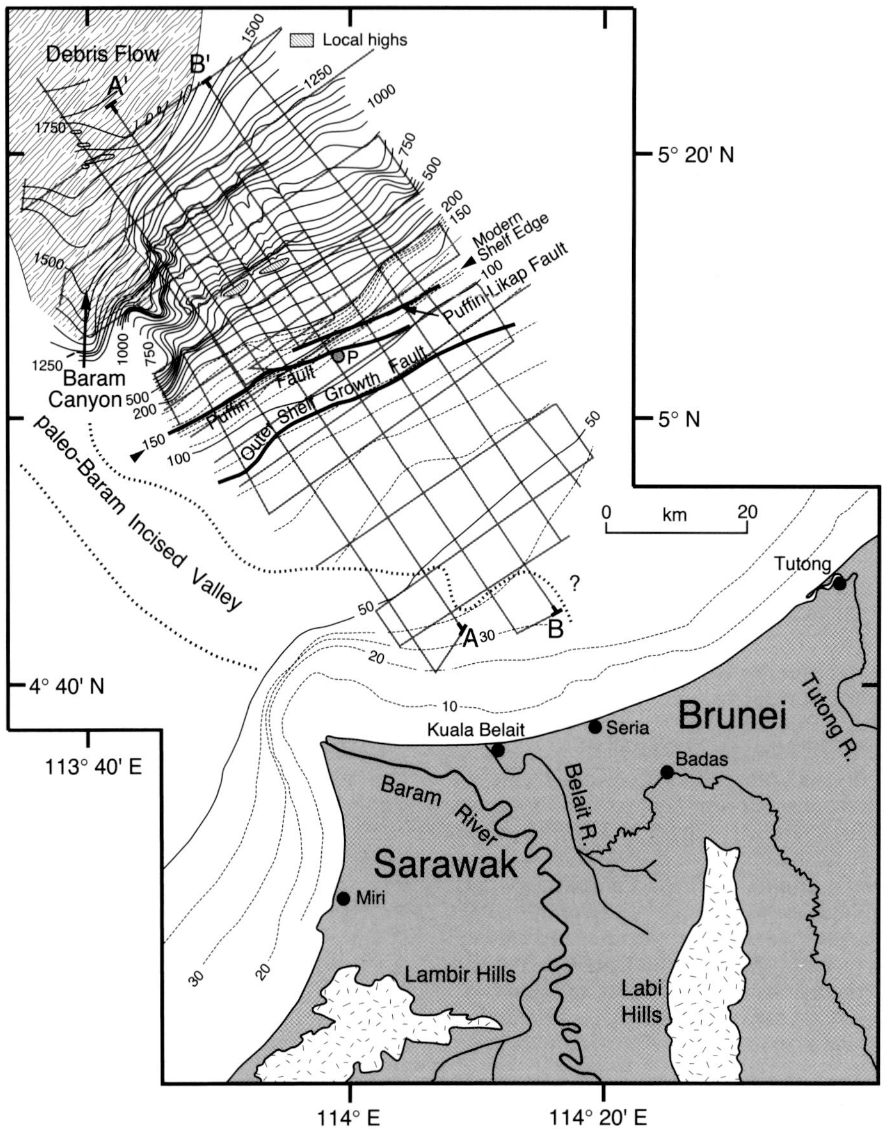

FIG. 2.—Bathymetric map compiled from *M.V. Armada Hydro* echo-sounder data, calculated depths below ~ 500 m (explained in text), and contours shown by Sandal (1996, his figure 1.2) outside the survey grid. Bathymetric contours each 10 m in depths < 200 m, and each 50 m elsewhere. Edges of incised valley on the middle and outer shelf were determined from examination of 2D and 3D seismic data of Sarawak Shell Berhad. P = Puffin-1 well location. AA' and BB' show the location of dip lines summarized in the line drawings of Figure 6.

having formed when sea level was high or slowly falling; they are therefore similar in their genesis to some prograding Quaternary shelf deltas in the northeastern Gulf of Mexico (Sydow and Roberts, 1994). In contrast, the forced regression and lowstand associated with the last glacial maximum (~ 24–16 ka) resulted in extensive development of fluvial channels all across the Brunei shelf, terminating at sand-prone shelf-edge deltas (e.g., Fig. 4). $\alpha 0$, the Holocene ravinement surface described by Hiscott (2001), overlies fluvial channels that developed at the last lowstand. Similar networks of lowstand channels deeper in the stratigraphy are inferred to coincide with earlier glacial lowstands, including those of oxygen isotope stages 6, 8, and 10 (Hiscott, 2001).

Sea-level variation during the late Quaternary closely parallels peaks and troughs in the oxygen isotope record (Imbrie et al., 1984). Calibration of the Quaternary sea-level curve is provided by Skene et al. (1998), who show that sea level remained near its modern elevation from 125 to 80 ka, then fell in one or two steps to ~ 30 m below its present height from 80 to 25 ka (Fig. 5). These conditions—an early extended highstand and then a period of slowly falling sea level—are consistent with other sea-level curves for the late Quaternary (Chappell and Shackleton, 1986) and with the thicknesses and landward limits of the highstand deltas on the Brunei shelf (Hiscott, 2001). The switching that occurred to form the separate delta lobes may have been entirely autocyclic in response to filling of accommodation space, or it may have been triggered by small-magnitude transgressions that characterize the sea-level curve (Fig. 5; Skene et al., 1998). When global sea level fell precipitously to about -115 m during oxygen isotope stage 2, the Baram River began to bypass the shelf through the

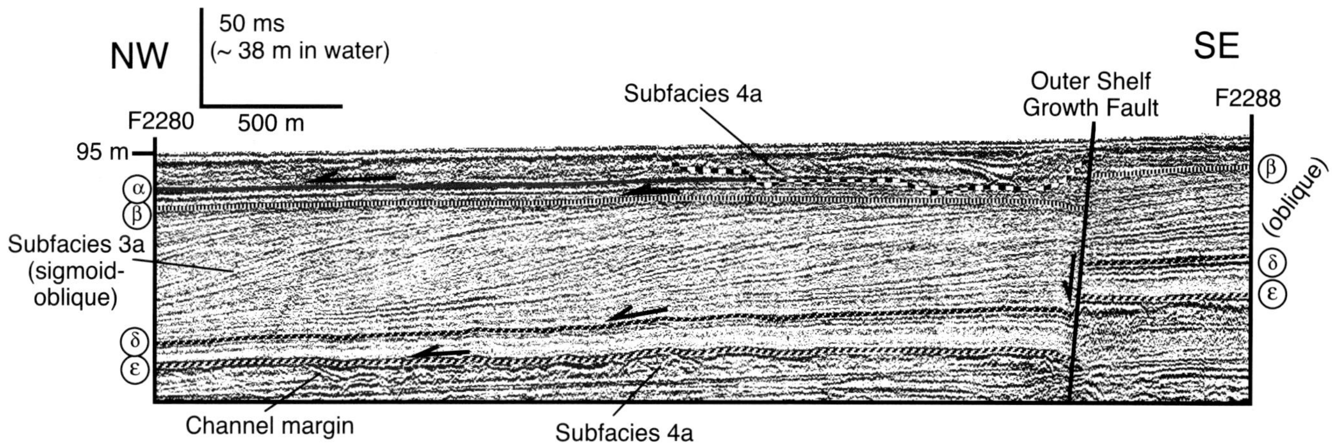

FIG. 3.—10 inch^3 airgun profile showing characteristics of key reflections on the middle and outer shelf. α, β, δ, and ε are downlap surfaces; in this area, γ = δ. The offlap break (topset–foreset transition in the β–δ unit) is preserved on the NW side of the Outer Shelf Growth Fault but has been eroded at the β reflector SE of the fault, where there was less accommodation space. Downlap (half arrows) in the very thinnest units is subtle.

Baram Canyon (Fig. 2), and smaller coastal rivers of limited discharge deposited small shelf-edge deltas across and seaward of a series of prominent growth faults. The sediment deposited in the vicinity of the shelf edge at this time is volumetrically much less than during the previous highstand (Fig. 6), consistent with bypassing of the much greater sediment load of the Baram River (presently ~ 12 million metric tons / yr) to the depths of the Borneo Trough.

A combination of deep-seated thrust faulting (Sandal, 1996; Fig. 1) and gravity-driven, down-to-the-basin growth faulting (Fig. 4) has resulted in profound thickness changes of time-equivalent strata from the inner shelf to the middle slope (Fig. 6). In particular, at the shelf edge, deposits thicken by a factor of 2–5 across a set of *en echelon* growth faults, with highest accumulation rates in the hangingwalls of these faults. The accentuated accumulation of deltaic sediments just seaward of the shelf edge, in the upper slope, distinguishes the Brunei shelf-to-basin profile from many other Quaternary deltas, for example the Rhône Delta (Torres et al., 1995, their figure 10).

This paper concerns the deposits of the Baram prodelta, mainly those associated with the 125–24 ka highstand delta lobes. The shelf stratigraphy is discussed elsewhere (Sandal, 1996; Hiscott, 2001). Because of a narrow shelf, high sediment yield from the Baram River, and an irregular slope along what was formerly a convergent plate boundary, the prodelta area is characterized by thick deposits, synsedimentary faults, slope failures, and mud diapirs. Most sediments were deposited on the prodelta slope when sea level was high or gently falling, and the slope became relatively starved at the last lowstand when the Baram River bypassed the shelf.

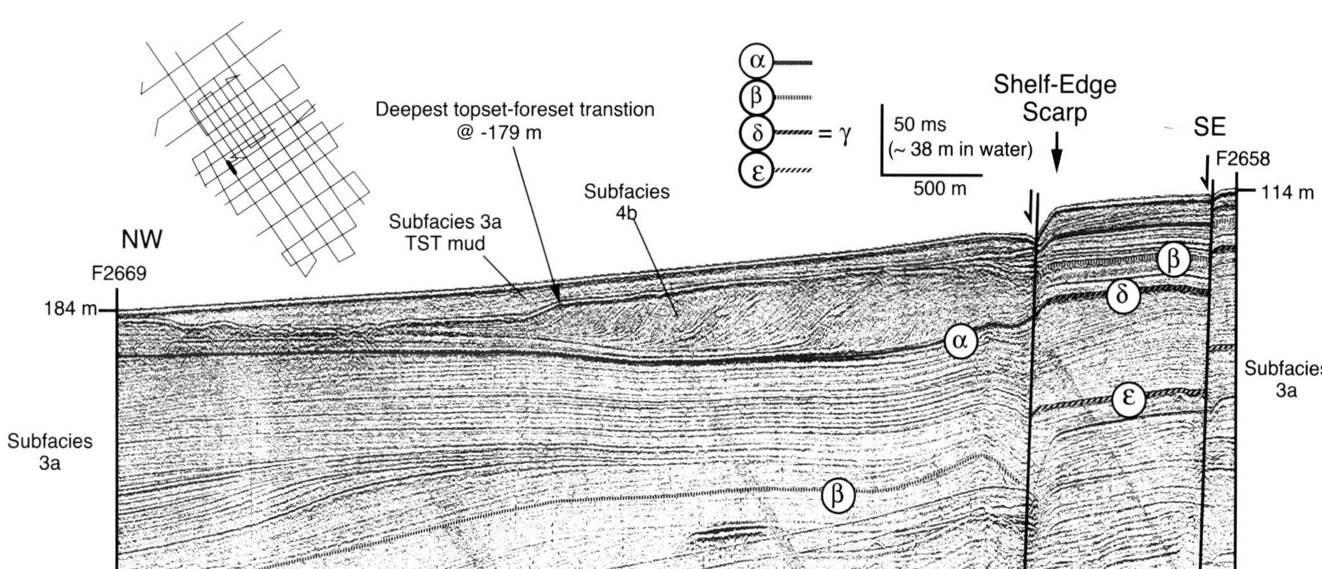

FIG. 4.—10 inch^3 airgun profile across the large-offset Puffin Fault at the shelf edge. The steep clinoforms of the post-α delta thicken by a factor of ~ 5 across the shelf edge.

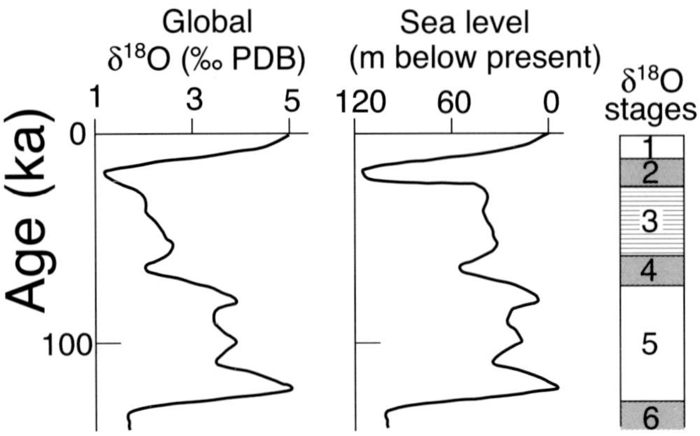

FIG. 5.—Global sea-level curve for the last ~ 140,000 yr (Skene et al., 1998), compared with the stacked, smoothed global oxygen isotope ($\delta^{18}O$) record and SPECMAP isotopic stage boundaries (from Imbrie et al., 1984). Glacial isotope stages 2, 4, and 6 are shaded, and the interstadial (not fully interglacial) stage 3 is ruled. The lowest sea levels occurred during glacial isotope stages 6 and 2. A slow sea-level fall occurred throughout stages 5–3.

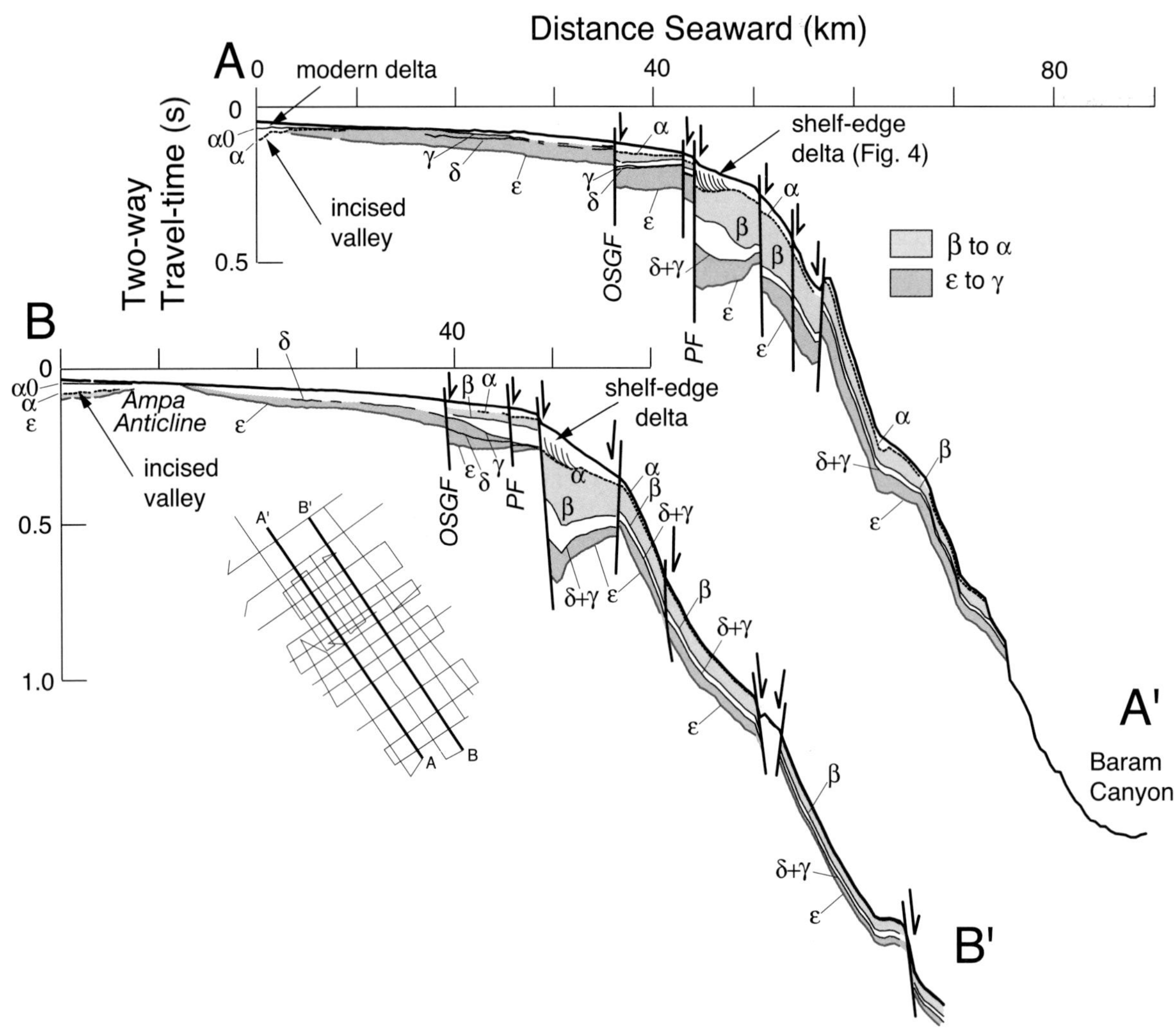

FIG. 6.—Vertically exaggerated (~ 40x) cross sections along two dip lines located in Figure 2. Each section shows the depths (in two-way travel time) of the seafloor and key reflections (from top: α0, α, β, γ, δ, ε). Strata between ε and γ, and between β and α, are shaded for clarity. γ is not recognized as a separate reflector beyond the shelf edge because it coincides with δ beneath the slope. γ is only shown on the shelf where it does not coincide with either the β reflection or the δ reflection. The Holocene ravinement surface, α0, is shown only near the landward end of each line where it defines the base of the modern Baram Delta. ε is arched over the actively growing Ampa Anticline on line BB'. At the landward end of each cross section, the paleo-Baram incised valley (Hiscott, 2001) occupies the synclinal depression landward of the Ampa Anticline. Only the Outer Shelf Growth Fault (OSGF) and the Puffin Fault (PF) extend between the cross sections. PLF = Puffin–Likap Fault (Fig. 2).

METHODS

In March 1998, ~ 1500 line-km of tracks were surveyed by *M.V. Armada Hydro* over the western shelf and slope of Brunei. Positions were determined using satellite navigation. Two seismic systems were used: a small airgun and a 16-transducer hull-mounted 3.5 kHz array. Bathymetry was recorded using a conventional narrow-beam echo sounder, with heave compensation, for water depths shallower than a few hundreds of meters to almost 1000 m, with the limiting depth dependent on sea state (i.e., noise level). All water depths were corrected for tidal water-level variations. For fix positions in deeper water, where the echo sounder could no longer track the rugged bottom, water depths were calculated from the last reliable depth determination, the difference in two-way travel-time between this point and the position with unknown depth, as measured from airgun profiles, and an acoustic velocity of 1530 m/s. Two "sound velocity profiles" were acquired beyond the shelf break during the survey, giving average velocities of 1530 m/s and 1522 m/s in the upper 110 m and 180 m of the water column, respectively.

The directly measured depths (echo sounder) are accurate to ± 10–20 cm. The calculated depths are assumed accurate to ± 0.2% (i.e., ± 2 m at 1000 m depth; ± 3 m at 1500 m depth). No compensation for ship's heave could be included in the calculations, and the precision of measurements of two-way traveltimes on the airgun profiles is ~ ± 2 ms.

RESULTS

Facies Characteristics of the Prodelta Wedge

Hiscott (2001) described five acoustic facies from the Brunei shelf (Table 1). The corresponding sedimentary facies are precisely known from geotechnical borings that penetrate typical examples of each facies. On the upper and middle slope, one additional facies (Facies 6) and one additional subfacies (Subfacies 4c) are present (italics entries in Table 1). Much of the prodelta region is characterized by laterally continuous, moderate-amplitude reflections of Subfacies 3a that progressively thin with increasing water depth (Fig. 7); these form stacked lens-shaped and wedge-shaped units in proximal areas (Fig. 8). Much of the equivalent sediment blanket over the deep slope is reflection-free or weakly reflective mud.

The muddy prodelta deposits are interrupted and disrupted by what are interpreted as diapir-cored ridges beneath the slope (Figs. 9 and 10). At these locations, the seabed is irregular, and otherwise continuous reflections are replaced by hyperbolic and chaotic returns. These ridges may be cored by mud diapirs like those exposed in outcrops in Brunei (Morley et al., 1998) or simply by zones where overpressured deep fluids have escaped to the surface, disrupting strata during their rise (Van Rensbergen et al., 1999). At some of the slope-parallel ridges, the seabed and "key" reflections are offset by counter-regional (landward-side-down) growth faults that produce steep seabed relief (Fig. 11). Very

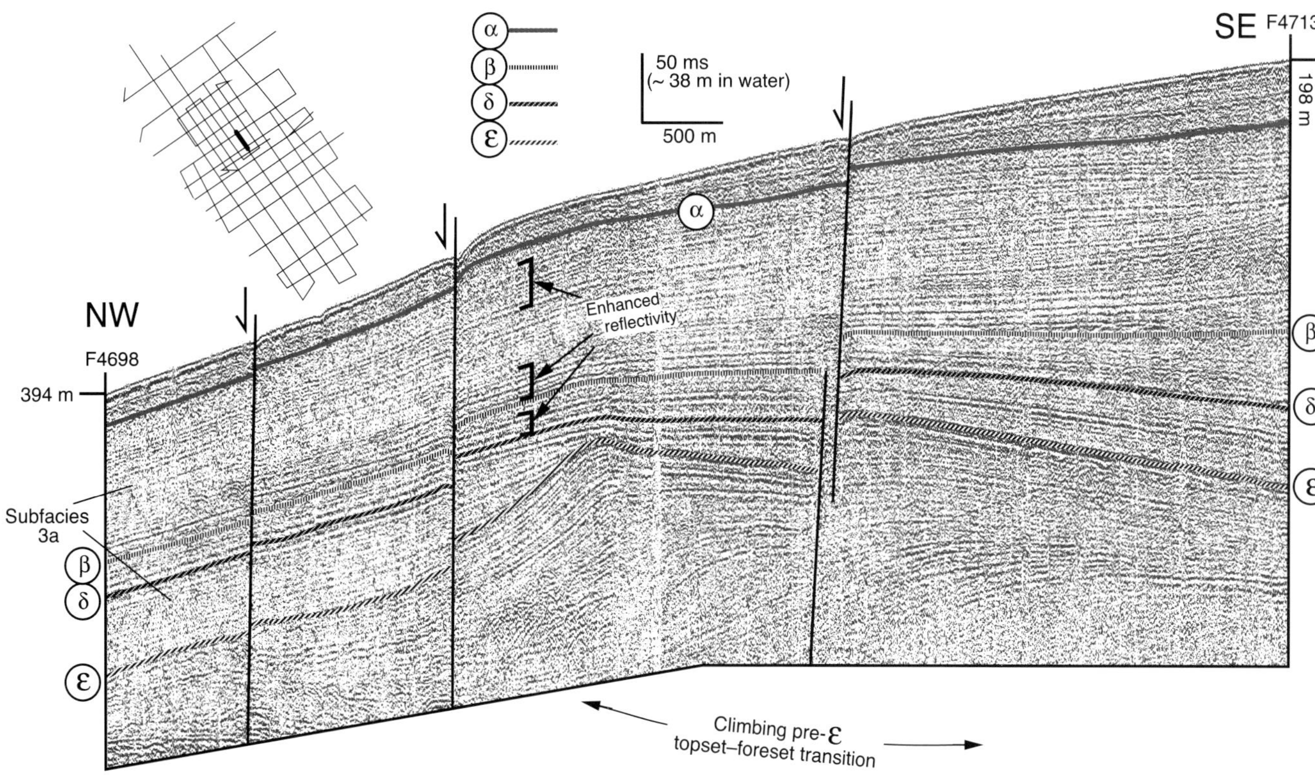

FIG. 7.— 40 inch^3 airgun profile showing progradational–aggradational clinoforms beneath the ε reflection and onlapping to slowly converging, subparallel reflections above. Parts of the ε–α succession with stronger reflections are indicated with bold brackets; these are interpreted to have thicker or more abundant sand and silt beds. This profile is located just downslope from the large-offset Puffin–Likap shelf-edge growth fault (Fig. 8), across which stratigraphic thicknesses decrease dramatically onto the shelf.

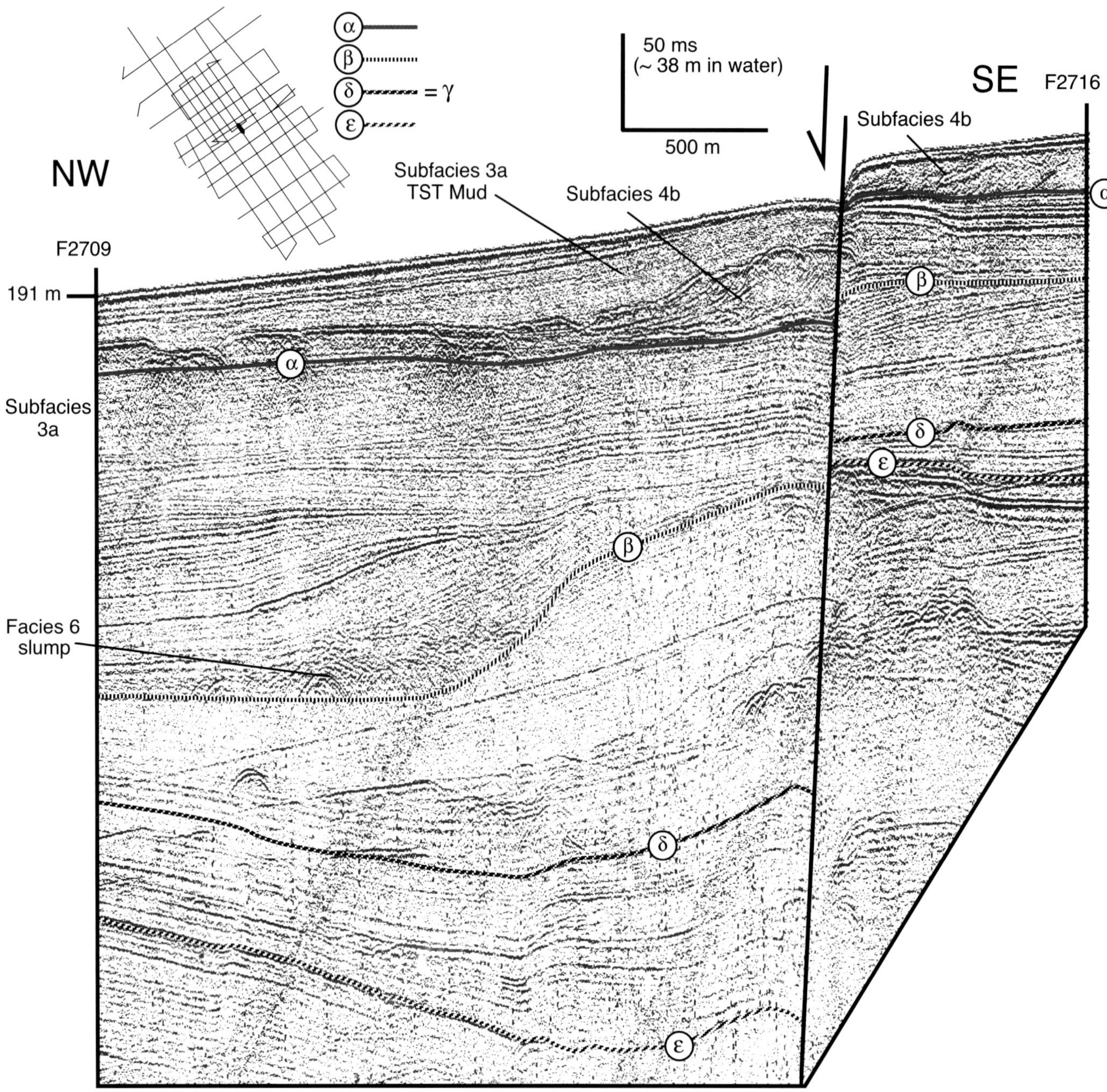

FIG. 8.—10 inch^3 airgun profile across a large-offset shelf-edge growth fault, showing dramatic thickening of deposits. Reflections were correlated across the fault elsewhere, using an industry 3D seismic dataset and reflection character. Offlap breaks of delta lobes in the δ–β and β–α units are located beyond the modern shelf edge. Figure 7 shows the succession just downslope of this profile.

similar features are present beneath the slope of the Niger Delta, where gravity tectonics deforms a thick and unstable sedimentary wedge (Damuth, 1994).

Subfacies 3a is interpreted as prodelta muds deposited from buoyant surface and mid-water plumes sourced from mud-load rivers. Wedge-shaped and lens-shaped geometries suggest progradation, aggradation, and lateral shifting of a number of muddy delta lobes. The lack of significant accommodation space on the shelf enhanced accumulation of these sediments in the moat seaward of the shelf-edge growth faults (Figs. 6 and 8).

Packets of Subfacies 3a strata with enhanced reflectivity (Fig. 7) are interpreted to have more or thicker silt–sand beds; these silt–sand beds may be plume deposits in proximal areas, or density-current deposits (turbidites) in distal areas. For Subfacies 3a as a whole, density-current deposits are believed to be far less important than plume deposits because thicknesses of the prodelta muds do not change significantly at sharp changes in slope (Fig. 12). Turbidite successions would instead infill bathymetric lows and thin dramatically onto basin slopes. Further indication of the importance of plume and hemipelagic

TABLE 1.— Facies characteristics and depositional environments.

Facies number, characteristics, and example Figures	**Subfacies letter and geometry (if applicable)**	**Interpreted depositional environment**
1. reflection-free deposits or parallel low-amplitude reflections forming a blanket over much of the modern seafloor; locally present in near-surface channel fills; borings recover soft clay (80–100% water) with local silt partings and silt-filled burrows		distal prodelta mud, plume deposits; contains maximum flooding surface
2. discontinuous, moderate-amplitude reflections at seafloor; more acoustic backscatter than Facies 1; borings recover sandy silty clays		current- and wave-reworked highstand muds and transgressive deposits in equilibrium with modern shelf hydrodynamics
3. continuous, low- to moderate-amplitude, dipping reflections (Figs. 3 and 7); borings in shelf clinoforms show laminated silty clays with silt and sand partings and thin beds, silt-filled burrows, a few shell fragments, and dispersed organic matter	(a) clinoforms (< 3°) and slope drape	muddy prodelta and plume deposits
	(b) draping channel fill	estuarine fill of gradually flooding channels during transgression
4. discontinuous to incoherent, locally concave upward (Figs. 4, 13, 15), moderate- to high-amplitude dipping reflections with maximum dips of 5–15°; dipping reflections resemble crossbedding, lateral accretion deposits in channels, or steep delta foresets. Borings through channel fills recover silty to silt-free, fine- to medium-grained sands with scattered decayed wood, rare shells, and lenses of clay	(a) channel fills with lateral accretion	meandering-river point bars
	(b) prograding wedges	coarse-grained forced-regressive and lowstand deltas
	(c) mounded lenses	*sandy turbidites from delta-front failures at lowstands*
5. continuous to discontinuous, moderate-amplitude, wavy, shingled to bundled reflections; restricted to the fill of the shelf-crossing incised valley; borings show moderate water contents (~ 40%) and silty clay with silt pockets and partings, some thin sand beds, local shell fragments, plant and wood fragments		shingled delta lobes filling a major incised valley during its flooding; also some small side-entry mini-deltas entering the valley from the adjacent shelf
6. chaotic reflection geometry, low backscatter; forms areally restricted lenses or mounds of disturbed strata (Figs. 17, 19)		*muddy to heterolithic debris-flow deposits, and sediment slides or slumps*

Italics entries were not described by Hiscott (2001)

sedimentation is found in the depositional architecture of a number of downslope-trending gullies and leveed channels in water depths of ~ 500–1000 m. The channels are spaced 1–2 km apart and have remained in the same place while 100 m or more of sediments have accumulated (Figs. 13 and 14). Their levees are largely formed of draping prodelta muds rather than spillover deposits from passing turbidity currents; the latter would be thicker at the levee crest and thinner in interchannel areas (e.g., Flood et al., 1991; Hiscott et al., 1997). The typical levee geometry, which is characterized by enhanced deposition at the levee crests, is seen only deep in the subsurface below one of the downslope-trending depressions (Fig. 13 near Fix4436, below reflector ε). Elsewhere, drape is the dominant geometry of the prodelta deposits.

Locally beneath the gullied slope, acoustic Subfacies 4c forms mounded deposits in depressions, mainly beneath the ε reflection (Figs. 13 and 15). These mounded deposits are possibly sandy turbidites ponded in slope depressions, like those described from nearby Sabah by Mohamad and Lobao (1997; their "orange" facies).

Areas of steep seabed are associated with lenses or mounds of disturbed strata assigned to acoustic Facies 6 (Table 1; Figs. 16 and 17). Low backscatter suggests muddy sediments. These are interpreted as mass-transport deposits (slumps, slides, and debris flows) on the basis of mounded, unstratified or disrupted reflec-

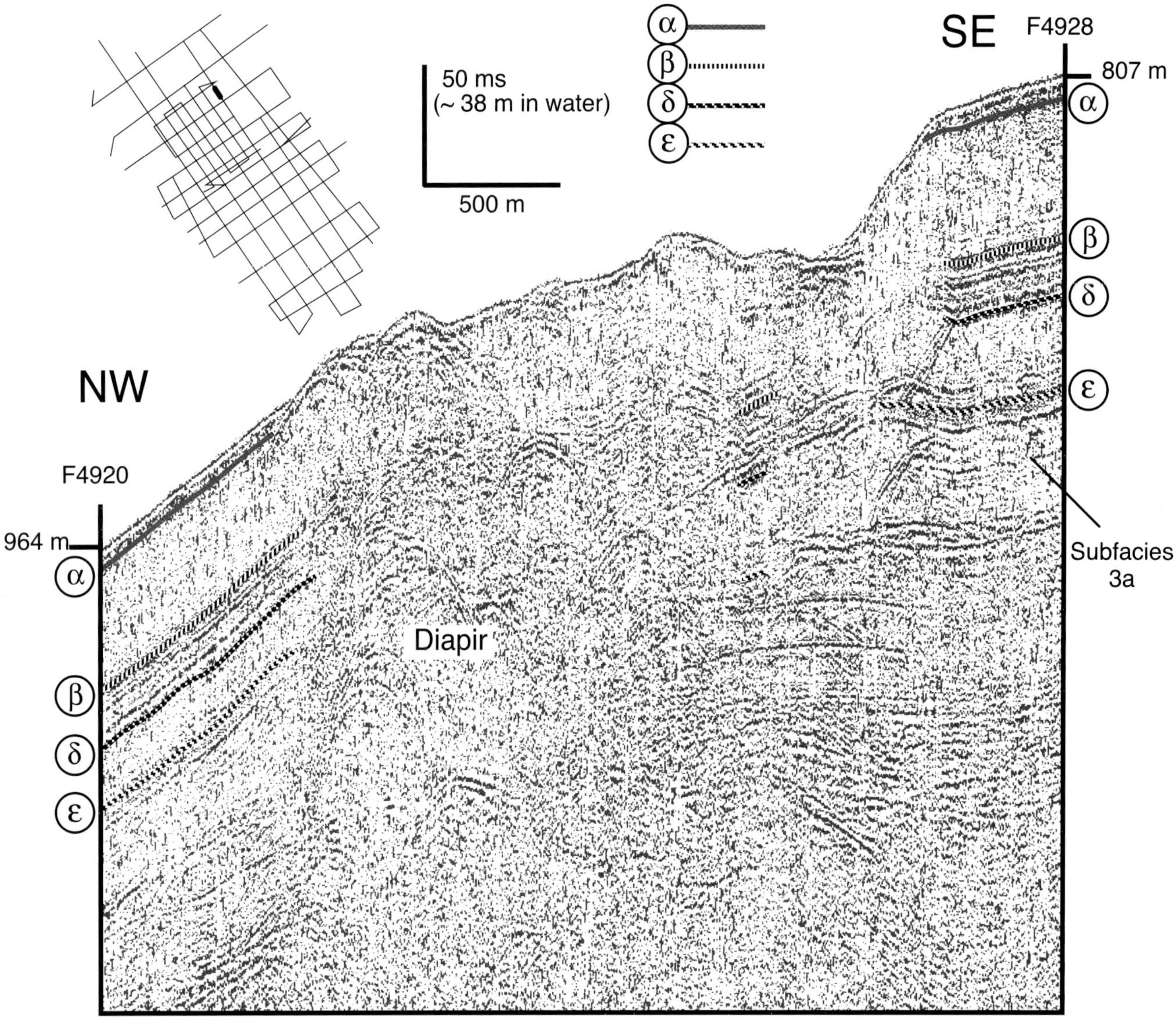

FIG. 9.—40 inch^3 airgun profile across an inferred mud-diapir-cored ridge. The diapir appears to have broken the seafloor and induced local collapse.

tion configurations and their association with slide scars (Fig. 18). Gradients beyond the shelf break are everywhere steep enough to promote the failure of water-rich, rapidly deposited sediments (Hein and Gorsline, 1981). The largest and thickest mass-transport deposit is confined to the floor of the Baram Canyon (Fig. 19). The stepped walls of the canyon are locally as steep as 15–20°, cutting through > 400 m of section.

Relationship of Slope Deposits to Phases of Deltaic Sedimentation

At water depths of < 500 m beyond the edge of the modern shelf, the e reflection separates a thick underlying unit characterized by clinoforms from overlying evenly bedded strata that onlap ε (Fig. 7). Below water depths of ~ 600 m, the ε reflection conformably underlies the same evenly bedded succession, but locally overlies depression-filling, mounded deposits (Figs. 13 and 15) or internally deformed sediment slides (Fig. 16). As explained by Hiscott (2001), ε represents a fourth-order cycle boundary on the shelf, separating lowstand fluvial and deltaic deposits below from mainly highstand delta lobes above. The thick clinoforms beneath ε on the upper slope (Fig. 7) are interpreted as a lowstand delta formed during glacial oxygen-isotope stage 6. Downslope from the front of this delta, leveed channels locally contain mounded deposits of acoustic Subfacies 4c, inferred to be sand-prone turbidites derived from delta-front failures (Figs. 13 and 15).

When relative sea level began to rise at the transition from glacial oxygen-isotope stage 6 to interglacial stage 5, the topset–foreset transition of the pre-ε delta (its offlap break) grew vertically upward. This lobe was then drowned as the Baram Delta shifted landward to the vicinity of the modern shoreline. When the rate of sea-level rise slowed, delta progradation began afresh, producing a set of delta lobes on the shelf (Hiscott, 2001). Once the margins of these delta lobes had reached the shelf edge and uppermost slope, reduced accommodation space on the

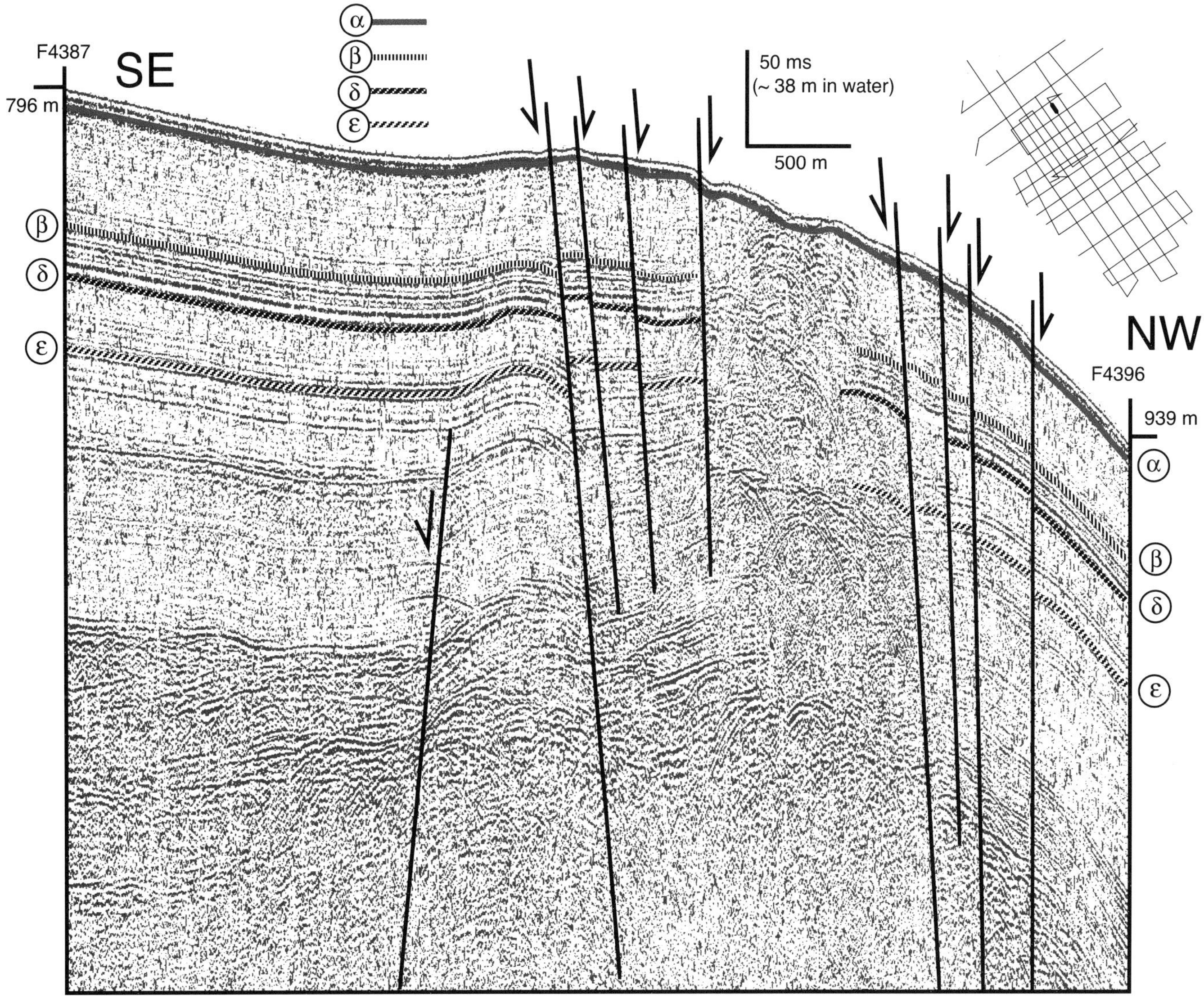

FIG. 10.—40 inch^3 airgun profile across an inferred mud-diapir-cored ridge. The diapir is surrounded by a swarm of subvertical faults with variable offsets.

shelf promoted the thick accumulation of sediments farther seaward (Fig. 6). The wedge-shaped deposits of Subfacies 3a between reflections ε and β (Fig. 8) are interpreted as the muddy plume deposits of these deltas. They drape the prodelta slope, and include only minor small-volume gravitational failures (Facies 6).

As sea level dropped to about -115 m during the last glacial maximum (Fig. 5), the Baram River cut a deep incised valley to the edge of the shelf (Hiscott, 2001), and deposited its load at the terminus of the valley, in water depths of ~ 130 m (outboard of shelf-edge growth faults). Here, 2D seismic profiles of Sarawak Shell Berhad (SSB) show a 120-ms-thick lens of disrupted reflections with an abrupt seaward limit at the steep headwall scarp of the canyon. This lens is interpreted as a remnant of the lowstand Baram Delta. Delta-front deposits that had accumulated on unstable slopes at the head of the canyon probably failed, on a regular basis, and generated efficient, mud-load turbidity currents that bypassed the slope and deposited their load in the Borneo Trough. Channel–levee deposits have been recognized in this deep-water area (BSP proprietary reports). At one point during the lowstand, a large mass of deltaic sediments is believed to have collapsed to form the thick, mounded debris-flow deposit that presently covers the floor of the canyon (Fig. 19) and extensive areas farther seaward (BSP proprietary reports). This proposed timing, during the last lowstand, is supported by the local presence of a veneer of stratified sediments on top of the debris flow. This veneer is inferred to be a Holocene drape.

During the prolonged period of lowstand bypass, most of the adjacent continental shelf and slope was relatively starved of sediment, except for a regional thin mud blanket that was likely still sourced from plumes entering the South China Sea at the delta front. Growth faulting, mud diapirism, and occasional sliding of unstable sediment masses all continued during this period, so that the seaward-tapering prodelta wedge is locally disrupted or partly missing where mass wasting has occurred.

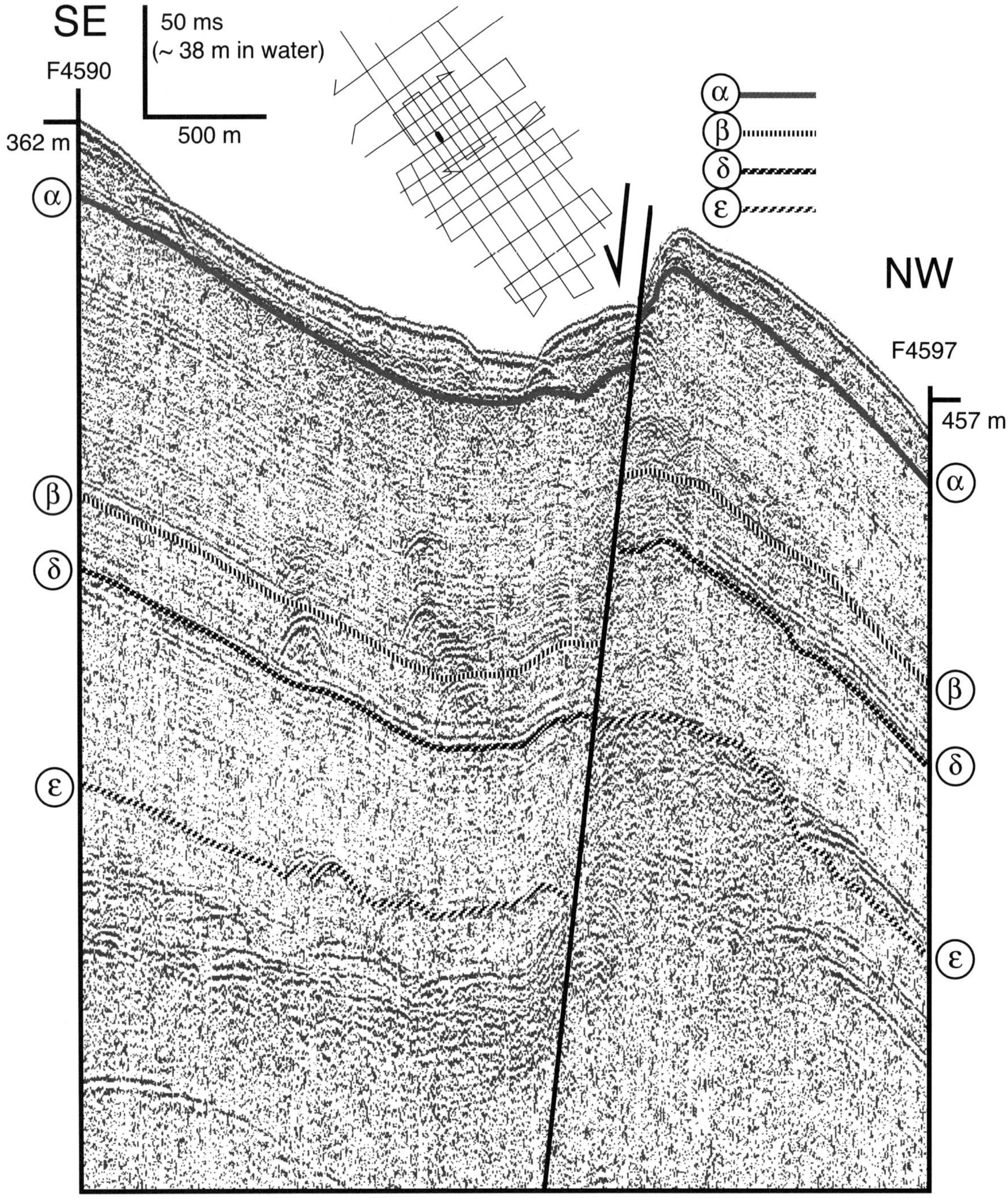

FIG. 11.—40 inch^3 airgun profile showing a counter-regional growth fault with some degradation of reflections in the footwall, perhaps indicating minor upward intrusion of wet muds or overpressured fluids.

DISCUSSION

On many passive continental margins, the thickest slope and basinal deposits accumulate during lowstands of sea level. This is equally true in the vicinity of passive-margin deltas like the Niger Delta (Damuth, 1994), shelf-edge deltas in the northeastern Gulf of Mexico (Winn et al., 1998), and the Rhône Delta (Torres et al., 1995). Along many active margins, however, a narrow shelf is quickly buried even at highstands by the deposits of rivers with high sediment discharges. In these situations, deltas can prograde sufficiently during highstands to reach the shelf edge in less than 10,000 years (Burgess and Hovius, 1998), after which mud and sand are supplied in large quantities to slope and basinal settings. As a result, the deep-water area farther seaward can receive substantial thicknesses of sediments that can no longer be accommodated on the shelf. In the case of the Baram Delta, at least during the last sea-level cycle (Fig. 20), accumulation rates of muddy prodelta deposits were greatest on the upper and middle slope at the highstand.

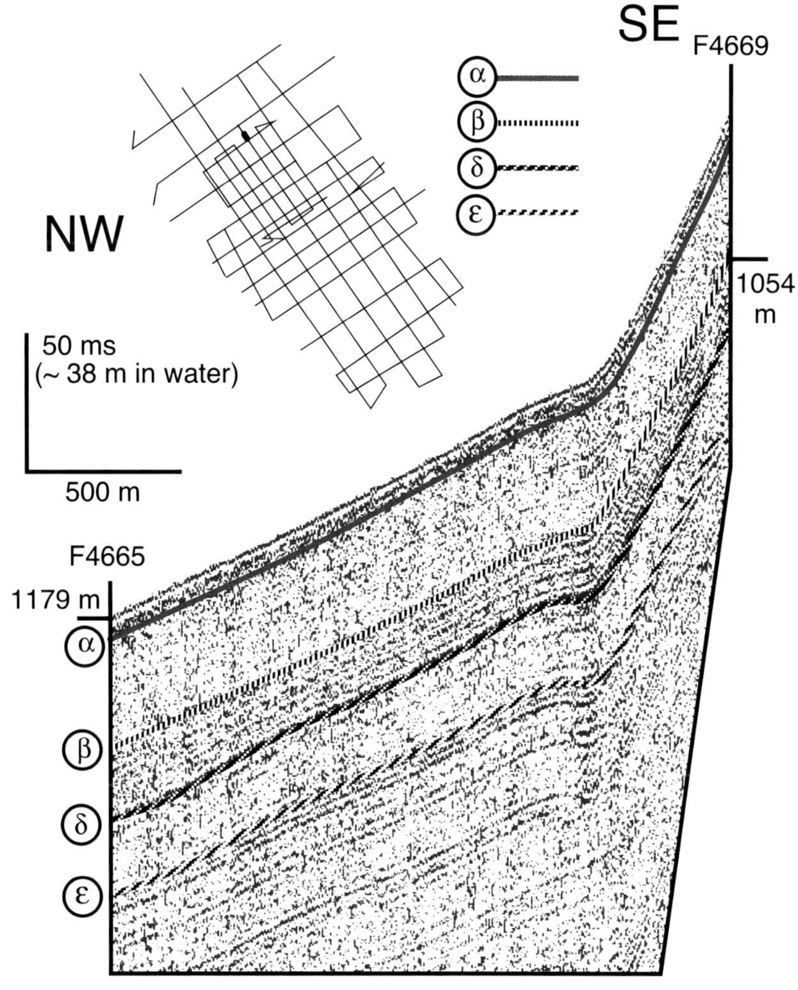

FIG. 12.—40 inch^3 airgun profile showing the behavior of Subfacies 3a silty muds across pronounced changes in bottom slope. The spacings between key reflections, measured vertically, show little decrease on the steeper slope, indicating that the sediments were deposited largely from suspension plumes rather than from turbidity currents.

The dominance of highstand deposition offshore Brunei is only partly a result of rapid delta-front advance to the shelf edge. A second important factor is that the outflow of the Baram River largely bypassed the shelf and upper–middle slope during the last lowstand through an incised valley and associated canyon system (Hiscott, 2001). Some of these bypassed sediments are present as thick mass-transport complexes (e.g., debris-flow deposits) in the canyon and beyond. Another part of the sediment load that bypassed the shelf is present as channel–levee systems in the deep-water Borneo Trough (BSP proprietary data), and probably as sheet-like turbidites on the floor of this trough.

The prodelta slope seems never to have been the site of significant turbidite deposition, except as minor wedge-shaped and lens-shaped deposits in slope channels. Over most of the slope, the position of leveed channels and gullies seems to be controlled by the location of underlying ridges, troughs, and saddles in deformed Tertiary deltaic deposits which have been thrusted, and locally intruded by mud diapirs. The channels have persisted as morphological features because of blanketing beneath inferred prodeltaic deposits of nearly even thickness, regardless of local seabed gradient or shape. This rather uniform blanket resembles the muddy successions that drape upper-slope gullies offshore from the Tiber Delta in the Tyrrhenian Sea (Chiocci and Normark, 1992) and likely consists of the deposits of buoyant plumes, and / or thick muddy gravity currents that inundated large areas of the slope rather than following the courses of individual slope channels. These deposits show only gradual changes in thickness from the upper to the middle slope, except where subsequent sediment sliding has removed part of the prodelta sedimentary wedge (e.g., Fig. 18).

Morphological channels and levees on the Baram prodelta are therefore downslope-elongated depressions and ridges with inherited relief. The term "channel" is perhaps inappropriate, because these depressions are *not* conduits for sediment supply to deep-water areas during most of their history. However, they have a morphology indistinguishable from that of active channels, and they could become active again if supplied from a shelf-edge delta lobe. Such morphological channels might be identified and misinterpreted as active transport pathways in low-resolution seismic data.

Burgess and Hovius (1998) have pointed out the danger of misinterpreting sandy slope and submarine fan deposits as lowstand indicators. If rate of delta progradation is sufficiently high, and the shelf sufficiently narrow, then coarser-grained

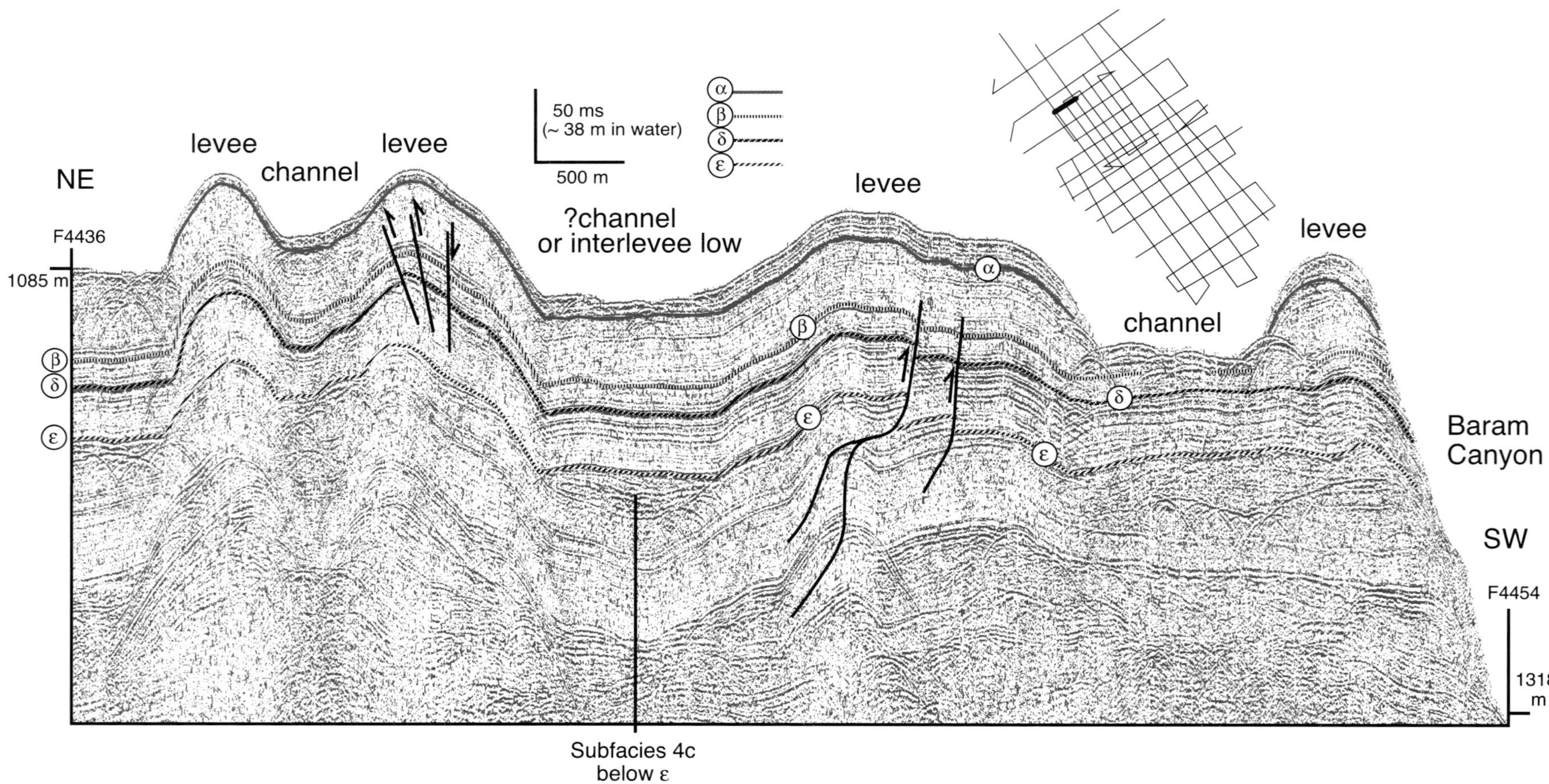

FIG. 13.—40 inch^3 airgun profile across the middle slope, where a number of levees have grown vertically during the Quaternary, principally through suspension fallout to form an even-thickness muddy drape. The channel to the SW, adjacent to the steep wall of the Baram Canyon, has been most recently active—its levees have enhanced sediment thicknesses and the floor of the channel appears to have a central thalweg channel flanked by low ?sandy inner levees (cf. Piper et al., 1999). There are mounded depression-filling deposits below the ε reflection in the southwestern channel and the central depression, which may be an interlevee low rather than a channel. A downslope crossing of this central depression is shown in Figure 15. Some of the faults in this part of the slope have a component of thrusting in this strike view.

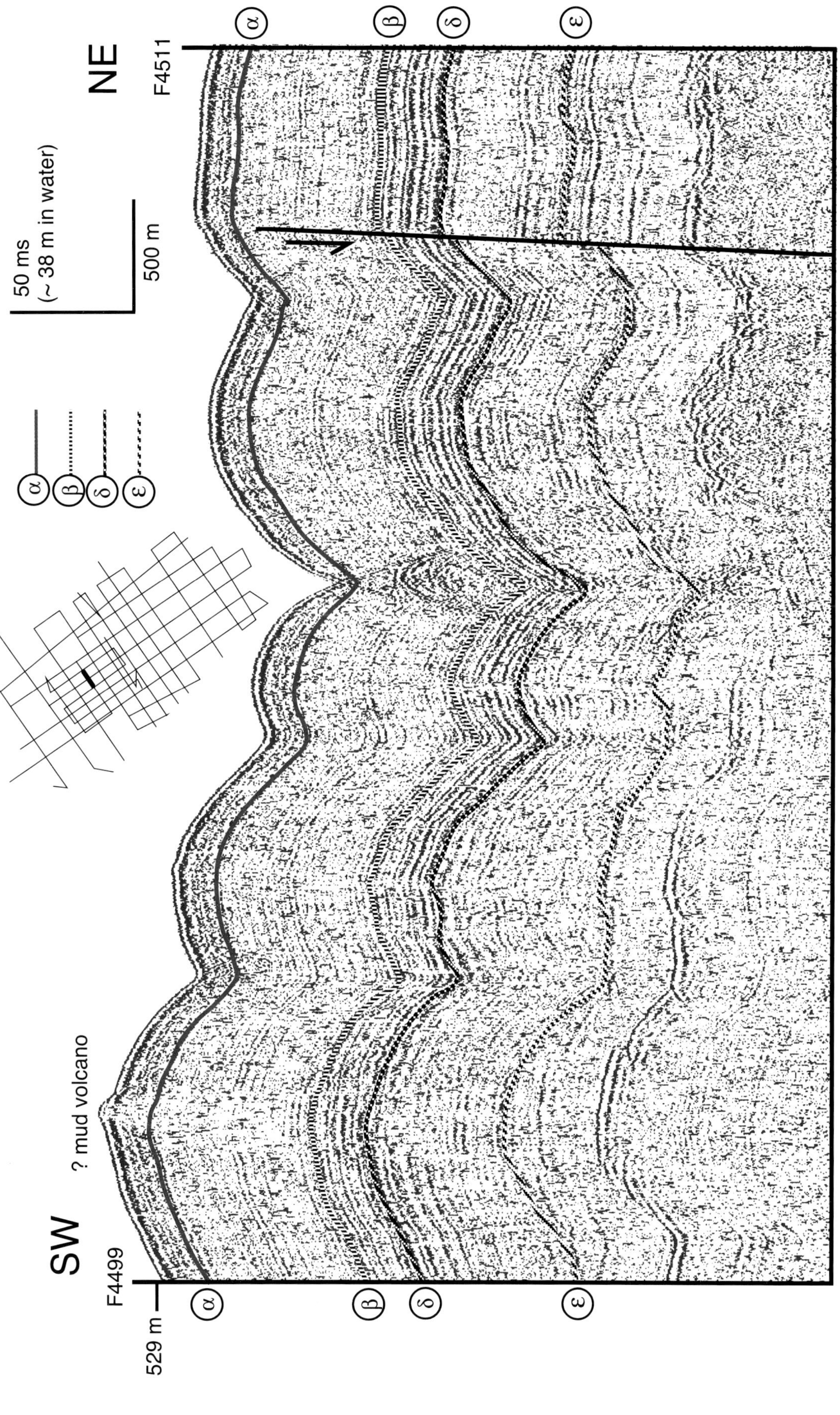

FIG. 14.—40 inch^3 airgun profile showing the smoothly draped middle slope. All deposits are Subfacies 3a prodelta muds, except surficial thin transparent mud of Facies 1 at the seafloor. The α, β, δ, and ε reflections occur in Subfacies 3a.

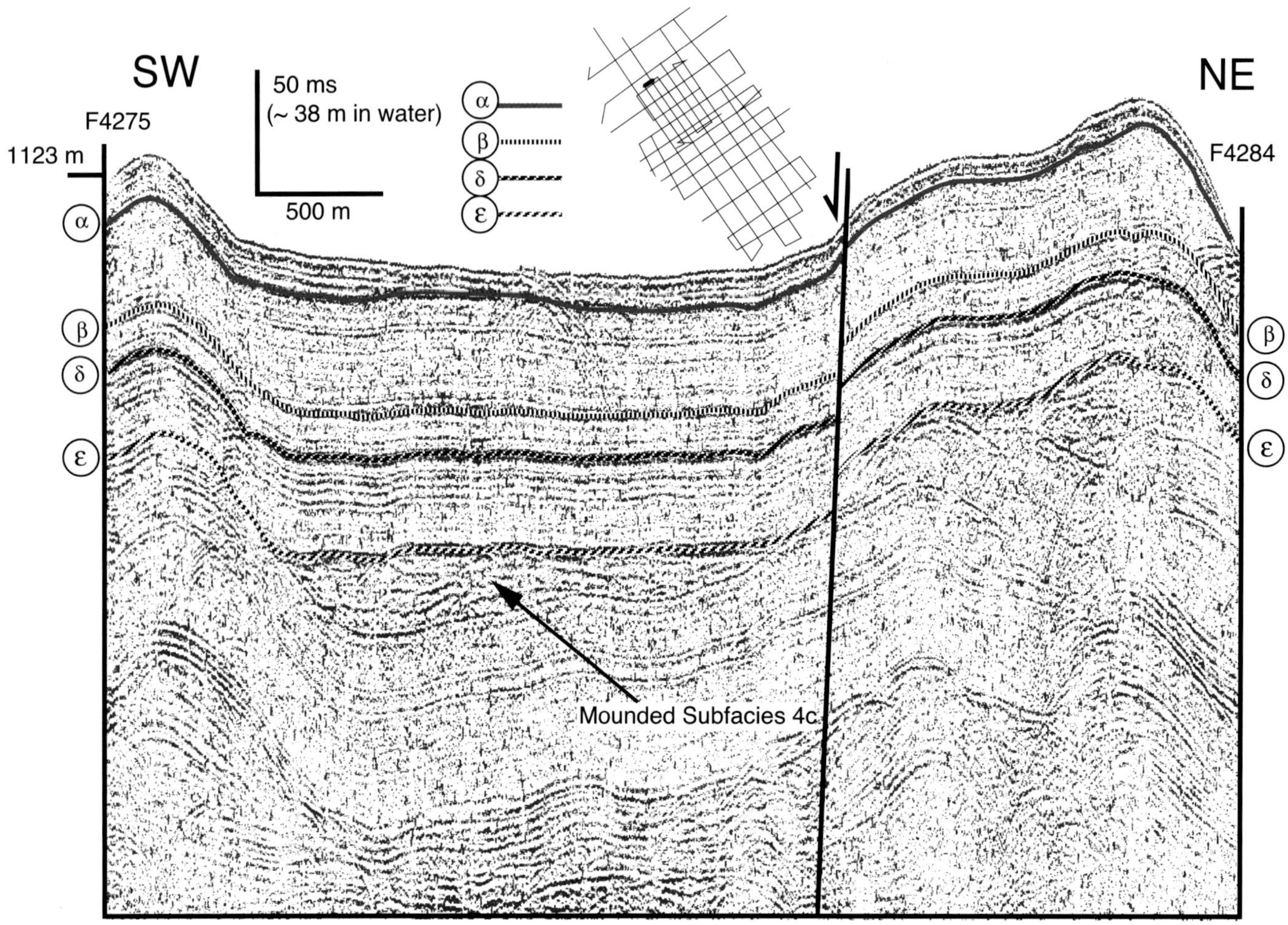

FIG. 15.—40 inch^3 airgun profile showing largely draping deposits above the ε reflection, and irregularly mounded deposits of Subfacies 4c in a broad depression (?channel) just beneath ε. Figure 13 crosses the same depression 1.5 km upslope.

sediments can be supplied to the shelf edge and beyond during most of a highstand cycle. The modern Baram Delta has prograded ~ 55 km in the last ~ 5,400 years (Caline and Huong, 1992). At this rate, and allowing for distal deepening of the shelf, it could reach the shelf edge in ~ 7,000–10,000 years. This would presumably result in renewed deposition of thick plume and turbidity-current deposits on the prodelta slopes and in the Borneo Trough (e.g., Fig. 8, ε–α). This style of rapid deposition of water-rich prodelta muds promotes mud diapirism, local sediment failures, and gravity tectonics across the entire margin (cf. Damuth, 1994).

CONCLUSIONS

1. A number of mostly highstand delta lobes accumulated in the area offshore from the Baram Delta during the Quaternary. Stratigraphic thicknesses of single delta lobes increase by a factor of 2–5 across growth faults, particularly at the edge of the modern shelf.

2. As glacio-eustatic sea level approached lowstand, the Baram River cut a major incised valley and then bypassed the shelf to the head of the Baram Canyon. The paleo-Baram lowstand delta thereafter provided muddy sediment directly to the deep-water Borneo Trough, in some cases as large debris flows generated by failures of oversteepened slopes.

3. Gravity-controlled growth faulting created a structural moat just beyond the modern shelf edge, where exaggerated thicknesses of mainly highstand muds and silts accumulated. Ridges cored by mud diapirs on the deeper slope locally trapped sandy turbidites in depressions to their landward side. Relict rugged topography on the slope has subsequently been draped by prodelta muds. The inherited relief may largely have been produced by the widespread faulting and diapirism.

4. Unlike many passive-margin deltas, a narrow shelf and high rate of sediment supply in the Baram system have promoted the development of thick highstand deposits, both as mud-prone delta lobes on the middle to outer shelf and as draping prodelta muds on the slope. During the last lowstand, in contrast, slope deposition was limited because river outflow bypassed the shelf through the Baram Canyon.

ACKNOWLEDGMENTS

This research was carried out during the tenure of a *Brunei-Shell Fellowship in Environmental Sciences* at Universiti Brunei Darussalam.

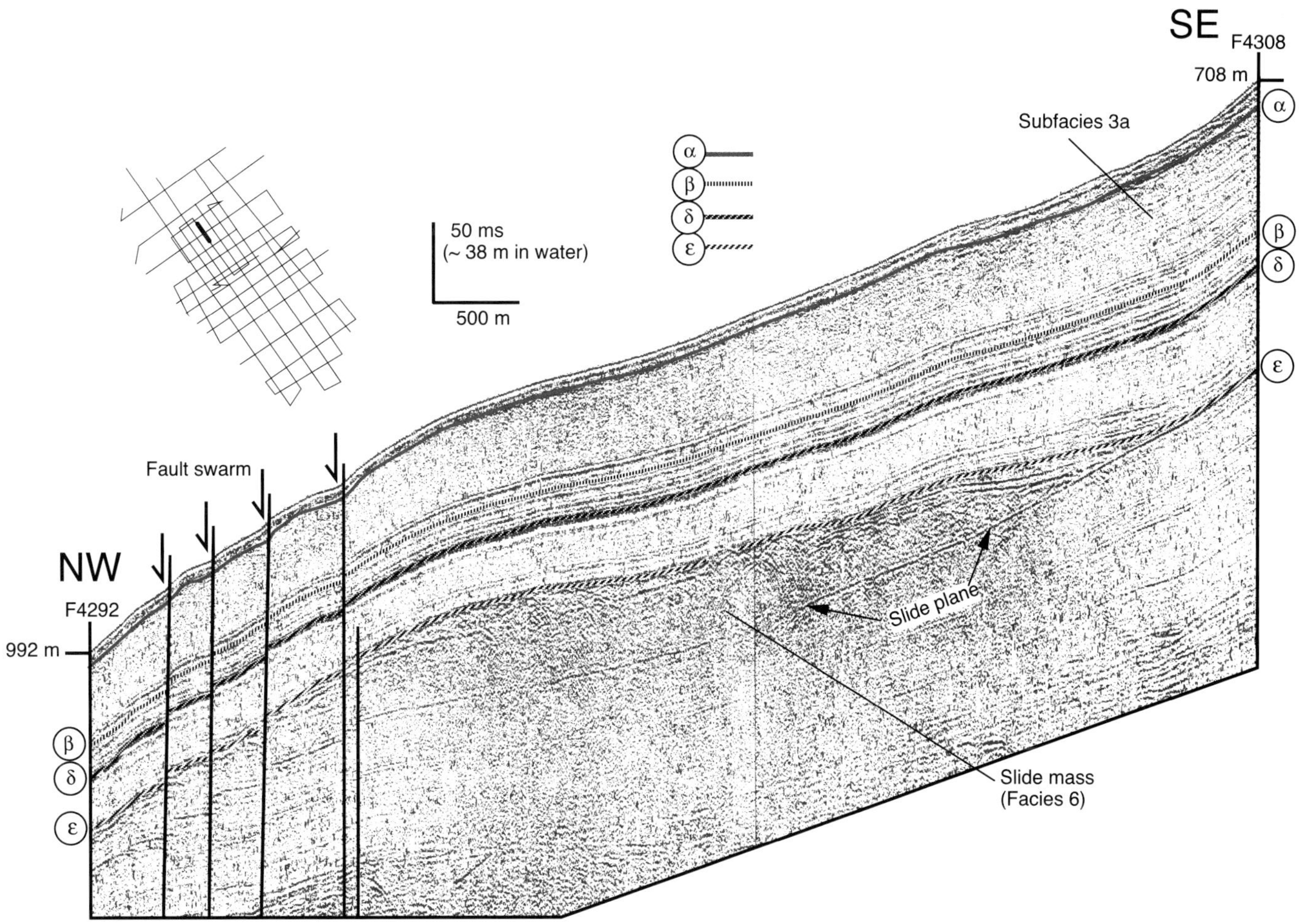

FIG. 16.—40 inch^3 airgun profile showing largely draping and seaward-thinning deposits above the ε reflection, and a large slide mass just beneath ε.

Colleagues in the Department of Petroleum Geoscience at Universiti Brunei Darussalam provided a critical sounding board as the research progressed, and offered invaluable suggestions for improvement of this paper. The marine survey and follow-up examination of industry seismic and borehole data were financially supported and encouraged by many geologists and technical experts from Brunei Shell Petroleum Company Sendirian Berhad. Professional data acquisition onboard *M.V. Armada Hydro* was provided by employees of Racal Surveys (Singapore) and Captain Rafael Misola and his crew (Nautika Support Services). Study of seismic data covering the edges of the paleo-Baram incised valley was generously permitted by Sarawak Shell Berhad, Malaysia. Early drafts of the manuscript were improved by critical comments and suggestions of Joseph Lambiase, Jeroen Peters, Patrick Allman-Ward, Pieter Van Rensbergen, and Ali Aksu. Reviewers P. Imbert and J.R. Suter are thanked for their constructive criticisms. Permission to publish was kindly granted by Brunei Shell Petroleum and the Petroleum Unit, Negara Brunei Darussalam.

REFERENCES

BURGESS, P.M., AND HOVIUS, N., 1998, Rates of delta progradation during highstands: consequences for timing of deposition in deep-marine systems: Geological Society of London, Journal, v. 155, p. 217–222.

CALINE, B., AND HUONG, J., 1992, New insight into the recent evolution of the Baram Delta from satellite imagery: Geological Society of Malaysia, Bulletin, v. 32, p. 1–13.

CHAPPELL, J., AND SHACKLETON, N.J., 1986, Oxygen isotopes and sea-level: Nature, v. 324, p. 137–140.

CHIOCCI, F.L., AND NORMARK, W.R., 1992, Effect of sea-level variation on upper-slope depositional processes offshore of Tiber Delta, Tyrrhenian Sea, Italy: Marine Geology, v. 104, p. 109–122.

DAMUTH, J.E., 1994, Neogene gravity tectonics and depositional processes on the deep Niger Delta continental margin: Marine and Petroleum Geology, v. 11, p. 320–346.

FLOOD, R.D., MANLEY, P.L., KOWSMANN, R.O., APPI, C.J., AND PIRMEZ, C., 1991, Seismic facies and late Quaternary growth of Amazon submarine fan, *in* Weimer, P., and Link, M.H., eds., Seismic Facies and Sedimentary Processes of Submarine Fans and Turbidite Systems: New York, Springer-Verlag, p. 415–433.

HAMILTON, W., 1979, Tectonics of the Indonesian region: U.S. Geological Survey, Professional Paper 1078, 345 p.

HEIN, F.J., AND GORSLINE, D.S., 1981, Geotechnical aspects of fine-grained mass flow deposits: California Continental Borderland: Geo-Marine Letters, v. 1, p. 1–5.

HISCOTT, R.N., 2001, Depositional sequences controlled by high rates of sediment supply, sea-level variations, and growth faulting: the Quaternary Baram Delta of northwestern Borneo: Marine Geology, v. 175, p. 67–102.

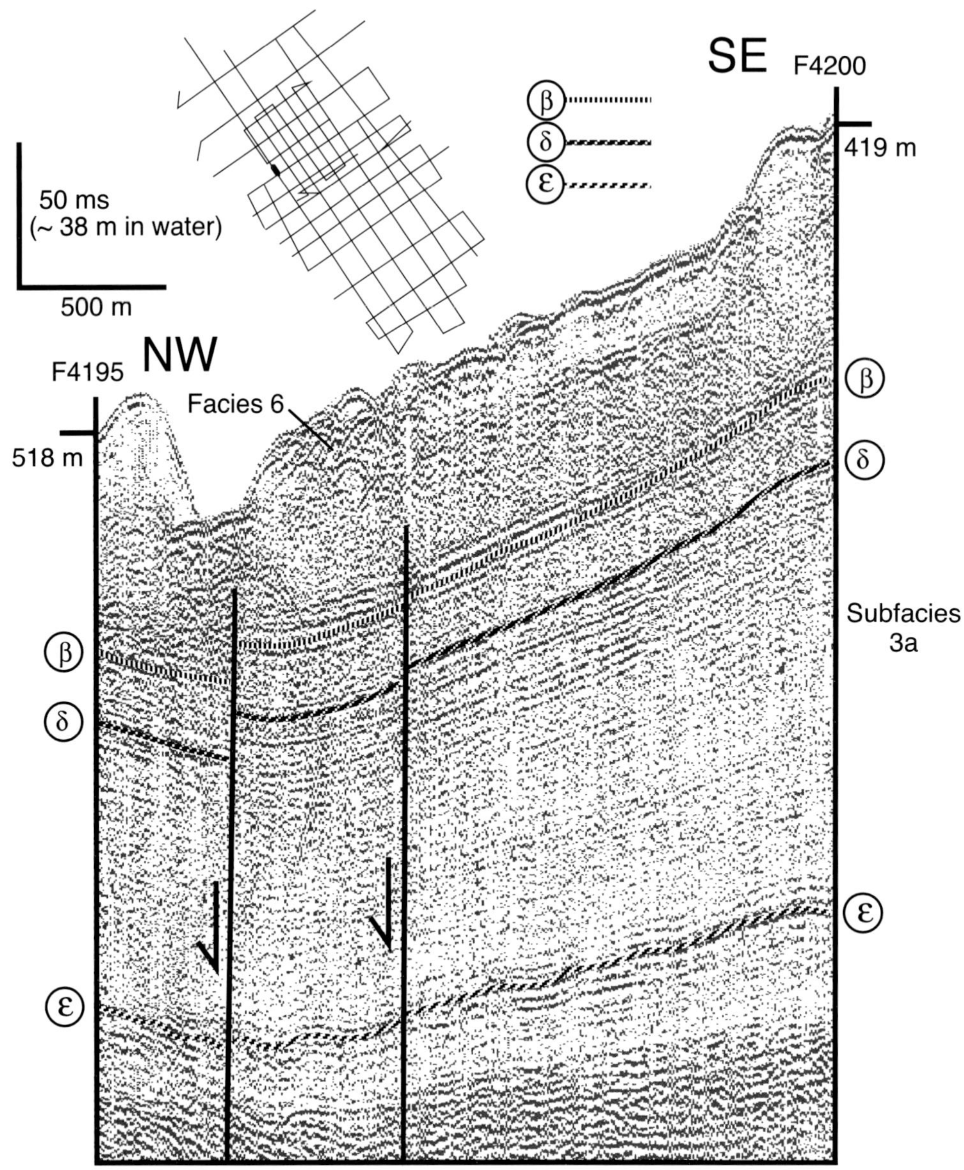

FIG. 17.—40 inch^3 airgun profile showing mounded, acoustically chaotic mass-transport deposits on the middle slope.

HISCOTT, R.N., HALL, F.R., AND PIRMEZ, C., 1997, Turbidity-current overspill from the Amazon Channel: texture of the silt/sand load, paleoflow from anisotropy of magnetic susceptibility and implications for flow processes, *in* Flood, R.D., Piper, D.J.W., Klaus, A., and Peterson, L.C., eds., Proceedings of the Ocean Drilling Program, v. 155, p. 53–78.

HUTCHISON, C.S., 1996, Geological Evolution of South-East Asia: Kuala Lumpur, Geological Society of Malaysia, 368 p.

IMBRIE, J., HAYS, J.D., MARTINSON, D.G., MCINTYRE, A., MIX, A.C., MORLEY, J.J., PISIAS, N.G., PRELL, W.L., AND SHACKLETON, N.J., 1984, The orbital theory of Pleistocene climate: support from a revised chronology of the marine $\delta^{18}O$ record, *in* Berger, A., Imbrie, J., Hays, J., Kukla, G., and Saltzman, B., eds., Milankovitch and Climate, Part 1: Dordrecht, The Netherlands, Reidel, p. 269–305.

MOHAMAD, M., AND LOBAO, J.J., 1997, The Lingan Fan: late Miocene/early Pliocene turbidite fan complex, north-west Sabah, *in* Howes, J.V.C., and Noble, R.A., eds., Petroleum Systems of SE Asia and Australia: Jakarta, Indonesian Petroleum Association, p. 787–798.

MORLEY, C.K., CREVELLO, P., AND HJ. AHMAD, Z., 1998, Shale tectonics and deformation associated with active diapirism: the Jerudong Anticline, Brunei Darussalam: Geological Society of London, Journal, v. 155, p. 475–490.

PIPER, D.J.W., HISCOTT, R.N., AND NORMARK, W.R., 1999, Outcrop-scale acoustic facies analysis and latest Quaternary development of Hueneme and Dume submarine fans, offshore California: Sedimentology, v. 46, p. 47–78.

SANDAL, S.T., ed., 1996. The Geology and Hydrocarbon Resources of Negara Brunei Darussalam: Brunei Darussalam, Brunei Shell Petroleum Company Sendirian Berhad and Brunei Museum, 243 p.

SKENE, K.I., PIPER, D.J.W., AKSU, A.E., AND SYVITSKI, J.P.M., 1998, Evaluation of the global oxygen isotope curve as a proxy for Quaternary sea level by modelling of delta progradation: Journal of Sedimentary Research, v. 68, p. 1077–1092.

SYDOW, J., AND ROBERTS, H.H., 1994, Stratigraphic framework of a late Pleistocene shelf-edge delta, northeast Gulf of Mexico: American Association of Petroleum Geologists, Bulletin, v. 78, p. 1276–1312.

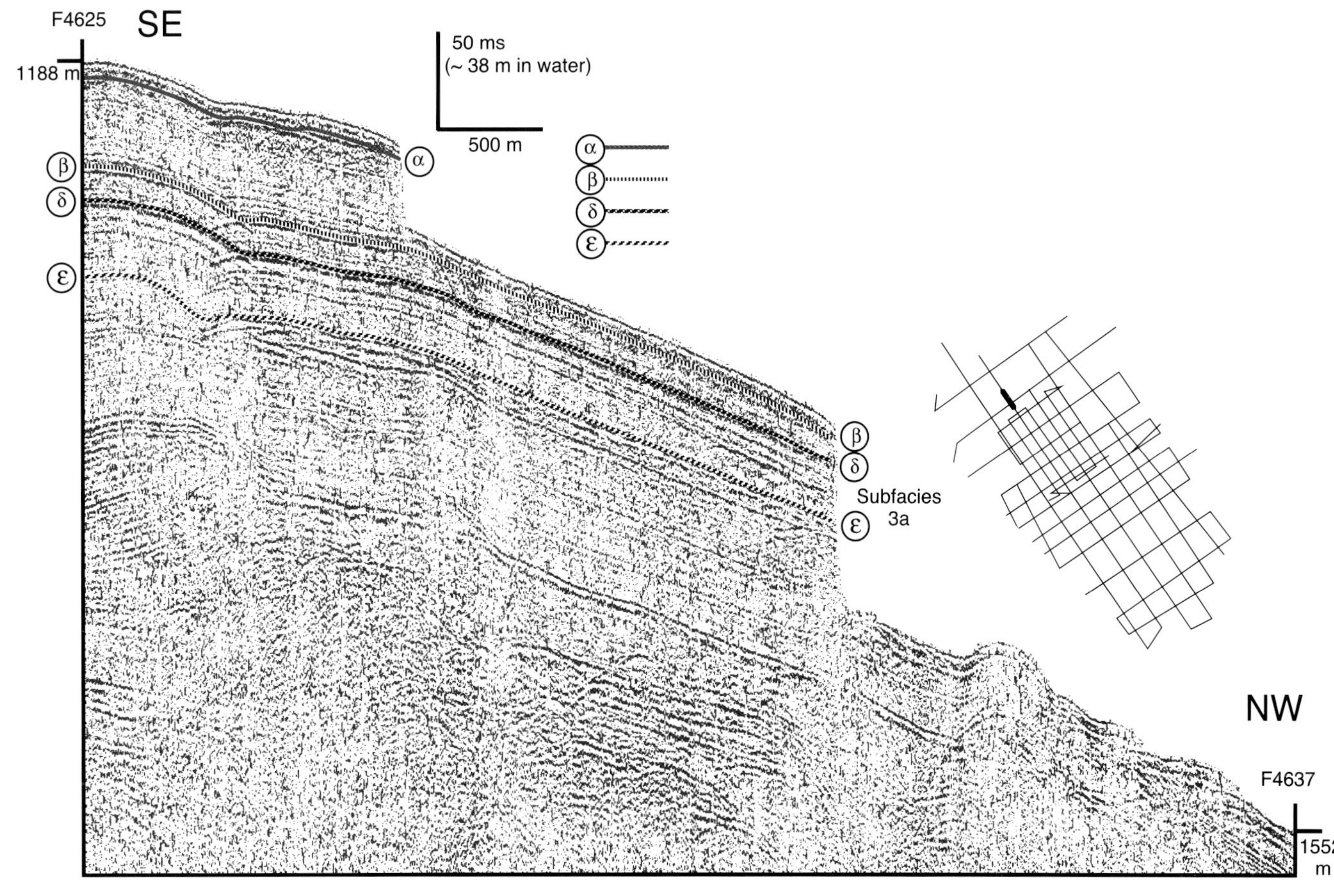

FIG. 18.—40 inch^3 airgun profile showing 30–70 m-high headwall scarps of slides on the middle slope, just landward of the thick and widespread debris-flow unit on the floor of the Baram Canyon (compare inset map location with Figure 2).

TORRES, J., SAVOYE, B., AND COCHONAT, P., 1995, The effects of Late Quaternary sea-level changes on the Rhône slope sedimentation (northwestern Mediterranean), as indicated by seismic stratigraphy: Journal of Sedimentary Research, v. B65, p. 368–387.

VAN RENSBERGEN, P., MORLEY, C.K., ANG, D.W., HOAN, T.Q., AND LAM, N.T., 1999, Structural evolution of shale diapirs from reactive rise to mud volcanism: 3D seismic data from the Baram delta, offshore Brunei Darussalam: Geological Society of London, Journal, v. 156, p. 633–650.

WINN, R.D., JR., ROBERTS, H.H., KOHL, B., FILLON, R.H., CRUX, J.A., BOUMA, A.H., AND SPERO, H.W., 1998, Upper Quaternary strata on the upper continental slope, northeast Gulf of Mexico: sequence stratigraphic model for a terrigenous shelf edge: Journal of Sedimentary Research, v. 68, p. 579–595.

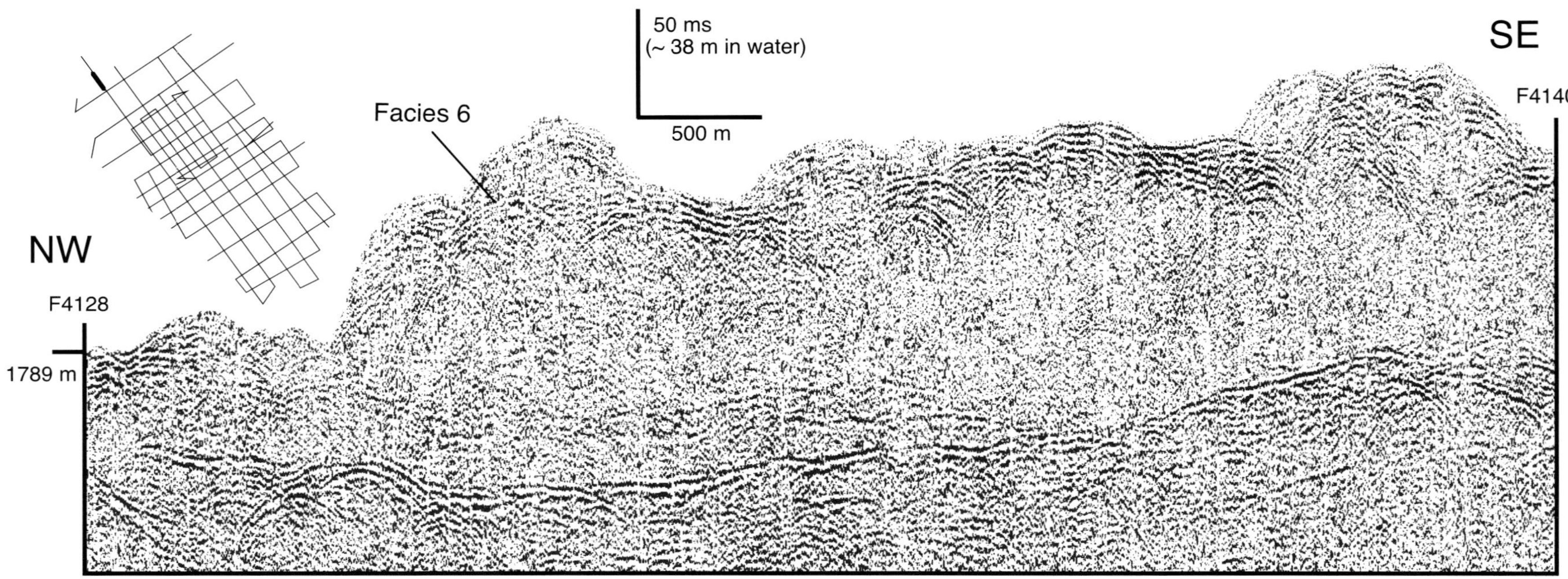

FIG. 19.—40 inch^3 airgun profile across the steep margin of an acoustically incoherent (chaotic) mega-debrite that fills the floor of the Baram Canyon and extends seaward into the Borneo Trough. Here, the deposit is > 100 m thick.

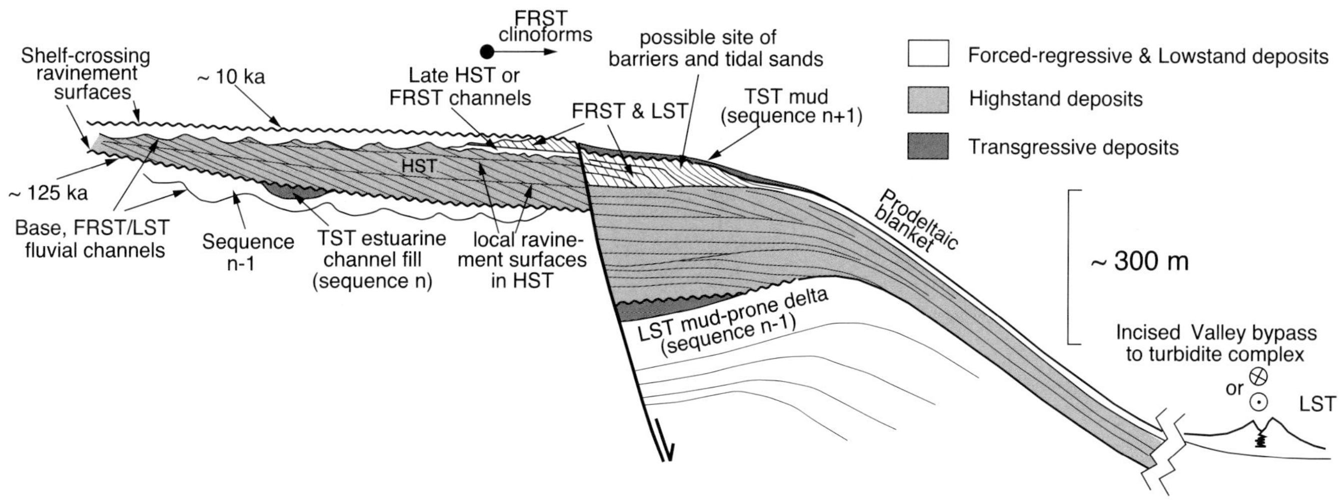

FIG. 20.—Schematic cross section of the youngest complete depositional sequence offshore Brunei (from Hiscott, 2001). FRST = forced regressive systems tract; LST = lowstand systems tract; TST = transgressive systems tract (estuarine valley fill); HST = highstand systems tract. Circles enclosing a cross and bold dot over the turbidite system in the deep-water Borneo Trough indicate turbidity-current flow in or out of the plane of the cross section. The LST mud-prone delta beneath the upper slope is shown in Figure 7.

A DEPOSITIONAL MODEL AND THE STRATIGRAPHIC DEVELOPMENT OF MODERN AND ANCIENT TIDE-DOMINATED DELTAS IN NW BORNEO

JOSEPH J. LAMBIASE
Department of Petroleum Geoscience, Universiti Brunei Darussalam, Tungku Link BE1410, Brunei
e-mail: lambiase@fos.ubd.edu.bn
ABDUL RAZAK DAMIT
Department of Geology and Petroleum Geology, University of Aberdeen, Aberdeen, AB24 3UE, United Kingdom
present address: Brunei Shell Petroleum, Seria 7082, Brunei
MICHAEL D. SIMMONS
Department of Geology and Petroleum Geology, University of Aberdeen, Aberdeen, AB24 3UE, United Kingdom
RADEN ABDOERRIAS
Department of Petroleum Geoscience, Universiti Brunei Darussalam, Tungku Link BE1410, Brunei
present address: Unocal Indonesia Company, Balikpapan, Kalimantan, Indonesia
AND
AZHAR HUSSIN
Department of Geology, University of Malaya, 50603 Kuala Lumpur, Malaysia

ABSTRACT: The modern deltas of NW Borneo have long been regarded as wave-dominated, as typified by the Baram River Delta. However, the sedimentary facies associated with several modern deltas within Brunei Bay are strongly tide-dominated. A notable example is the Trusan River Delta, which occurs within a subsiding sub-basin and has an intertidal morphology and facies distribution that indicates tidal dominance, although the shoreline geometry suggests significant fluvial influence. The succession becomes sandier upward from embayment muds through mixed sand and mud flats to tidal-channel and bar sands near mean sea level, and then fines upward to tidal sand flats and finally mud flats near the high-tide shoreline.

Tide-dominated sediments also are common in the outcropping strata of the mid-Miocene and younger Belait Formation. The Seven-Up Beach succession is interpreted as the distal ends of progradational lobes of a tide-dominated delta that coarsen upward from brackish-water mudstone through muddy and sandy tide-dominated mouth-bar deposits to tidal-channel and bar sandstones. The tidal-flat and tidal-channel and bar sandstones and interbedded tidal-flat mudstones of the Jalan Sungai Akar succession are more proximal deposits of a similar delta. Examples of distributary channels that eroded into lobe deposits during a relative sea-level fall and were backfilled during the subsequent relative sea-level rise are exposed in the Tanjong Batu succession. All three outcropping successions are strongly aggradational.

A model for tide-dominated deltas in NW Borneo derived from the modern facies distribution and outcrop stratigraphy indicates that stratigraphic architecture is controlled primarily by the interaction of tectonic and compaction-driven subsidence, a relatively high rate of sediment supply, and basin hydrodynamics. The result is a largely aggradational, sandy intertidal succession on the delta plain and muddy delta-front deposits, with aggradational stacking of individual delta lobes.

INTRODUCTION

The thick (> 10 km) mid-Miocene to Recent succession of clastic sediments that spans northern Sarawak, Brunei Darussalam, and southern Sabah is regarded as the product of three delta systems that are progressively younger to the northwest (Fig. 1; Koopman, 1996). The youngest of these, the modern Baram River Delta, has been characterized as wave-dominated primarily on the basis of its morphology (Galloway, 1975), and the presence of abundant wave-deposited shoreline sands in outcrop and in the subsurface led to the general view that all three delta systems are wave-dominated (e.g., James, 1984; Atkinson et al., 1986; Sandal, 1996). However, recent outcrop and subsurface studies have recognized that tidal deposits are an important component of all three deltas, with more than half of the sandstones deposited by tidal currents in shallow-water environments in some areas (Prosser and Carter, 1997). The three delta systems are now viewed as sedimentary complexes comprising three distinct depositional settings, each including several depositional environments.

Ongoing studies indicate that tidal sedimentation is prevalent on modern deltas within Brunei Bay (Fig. 1). One of these, the Trusan River Delta, has been studied in detail, and it consists of a variety of tide-dominated sedimentary environments. It also appears to be a modern analogue for outcropping tidal deposits, notably at Seven-Up Beach, Jalan Sungai Akar, and Tanjung Batu in coastal Brunei (Fig. 1). This paper discusses the controls on the occurrence of tide-dominated deltas in NW Borneo and examines the facies distribution of the Trusan River Delta. Predicted successions from the Trusan are compared to outcropping examples to derive a stratigraphic model for tide-dominated deltas in NW Borneo that can be used for reservoir prediction and modeling.

An important observation is that the Trusan Delta, and others on the NW Borneo margin, are tide-dominated with respect to their sedimentary features and the stratigraphic architecture of their deposits and, consequently, the signature that would be preserved in the rock record, yet their morphology varies significantly from that predicted by models derived from other tide-dominated systems, e.g., the Fly River (Harris et al., 1993) and Ord River (Wright et al., 1975) deltas. Consequently, commonly used delta classification schemes that rely heavily on surface morphology (e.g., Galloway, 1975; Coleman, 1982) to define hydrodynamic dominance and predict stratigraphic architecture are inappropriate for the NW Borneo examples. The local hydrodynamic environment is the cause of this apparent mismatch; one implication is that many models may be strictly applicable only to hydrodynamic settings that are similar to those from which they were derived.

Tropical Deltas of Southeast Asia—Sedimentology, Stratigraphy, and Petroleum Geology
SEPM Special Publication No. 76, Copyright © 2003
SEPM (Society for Sedimentary Geology), ISBN 1-56576-086-7, p. 109–123.

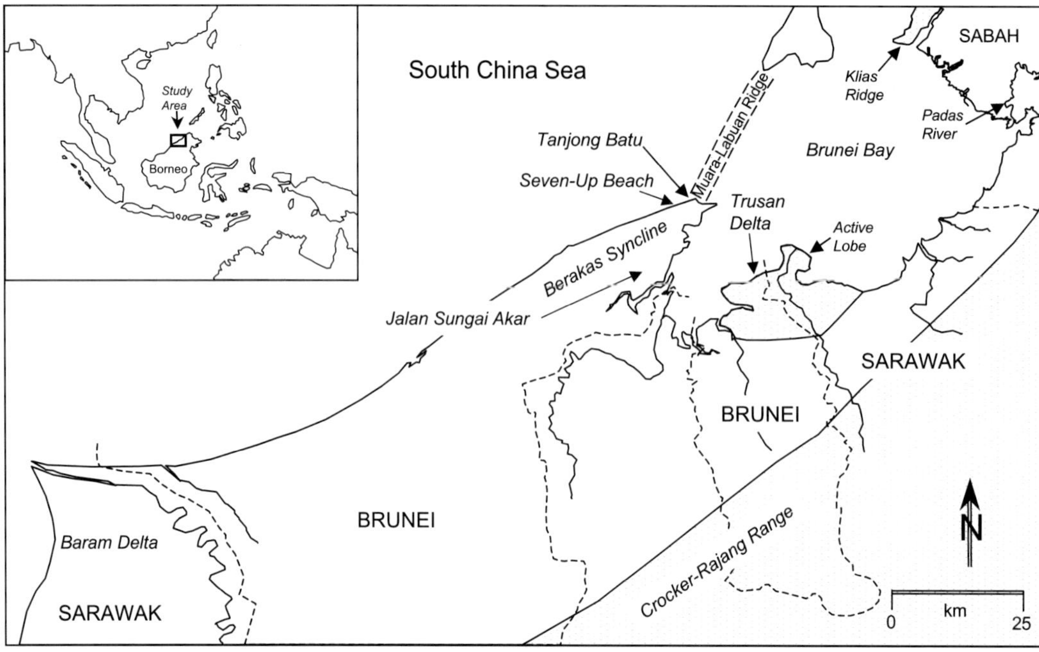

FIG. 1.—Location map of the study area in Brunei and Sarawak and the structural setting of Brunei Bay.

MODERN TIDE-DOMINATED DELTAS

Brunei Bay

There are four tide-dominated deltas on the NW Borneo margin. All four occur within Brunei Bay, where they are the locus of sedimentation. Brunei Bay is a structurally controlled coastal embayment that is bounded to the NW by the Muara–Labuan Ridge and to the SE by the foothills of the Crocker–Rajang Range (Fig. 1). The bay is an estuary in the hydrodynamic sense according to the definition of Cameron and Pritchard (1963) because it has salinities of < 30‰ throughout and is semi-enclosed by a sill at its mouth, where water depths are < 10 m. The sill apparently prevents sand from being transported into the bay from offshore, because the bay is floored by mud, except near the mouths of the major rivers. Thus, Brunei Bay is not an estuary by the Dalrymple et al. (1992) definition, which requires an offshore component of sediment supply, and the sand bodies within it are, therefore, bayhead deltas.

The South China Sea margin of NW Borneo is a low-energy coast. It is microtidal to mesotidal, with a maximum tidal range of 2.3 m and a median wave height of only 0.7 m (Schreurs, 1996a). Most of the margin appears to be wave-dominated, but the geometry of Brunei Bay shelters it from the dominant SW and NE monsoon wind–driven waves. The hydrodynamics of the bay are not well known, but observed wave heights are consistently < 0.5 m. Tidal currents have been measured in 24 locations and, generally, are capable of transporting fine sand in water depths up to 1 m above and 2 m below mean sea level, but not in greater or lesser depths. Therefore, Brunei Bay is perceived as an even lower-energy environment than the South China Sea coast, with hydrodynamics, and sedimentation, dominated by tidal currents.

Annual rainfall in the Brunei Bay catchment area ranges from 3250 to > 4000 mm and generally increases inland as a result of topography (Wee and Schreurs, 1996). Rainfall is highly variable throughout the year, although there are two distinctly wetter seasons. Discharge data are unavailable for the rivers that drain into Brunei Bay. However, the nearby Belait, Baram, and Tutong rivers all have highly variable discharge in response to similar rainfall patterns in their catchment areas (Watson, 1979; Schreurs, 1996b), and the same is expected for those rivers draining into Brunei Bay.

Brunei Bay appears to be actively subsiding because maximum water depth exceeds 50 m in the NW part of the bay despite a very high sediment transport rate in the Padas River (R. Hall, personal communication, 1999, Fig. 1). Subsidence is probably caused by shale withdrawal into the Klias Ridge, where there are active mud volcanoes. The combination of high rates of sediment supply and subsidence has implications for the sequence stratigraphic signature of the Brunei Bay successions. Accommodation space is created and filled quickly, so that most of the stratigraphy at the outcrop and reservoir scale reflects high-order events that are more commonly the product of autocyclic processes rather than eustasy.

Trusan River Delta

The Trusan River drains an area of approximately 1700 km^2 and has deposited a delta with an area of approximately 400 km^2. The delta consists of an active lobe and several abandoned lobes and has a shoreline morphology similar to that of a fluvial-dominated delta (Fig. 2; Galloway, 1975). It has a single distributary channel system (Coleman and Wright, 1975) that averages 5 to 8 m deep, and there are abandoned distributary channels on the delta plain. However, mangrove swamps, rather than bays, occupy the interdistributary areas, and nearly the entire delta consists of marginal marine sedimentary environments (Fig. 2). The active lobe of the Trusan Delta has an area of approximately 14 km^2 and includes several intertidal and shallow subtidal sedimentary environments (Fig. 3). As discussed below, the lobe is interpreted as a complex of abandoned small-scale distributary channels and mouth bars, plus a mouth bar that is associated with the active distributary channel.

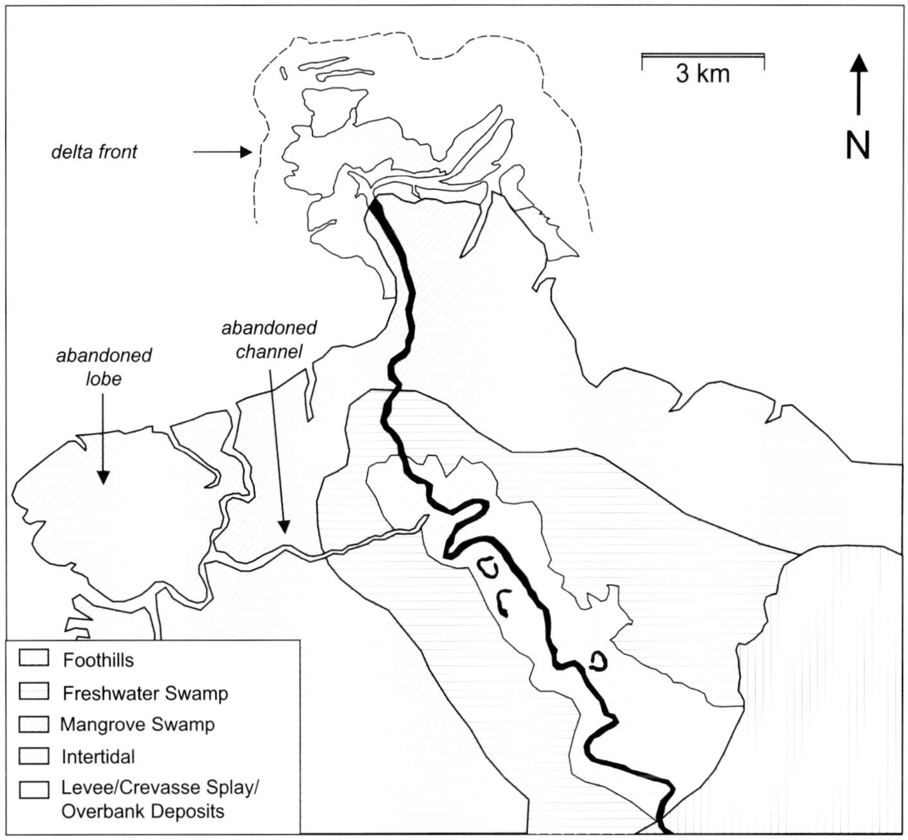

FIG. 2.—Morphology and sedimentary environments of the Trusan Delta. The boundary between the freshwater and mangrove swamps is approximate.

Facies distribution.—

The intertidal and subtidal lithofacies on the Trusan Delta were investigated with surface observations, trenches, and 36 shallow (0.5 m) cores. Generally, the coarsest sediments occur near the mean tide line near the seaward edge of the delta plain, with a gradual fining upward across the delta plain to a supratidal mangrove swamp as well as a fining downward on the delta front into pro-delta Brunei Bay muds. All of the various facies are burrowed to some degree by marine organisms, and marine microorganisms occur in all the fine-grained facies. The facies distribution within the different parts of the delta is as follows:

Delta plain.—The delta plain extends for approximately 10 km along the lower course of the Trusan River to the delta front and includes a supratidal swamp and several intertidal environments. The supratidal swamp is covered by mangrove, dominated by *Rhizophora* species, except for areas adjacent to the distributary channel that are lined by *Nipa* palm. The substrate is very dark gray mud with abundant plant debris and is highly bioturbated by a dense network of mangrove roots and crustacean burrows. A narrow, 0.3-m-thick beach of fine sand occurs on the eastern shoreline of the delta (Fig. 3). In a few locations, plant debris up to a meter thick covers the beach. There are no beaches on the western flank of the delta.

The intertidal zone extends seaward from the shoreline for a distance of approximately 4 km almost to the top of delta front (Fig. 3). Over that distance, there is an elevation drop of 2 m and the delta front slopes into the basin at an average of 1°. Near the shoreline, tidal flats consist of dark brownish-gray clay with very thin sandy or silty streaks and starved, low-relief ripples. Plant debris is very abundant and locally forms discontinuous layers up to 1 cm thick; rootlets are common in some areas. The muds are highly to very highly bioturbated; burrows range in diameter from a few millimeters to 2 cm and are unlined (*Thalassinoides*, *Chondrites* type). Many larger horizontal burrows are flattened and exhibit faint, concave-downward concentric laminae (*Teichichnus* type).

The mud flats grade seaward into an area of mixed flats that represent abandoned mouth bars and consist of interbedded very fine to fine sand and mud (Fig. 4). Sand beds thicken from < 1 cm to > 15 cm seaward, and mud-layer thickness varies inversely. Sandy areas are covered by straight-crested wave ripples and small current ripples, commonly with thick mud drapes; the fine sands are ripple cross-laminated with local mud and carbonaceous laminae. Interbedded muds often contain carbonaceous layers up to 1 cm thick; starved ripples are common on the surfaces of muddy areas. Plant fragments are common, and in places the content of plant debris is so high that the mud has a peat-like appearance. Bioturbation intensity is relatively high, with a wide variety of burrows, ranging from 5 to 20 mm in diameter and including *Thalassinoides*, *Teichichnus*, and *Chondrites* types, and occasional *Cylindrichnus* type. In muddy areas, burrows are unlined, whereas organic-lined vertical tubes are very common in the sandy areas, although sand-lined tubes are rare.

An extensive area of sand flats lies seaward of the mixed flats (Fig. 3). These generally consist of very fine to fine sand and are

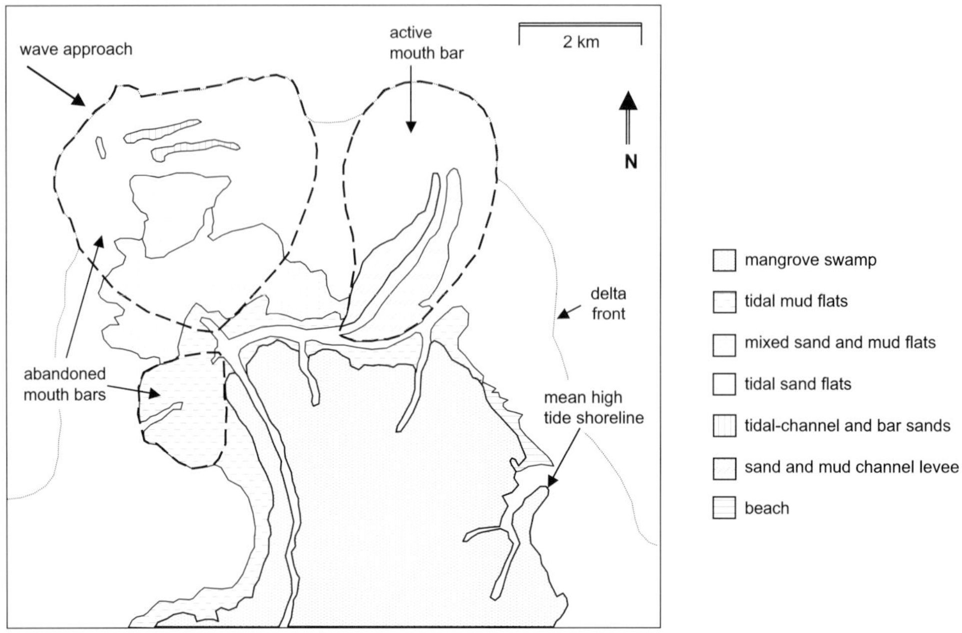

FIG. 3.—Intertidal facies distribution on the active lobe of the Trusan Delta.

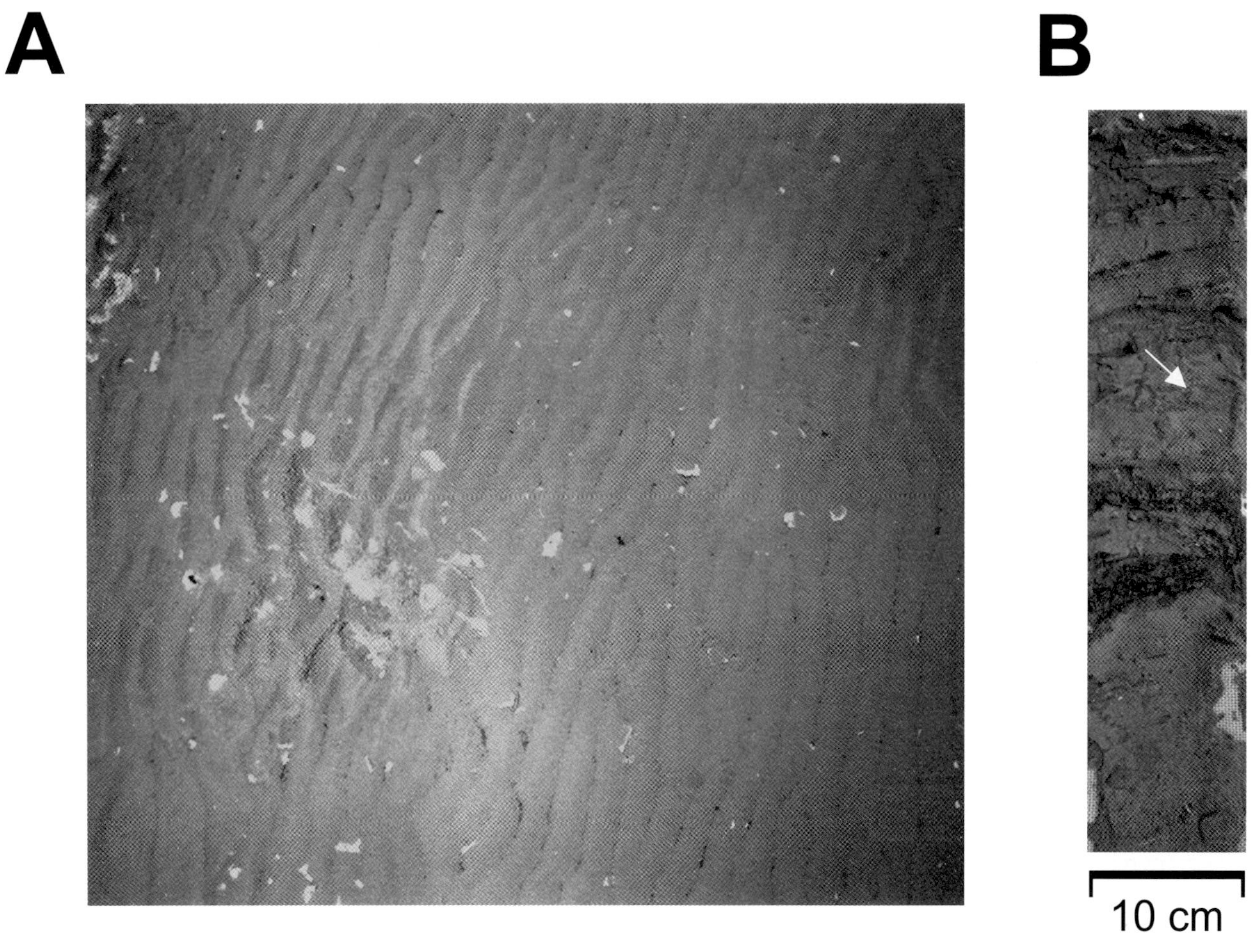

FIG. 4.—Mixed tidal flats on the Trusan Delta. **A)** Thick mud drapes on straight-crested ripples with an approximate wavelength of 0.1 m. **B)** Cross-laminated fine sand with mud and organic drapes in core. The arrow points to an *Ophiomorpha* burrow.

covered by straight-crested wave ripples and straight-crested and linguoid current ripples. Rill marks, ladder ripples, and double-crested ripples are common, and shallow depressions are mud draped; mud drapes are more widespread after periods of heavy runoff (Fig. 5). The sands are cross-bedded with set thicknesses of 0.1–0.3 m and ripple cross-laminated with local flasers, mud clasts, and thin (1–2 cm) mud layers. Mollusc shells and plant fragments are common. Relative bioturbation intensity is moderate to very high. Many of the burrows are small and unlined, including *Planolites*, *Arenicolites*, and *Skolithos*, whereas larger, mud-lined traces, such as *Ophiomorpha*, also are common. Clusters of vertical or subvertical tubes lined by organic and/or shell detritus or sand grains are common.

Several long, slightly lunate sand bars are found at the seaward edge of the delta plain (Figs. 3, 6). They are 1 to 2 km long and 200 to 300 m wide and oriented parallel to the shoreline. Swash-formed flat beds and wave ripples dominate the windward side, and the crest and leeward side are usually covered by ebb-oriented current ripples, and small three-dimensional lunate dunes during spring tide. Sedimentary structures are limited to parallel lamination and thin flasers; burrows are not common and are mud-lined.

Interbedded trough cross-bedded sands and muds with common plant fragments floor the active distributary channel. The sands have very sharp bases and tops; some are ripple cross-laminated, and mud drapes are common. Burrows are limited to a few small horizontal and subvertical *Chondrites* type and larger, but less common, *Thalassinoides*. Thin (less than 1 to 10 cm) interbedded sands and muds form the active distributary channel mouth bar and levees. The sands are wavy bedded and ripple cross-laminated with occasional flasers, and plant fragments are common in the muds.

Delta front.—From the top of the delta front, there is a rapid downslope facies change from sand to mud such that virtually all the delta-front sediments more than 1.5 m below mean low water are mud. Generally, the mud is slightly compacted and dark gray with thin (1–2 mm) streaks or lenses of silt or very fine sand usually spaced a few centimeters apart (Fig. 7A). In places, the muds have alternating darker and lighter layers indicating higher and lower organic content, respectively, and there are some mottled, silty and sandy intervals. Plant fragments are common, and mollusc shells occur occasionally. Bioturbation intensity is low to moderate, and burrows are mostly horizontal and unlined, with subvertical tubes less common. Most burrows are small with a diameter of 2 to 4 mm, although larger burrows (15 to 20 mm) do occur. Some small tubes form clusters and may be a single branching network of *Chondrites* type.

In some areas, there are intervals within the muds that have relatively common thin (0.5 to 3 cm) beds and laminae (< 0.5 cm) of fine light yellowish-brown sand and/or silt (Fig. 7B). The thicker sand beds typically have sharp bases and undulating or rippled tops. Wavy and ripple cross lamination is common, and some sand beds are normally graded. Plant fragments are often found and, more rarely, thin mud drapes and small mud clasts. Bioturbation intensity is relatively low to moderate and consists mostly of unlined, sand-filled semicircular horizontal tubes 3 to 5 mm in diameter and occasional

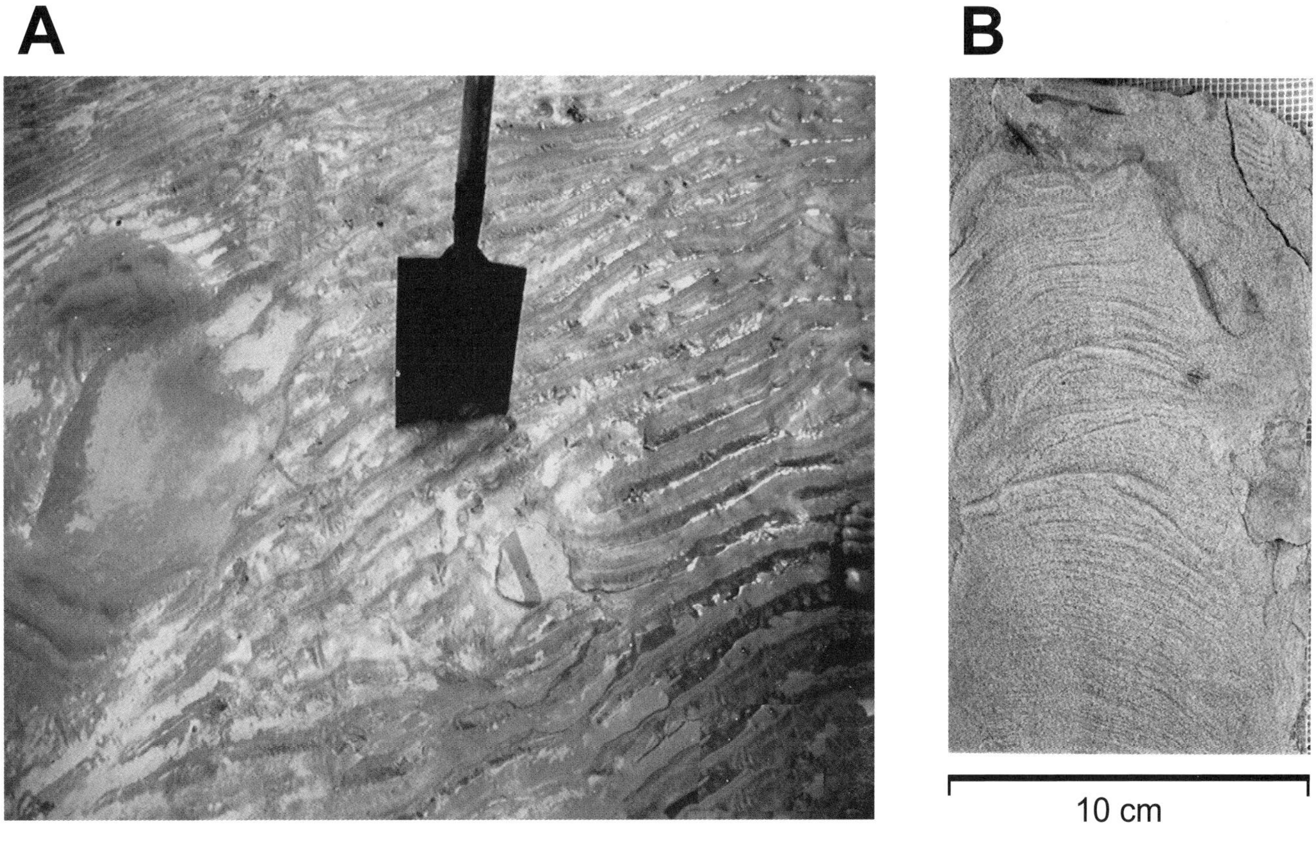

FIG. 5.—Sandy tidal flats on the Trusan Delta. **A)** Straight-crested current ripples with mud drapes. **B)** Ripple cross-laminated and parallel-laminated fine sand with a single mud flaser near the top of the core. The distorted bedding is a product of the coring process.

FIG. 6.—**A)** Lunate dunes with superimposed current ripples covering the surface of a sand bar at the seaward edge of the delta plain. **B)** Cross-laminated and parallel-laminated sand in a core from the same environment.

larger tubes (up to 2 cm). Small, vertical sand-filled tubes terminate at the base of some sand beds. V-shaped depressions on the slope, 50–100 m wide, suggest gravity-induced failure (Coleman, 1982; Hoekstra, 1993).

Prodelta.—Prodelta sediments occur below the base of the delta front in 20–50 m of water and consist of unconsolidated to moderately compacted, dark gray mud with abundant plant fragments.

Depositional Model.—

The morphology and facies distribution on the active lobe of the Trusan River Delta suggest that sediments assume their ultimate character after deposition by fluvial processes and subsequent reworking by tidal currents and, to a minor extent, waves. Sediment supplied to the active delta lobe by the Trusan River accumulates in a relatively small, lobate mouth bar and levees associated with the active distributary channel (Fig. 3). Observations suggest that the distributary channel carries sand only when the river is in a relatively high flow stage because it is floored by mud at other times. The deposits are mostly mud, with sand and mixed sand facies restricted to the levees and water depths of less than 2 m below mean sea level on the mouth bar. Sand does not penetrate farther basinward because the low-energy hydrodynamic environment of Brunei Bay cannot transport sand into greater water depths. Also, slumping does not appear to be a significant process for distributing sand into deeper water; no sand was recovered in bottom samples from the delta front adjacent to the active mouth bar, and all slump scars noted occur exclusively within mud. Consequently, sand is trapped on the lower delta plain so that the mouth bar is essentially a mound of mud capped by a relatively thin (~ 3 m) veneer of sand and mixed sand and mud.

As sand continues to be supplied to the delta and trapped in shallow water, it lowers the gradient of the distributary channel by quickly filling some of the small available accommodation space and precipitates small-scale channel switching on the surface of the active lobe and abandonment of the mouth bar. Indeed, the lobe apparently is constructed from the successively abandoned mouth bars as the distributary channel system swings back and forth across its surface. Two abandoned mouth-bar and channel systems are easily recognized on the active lobe; the abandonment was caused by channel switching that occurred, in both cases, just seaward of the mean-high-water shoreline (Fig. 3). Deposition of the active mouth bar apparently began immediately after the younger of the two lobes was abandoned.

Sediment deposited by fluvial processes on mouth bars is reworked after abandonment by tidal currents and, to a lesser degree, by waves. The effect of waves is seen exclusively in the part of the delta front that faces the NW direction of wave approach (Fig. 3). There, the outer edge of the large abandoned mouth bar is the site of the wave ripples and swash-formed flat beds noted above, plus the morphology of the bar appears modified by waves. In contrast to the elongate shape of the active mouth bar, the abandoned mouth bar has a broad shape with a straight seaward edge, nearly perpendicular to wave approach, that probably results from wave reworking.

Tidal currents have a significant effect on the sediment of the active lobe of the Trusan Delta. The channel–levee topography of the active mouth bar is maintained by ongoing deposition whereas levees on the abandoned mouth bar systems have been modified and are no longer topographic features; their associated channels are nearly filled with sediment (Fig. 3). However, with ongoing subsidence, there apparently is insufficient time for the relatively weak tidal currents to rework sand into the shoreline-perpendicular bars commonly observed on other tide-dominated deltas, such as the Mahakam Delta (Allen et al., 1979).

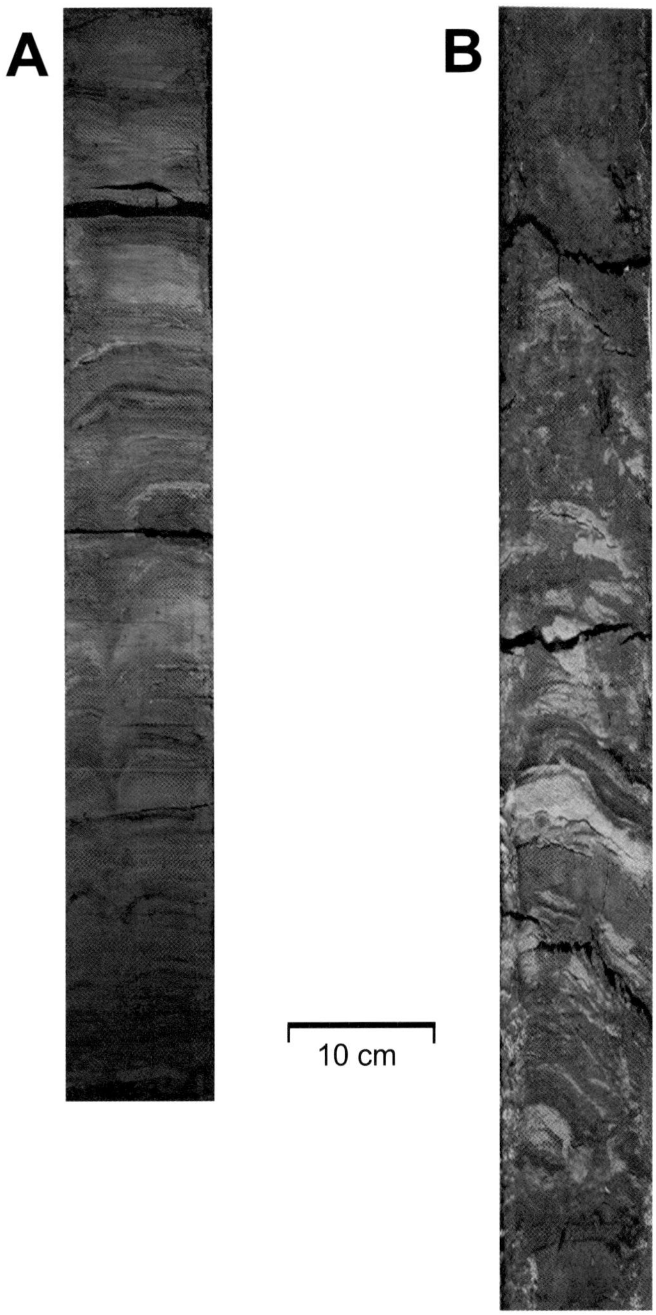

FIG. 7.—Moderately consolidated delta front mud with **A)** thin sand streaks and plant fragments; there are several unlined, mud-filled burrows of *Planolites* and *Chondrites* type; **B)** thin sand beds with wavy or ripple cross-lamination and unlined, mud-filled traces.

More significantly, the lithofacies distribution on all but the active mouth bar no longer reflects that of channel-levee systems but has been reworked into a single, typically fining-shoreward, tidal shoreline (e.g., Dalrymple, 1992; Fig. 3). Combined with the sedimentary structures that indicate tidal deposition and the occurrence of marine burrowing organisms, as described above, this gives the sediments of the active lobe an overall tidal character that would be the dominant signature preserved in the rock record. The tidal character of the Trusan Delta deposits suggests, therefore, that shoreline geometry is not necessarily indicative of hydrodynamic dominance.

Depositional lobes are abandoned by switching of the river's single distributary channel system on the inner delta plain (Fig. 2). The morphology of the delta is a product of deposition and then abandonment of successive lobes; the abandoned lobes are then covered by mangrove swamps. Some sediment accumulates in levees and fills the oxbow lakes that are common on the inner delta plain and crevasse-splay, and overbank deposits are expected in the freshwater swamps that border the distributary channel (Fig. 2).

Stratigraphic Succession.—

The facies distribution on the Trusan Delta indicates that the stratigraphic succession generated by progradation of a delta lobe would be heavily dominated by mudstone. Out of a total uncompacted thickness of approximately 25 m, the sandy and mixed sand and mud facies are limited to a unit 3.5 m thick that straddles the mean tide line, with clean sands comprising only about 2 m (Fig. 8). The overall succession coarsens upward from prodelta muds through mixed delta-front sediments and sands deposited at the seaward edge of the delta plain. Overlying deposits then fine upward through sandy, mixed, and muddy tidal-flat sediments to the mean high tide shoreline (Fig. 8).

The very high proportion of mud in the succession illustrated in Figure 8 suggests that synsedimentary compaction is likely to be relatively fast. When this compaction is combined with the active tectonic subsidence of Brunei Bay, accommodation space probably is generated relatively quickly. This introduces a strong aggradational component to the stratigraphic development, which causes significant stacking of the various facies, as is seen in outcropping examples of tide-dominated deltas and discussed below. Consequently, the complete succession of a delta lobe is expected to be several times thicker than that generated by simple progradation.

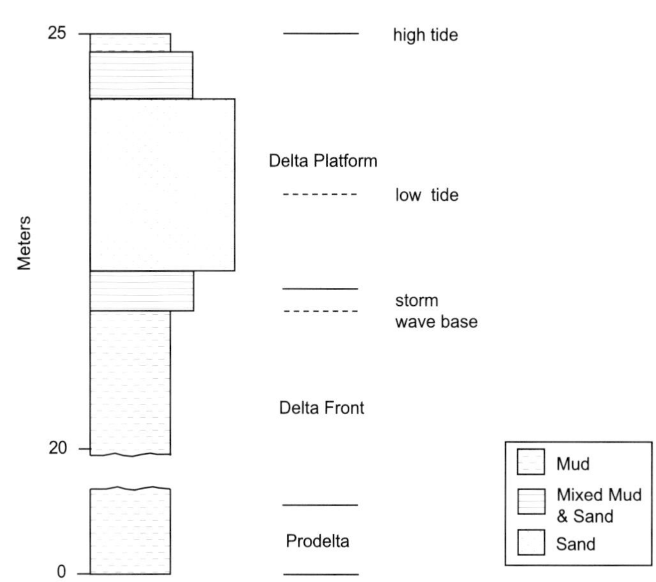

FIG. 8.—Predicted stratigraphy from the progradation of one delta lobe on the Trusan Delta.

OUTCROP EXAMPLES

A number of outcrops in Brunei and adjacent Malaysia include tidal sediments in the middle Miocene and younger Belait Formation and equivalents. Generally, these consist of cycles of mudstones and tidal sandstones that are interpreted to have been deposited in coastal embayments, probably on or near tide-dominated deltas. The outcropping strata include examples that bear close resemblance to expected successions for the distal and proximal parts of delta lobes, and also for distributary channels.

Distal Delta Lobes

The middle Miocene Seven-Up Beach succession crops out along the Brunei coast on the eastern limb of the Berakas Syncline (Fig.1). Approximately 400 m of steeply dipping, interbedded fine- to medium-grained sandstones and shales are exposed in a succession that is best characterized as 11 coarsening-upward sedimentary cycles that vary in thickness from 9 to 53 m (Fig. 9). Each cycle begins with a shale, which grades upward into a unit of interbedded thin mudstones and sand-

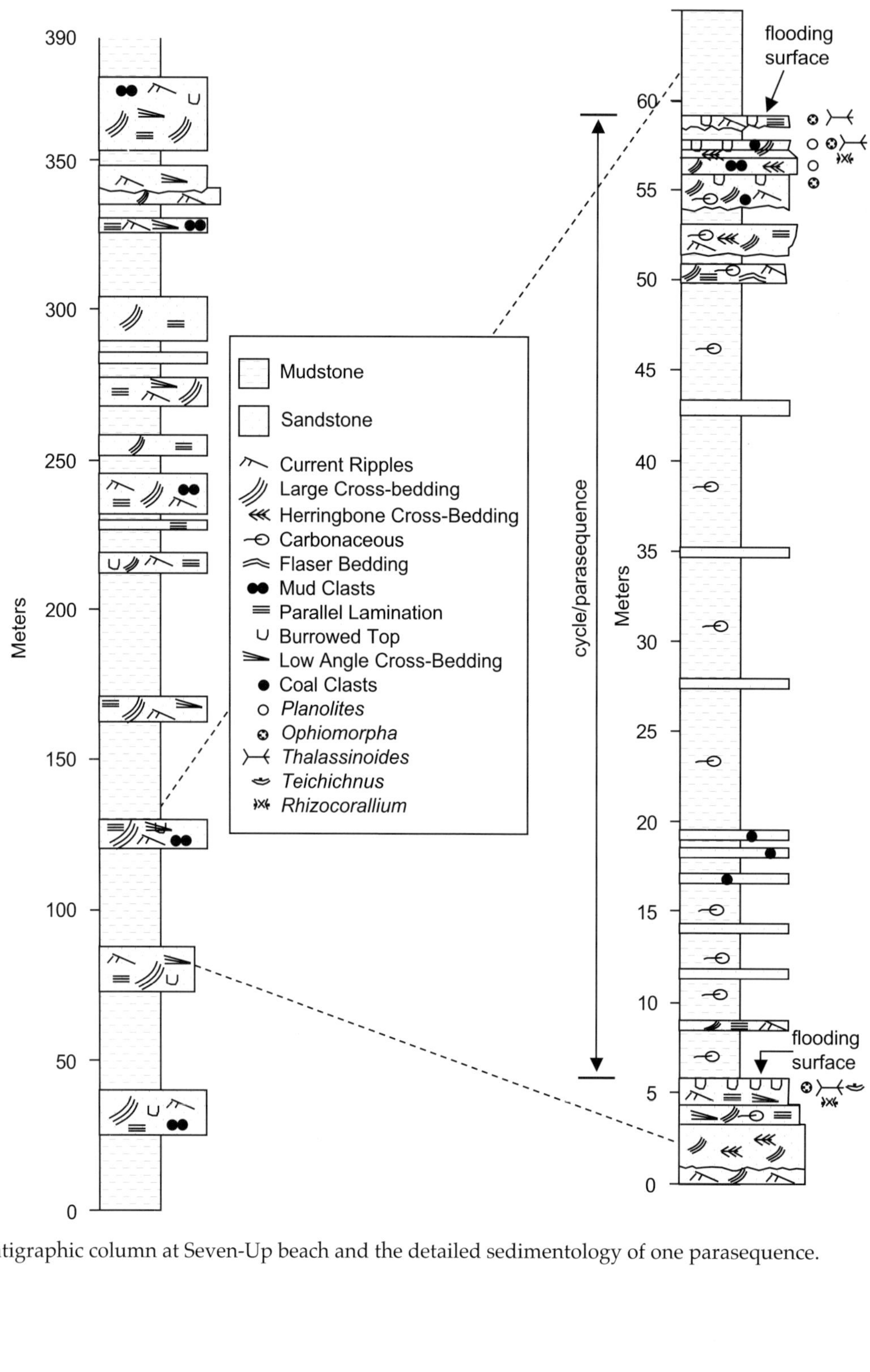

FIG. 9.—The stratigraphic column at Seven-Up beach and the detailed sedimentology of one parasequence.

stones up to several meters thick and then into several more meters of sandstone (Fig. 9). The top of each cycle is an abrupt contact between the top of the sandstone and the overlying mudstone at the base of the succeeding cycle.

Facies Distribution.—

Mudstones from the Seven-Up Beach succession contain a microfossil assemblage that includes coarse *Trochammina*, *Recurvoides*, and *Miliammina* with abundant mangrove pollen and is indicative of a brackish-water environment (Simmons et al., 1999). The mudstones are interpreted to be coastal embayment deposits rather than lagoonal sediments, because of their thickness and their stratigraphic relationship with the other facies. (See the discussion below.)

The cycle illustrated in Figure 9 typifies the depositional history of the succession. It begins with 42 m of embayment mudstone containing a few thin, highly bioturbated sandstone interbeds, which are interpreted to be storm beds. This facies grades into 4.5 m of interbedded thin, highly bioturbated sandstones and carbonaceous mudstones. The sandstones are ripple cross-laminated with abundant mud drapes and flaser bedding near the base, and with an upward increase in rare herringbone and large-scale tabular cross-bed sets (set thickness up to 0.2 m), which are also mud draped (Fig. 10). Coal clasts and plant fragments are common. The unit is interpreted to be a tidal sand bar deposit. Palynological studies of a few of the thin (1–2 cm) mudstones reveals that they all contain a *Dipterocarpus* forest-derived assemblage, suggesting that they were deposited during a flooding event on a nearby river.

The succeeding 3.5 m of strata were deposited in tidal channels. Channel sands are relatively thick (1 m) with erosional bases, and there is little interbedded mudstone and few mud drapes; mud rip-up clasts are common locally (Fig. 11). Large-scale cross-stratification is common, as is lateral accretion. Bioturbation intensity is moderate, with the ichnofossil assemblage including *Planolites*, *Ophiomorpha*, *Thalassinoides*, and *Rhizocorallium*.

The sands in the top 0.5 m of the cycle form one bed that is transitional from the underlying sands and contains no mud (Fig. 12A). Trough cross-bedding is common, as is low-angle cross-bedding, wave ripples, and coalified wood clasts up to 0.2 m in diameter; equivalent beds at the top of other cycles often are swaly bedded. There is a very sharp contact at the top of the bed

Fig. 10.—Ripple cross-laminated sands with mud drapes at Seven-Up Beach. Lens cap is 7.5 cm in diameter.

Fig. 11.—Cross-bedded tidal channel sandstones with mud pebble laminae at Seven-Up Beach. Lens cap is 7.5 cm in diameter.

A

B

Fig. 12.—**A)** Trough cross-bedded sandstone with a wave-rippled surface at the top of a parasequence. **B)** A heavily burrowed surface at the top of another parasequence. Both A and B are from the Seven-Up Beach succession. The burrows are dominantly *Thalassinoides* and *Ophiomorpha*, and the lens cap is 7.5 cm in diameter.

with an overlying mudstone that forms the base of the next cycle. The bed top is very heavily burrowed with *Thalassinoides* and *Rhizocorallium* (Fig. 12B). *Planolites* and *Ophiomorpha* also are abundant on the bed top, with the latter penetrating downward as much as 0.25 m. All of the burrowing apparently occurred during deposition of the overlying mudstone. The wave-generated sedimentary structures with no mud and syndepositional burrowing indicate a high-energy wave-dominated environment. Consequently, the bed is interpreted as a sand bar at the outer edge of a delta plain.

Stratigraphic Development.—

The vertical facies distribution of all eleven cycles in the Seven-Up Beach succession is similar to that described above. It represents a progressive change in depositional environment that begins with embayment mudstones and shallows through tide-dominated environments, often to a wave-dominated clean sandstone, and constitutes a progradational parasequence that is capped by a flooding surface (Fig. 9). Generally, the parasequences become thinner and proportionately more sandy upward so that the whole succession is a progradational parasequence set; the thinning and increase in sand is achieved by thinning of the embayment mudstones at the base of each parasequence.

The total thickness of nearly all the sandstone and mixed sandstone and mudstone facies varies between 8 and 15 m in each cycle (Fig. 9), which is significantly thicker than the water-depth range of approximately 3 m for the same facies on the Trusan Delta (Fig. 8). It is likely that the hydrodynamics of Miocene coastal embayments in NW Borneo were essentially equivalent to the hydrodynamics in their modern counterparts because the size and shape of the South China Sea have not changed significantly (Lee and Lawver, 1995). Consequently, waves probably had little influence on sedimentation in Miocene embayments because their size and approach were similar to the present day. Tidal range may have varied somewhat from embayment to embayment as a function of local geometry, but probably not enough to account for a significantly larger range of water depth for the intertidal facies, suggesting that the thick intertidal succession at Seven-Up Beach is strongly aggradational. This is supported by the fact that the tidal-channel facies in each parasequence consists of stacked 1-m-thick channel sandstones. In the Seven-Up Beach succession, individual facies can be as much as 3–5 times as thick as their modern equivalents.

Each parasequence in the Seven-Up Beach succession is interpreted as the distal end of a prograding, tide-dominated delta lobe within a coastal embayment. The observed stratigraphy at Seven-Up Beach is the same as that predicted for the prodelta and delta-front environments on the Trusan Delta. The basal mudstones are carbonaceous, plant fragments and coal clasts occur in the same facies, and microfossils and, to a lesser extent, ichnofossils indicate shallow, brackish-water environments for the Seven-Up Beach succession. Sedimentary structures and ichnofossils in the sandy facies are consistent with a coarsening-upward succession representing upper delta-front and outer delta-plain environments. The presence of thick embayment mudstones, plus the strongly aggradational character of the succession, suggests that the depositional setting of the Seven-Up Delta was an actively subsiding sub-basin at the time of deposition, much as Brunei Bay is today.

Proximal Delta Lobes

The Belait Formation also crops out at Jalan Sungai Akar (Fig. 1). Within the succession are 60 m of interbedded sandstones and mudstones that become sandier upward and form ten cycles. The individual cycles fine upward and range in thickness from 3 to 15 m (Fig. 13).

Facies Distribution.—

Generally, the facies at the base of each cycle is tabular and trough cross-bedded fine- to medium-grained sandstone. Cross-bed sets are up to 30 cm thick, and many of the sandstones consist of stacked channels, 0.5–1 m thick and 20–30 m wide, with lateral accretion surfaces and erosional bases (Fig. 14). Mud drapes and flasers are rare, and the channel sandstones are not appreciably bioturbated. The sandstones are interpreted as tidal-channel and bar deposits that gradually pass upward into ripple cross-laminated, fine-grained sandstones with common mud drapes and flaser bedding and occasional wave-rippled tops of sandy tidal-

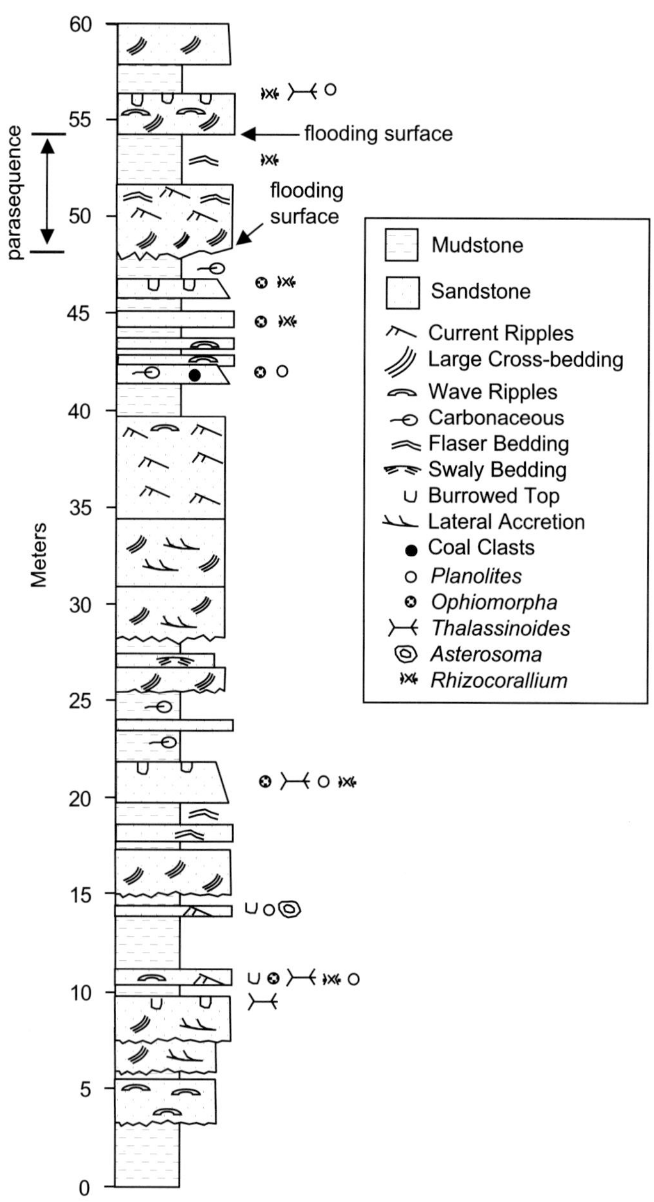

FIG. 13.—Stratigraphic log of delta-plain parasequences in the Jalan Sungai Akar succession.

FIG. 14.—Stacked tidal-channel and bar sandstones passing upward (toward the right of the photo) into tidal-flat sandstones and then tidal-flat mudstones at Jalan Sungai Akar. The overlying stacked tidal-channel and bar succession at the top right of the photo forms the base of next fining-upward parasequence. Hammer (center of photo) for scale.

flat origin. Bioturbation is moderate to intense in the tidal-flat sandstones, with a variety of trace fossils, including abundant *Ophiomorpha* and *Rhizocorallium*, common *Thalassinoides* and *Planolites*, and less common *Asterosoma*.

The tidal-flat sandstones become muddier upward and gradually pass into laminated mudstones that mark the top of a parasequence and that also were deposited on tidal flats (Fig. 14). Plant fragments are common in the mudstones, and bioturbation varies from moderate to intense. Individual burrows are often difficult to identify, although *Ophiomorpha* is relatively common and *Rhizocorallium*, *Thalassinoides*, and *Planolites* all occur.

There is a fine-grained sandstone 1 m thick at the 27 m level of the Jalan Sungai Akar succession that is swaly cross-stratified (Fig. 13). There are no mud drapes or flasers in this sandstone, and it is not bioturbated except for a few trace fossils on its upper surface. The bed clearly is a wave deposit and is interpreted as a subtidal, upper-shoreface sandstone.

Stratigraphic Succession.—

Each cycle in the Jalan Sungai Akar succession corresponds to a progradational cycle on a tide-dominated delta plain. The transition from tidal-channel and bar sediments, and in one case upper-shoreface sandstones, through sandy tidal-flat and muddy tidal-flat deposits represents a shallowing from environments at the seaward edge of the delta plain to the high-tide shoreline. The thickness of the individual facies suggests that most of the cycles have a strong aggradational component. As in the Seven-Up Beach succession, the tidal-channel and bar facies at Jalan Sungai Akar includes stacked channels, and the tidal-flat facies also are several times thicker than their modern equivalents (Fig. 13).

The cycles in the Jalan Sungai Akar succession are parasequences according to the Van Wagoner et al. (1990) definition because each represents a conformable succession of genetically related strata bounded by flooding surfaces. The fining-upward character of the parasequences causes the flooding surfaces to be recorded as erosional contacts between tidal-flat mudstones and the overlying tidal-channel sandstones at the base of the succeeding parasequence (Figs. 13, 14). Similarly, the overall upward increase in sandstone abundance through the exposed strata reflects a progressively more seaward position and indicates that the Jalan Sungai Akar succession is a retrogradational parasequence set deposited on the proximal part of a tide-dominated delta. A progradational succession from equivalent environments would have an overall fining-upward character due to an increasingly higher proportion of tidal-flat mudstones toward the top of a parasequence set.

Distributary Channels

The middle Miocene outcrops at Tanjung Batu (Fig. 1) consist of a complex of tidal-flat and tidal-channel deposits that are cut by large channels. Individual large channels cut into the underlying strata as much as 15 m and can be at least 40 m wide.

Facies Distribution.—

The large, incised channels are filled mostly with trough cross-bedded fine to medium sandstone that occurs as stacked small channels and bars, 1 to 2 m thick, with lateral accretion surfaces (Fig. 15). In the example illustrated in Figure 16, stacked channel sandstones account for nearly all the fill of the large channel except for 2 m of mudstones with thin interbedded rippled sandstones near the base. The thin sandstones have erosional bases, and some of the ripples are mud draped. Trace fossils are restricted to rare *Planolites* in the mudstone unit.

The channel-fill sandstones represent the basal part of a fining-upward succession that continues above the channel top with 7 m of very thin interbedded fine rippled sandstones and mudstones with a 2 m interbed of mudstone (Fig. 16). This is followed sequentially by a thin channel sandstone with an erosional base, 5 m of homogeneous claystone, 2 m of coaly claystone, and 6 m of coal. Palynological analysis of the claystone, coaly claystone, and coal indicates an upward increase in mangrove pollen from near zero in the lower part of the claystone to 99% in the coal.

Stratigraphic Development.—

The deeply incised channels apparently were cut during a relative sea-level fall. They indicate erosion to a depth of up to 15 m into tidal sediments that were deposited near a tide-dominated shoreline, and are, therefore, similar to drowned, abandoned distributary channels on the Trusan Delta that are up to 20 m deep. The abandoned Trusan channels are more than 10 m deeper than the active distributary channel and probably were cut during the 5000–2000 yr BP relative sea-level fall recorded on the nearby Baram Delta (Caline and Huong, 1992). The channels are presently accumulating mud, suggesting that the present-day energy regime is significantly lower than when they were active.

The channels outcropping at Tanjong Batu were backfilled with fluvial and tidal sediments during the subsequent relative sea-level rise. In the example illustrated in Figure 16, the cross-bedded sandstones at the base of the channel fill and the overlying mudstone are interpreted to be fluvial channel sands and overbank deposits, respectively, because trace fossils are simple and rare and no marine microfossils were recovered. The sandstones and mudstones probably were deposited soon after the

FIG. 15.—Cross-bedded sandstone with lateral accretion surfaces filling the large channel at Tanjong Batu. The top of the succession is to the right of the photo, and the lens cap is 7.5 cm in diameter.

channel was incised, the sandstones by an active fluvial channel and the mudstones by overbank processes following later channel abandonment.

The overlying cross-bedded sandstones were deposited in small channels within the larger channel cut. These shallow, relatively broad channels with significant lateral accretion are much more similar in shape and character to observed modern and ancient tidal channels than they are to fluvial channels seen in outcrop and in modern rivers in NW Borneo. Also, the base of the unit has an erosional contact with the underlying mudstones and passes conformably into tidal strata above (Fig. 16). Consequently, the thick cross-bedded sandstones are interpreted as tidal-channel deposits that aggraded during a relative sea-level rise. As sea level continued to rise, the incised channel eventually was filled, and tidal-current speed decreased because currents were no longer confined by the incised geometry. This resulted in tidal-flat sedimentation, as indicated by the very thin interbedded sandstones and mudstones that immediately overlie the cross-bedded channel-fill sandstones. As the rate of relative sea-level rise slowed, the area was colonized by mangroves, which led to the deposition of the swamp claystones and mangrove coal that cap the succession (Fig. 16; Lambiase et al., in press).

The entire succession from the basal fill of the large channel through the coal was deposited during a continuous relative sea-level rise that probably was generated by subsidence of an abandoned distributary channel. Consequently, the fining-upward succession of fluvial, tidal, and paludal deposits is the product of an autocyclic process and is expected to be common in tide-dominated delta successions. Conversely, abandoned distributary channels are expected to be mud-filled whenever a rapid relative sea-level rise occurs.

Stratigraphic Model

Integration of the modern facies distribution and depositional model for the Trusan Delta with the stratigraphic architecture of the outcrop examples allows construction of a model for the stratigraphic development of tide-dominated-delta deposits in NW Borneo. Although the tectonics, morphology, and hydrodynamics of Brunei Bay allow tide-dominated deltas to develop, an important consideration for any model that attempts to explain the common occurrence of ancient tide-dominated delta deposits is the frequency with which a suitable setting may have developed previously.

In NW Borneo, the key factor for generating tide-dominated hydrodynamics is the presence of an embayed coastline that reduces wave influence. This morphology apparently has occurred frequently since the middle Miocene because of local tectonics. The NW Borneo margin is segmented into a series of fault-bounded synclines, of which Brunei Bay is one segment (Fig. 17). Many of the bounding fault systems are cored by shale diapirs that experience episodic growth as the result of shale withdrawal from beneath adjacent synclines (Van Rensbergen et al., 1999). Consequently, the segments subside quasi-independently, which provides a mechanism for one segment to be a relative topographic low and therefore to form a coastal embayment during a relative sea-level highstand. The episodic nature of the tectonism and sea-level change suggests that each segment could become an embayment at several different times in its history.

A stratigraphic model for tide-dominated deltas in NW Borneo is illustrated in Figure 18. Most sediments are deposited in aggradational lobes that extend laterally parallel to the shoreline for approximately 5 km. A deltaic succession is composed of stacked lobes. The thickness of the succession, and of the individual lobes, is strongly influenced by the rate of basin subsidence, and hence the degree of aggradation. Judging by the Jalan Sungai Akar and Seven-Up Beach examples, the deposits of one delta lobe can reach 50–100 m thick, with the complete succession of stacked lobes totaling several hundred meters.

Stratigraphic development is controlled by tectonic and compaction-generated subsidence, a relatively high rate of sediment supply, and, to a large degree, basin hydrodynamics. Ongoing subsidence generates accommodation space on the delta plain, which is filled relatively quickly by a strongly aggradational and weakly progradational parasequence (Fig. 18). The rate of sediment supply is high enough that excess material is available for delta-front progradation, but hydrodynamics prevents sand from penetrating into the basin, resulting in small-scale channel switching as observed on the Trusan Delta.

Progradation of the delta front occurs solely by the deposition of mud and proceeds relatively slowly. Consequently, the process of subsidence, filling of accommodation space on the delta plain, and lateral shifting can be repeated several times, resulting in a parasequence set composed of 5–15 m thick

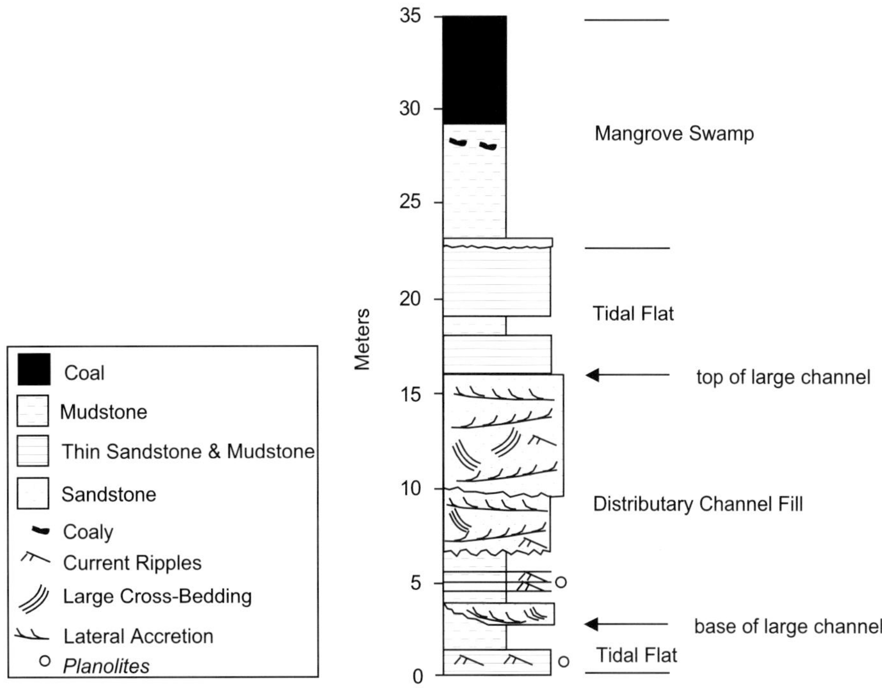

FIG. 16.—Stratigraphic log at Tanjung Batu from below the base of a distributary channel to the coal bed at the top of the exposed succession.

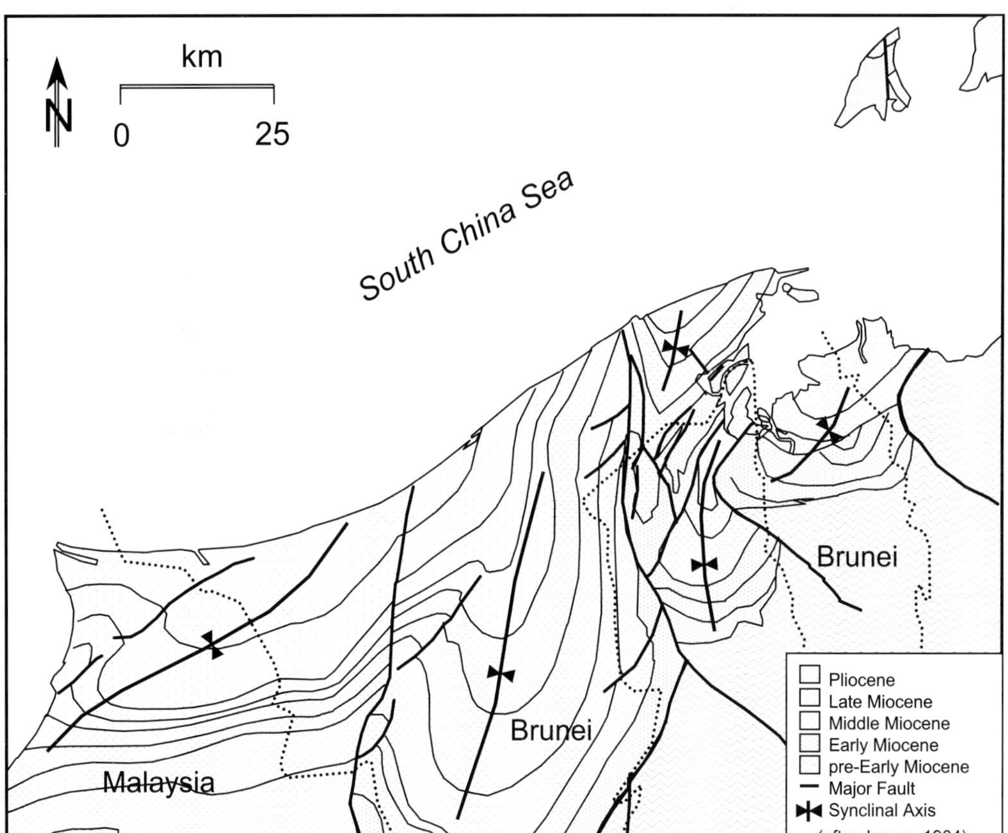

FIG. 17.—Chronostratigraphic map of Brunei and adjacent Malaysia showing segmentation of the margin into fault-bounded synclines.

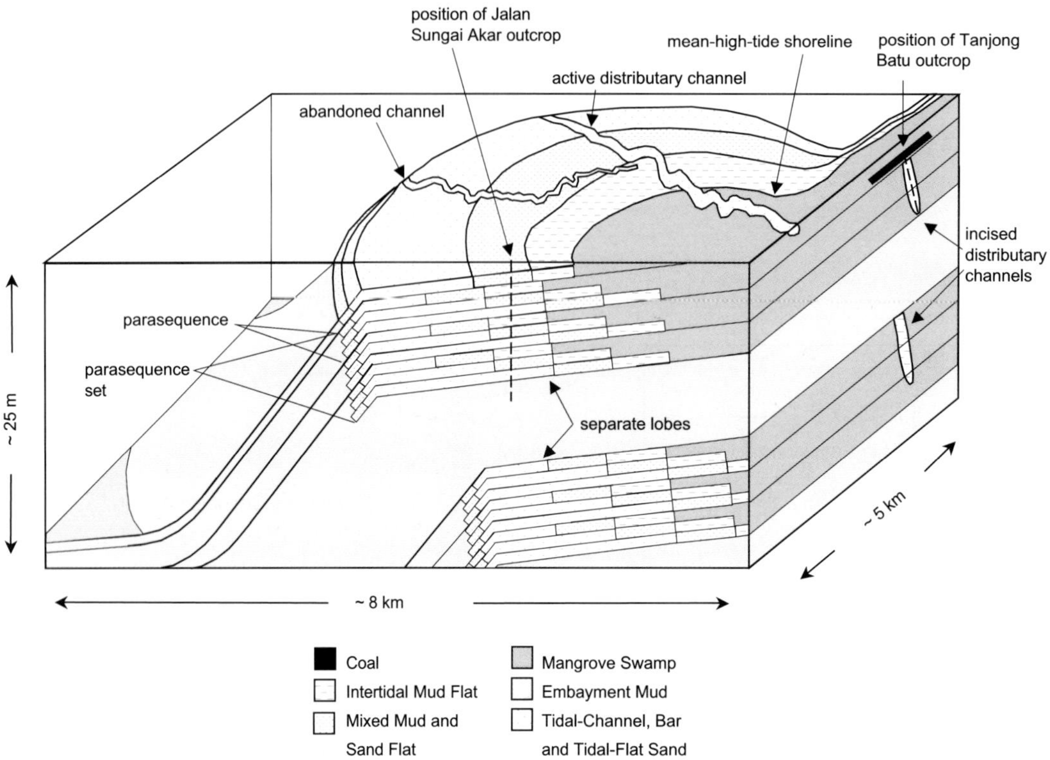

FIG. 18.—Stratigraphic model for tide-dominated deltas in NW Borneo showing facies distribution and stratigraphic architecture of proximal parasequences deposited on the delta plain.

parasequences of delta-plain facies similar to those exposed at Jalan Sungai Akar (Figs. 13, 18). As demonstrated by the Trusan Delta, the mangrove-swamp environment often progrades to the seaward edge of the delta plain after a lobe is abandoned (Fig. 2); continued subsidence may lead to the accumulation of mangrove peat and eventually a coal deposit (Lambiase et al., in press).

The deposits of individual delta lobes are vertically separated by embayment mudstones deposited during relative sea-level rises probably caused by lobe switching (Fig. 18). Each lobe constitutes a coarsening-upward succession at a larger scale (tens of meters) than the delta-plain parasequences, as exemplified by the Seven-Up Beach succession (Fig. 9).

CONCLUSIONS

A comparison of the modern tide-dominated Trusan River Delta and outcrops of the mid-Miocene and younger Belait Formation in Brunei indicates that:

1. The geographic setting and hydrodynamic environment that is responsible for the tidal dominance of the Trusan Delta probably has occurred frequently since the middle Miocene, suggesting that tide-dominated deltas also have occurred frequently.

2. The facies distribution and depositional system of the Trusan Delta establish it as a modern analogue for the Seven-Up Beach, Jalan Sungai Akar, and Tanjong Batu outcrops.

3. Tide-dominated delta successions are strongly aggradational, largely because of the interplay of ongoing subsidence and basin hydrodynamics.

4. The distal deposits of one delta lobe coarsen upward from prodelta and delta-front mudstones to wave-deposited sands at the seaward edge of the delta plain.

5. Proximal parasequences are deposited on the delta plain and fine upward from tidal-channel and bar sandstones through tidal-flat sandstones to tidal-flat mudstones.

6. Distributary channels erode into lobe deposits during relative sea-level falls and are backfilled during the next relative sea-level rise.

ACKNOWLEDGMENTS

The authors wish to thank their colleagues Chris Morley, Pieter Van Rensbergen, and Stefan Back for valuable discussion. Abdul Razak Damit thanks Brunei Shell Petroleum for financial support of his Ph.D. studies at the University of Aberdeen. Comments by Bob Dalrymple and an anonymous reviewer improved the manuscript significantly. Mike Bidgood and Patrice Brenac assisted in the biostratigraphic studies.

REFERENCES

ALLEN, G.P., LAURIER, D., AND THOUVENIN, J., 1979, Etude sédimentologique du delta de la Mahakam, Notes et Mémoires no. 15, Total Compagnie Français des Pétroles, 156 p.

ATKINSON, C.D., GOESTEN, M.J.B.G., SPEKSNIJDER, A., and VAN DER VLUGHT, W., 1986, Storm-generated sandstone in the Miocene Miri Formation, Seria Field, Brunei (N.W. Borneo), *in* Knight, R.J., and McClean, R.J., eds., Shelf Sands and Sandstones: Canadian Society of Petroleum Geologists, Memoir 11, p. 213–240.

CALINE, B., and HUONG, J., 1992, New insights into the recent evolution of the Baram Delta from satellite imagery: Geological Society of Malaysia, Bulletin, v. 32, p. 1–13.

CAMERON, W.M., and PRITCHARD, D.W., 1963, Estuaries, *in* Hill, M.N., ed., The Sea... New York, Wiley Interscience, v. 2, p. 306 – 324.

COLEMAN, J.M., 1982, Deltas, Processes of Deposition and Models for Exploration: Boston, International Human Resources Development Corporation, 124 p.

COLEMAN, J.M., and WRIGHT, L.D., 1975, Modern river deltas: variability of processes and sand bodies, *in* Broussard, M.L., ed., Deltas: Models for Exploration: Houston, Houston Geological Society, p. 99–150.

DALRYMPLE, R.W., 1992, Tidal depositional systems, *in* Walker, R.G., and James, N.P., eds., Facies Models: Response to Sea Level Change: Geological Association of Canada, p. 195–218.

DALRYMPLE, R.W., ZAITLIN, B.A., and BOYD, R., 1992, A conceptual model of estuarine sedimentation: Journal of Sedimentary Petrology, v. 62, p. 1130–1146.

HOEKSTRA, P., 1993, Late Holocene development of a tide-induced elongate delta, the Solo delta, East Java: Sedimentary Geology, v. 83, p. 211–233.

GALLOWAY, W.E., 1975, Process framework for describing the morphologic and stratigraphic evolution of the deltaic depositional systems, *in* Broussard, M.L., ed., Deltas, Models for Exploration: Houston, Houston Geological Society, p. 87–98.

HARRIS, P.T., BAKER, E.T., COLE, A.R., and SHORT, S.A., 1993, A preliminary study of sedimentation in the tidally dominated Fly River Delta, Gulf of Papua: Continental Shelf Research, v. 13, p. 441–472.

JAMES, D.M.D., ed., 1984, The Geology and Hydrocarbon Resources of Negara Brunei Darussalam: Singapore, Kok Wah Press, 169 p.

KOOPMAN, A., 1996, Regional geological setting, *in* Sandal, S.T., ed., The Geology and Hydrocarbon Resources of Negara Brunei Darussalam: Bandar Seri Begawan, Syabas, p. 49–63.

LAMBIASE, J.J., DAMIT, A.R., and SIMMONS, M.D., in press, A depositional model for mangrove-derived coals in NW Borneo: Proceedings of the Southeast Asian Coal Conference, Bandung.

LEE, T.Y., and LAWVER, L.A., 1995, Cenozoic plate reconstruction of Southeast Asia: Tectonophysics, v. 251, p. 85–138.

PROSSER, D.J., and CARTER, R.R., 1997, Permeability heterogeneity within the Jerudong Formation: an outcrop analogue for subsurface Miocene reservoirs in Brunei, *in* Fraser, A.J., Matthews, S.J. and Murphy, R.W., eds., Petroleum Geology of Southeast Asia: Geological Society of London, Special Publication 126, p. 195–235.

SANDAL, S.T., ed., 1996, The Geology and Hydrocarbon Resources of Negara Brunei Darussalam: Bandar Seri Begawan, Syabas, 243 p.

SCHREURS, J., 1996a, Coastal processes and shelf dynamics, *in* Sandal, S.T., ed., The Geology and Hydrocarbon Resources of Negara Brunei Darussalam: Bandar Seri Begawan, Syabas, p. 32–37.

SCHREURS, J., 1996b, Geographical Overview, *in* Sandal, S.T., ed., The Geology and Hydrocarbon Resources of Negara Brunei Darussalam: Bandar Seri Begawan, Syabas, p. 22–27.

SIMMONS, M.D., BIDGOOD, M.D., BRENAC, P., CREVELLO, P.D., LAMBIASE, J.J., and MORLEY, C.K., 1999, Microfossil assemblages as proxies for precise palaeoenvironmental determination—an example from Miocene sediments of north-west Borneo, *in* Jones, R.W., and Simmons, M.D., eds., Biostratigraphy in Production and Development Geology: Geological Society of London, Special Publication 152, p. 219–241.

VAN RENSBERGEN, P., MORLEY, C.K., ANG, D.W., HOAN, T.Q., and LAM, N.T., 1999, Structural evolution of shale diapirs from reactive rise to mud volcanism: 3D seismic data from the Baram Delta, offshore Brunei Darussalam: Geological Society of London, Journal, v. 156, p. 633–650.

VAN WAGONER, J., MITCHUM, R.M., CAMPION, K.M., and RAHMANIAN, V.D., 1990, Siliciclastic sequence stratigraphy in well logs, cores and outcrops: American Association of Petroleum Geologists, Methods in Exploration Series, no. 7, 55 p.

WATSON, 1979, State of Brunei water resources study report: Watson SE Asia Ltd., report to Government of Brunei.

WEE, T.Y., and SCHREURS, J., 1996, Climate, *in* Sandal, S.T., ed., The Geology and Hydrocarbon Resources of Negara Brunei Darussalam: Bandar Seri Begawan, Syabas, p. 27–31.

WRIGHT, L.D., COLEMAN, J.M., and THOM, B.G., 1975, Sediment transport in a macrotidal river channel, Ord River, Western Australia, *in* Cronin, L.E., ed., Estuarine Research, v. 2: New York, Academic Press, p. 309–322.

LATE QUATERNARY STRATIGRAPHY AND SEDIMENTOLOGY OF THE OFFSHORE MAHAKAM DELTA, EAST KALIMANTAN (INDONESIA)

HARRY H. ROBERTS
Coastal Studies Institute and the Department of Oceanography and Coastal Sciences, Louisiana State University, Baton Rouge, Louisiana 70803, U.S.A.
and
JOHAN SYDOW
BP-AMOCO, Trinidad and Tobago Exploration Business Unit, P.O. Box 714, Port of Spain, Trinidad, West Indies

ABSTRACT: Late Quaternary stratigraphy of the shelf over which the modern tropical Mahakam River delta is prograding indicates an alternating dominance of shelf carbonate sediments and siliciclastic deltaic deposits. Control of these major facies shifts is in response to high-frequency cyclic relative sea-level changes. Seismic facies and stratigraphy of the two latest Pleistocene depositional episodes reflect interplay of deltaic and shelf processes as well as accommodation during the last two glacio-eustatic cycles. Over 3000 line km of high-resolution seismic and side-scan sonar data calibrated to 380 bottom samples and 97 vibracores amd piston cores help characterize seismic facies of the Holocene and latest Pleistocene shelf deposits, which have been influenced by a strong north-to-south flowing current from the Makassar Strait. Holocene prodelta sediments are confined to the inner shelf of the northern and central sectors of the delta front, but they are skewed to the south, creating a broad facies tract in the southern sector. The middle and outer parts of the shelf are dominated by mounded topography characterized by both individual and aggregate bioherms. Mound relief varies from a few meters to over 30 m, but the average is about 20 m. This topography is locally accentuated by intervening erosional lows. Cores through the upper 6 m of these features indicate that the mounds are constructed of flakes of the calcareous green alga *Halimeda* in a matrix of foraminifera-rich terrigenous mud. The *Halimeda* bioherms are established on a ravinement surface formed during the Pleistocene to Holocene transgression. Inner-shelf bioherms are now being slowly buried by the Holocene advance of the Mahakam delta.

Below the Pleistocene to Holocene ravinement surface are the deposits associated with the preceding relative sea-level fall. Deposits defining this falling-stage systems tract and associated lowstand systems tract consist of entrenched fluvial networks, incised-alluvial-valley fill, aggraded delta-plain and platform deposits, and prograded delta lobes. Pleistocene deltaic deposits have prograded to the shelf edge along the entire shelf–slope break. Over 50% of the stratigraphy in the lowstand systems tract is related to aggradation during the lowstand turnaround. The Pleistocene deltaic deposits bury massive carbonate bioherms and aggregates of bioherms that mimic their Holocene counterparts in morphology and scale. A prominent ravinement surface underlies the bioherm facies, and a maximum flooding surface runs through these features.

INTRODUCTION

The modern Mahakam River of East Kalimantan, Indonesia, has produced a river–tide dominated delta that is building across a narrow shelf. This shelf borders the north–south oriented Makassar Trough, which acts as a throughflow pathway for Pacific water to the Indian Ocean (Fig. 1; Wyrtki, 1987). The sedimentological characteristics and facies architecture of the modern highstand delta provide an analog model for hydrocarbon exploration and production in Tertiary deltaic sediments representing ancient Mahakam delta counterparts. The project described in this paper improves our understanding of both the modern Mahakam delta and its latest Pleistocene counterparts by analysis of high-resolution seismic and side-scan sonar data as well as sediment cores from the shelf setting. Results of the offshore Mahakam delta study help upgrade models for deeper subsurface stratigraphy by defining the relationships between important stratigraphic elements with an entire sea-level cycle of deposition. In addition, this study provides new insight into the distribution and preservation potential of deltaic and shelf facies. Many of the important characteristics of these high-order depositional cycles are below the resolution of conventional exploration-scale seismic, and therefore the importance of such high-order cycles as stratigraphic elements and sites of reservoir and source facies has not been fully appreciated to date.

Following the thorough investigation of the onshore modern Mahakam delta by Allen et al. (1979) and Allen (1994), the offshore was studied using high-resolution seismic and side-scan sonar (over 3000 line km) calibrated to bottom samples (380 samples, some of which were from delta distributaries and tidal channels) and 97 piston cores and vibracores (both in the modern delta and on the shelf). These data sets provided the foundation for developing the first detailed sedimentological and stratigraphic appraisal of the modern offshore Mahakam delta and shelf, as well as insight into the fundamental role of relative sea-level change in defining the stratigraphic framework that is actually preserved in the geologic record.

Stratigraphy of the shelf seaward of the modern Mahakam delta has developed under a complex interplay of rapid tectonic subsidence, marine shelf tidal transport processes, high-order relative sea-level cyclicity, significant terrigenous sediment input by the Mahakam River, and development of biohermal carbonates. The objective of this paper is focus on the modern offshore Mahakam delta and shelf deposits as prime components of a depositional model for understanding lowstand deltas that dominate the Pleistocene stratigraphic architecture of shelf.

RIVER-SYSTEM CHARACTERISTICS AND PROCESS SETTING

The Mahakam delta is located approximately 50 km south of the equator along the east coast of Borneo (Fig. 1). Its drainage basin covers about 75,000 km^2 and captures an annual rainfall that varies between 2,000–3,000 mm near the coast and 4,000–5,000 mm in the central highlands (TAD report, 1982). Two wet seasons occur annually. They correspond to the beginning of the northwest monsoon (November–December) and a transitional

Tropical Deltas of Southeast Asia—Sedimentology, Stratigraphy, and Petroleum Geology
SEPM Special Publication No. 76, Copyright © 2003
SEPM (Society for Sedimentary Geology), ISBN 1-56576-086-7, p. 125–145.

FIG. 1.—Location map for the Mahakam delta study area along the East Coast of Borneo (Indonesia).

period to the southeast monsoon (March–May). Although river discharge is poorly documented, rainfall and limited discharge measurements suggest that an annual average fluvial discharge of about 3000 m^3/s occurs, when measured approximately 20 km up-river from "head of passes", where the Mahakam River breaks into its distributaries (Sydow, 1996). For scale, this discharge is about one-third of the Mississippi River discharge. The average annual sediment load is estimated at 8×10^6 m^3/yr (Allen et al., 1979). The sediment load is about 70% mud and 30% sand, with silt and clay carried as suspended load, and fine to medium sand as bedload (Allen et al., 1979).

The Holocene delta and its submerged platform create an arcuate protrusion of the coastline, which has prograded nearly 50 km over approximately the last 7 ka (Fig. 3). The subaerial delta covers about 5,000 km^2, contains both active bifurcating distributaries and tidal channels, and is distinguished by plant communities that range from tropical rain forest at the supratidal delta apex to *Nipa* palm and mangrove swamp on the intertidal parts of the delta plain (Allen et al., 1979; Dutrieux et al., 1990; Dutrieux, 1991). The delta front has a relatively steep slope (1.0–2.5°E), which appears to be very stable, with no evidence of gravity-driven mass-movement features. Instead, a thin prodelta sheet extends seaward from the base of the prograding delta front. This fine-grained facies is widespread on the southern shelf (> 40 km wide) and to a lesser extent on the northern shelf, but is confined to a narrow belt along the central part of the delta (10–15 km wide). The distribution pattern of this prodelta facies is partially explained by the southern and northern positions of active distributaries, but a dominant controlling factor is the strong north-to-south flowing Indonesian throughflow current that originates in the Pacific Ocean, flows southwards through the Makassar Strait between Borneo and Sulawesi, and eventually exits through the Indonesian Archipelago to the Indian Ocean (Wyrtki, 1987; Murray and Arief, 1988; Arief and Murray, 1996). This current sweeps across the shelf opposite the delta, deflecting prodelta sediments to the south. Current speeds in excess of 80 cm/s have been recorded by meters moored near the central delta front (TOTAL internal report, 1986). In addition to transporting sediment, this current helps transport upwelled and nutrient-rich water across the outer and central parts of the shelf, as appears to also be the case along the shelf and over carbonate platforms in the Java Sea south of the Mahakam study area (Samudra, 1975; Roberts et al., 1987; Roberts et al., 1988).

Because the delta is located near the equator, storms are rare events, and limited fetch between Borneo and Sulawesi restricts the wave field. Wave energy generated by the dominant southeast and northwest monsoonal winds is minimal. Significant wave heights exceed 80 cm less than 5% of the time (TOTAL internal report, 1986). Tides are semidiurnal and range from about 0.5 m for neap tide, 1.7 m for high tide, and up to 3 m during equinox high tide, with an overall average tide of 1.2 m (TOTAL internal report, 1986; offshore Mahakam delta project data).

SEDIMENTS AND STRATIGRAPHY: MODERN DELTA AND SHELF

Plant debris is an important component of nearly all sedimentary facies of the Mahakam delta. The modern delta has been divided into zones based on vegetation types and salinity (Allen et al., 1979; Dutrieux et al., 1990) and on delta morphology (Allen et al., 1979). These zones, as illustrated in Figure 4, consist of upper delta plain (tropical lowland rain forest and intermediate mixed hardwoods–palms) and lower delta plain (*Nipa* palm and mangrove swamps). The upper delta plain (interior zone of Allen, 1979) is characterized by terrestrial and supratidal conditions. This apex or terrestrial part (supratidal) of the delta is vegetated by a tropical lowland rain forest, whereas intermediate mixed hardwood and palm (*Heritiera littoralis*) forest occupies the rest of the upper delta plain (Dutrieux et al., 1990). The tidally influenced lower deltaic plain is inhabited almost entirely by *Nipa*

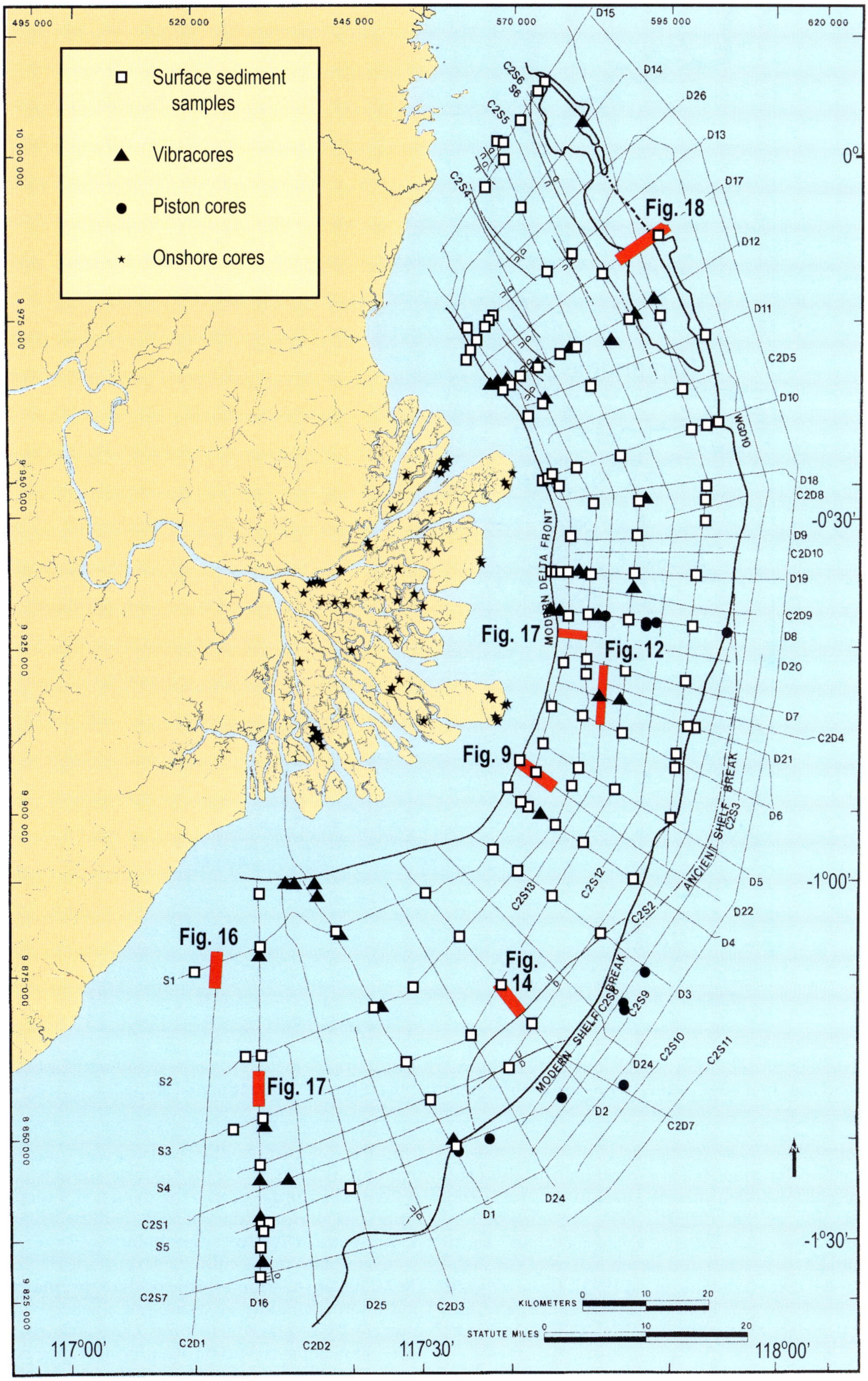

FIG. 2.—Map of the study area showing the grid of high-resolution seismic lines and side-scan sonar swaths, sample sites, and locations of seismic line segments used as illustrations in this paper.

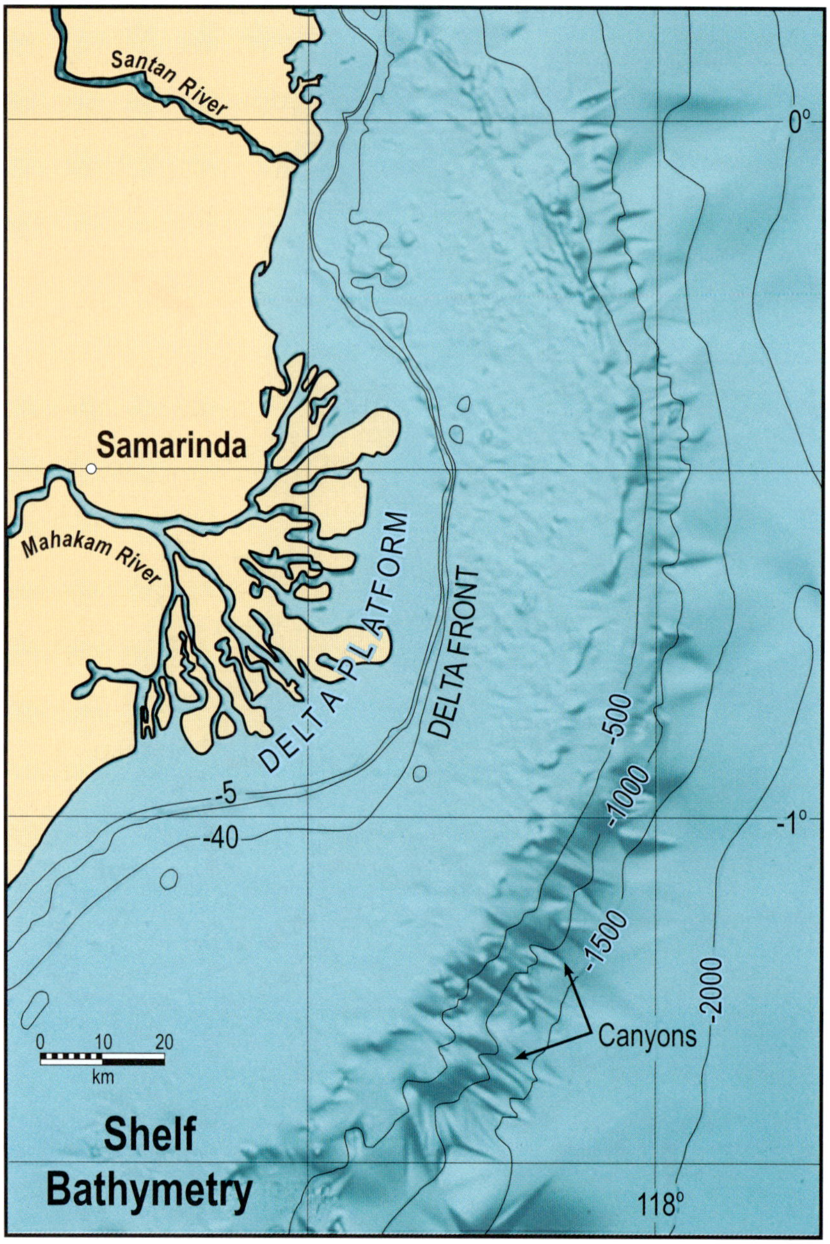

FIG. 3.—Shaded relief representation of shelf and slope bathymetry. Contours are in meters. Slope bathymetry is limited by low data density beyond the shelf edge, but canyons are clearly identifiable on the upper continental slope.

palm swamp vegetation (*Nipa fructicians*). This plant species, which has large vertically oriented leaves that attain heights of up to 8 m, dominates the lower delta plain (Gastaldo and Huc, 1992). The seaward margins of the subaerial delta are inhabited by coastal mangroves (*Avicennia, Rhizophora, and Sonneratia*). Detrital peats accumulate as beaches and beach ridges at the distal ends of the subaerial delta (Gastaldo et al., 1993). These deposits may attain a thickness of over 2 m, are observed 3 km inland, and may cover a total surface area of up to 50 km^2 (Allen and Pizon, 1986).

The delta plain is segmented by distributary and tidal channels. Figure 4 shows that three major distributaries branch from the Mahakam River from the "head of passes" at the apex of the fan-shaped delta. Two of these distributaries bifurcate further to form the dominant southern distributary network, and the remaining distributary branches form the northern distributary system. The central part of the delta is tide-dominated and has numerous sinuous tidal channels with bell-shaped mouths but no presently active river distributaries.

Delta-plain topography is very subtle. Unlike many deltas, the Mahakam distributaries do not have prominent natural levees and interdistributary areas are not affected by levee breaching and crevasse-splay development. Because natural levees and levee breaching are primarily products of a pronounced river flood cycle, the rather constant discharge of the Mahakam River does not favor levee development or crevasse-splay formation. Interdistributary areas are flooded by tidal action. This process creates a low-relief delta plain.

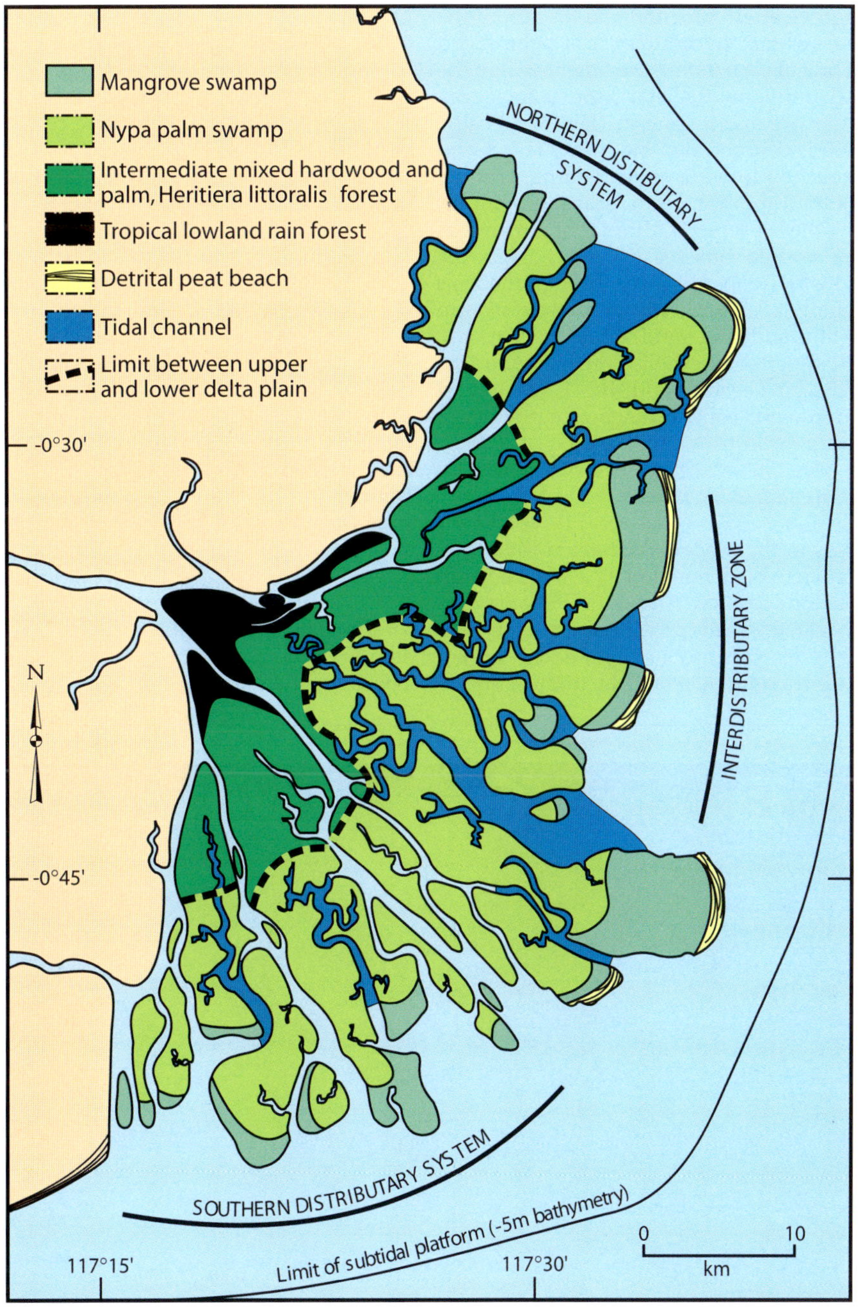

FIG. 4.—Sedimentary environments of the subaerial delta (modified from Allen et al., 1979; Dutrieux, 1991).

Tropical Lowland Rain Forest (Delta Apex)

Vibracores taken in this environmental setting indicate a shallow sediment column dominated by 1.5–2.0 m of root-burrowed and locally organic silty clay over couplets of sand, silty sand, and mud. The couplets frequently illustrate the disrupting effects of root burrowing. The sand may be cross-stratified, and detrital organic particles are deposited with the clays and form organic partings. As the X-ray radiographs of Figure 5 illustrate, the fine- to medium-grained sand can be quite massive. Both sand and sand–mud couplets illustrate the emplacement of early diagenetic iron carbonates (Bailey and Roberts, 1998). Plant fragments are found throughout the sand and silt units in this depositional setting.

Swamps

Nipa palm swamps dominate the delta plain (Fig. 4). Like their counterparts in the tropical rain forest at the apex of the delta and the mixed hardwood and palm forest environment of the upper delta plain, the *Nipa* swamp deposits contain highly organic and root-burrowed sediment in the top 1.0–1.5 m. Because the elevation of the lower deltaic plain is approximately at the mean tide level (Allen et al., 1979), the seaward parts are submerged on a semidiurnal basis (mangrove environments) and the rest of the lower delta plain (*Nipa* swamp environment) is submerged semidiurnally for about half of each month. Sediment within the *Nipa* swamp is composed of stiff, mottled, dark gray to light gray mud containing abundant organic fragments (Fig. 5). Locally, particulate organic matter is concentrated into peaty layers and clots. Root burrows disrupt subtle evidence of bedding, producing a near-massive-to-mottled appearance on X-ray radiographs. Oxidation rims are sometimes observed around root burrows as well as development of nodular masses by early diagenetic processes (Bailey et al., 1998). Deeper in cores through the *Nipa* swamp environment, usually 1.5–2.0 m below the present swamp surface, organic-rich silt–mud and sand–mud couplets are present. Environmentally specific invertebrate shell remains are rare in these *Nipa* swamp deposits, but Gastaldo and Huc (1992) suggest that all delta-plain environments contain diagnostic phytoclast compositions.

Fluvial Channels

Sedimentary processes in the modern Mahakam River distributaries have been shown to be dominated by active bedform migration, erosion, and lateral accretion (Allen et al., 1979; Siemers et al., 1994). Distributary channels in the proximal part of the delta are approximately 7–10 m deep (locally to 15 m) and 1–1.5 km wide. Sand waves and dunes migrate along the thalweg. Elongate lateral accretion bars are common. These bars have erosional bases and are composed of cross-laminated to massive sands containing plant fragments (Allen et al., 1979). The lateral bars are thickest adjacent to the channel margins, and they slope toward the thalweg (Gastaldo and Huc, 1992). Cores from this study penetrated the upper 5–7 m of these features. These cores from lateral bars document a sedimentary succession starting in cross-bedded sands (medium sand size) to flaser bedding followed by mud–sand couplets, lenticular bedding, and finally organic-rich muds typical of swamp deposits at the present bar surface (Fig. 5).

Mid-channel bars and distributary-mouth bars in a distal distributary setting are influenced by a stratified water column where net flow along the channel bed is upstream (Allen et al., 1979). These sand bodies are generally muddier than their lateral channel bar counterparts. They have gradational bases and are accreting over mud-dominated subtidal delta platform sediment. Mud–sand couplets in the basal parts grade upward into sand-rich lenticular bedding and finally into a mud drape. Benthic communities of infaunal organisms burrow these sediments, and shell fragments are frequently encountered in distal areas.

Tidal Channels

The lower delta plain is intricately incised by numerous sinuous tidal channels, especially in the central sector of the delta. These channels are typically 20–100 m wide, may incise up to 20

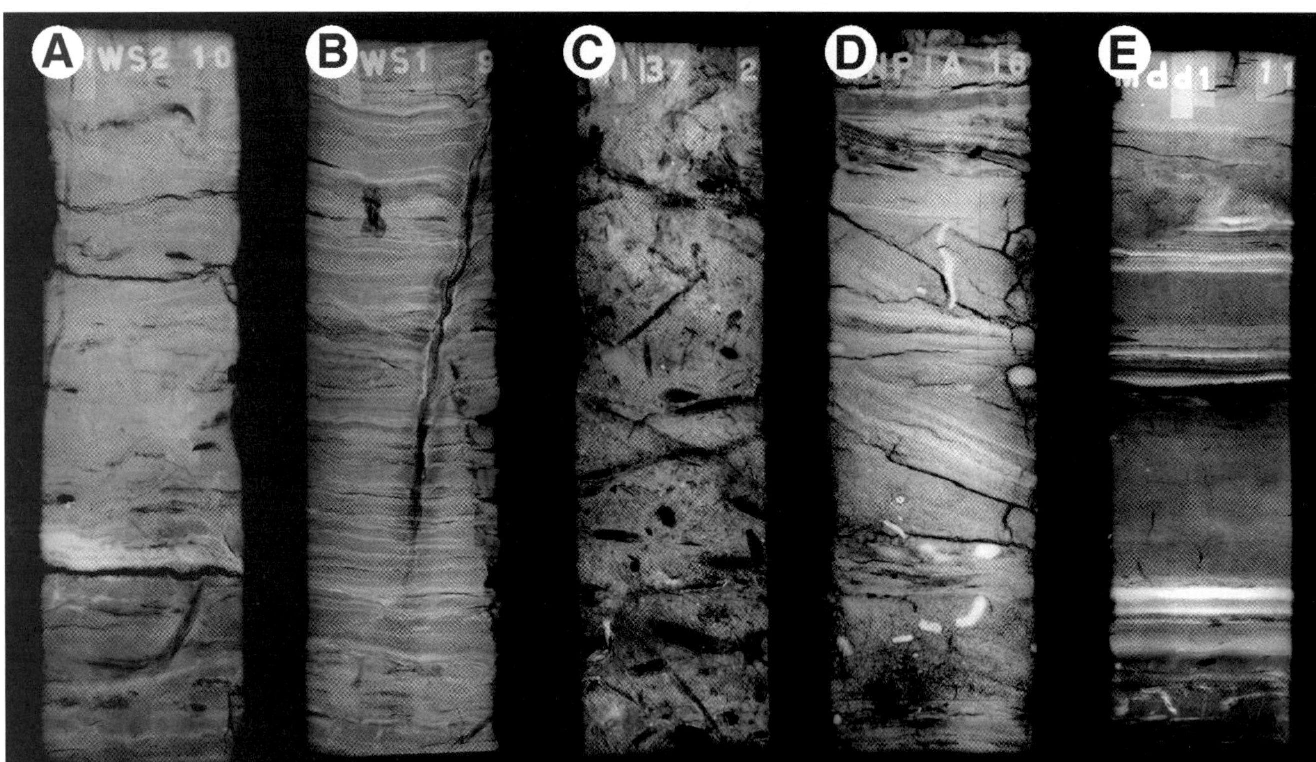

FIG. 5.—X-ray radiographs of selected core sections from the onshore modern Mahakam delta. The diameter of each core is 7.5 cm. The pictures are of the X-ray radiograph negative, so lighter tones represent areas that are denser to the penetration of X-radiation. These cores represent the following depositional environments: **A)** root-burrowed massive sand of the tropical lowland rain forest at the delta apex, **B)** sand–mud couplets with a long root burrow that occur approximately 2.0 m below the root-burrowed massive sand of the delta-apex region, **C)** *Nipa* swamp deposits of the lower delta plain, illustrating the inclusion of large amounts of organic debris (black areas), **D)** cross-stratified sand of the fluvial-channel facies (note diagenetic inclusions), and **E)** alternating silty clay, clay, and organics of the intertidal mudflat facies.

m into underlying deltaic deposits (Hardy, 2000), and are often deeper than their fluvial counterparts. As Allen et al. (1979) point out, some tidal channels in the central interdistributary areas are former fluvial channels that have been abandoned. Sediments within the channels are now being reworked by tide-driven processes.

Because input of sediment to tidal channels does not involve direct fluvial contribution, the channel-fill sediments are mud-dominated and have a high content of plant debris. Vibracores taken in tidal-channel deposits contain weakly bedded to laminated organic mud with thin layers of detrital organic fragments and silt. Bedding is highly distorted by root burrowing from *Nipa* palms over most of the lower delta plain and from mangroves toward the distal parts of the delta.

Intertidal Mudflats and Detrital Organic Beaches and Beach Ridges

Extensive mudflats form at the transition between the vegetated delta plain and the broad subtidal platform. In cores, these deposits consist of parallel-bedded muddy peat to organic-rich mud (Fig. 5). Although wave action along the seaward perimeter of the delta is extremely low, it is sufficient to concentrate detrital organic debris along the inner parts of the mudflats to form beaches and beach ridges. These organic deposits occur at the distal ends of many protruding lobes of the delta, especially in the central interdistributary zone (Fig. 4). The organic deposits may be up to 2.5 m thick and have been observed to cover broad cover areas of the distal subaerial delta (Allen and Pizon, 1986; Gastaldo et al., 1993).

Subtidal Delta Platform

In accordance with the most widely accepted deltaic literature, the shallow transitional area between the subaerial delta and the relatively steep slope to the shelf is termed the subtidal delta platform, rather than the delta front of Allen et al. (1979). The bathymetry of the delta platform, steeper delta front, and adjacent shelf clearly distinguish these features as being distinct geomorphic components of the prograding delta (Fig. 3). The shallow water depth of the subaqueous delta platform (< 5 m, 5–10 km wide) makes this province difficult to sample. Most research in this part of the delta has been concentrated opposite the major distributary mouths (Allen et al., 1979; Gastaldo and Huc, 1992). Samples from areas other than distributary mouths have been analyzed by Carbonel et al. (1987) as part of the MISEDOR drilling program. A wide variety of sedimentary facies have been reported from the subtidal platform, including organic-rich mud, interbedded sand and mud, bioclastic sand, and massive sand. Thorough burrowing by benthic organisms and locally abundant shell debris distinguish these sediments from delta-plain deposits. Sand bodies typically occur near the distributary mouths as elongate, shore-normal tidal bars and distributary-mouth bars. Most of the platform could be characterized as bioturbated organic-rich mud containing shell debris (Fig. 6).

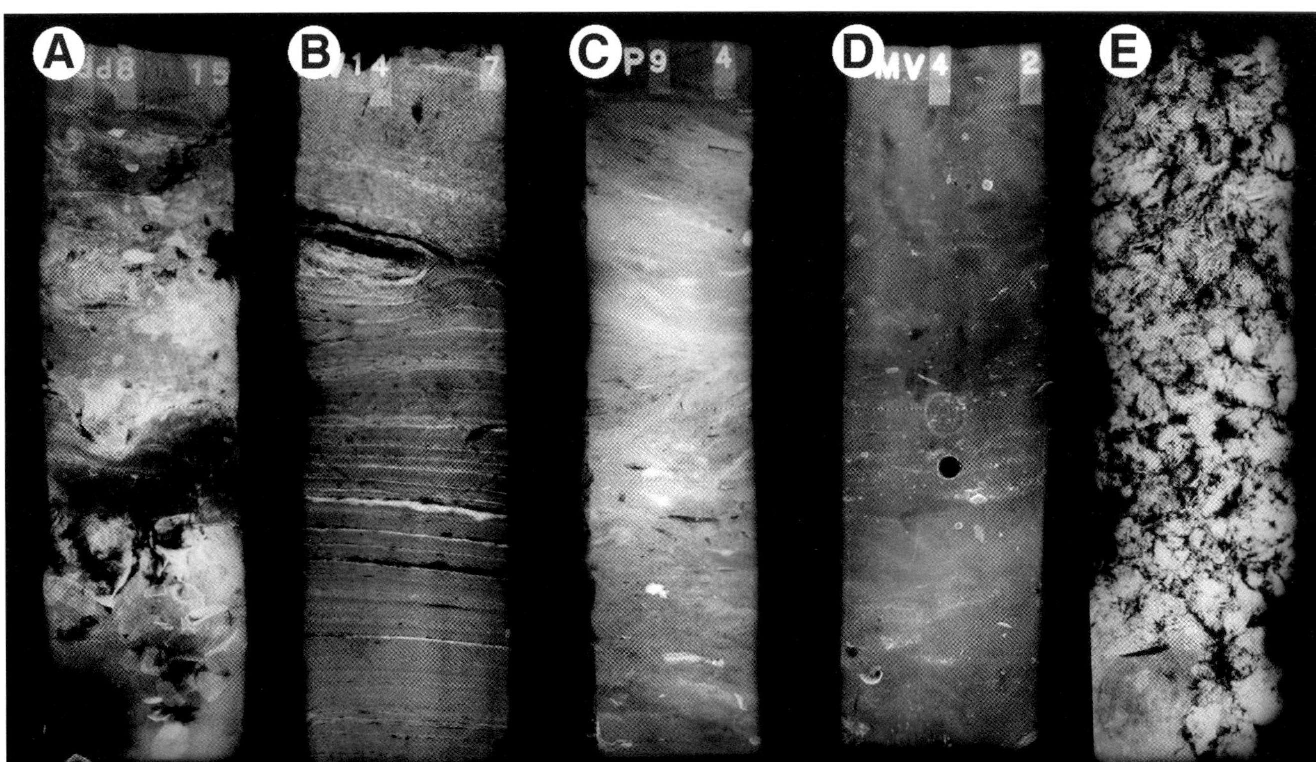

FIG. 6.—X-ray radiographs of selected core sections from the modern offshore Mahakam delta and shelf. The diameter of each core is 7.5 cm. The pictures are of the X-ray radiograph negative, so lighter tones represent areas that are denser to the penetration of X-radiation. These cores represent the following depositional environments: A) shell-rich subtidal-delta-platform mud, B) organic-rich bedded mud of the transition between the subtidal delta platform and the delta front, C) rather massive but organic-rich mud of the actively prograding delta front (dark spots are organic fragments), D) massive, shell-rich mud of the prodelta facies, and E) packed aragonitic flakes of *Halimeda* in a matrix of foraminifera-rich mud from a mid-shelf *Halimeda* bioherm.

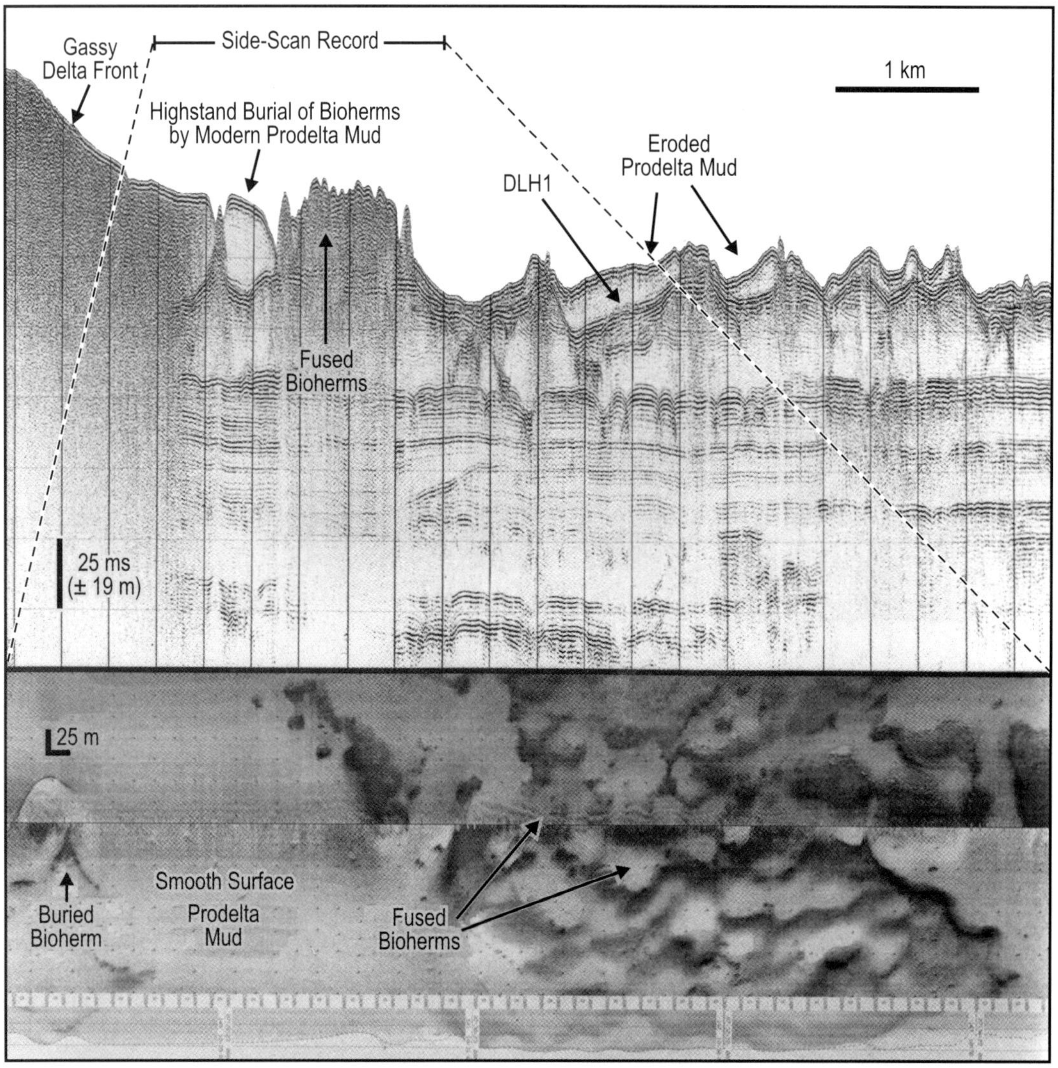

FIG. 7.—A high-resolution seismic profile (top) and side-scan sonar data (bottom) acquired across the inner shelf off the central sector of the modern Mahakam delta. Note the acoustically opaque delta front (gassy sediments) and the prodelta mud in the process of burying inner-shelf *Halimeda* bioherms. The side-scan sonar record (bottom) shows the prodelta sediment burying a single bioherm and the multimounded surface of an exposed aggregate of bioherms. See Figure 2 for the location of this seismic profile.

Delta Front

The relatively steep (1–2.5°) seaward transition of the platform to the shelf, the delta front, is composed of organic-rich mud (Fig. 6). This area of the prograding delta expresses itself on high-resolution seismic profiles as an acoustically opaque zone, presumably owing to scattering of the acoustic energy by bubble-phase gas (Fig. 7). The dark, organic-rich reducing character of cored sediments from this part of the delta is consistent with their acoustic signature. Inspection of X-ray radiographs from this setting indicate an abundance of detrital organic particles, sparse burrowing (only microburrowing), sparse faunal remains, and nodular masses of early diagenetic siderite (Bailey and Roberts, 1998). The high organic content of these sediments (TOC usually > 5%; unpublished project data) is in direct contrast to that of the prodelta facies that extends seaward from the base of the delta in water depths of 30–40 m and is more closely aligned with organic-rich delta-plain sediment (Hardy, 2000).

Prodelta

The transition from delta-front to prodelta facies is quite dramatic as observed in both seismic profiles (Fig. 7) and cores (Fig. 6). Sediments change from dark, reducing, organic-rich mud containing few shell fragments to olive green mud containing an abundance of shell fragments and foraminifera tests (Fig. 6). Seismically, the prodelta facies has little internal character, but it is not an acoustic "wipe-out zone" as is the case with the delta front. As the X-ray radiographs demonstrate (Fig. 6D), prodelta deposits generally have a massive character at the core-sample scale. Bedding is obscured because of the homogenizing effects of bioturbation in this slowly accumulating facies. Abundant calcareous foraminifera tests, gastropod shells, echinoid remains, and bivalves form the coarse fraction of the shelly muds of this facies. Because of the slow deposition rate and extensive bioturbation, the preservation potential of organic material in this facies is low (TOC 0.5–1.0%). On seismic profiles the prodelta facies forms a seaward-thinning wedge of sediment with subtle, conformable internal reflectors. This facies forms a smooth seafloor and in many places fills irregularities in underlying shelf topography (Fig. 7).

Shelf Bioherms and Shelf-Edge Buildups

The bathymetry of the shelf clearly distinguishes the rough bottom topography typical of the northern and central shelf from the smoother shelf to the south of the delta (Fig. 3). High-resolution seismic profiles, side-scan sonar swaths, bottom samples, and cores confirm this rough shelf topography to consist of individual and aggregate mound-like bioherms built mainly by the calcareous green alga *Halimeda* (Roberts and Sydow, 1996; Sydow, 1996). These bioherms attain over 30 m of relief in local parts of the northern shelf, but over most of the shelf their average relief is about 20 m. Cores acquired from the *Halimeda* bioherms indicate that they are composed of aragonitic flakes of *Halimeda* in a foraminifera-rich mud matrix during the early to middle parts of their accretion history (Fig. 6E). However, late-stage growth is mixed with progressively more Mahakam River mud as the modern delta progrades across the shelf.

Seismically, the bioherms either reveal subtle internal reflectors that mimic the mound external shape or they are acoustically transparent (Fig. 7). Shelf bioherms are established on a regional erosional surface that truncates clinoforms of the underlying latest Pleistocene deltaic deposits. On side-scan sonar data all bioherms, single and aggregate, exhibit a unique surface topography composed of numerous small-scale low-relief mounds (Fig. 7). Toward the edge of the shelf, more acoustically "hard targets" are scattered among these low-relief surface features and a higher-amplitude return is typical on high-resolution seismic profiles, both at the surface and from the mound interior. Cores and surface samples indicate a significant increase in coralline algal nodules and small buildups in this distal part of the shelf, which correlate with the higher-amplitude reflections observed on both side-scan sonar and seismic data.

Along the shelf edge, especially in the northern and central sectors, mounded and linear reef-like buildups occur with a relief of 80 to 100 m. These features have no internal reflectors on seismic profiles and give hard surface returns on side-scan sonar data. Coring into these features is difficult. However, short vibracores (3–4 m long) indicate no true reef-building corals. Coralline algal encrustations and nodules plus bioclastic sand composed of degraded *Halimeda* fragments and foraminifera tests (*Amphistegina–Tinoporus–Operculina*) similar to those found in association with *Halimeda* bioherms are observed (Dr. Barun Sen Gupta, personal communication).

Holocene Stratigraphic Summary

A long boring taken through the modern delta, the MISEDOR boring (Carbonel et al., 1987; Gayet and Legigan, 1987), indicates that Holocene deposits are approximately 50 m thick (Fig.

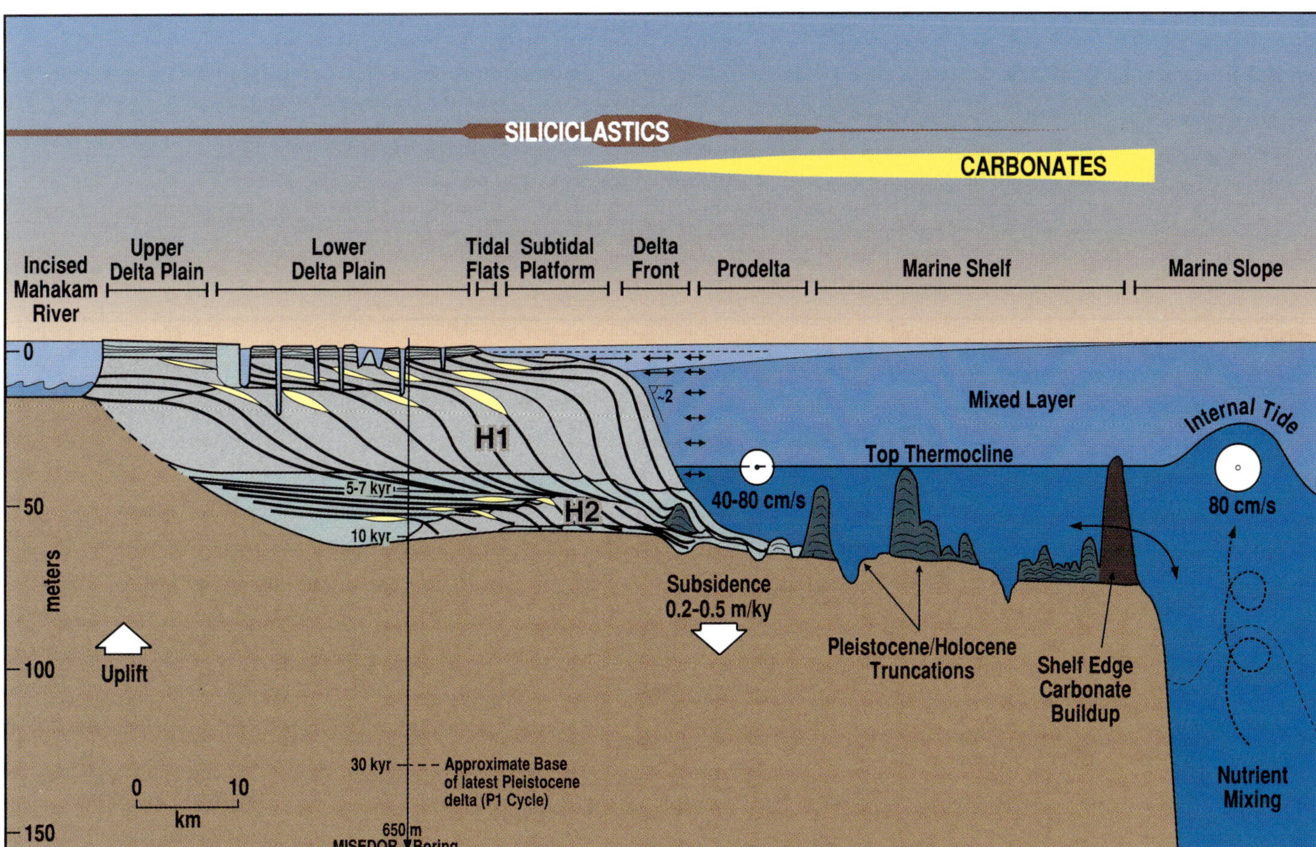

FIG. 8.—Model for Holocene stratigraphy of the Mahakam delta and adjacent shelf. Note the early Holocene delta (H2 delta) in relation to the modern highstand Holocene delta (H1 delta). Stratigraphic relationships are provided by both high-resolution seismic profiles and results from the MISEDOR boring. Dates are extrapolated from the MISEDOR boring (Carbonel and Moyes, 1987).

8). Seismic data collected as part of this study of the offshore Mahakam delta reveal that an early Holocene delta exists below the present delta. Sediments from this early Holocene delta-building episode protrude from beneath the presently active delta onto the southeast sector of the shelf. The clinoforms of this delta downlap on the regional Pleistocene–Holocene erosional surface and are themselves truncated by a planar surface interpreted to be a product of seafloor ravinement processes. Clinoforms of the present highstand Holocene delta downlap onto the truncated earlier clinoforms (Fig. 9).

The erosional surface that separates the two Holocene deltas occurs about 40 m beneath the present delta surface. The present sea-level highstand was attained about 5 ka (Aharon and Chappell, 1986; Matthews, 1990). When compared to dated horizons in the MISEDOR boring (Fig. 8), the base of the present delta (prodelta facies) is dated at 5–7 ka. Even though subsidence rates are reasonably high for the Mahakam delta (0.2–0.35 m/kyr; Sydow, 1996), the stratigraphic position of the early Holocene lobe suggests deposition during a time of substantially lower sea level relative to present. Delta-plain sediments that date 10 ka in the MISEDOR boring correlate to the early Holocene delta rather than the present delta (Fig. 8). Other deltas, like the Ganges (Goodbred and Kuehl, 2000), have early Holocene lobes formed at a lower-than-present sea level. Deltaic deposition at this time may be related to the Younger Dryas event (Broecker et al., 1988; Fairbanks, 1989), which potentially reversed or substantially slowed the rate of eustatic sea-level rise in the 10–12 ka time frame.

Overlying the early Holocene delta is the modern highstand Mahakam delta. Prodelta deposits of this presently active delta are 10–20 m thick. Volumetrically, delta-front deposits are much more important than their prodelta, delta-platform, or delta-plain counterparts (> 40 m thick beneath the outer delta platform). Approximately 10 m of subtidal-platform deposits overlie sediments of the delta front. The entire delta sedimentary succession is capped by 3–4 m of delta-plain swamp deposits (Fig. 8).

The seaward-thinning prodelta deposits attain their maximum offshore extent on the southern shelf where both the dominant fluvial distributaries and along-shelf currents are focused. Offshore from the arcuate delta front of the central shelf, prodelta deposits are confined to a narrow inner-shelf zone no more than 10 km wide. Seaward of the prodelta deposits on the northern and central shelf, Holocene topography, sedimentology, and stratigraphy are dominated by the carbonate deposits of *Halimeda* bioherms. These bioherms occur mainly on the middle to outer parts of the central shelf and mainly on the outer shelf of the northern sector. Both individual and aggregate bioherms that form extensive platforms are seated on a distinct Pleistocene–Holocene regional ravinement surface. The relationship of reef-like buildups at the shelf edge to the ravinement surface is not as clear because these features are usually acoustically opaque. Fragmentary evidence from high-resolution seismic profiles suggests that at least some of these features are seated on the same surface as *Halimeda* bioherms landward of the shelf edge. Figure 8 summarizes these stratigraphic relationships between the Pleistocene to Holocene ravinement surface and overlying shelf-edge buildups, *Halimeda* bioherms, and deltaic deposits.

SEISMIC STRATIGRAPHY—LATE PLEISTOCENE TO HOLOCENE DEPOSITS

Late Pleistocene to Holocene Stratigraphic Framework

Three Pleistocene depositional cycles, P1, P2, and P3, are identified in the present data. The last cycle, P1, is best resolved and has been mapped in detail. Representative dip and strike sections illustrate stratigraphic positions of depositional units and surfaces within the P1 and stratigraphically deeper cycles (Fig. 10). A two-way travel time (TWT) isochron map of P1, excluding carbonate bioherm facies, reveals two principal depocenters, a northern and a larger southern depocenter, referred to as P1N and P1S, respectively (Fig. 11).

Bounding Downlap Surfaces

Bounding discontinuity surfaces of depositional cycles P1 and P2 are defined on high-resolution seismic data by prominent

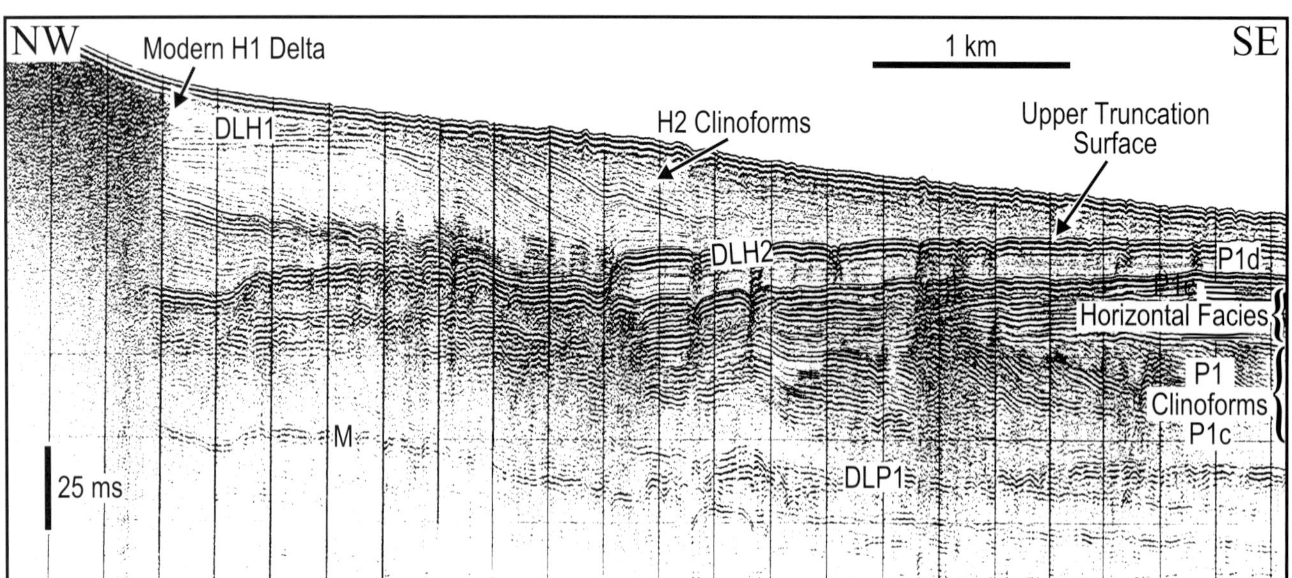

FIG. 9.—High-resolution seismic profile illustrating the truncated clinoforms of the early Holocene delta (H2 delta) downlapped by the prograding front of the modern delta (H1 delta).

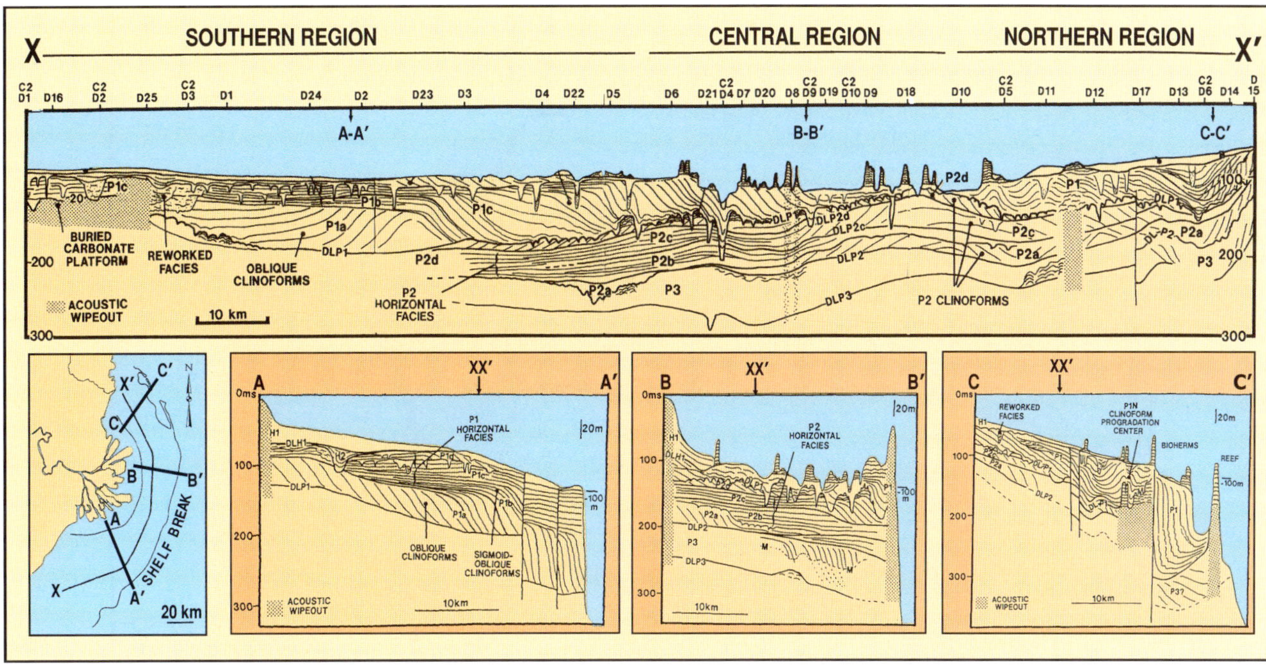

FIG. 10.—Selected strike and dip lines drawn from high-resolution seismic profiles illustrating the stratigraphic relationships between the Holocene and the three latest Pleistocene depositional cycles (P1, P2, and P3). Units P1a–P1d and P2a–P2d represent various progradational lobes of the P1 and P2 depositional cycles.

regional downlap surfaces such as DLP1 and DLP2. Although cycle P3 is partially imaged in the seismic data, this depositional cycle could not be mapped in detail. Downlap surfaces are generally higher-amplitude, regionally continuous seismic events, and they are prominent and practical surfaces to map within the complex shelf stratigraphy (Fig. 10). The correlative surfaces on the slope are similarly continuous, moderate- to high-amplitude reflections that are also obvious features to map. Surfaces having similar geometric and stratigraphic relationships penetrated in the Gulf of Mexico and elsewhere are considered to represent condensed intervals associated with major highstand shelf flooding events (Coleman and Roberts, 1988; Loutit et al., 1988; Sydow and Roberts, 1994; Anderson et al., 1996). These surfaces are the ancient equivalents of the modern, sediment-starved middle to outer shelf, which is presently being buried and downlapped by the presently prograding Holocene highstand (H1) Mahakam delta on the inner shelf (surface DLH1) (Fig. 12).

In the middle to outer shelf of the central sector, surface DLP1 is highly rugose, passing over the tops of buried bioherms, and is deeply incised in between carbonate buildups. Deepest incision occurs in shore-normal valleys 1–2 km wide, in some areas with as much as 50 m of erosional relief. These incisions are further accentuated by the positive topography of the adjacent biohermal buildups (Figs. 10 and 12). The incisions are prominent features on the two-way-travel-time (TWT) structure map of DLP1 (Fig. 13). Away from areas of carbonate deposition, downlap surfaces are less rugose and in some areas planar (Fig. 10).

Seismic Facies

Four seismic facies can be recognized within each depositional cycle: (1) parallel-to-subparallel facies (delta platform), (2) channelized facies (distributaries, tidal channels, and incised valleys), (3) clinoform facies (delta front to prodelta), and (4) rugose facies (*Halimeda* bioherms and shelf edge buildups). Volumetrically, clinoform and delta-platform facies are the most significant, followed by carbonate facies and finally channel facies.

Horizontal-to-Subhorizontal Facies (Delta Platform).—

Horizontal-to-subhorizontal, conformable reflections of the platform facies represent updip extensions of sigmoid clinoforms and lap out in a landward direction against an onlap surface (OLP1) (Figs. 10 and 14). Stratigraphic position and the horizontal, parallel character of this facies imply vertical aggradation in a delta-platform depositional setting, ranging from intertidal-delta-plain to subtidal environments. Four vibracores targeted this facies, but owing to the stiff nature of the clay at the top of this facies, penetration was poor. The clay recovered from the top was light gray, mottled, and root burrowed, and contained abundant red oxidation zones as well as scattered lignite particles. Macroscopic faunal remains were absent. The sediment appeared to be a more oxidized version of the mottled organic-rich clay of the modern delta-plain *Nipa* palm swamps. Thick, extensive delta-platform facies were best developed above the P1S depocenter, with significant aggradation (up to 30 m) all the way to the shelf edge (Figs. 10 and 15). The P1N depocenter was dominated by topset-truncated oblique to sigmoid-oblique clinoforms, with platform facies restricted to fault-controlled areas of expanded section.

Channelized Facies (Distributaries, Tidal Channels, and Incised-Valley Fills).—

The channelized facies is mainly associated with P1S platform deposits and defined by shore-normal erosional trends containing a complex subset of laterally discontinuous facies

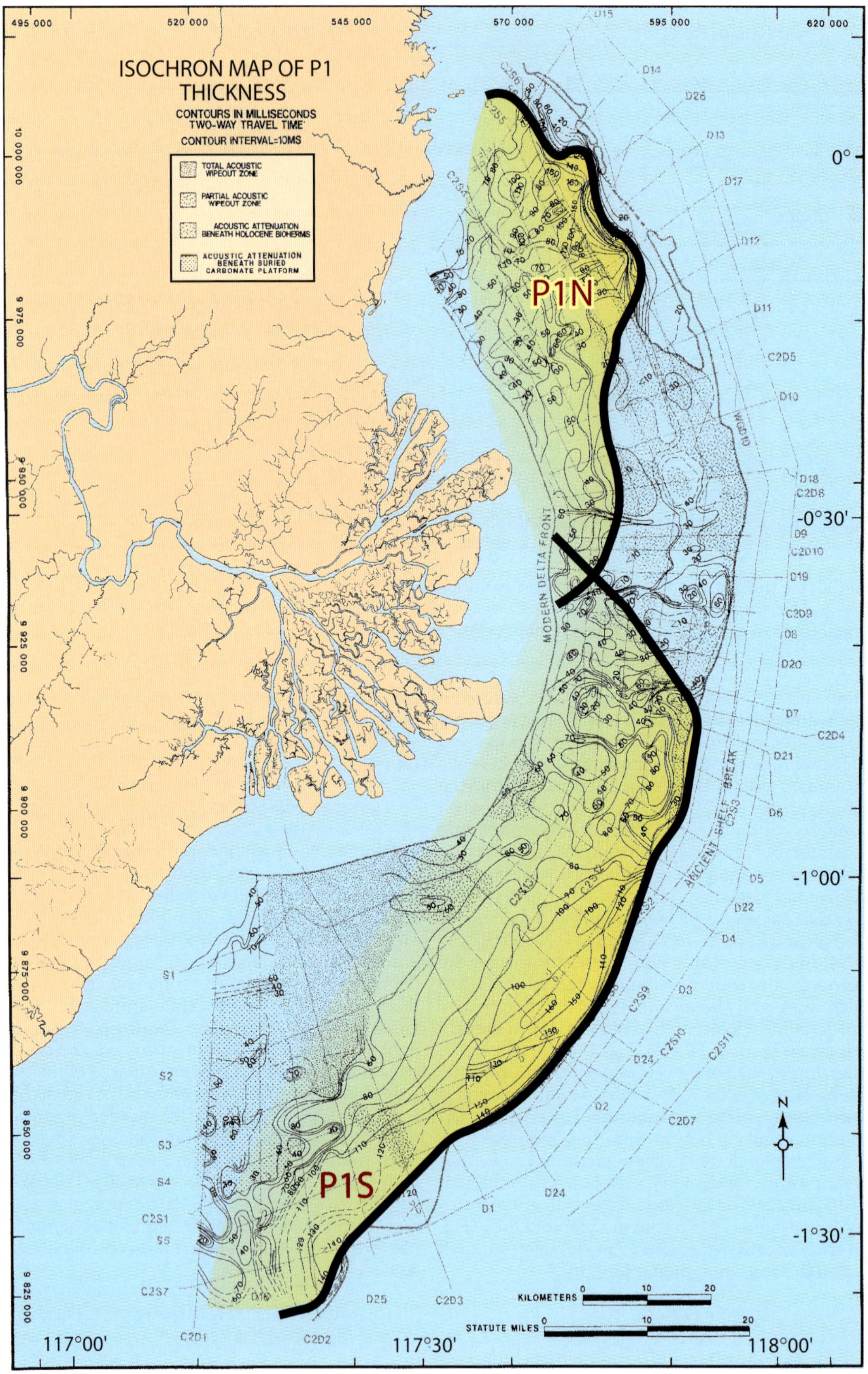

FIG. 11.—Isochron map of deltaic deposits related to the latest Pleistocene sea-level cycles (Cycle P1). Carbonates associated with the P1 Cycle are not included. Thickness is mapped in two-way travel time (10 ms ≈ 7.5 m).

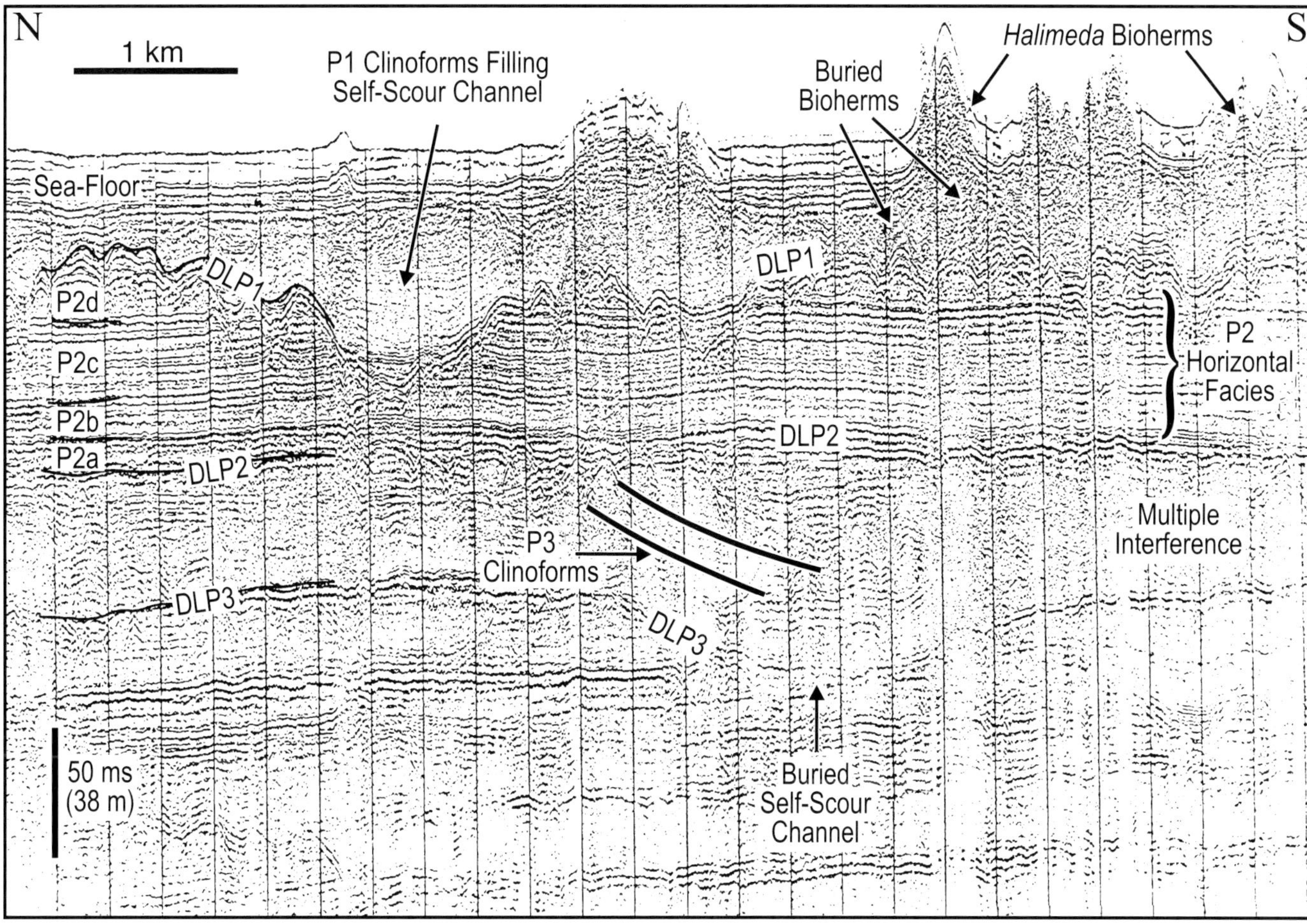

FIG. 12.—A high-resolution strike-oriented seismic profile of the central sector of the inner shelf. The P2 and P3 depocenters are located in this shelf sector. Stacked horizontal and subhorizontal seismic facies constitute much of the P2 seismic facies of this profile from the inner shelf. Units P2a–P2d represent deposits associates with individual progradational lobes of the P2 depositional cycle. These units are shown in the strike and dip sections of Figure 10. See Figure 2 for the location of Figure 12 seismic profile.

(chaotic, channel form, subhorizontal, lateral accretion reflections, and passive fill). Two types of channel trends are recognized. Volumetrically, the most significant is the shore-normal swath, 10–15 km wide and 30–40 m thick, located along the progradational axis of the P1S depocenter, interpreted as an incised-valley fill (Figs. 15 and 16). Encapsulated within the platform facies, on either side of the incised-valley fill, are narrow (< 1 km), linear channel trends, which may have passive fills, or a combination of reworked chaotic facies at the base and passive fill at the top. These trends are dimensionally similar to the major fluvial distributaries and large tidal channels of the modern delta plain (Fig 17). Narrow swaths (< 2 km) of channel facies are associated with progradational axes of the P1N clinoform depocenter.

Clinoforms Facies (Delta Front to Prodelta).—

Oblique to sigmoid clinoforms, with gradients of 2 to 5°, represent the prograding delta front to prodelta of the ancestral Mahakam River. Cores in clinoform topsets at the shelf edge, and a core in clinoform toesets exposed by a modern erosional scour channel, all contain dark gray organic-rich mud and silt, interbedded with sand and lignite laminae, similar to the Holocene distal-delta-platform and delta-front sediments. Oblique clinoforms occur on the inner to middle shelf, with topset truncation against a planar onlap surface. Oblique clinoforms transform into sigmoid clinoforms on the middle to outer shelf (Fig. 10).

Rugose Facies (Halimeda Bioherms).—

Carbonates overlying the P1 cycle are related to the Pleistocene to Holocene transgression and have previously been described in the Holocene section (Fig. 7). Pleistocene equivalents of the Holocene *Halimeda* bioherms and shelf-edge reef-like trends are also found overlying the P2 cycle, and earlier cycles. Bioherms occur above a planar ravinement surface, are commonly established on topographic highs (tops of abandoned delta lobes, footwalls of normal faults, edges of erosional channels), and may be solitary pinnacles, aggregate pinnacles that form platforms, or reef-like walls up to 100 m high at the shelf edge (Fig. 18). Regions in between bioherms are usually scoured to form deep shore-normal incisions such as associated with the DLP1 surface (Fig. 13). This scour also includes erosional "moats" around solitary pinnacles. The equivalent to the erosional surface beneath the P1 bioherms has been penetrated by vibracores outside of carbonate accretion areas, and represents a transgressive ravinement sur-

Fig. 13.—Isochron structure map of the Cycle P1 downlap surface (DLP1), the basal boundary of the P1 deposits. Similarly to the modern seafloor, the surface represents highstand sea-level flooding of the shelf.

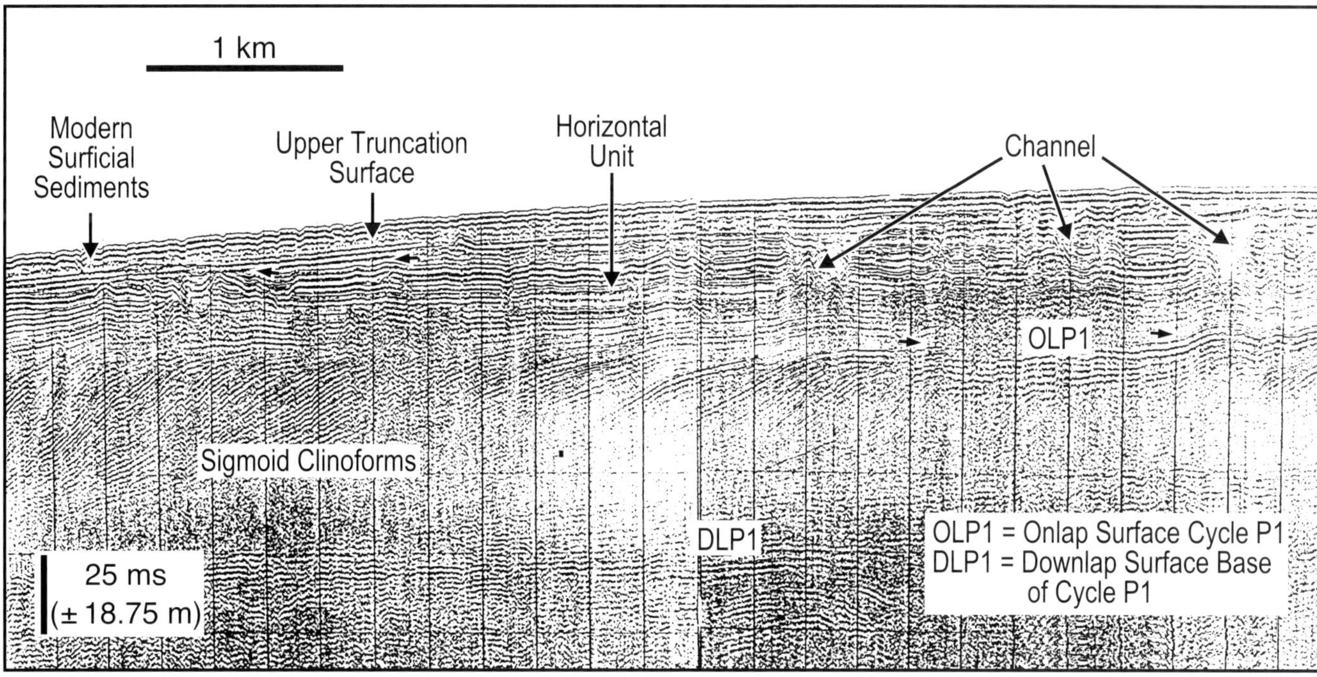

FIG. 14.—High-resolution seismic profile illustrating the transition from oblique to sigmoid clinoforms. Note that onlapping strata of horizontal to subhorizontal seismic facies represent the updip extensions of sigmoid clinoforms. See Figure 2 for the location of this seismic profile.

face, with deltaic and nonmarine sediment below and fossiliferous marine shelf mud above.

P1 Stacking Patterns and Systems Tracts

Reconstruction of detailed stratigraphy in ancient delta and shelf deposits is frequently made difficult by the limited resolution of conventional seismic data. The high resolution of the seismic data used in this study (bed resolution of < 2 m) provides an opportunity to construct a detailed stratigraphic framework for the cyclic late Pleistocene deposits that underlie the modern sediments of the shelf seaward of the Holocene Mahakam River delta. Seismic stratal geometries, surfaces, and stacking patterns are used to define the stratigraphic architecture of the latest Pleistocene depositional cycle (P1) and its relation to relative sea level in accordance with the basic principles described by Mitchum et al. (1977) and later additions by Posamentier et al. (1992), Nummedal et al. (1995), and Plint and Nummedal (2000). In this study we accept the arguments of Nummedal et al. (1995) and Plint and Nummedal (2000) that deposition during relative sea-level fall results in a distinctive "falling-stage systems tract" (FSST); this terminology has been adopted for describing the late Pleistocene deposits of the Mahakam shelf.

Falling-Stage Systems Tract (Forced Regression).—The transition from oblique clinoforms on the inner and middle shelf to sigmoid clinoforms and updip aggradation of delta-platform facies on the outer shelf implies a significant change in shelf accommodation space. Oblique clinoforms with truncated topsets are interpreted to reflect progradation during limited to decreasing accommodation, closely followed by subaerial exposure and erosion (falling relative sea level). Elevations of truncated topsets decrease steadily in a seaward direction, in large part owing to successively lower eustatic levels, but also owing to modification by differential subsidence, erosion, and structural deformation (Sydow, 1996). This surface of erosion represents the top of the FSST and is correlative with a submarine conformity that forms the base of the lowstand systems tract (LST). The surface created by this subaerial erosional truncation and sediment bypass is defined by Plint and Nummedal (2000) as a sequence boundary (SB). This FSST, as delineated by Plint and Nummedal (2000), has been recognized elsewhere, although a variety of different terms have been used to describe it (Posamentier et al., 1992; Hunt and Tucker, 1992; Sydow and Roberts, 1994). In early sequence stratigraphic literature, the FSST was considered to be part of the highstand systems tract (HST) as described by Posamentier and Vail (1988) and Van Wagoner et al. (1990). Recent studies using high-resolution seismic data indicate that in some cases highstand deltas prograde to the outer shelf (Anderson et al., 1996). However, most HST deposits occur on the inner shelf and have poor preservation potential because of subsequent erosion during relative sea-level fall. Because of this erosion it is very difficult to differentiate true HST from FSST deposits.

Lowstand Systems Tract.—Following lowstand sea-level turnaround, accommodation is created landward of the prograding delta front, allowing space for aggradation and expansion of the tidally influenced delta platform. The resulting horizontal delta-platform facies onlap the surface that truncates the oblique clinoforms, and is termed the onlap surface (Figs. 14 and 15). The shift from oblique to sigmoid clinoforms represents this lowstand turnaround. Sigmoid clinoforms and updip delta-platform facies are therefore considered to be part of the LST. Distribution of oblique and sigmoid clinoforms (FSST vs. LST) are shown in Figure 19. Most of the P1N depocenter is composed of oblique clinoforms. Sigmoid-oblique clinoforms on the outer shelf may

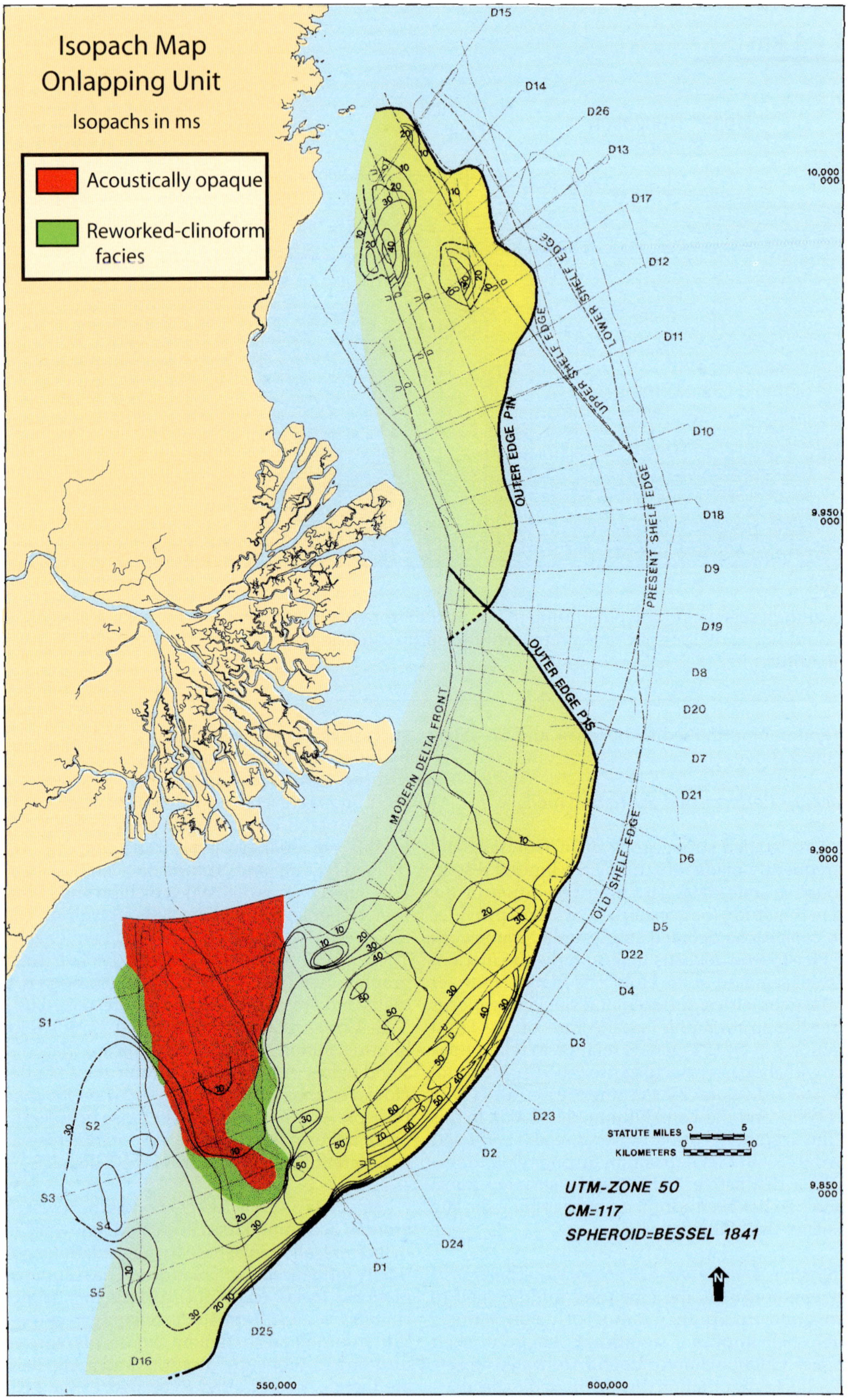

FIG. 15.—Isochron map of the distribution and thickness of horizontal to subhorizontal seismic facies (interpreted as delta-plain deposits) from the latest Pleistocene sea level cycle (P1 cycle). Thickness is mapped in two-way travel time (10 ms ≈ 7.5 m).

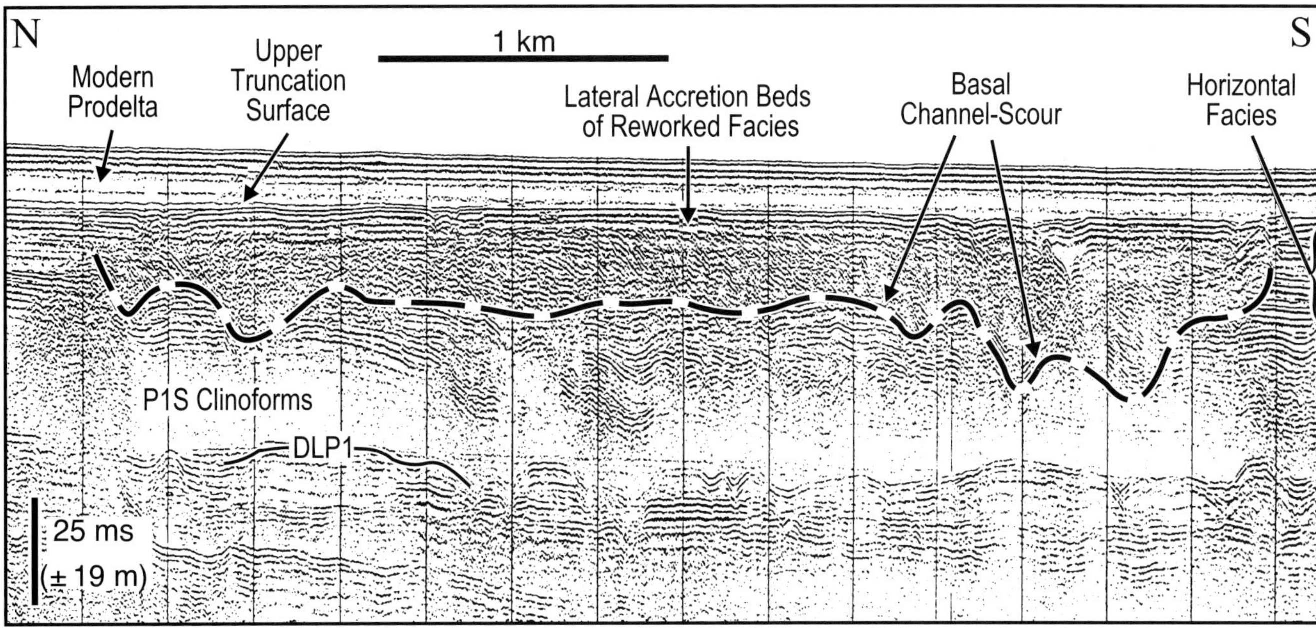

FIG. 16.—A high-resolution seismic profile across a reworked unit on the inner shelf associated with the southern extension of the latest Pleistocene sea-level cycle (the P1S depocenter). Note the erosional base on high-amplitude infilling units. See Figure 2 for the location of this seismic profile.

represent the lowstand relative sea-level turnaround. Poor preservation of delta platform facies probably reflects an overall smaller sediment supply from the Santan River (Fig. 3) and early abandonment of the Mahakam River supply (avulsion to the southern shelf).

The surface at the base of the incised-valley fill was clearly eroded during relative sea-level fall, but most of the fill thickness is likely to be the result of aggradation along with adjacent delta-platform facies accretion. Expansion of the tidal delta platform during this stage resulted in a large tidal prism, and development of an extensive network of distributaries and tidal channels, similar to the modern Mahakam delta (narrow channel trends, Fig. 17). The incised-valley fill and narrow channel facies are therefore also considered part of the LST.

Transgressive Systems Tract (delta abandonment and carbonate accretion).—Accelerating rates of relative sea-level rise eventually outstripped the ability of Mahakam River sediment to fill the new shelf accommodation. The delta front was abandoned, and transgression of the coastline began. Transgression of P1S

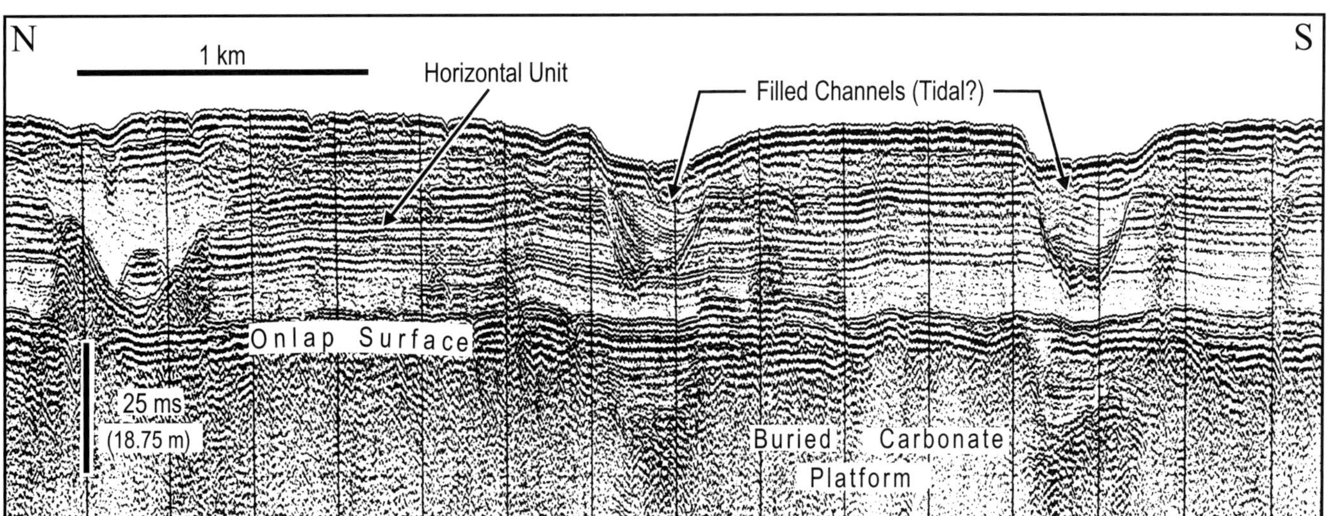

FIG. 17.—High-resolution seismic profile illustrating the narrow, steep-sided channels that cut through the horizontal to subhorizontal seismic facies associated with the latest Pleistocene sea-level cycle (P1 Cycle). See Figure 2 for the location of this seismic profile.

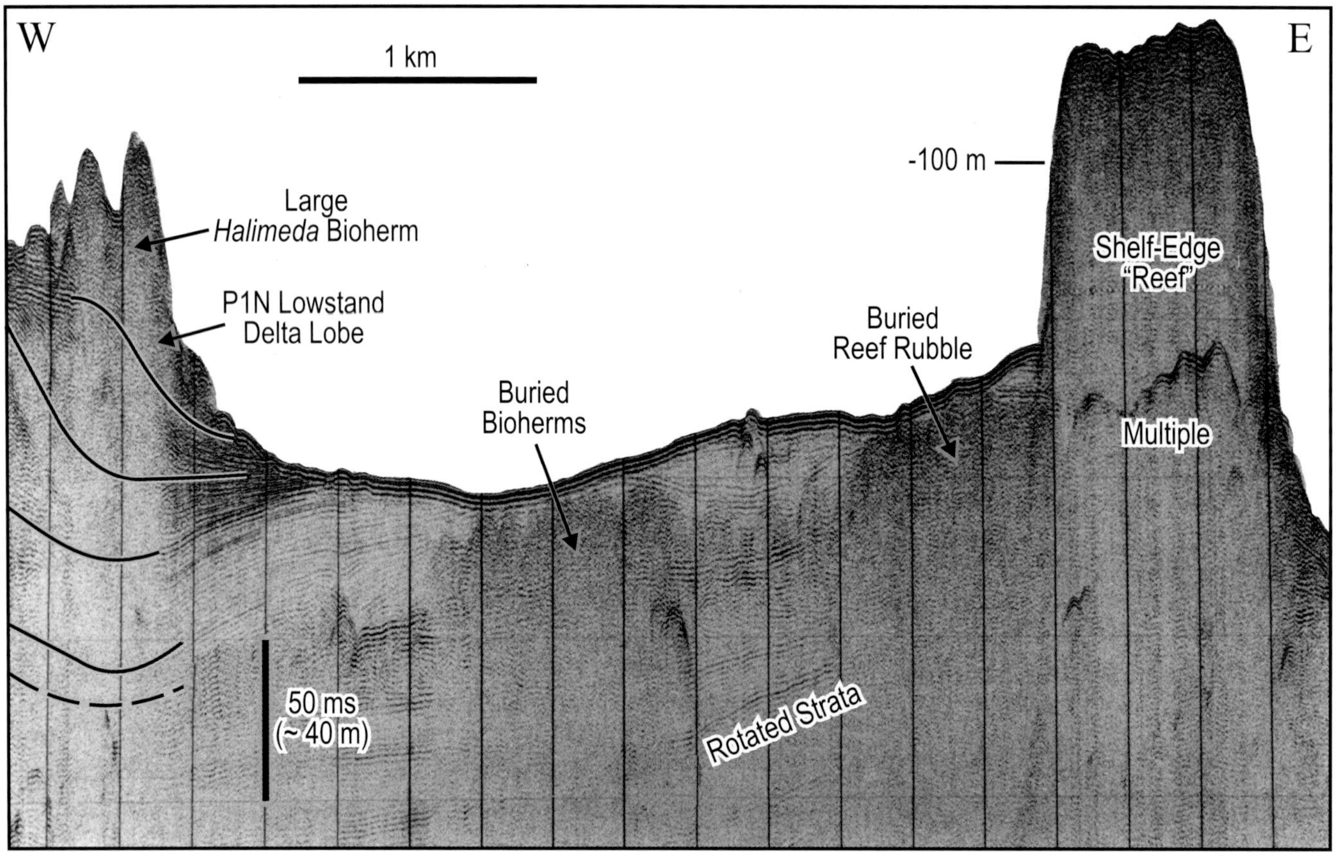

FIG. 18.—A high-resolution seismic profile across the shelf edge of the northern shelf sector illustrating a large (linear) reef-like buildup on the outer edge of a downfaulted and rotated fault block. See Figure 2 for the location of this seismic profile.

occurred in at least three backstepping increments, interrupted by a progradational pulse P1d (Fig. 10). Sediments constituting P1d are associated with significant delta-platform aggradation. Limited delta-plain deposition associated with P1N resulted in amalgamation of the subaerial exposure surface and transgressive ravinement surface. Vibracores penetrating the P1N ravinement reveal a nonmarine–marine contact. The physical processes that caused erosion at the ravinement surface were dominated by shore-normal tidal (surface tides nearshore and internal tides on outer shelf) and shore-parallel throughflow currents. Ravinement was therefore not restricted to nearshore marine areas (above fair-weather wave base), as is the case for most wave-influenced settings. Erosion on the shelf is active even today, in water depths of up to 100 m, and is amplified by the constricting and focusing effects of carbonate bioherms (Sydow, 1996).

Halimeda bioherms of the present shelf established themselves on top of the time-transgressive Pleistocene to Holocene ravinement surface and are therefore considered a transgressive phenomenon. The tops of the bioherms are covered by a thin drape of fossiliferous shelf mud with abundant flakes of the calcareous green algae *Halimeda* and reworked fauna. This drape represents a dramatic slowing of carbonate accretion. Radiocarbon dates of the *Halimeda* deposits beneath the drape indicate bioherm accretion rates from 72 mm to 170 mm/100 yr (unpublished project data from three piston cores taken from the inner shelf to the shelf edge). Bioherms were established sometime after P1 delta transgression, when the Mahakam River sediment plume was restricted to the inner shelf, and the shelf was sufficiently flooded for the Indonesian throughflow current to dilute and deflect the plume to the south. There is a clear link between distribution of bioherms and erosion related to the throughflow current and possibly internal tides. *Halimeda* communities are favored over more common framework-building coral communities under nutrient-enriched photic zone conditions (Hallock and Schlager, 1986). Nutrient-enriched waters are present beneath the mixed surface layer, and a distinct thermocline presently occurs at about 40 m water depth. Relative sea-level rise sufficient to allow flooding of the shelf by the nutrient-enriched waters, aided by internal tides and dynamic upwelling, is a further constraint on the onset of *Halimeda* bioherm accretion. Prolific rates of carbonate accretion result in buildups with 20 m to 50 m relief in the central and northern shelf sectors (Roberts and Sydow, 1996). Soon after onset of modern delta progradation (7 ka; Allen et al., 1979) bioherm communities were detrimentally affected by terrigenous sediment influx, and carbonate production systematically began shutting down. Outer-shelf reef trends in the northern sector accreted to water depths less than 20 m, and they are still actively accreting today.

DISCUSSION AND CONCLUSIONS

Results of the study highlighted in this paper provide a new perspective on the depositional linkage between deltaic and shelf

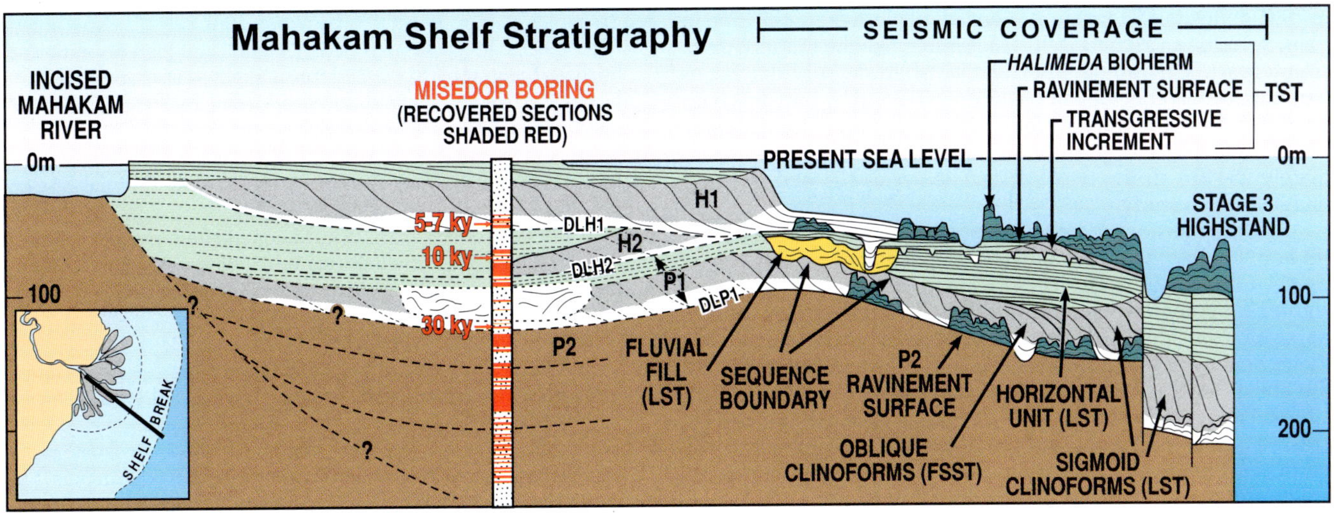

FIG. 19.—A schematic representation of Holocene and P1 cycle (deposits associated with the latest Pleistocene sea-level cycle) stratigraphy. Timing of important surfaces is extrapolated from the MISEDOR boring (Carbonel et al., 1987). In order to illustrate both interfluve and feeder systems, the alluvial valleys are shown schematically as axial-transverse views.

facies in the framework of the modern Mahakam delta. These results also underscore the importance of cyclic relative sea-level fluctuations in controlling the sedimentary architecture of deposits underlying the present shelf.

In the Holocene section, results defined an early pulse of deltaic progradation that underlies the present arcuate delta front. Delta-plain sediments from this early progradational event date to approximately 10 ka, as determined from the MISEDOR boring, whereas prodelta deposits of the present delta that downlap onto the truncated clinoforms of the earlier delta are dated at 5–7 ka. The reason for the development of this early delta lobe that formed at a lower-than-present sea level is not clear. Climatic conditions may have favored more water discharge and sediment discharge around 10 ka, relative sea level may have experienced a short-term fall or a dramatic decrease in rate of rise (perhaps the Younger Dryas event), or a combination of climate change and relative sea-level change encouraged this depositional pulse. Regardless, there is no question that the delta prograded in the early Holocene, retreated, and prograded again starting in the middle to late Holocene. Within these latest Holocene deltaic deposits, which are approximately 50 m thick, prodelta deposits are 10–20 m thick, but delta-front deposits are volumetrically the most important (about 40 m thick beneath the outer delta platform). Subtidal-platform deposits are approximately 10 m thick, and delta-plain sediments (*Nipa* swamp deposits) that cap the deltaic sedimentary sequence are typically 3–4 m thick. Nested within these deltaic facies are coarse-grained channel deposits, lateral channel bars, and tidal bars–distributary mouth bars that have variable thicknesses but do not exceed 15 m thick. There is an inverse relationship between TOC and sand percentage, with the highest TOC values occurring in the delta-plain and delta-front environments.

Holocene shelf sediments are dominated by carbonates in the northern and central sectors, whereas suspended-load sediments from the Mahakam River are advected to the southern shelf by north-to-south flow that originates from the Makassar Strait. The carbonates are dominated by individual and aggregate bioherms built by accumulated aragonitic plates of the calcareous green alga *Halimeda* in a mud matrix rich in foraminifera tests. These bioherms are seated on a regional ravinement surface formed as sea level transgressed the shelf following the latest Pleistocene glacial maximum. Initial radiocarbon dating of *Halimeda* plates from three piston cores (up to 6 m long) into these features suggests accretion rates of 72 mm to 170 mm/100 yr. Cores on which these accretion rates were derived are located from the inner shelf to the outer shelf. The outer-shelf core shows the highest accumulation rate. These rates are calculated from piston cores collected from the tops of bioherms. Increasing proportions of terrigenous mud in the core top suggest that suspended sediments from the rapidly prograding delta are negatively affecting *Halimeda* productivity. Carbonate accretion rates from the middle to basal parts of bioherms may have been substantially higher, but they could not be sampled with coring equipment available to this study.

At the shelf edge, reef-like buildups composed of coralline algae, *Halimeda*, and foraminifera tests form relatively hard and acoustically opaque pinnacles, mounds, and ridges. The dominance of calcareous and coralline algal species in place of reef-building hermatypic corals is interpreted to be a response to nutrient loading. On the Mahakam shelf this phenomenon is caused by upwelling of nutrient-rich Pacific intermediate water carried by the Indonesian throughflow current that flows north to south through the Makassar Strait. True coral reefs on the Mahakam shelf are limited to local areas of nearshore patch-reef development on the northern shelf.

Below the shelf ravinement surface over which the modern *Halimeda* bioherms have developed are falling-to-lowstand deposits of the latest Pleistocene sea-level cycle. These deposits contain surprising differences in stratigraphic elements when compared to the modern highstand delta and shelf. Falling-stage systems tract deposits of the northern shelf are represented by oblique clinoforms truncated by an interpreted exposure surface. These deposits when viewed in plan describe a multi-lobate delta that had a limited delta platform and was probably more river-dominated and with a smaller tidal prism than the modern Mahakam (Roberts and Sydow, 1996). Sediments deposited during lowstand and initial sea-level rise, especially those associated with the southern shelf, are represented by sigmoid clinoforms. Their subhorizontal updip extensions, interpreted as delta-plain deposits, onlap the early lowstand exposure surface. In contrast to FSST deposits, LST progradational deposits, when analyzed in plan, describe an arcuate delta front much like the modern delta counterpart. Onlapping intertidal to shallow subtidal facies (subhorizontal reflectors on seismic profiles cut by narrow channels) become quite thick in the lowstand to early-rise deposits. These organic-rich deposits volumetrically account for approximately 25–50% in the two Pleistocene cycles (P1 and P2) best imaged in this study.

The late Pleistocene deltaic deposits contain incised-valley-fill and smaller incised-channel-fill deposits not present in the modern highstand delta. These features are perhaps the most significant potential reservoir trends in these Pleistocene deltaic deposits. Considering deposits from the two best-imaged late Pleistocene cycles on the Mahakam shelf, lowstand deltas reached the abrupt shelf–slope transition along all sectors of the shelf. Clinoform aggradation and onlap of the FSST exposure surface created a thick package of subhorizontal delta-plain facies. The vertical accretion of a thick delta-plain facies suggests a reasonable residence time for the delta at the shelf edge.

This study of the modern Mahakam delta and adjacent shelf deposits, using high-resolution seismic and side-scan sonar data calibrated to sediment cores, has provided the framework for understanding the cyclic late Pleistocene stratigraphy of the tropical Mahakam shelf. It is now clear that the alternating dominance of deltaic deposition and carbonates so evident in the stratigraphic architecture of the shelf is the product of cyclic relative sea-level fluctuations typical of the Pleistocene.

ACKNOWLEDGMENTS

The offshore Mahakam delta study was funded Virginia Indonesia Company (VICO), UNOCAL, and TOTAL. Our thanks go to these companies, especially VICO, not only for project funding but also for logistics assistance and supporting data sets. Pertamina is gratefully acknowledged for permitting us to conduct this investigation. Our host company, P.T. Geoservices, provided excellent logistics support and help with the permitting process. Dr. H.L. Ong is cited for his individual interest in our research and personal help with governmental permits necessary to conduct our studies. Special thanks go to Walker Winans, Floyd DeMers, and Bill Gibson of the Coastal Studies Institute Field Support Group for their excellent technical assistance in preparing for and executing our data collection program. Mr. Richard Chua is gratefully acknowledged for his excellent help during the mob-demob phases of the project in Singapore. We thank Mary L. Eggart for preparation of the figures and April Hawkins for helping with manuscript preparation. Finally, we thank the reviewers who made suggestions that substantially improved this paper.

REFERENCES

Aharon, P., and Chappell, J., 1986, Oxygen isotopes, sea level changes and temperature history of a coral reef environment in New Guinea over the last 10^5 years: Palaeogeography, Palaeoclimatology, Palaeoecology, v. 56, p. 337–379.

Allen, G.P., 1994, Sediment patterns and facies in the Modern Mahakam Delta: TOTAL Field Guide, Paris, 36 p.

Allen, G.P., and Pizon, J., 1986, Detrital peat deposits in the modern Mahakam (Indonesia): Environmental significance and fossil analogues in adjacent Miocene delta deposits (abstract): 12th International Sedimentological Congress, Canberra, Australia.

Allen G.P., Laurier, D., and Thouvenin, J., 1979, Etude sédimentologique du delta de la Mahakam. Compagnie Française des Pétroles, TOTAL, Paris, Notes et Mémoires, no. 15, 154 p.

Anderson, J.B., Abdulah, K.C., Sarzalejo, S., Siringan, F., and Thomas, M.A., 1996, Late Quaternary sedimentation and high-resolution sequence stratigraphy of the east Texas shelf, *in* De Baptist, M., and Jacobs, P., Geology of Siliciclastic Seas: Geological Society of London, Special Publication 117, p. 95–124.

Arief, D., and Murray, S.P., 1996, Low-frequency fluctuations in the Indonesian throughflow through Lombok Strait: Journal of Geophysical Research, v. 101, p. 12,455–12,464.

Bailey, A.M., and Roberts, H.H., 1998, Minor and trace element compositions of carbonates formed during burial diagenesis of deltaic sediments: 8th Annual V.M. Goldschmidt Conference, August 30–September 3, Toulouse, France, p. 1–3.

Broecker, W.S., Andree, M., Wolfli, W., Oeschger, H., Bonani, G., Kennett, J., and Peteet, D., 1988, The chronology of the last deglaciation: Implications to the cause of the Younger Dryas event: Paleoceanography, v. 3, p. 1–19.

Carbonel, P., Hoibian, T., and Moyes, J., 1987, Ecosystems et paléoenvironments de la zone deltaique de la Mahakam depuis la fin de Neogène: LeSondage Misedoe, Editions Technip, Paris, p. 85–135.

Coleman, J.M., and Roberts, H.H., 1988, Sedimentary development of the Louisiana continental shelf related to sea level cycles, part 1: Sedimentary sequences: Geo-Marine Letters, v. 8, p. 63–108.

Dutrieux, E., 1991, Study of the ecological functioning of the Mahakam Delta (East Kalimantan, Indonesia): Estuarine, Coastal and Shelf Science, v. 32, p. 415–420.

Dutrieux, E., Denis, J., and Populus, J., 1990, Application of SPOT data to a base-line ecological study of the Mahakam Delta mangroves (East Kalimantan, Indonesia): Oceanologica Acta, v. 13, p. 317–326.

Fairbanks, R.G., 1989, A 17,000 year glacio-eustatic sea level record: Influence of glacial melting rates on the Younger Dryas event and deep ocean circulation: Nature, v. 342, p. 637–642.

Gastaldo, R.A., Allen, G.P., and Huc, A.Y., 1993, Detrital peat formation in the tropical Mahakam River Delta, Kalimantan, eastern Borneo: Sedimentation, plant composition, and geochemistry, *in* Cobb, J.C., and Cecil, C.B., eds., Modern and Ancient Coral-Forming Environments: Geological Society of America, Special Paper 286, p. 107–118.

Gastaldo, R.A., and Huc, A.Y., 1992, Sediment facies, depositional environments, and distribution of phytoclasts in the recent Mahakam River Delta, Kalimantan, Indonesia: Palaios, v. 7, p. 574–590.

Gayet, J., and Legigan, P., 1987, Etude sédimentologique du Sondage Misedor (delat de la Mahakam, Kalimantan, Indonesia): LeSondage Misedor, Editions Technip, Paris, p. 23–71.

Goodbred S.L., and Kuehl, S.A., 2000, Enormous Ganges–Brahmaputra sediment discharge during strengthened early Holocene monsoon: Geology, v. 20, p. 1083–1086.

Hallock, P., and Schlager, W., 1986, Nutrient excess in the demise of coral reefs and carbonate platforms: Palaios, v. 1, p. 389–398.

Hardy, M.J., 2000, Origin, distribution, and degradation of sedimentary organic matter in a modern tropical deltaic system (Mahakam delta, Borneo, Indonesia): Ph.D. dissertation, Department of Geology and Geophysics, Louisiana State University, 368 p.

Hunt, D., and Tucker, M.E., 1992, Stranded parasequences and the forced regressive wedge systems tract: deposition during base-level fall: Sedimentary Geology, v. 81, p. 1–9.

Loutit, T.S., Hardenbol, J., Vail, P.R., and Baum, G.R., 1988, Condensed sections: the key to age determination and correlation of continental margin sequences, *in* Wilgus, C.K., Hastings, B.S., Kendall, C.G.St.C., Posamentier, H.W., Ross, C.A., and Van Wagoner, J.C., eds., Sea-Level Changes: an Integrated Approach: SEPM, Special Publication 42, p. 183–213.

Matthews, R.K., 1990, Quaternary sea level change, *in* Sea Level Change: Geophysics Study Committee, Commission on Physical Sciences and Resources: National Research Council, National Academy Press, Washington, p. 88–103.

MITCHUM, R.M., VAIL, P.R., AND SANGREE, J.B., 1977, Seismic stratigraphy and global changes of sea level, part 6: stratigraphic interpretation of seismic reflection patterns in depositional sequences. *in* Payton, C.E., ed., Seismic Stratigraphy: Applications to Hydrocarbon Exploration: American Association of Petroleum Geologists, Memoir 26, p. 117–134.

MURRAY, S.P., AND ARIEF, D., 1988, Throughflow into the Indian Ocean through the Lombok Strait, January 1985–January 1986: Nature, v. 333, p. 444–447.

NUMMEDAL, D., GUPTA, S., PLINT, A., AND COLE, R., 1995, The falling stage systems tract: definition, character, and expression in several examples from the Cretaceous from the U.S. Western Inferior, *in* Hunt, D., Gawthorpe, R.L., and Docherty, M., convenors, Sedimentary Responses to Forced Regressions: The Geological Society of London, Abstract Volume, p. 45–48.

PLINT, G., AND NUMMEDAL, D., 2000, The falling stage systems tract: recognition and importance in sequence stratigraphic analysis, *in* Hunt, D., and Gawthorpe, R.L., eds., Sedimentary Responses to Forced Regressions: The Geological Society of London, Special Publication 172, p. 1–17.

POSAMENTIER, H.W., AND VAIL, P.R., 1988, Eustatic controls on clastic deposition II: sequence and systems tract models, *in* Wilgus, C.K., Hastings, B.S., Kendall, C.G.St.C., Posamentier, H.W., Ross, C.A., and Van Wagoner, J.C., eds., Sea Level Changes: An Integrated Approach: SEPM, Special Publication 42, p. 125–154.

POSAMENTIER, H.W., ALLEN, G.P., JAMES, D.P., AND TESSON, M., 1992, Forced regressions in a sequence stratigraphic framework: concepts, examples, and exploration significance: American Association of Petroleum Geologists, Bulletin, v. 76, p. 1687–1709.

ROBERTS, H.H., AND SYDOW, J., 1996, The offshore Mahakam Delta: stratigraphic response of Late Pleistocene-to-Modern sea level cycle: Indonesian Petroleum Association, Proceedings, v. 25, p. 147–162.

ROBERTS, H.H., AHARON, P., AND PHIPPS, C.V., 1988, Morphology and sedimentology of *Halimeda* bioherms from the eastern Java Sea (Indonesia): Coral Reefs, v. 6, p. 161–172.

ROBERTS, H.H., PHIPPS, C.V., AND EFFENDI, L., 1987, *Halimeda* bioherms of the Eastern Java Sea, Indonesia: Geology, v. 15, p. 371–374.

SAMUDRA, R.V., 1975, Hydrological, plankton, and pigment observations in the Makassar Strait, Madura Strait, and eastern Java Sea: Oceanographic Cruise Report 16, Institute of Marine Research, National Institute of Oceanography, Indonesian Institute of Science.

SIEMERS, C.T., CHAMBERS, J.L.C., AND ALLEN, G.P., 1994, Field seminar on the sedimentology and sequence stratigraphy of modern and ancient fluvial, deltaic, and shelf deposits: American Association of Petroleum Geologists and P.T. Geoservices, 13–20 November, Guidebook, 85 p.

SYDOW, J., 1996, Holocene to Late Pleistocene stratigraphy of the Mahakam Delta, Kalimantan, Indonesia: Ph.D. dissertation, Louisiana State University, Baton Rouge, 170 p.

SYDOW, J., AND ROBERTS, H.H., 1994, Stratigraphic framework of a Late Pleistocene shelf-edge delta, northeast Gulf of Mexico. American Association of Petroleum Geologists, Bulletin, v. 78, p. 1276–1312.

TAD—TRANSMIGRATION AREA DEVELOPMENT PROJECT REPORT, 1982, Rainfall records, East Kalimantan, report: analysis summaries and histograms: Samarinda, Kalimantan Timur, Indonesia, 21 p.

TOTAL, 1986, Meteorological and Oceanographical Campaign Balikpapan Final Report: TOTAL internal Report, 67 p.

VAN WAGONER, J.C., MITCHUM, R.M., CAMPION, K.M., AND RAHMANIAN, V.D., 1990, Siliciclastic Sequence Stratigraphy in Well-Logs, Cores, and Outcrops: American Association of Petroleum Geologists, Methods in Exploration Series 7, 55 p.

WYRTKI, K.E., 1987, Indonesian throughflow and the associated pressure gradient: Journal of Geophysical Research, v. 92, p. 12,941–12,946.

SEDIMENTOLOGY AND STRATIGRAPHY OF A TIDE-DOMINATED, FORELAND-BASIN DELTA (FLY RIVER, PAPUA NEW GUINEA)

ROBERT W. DALRYMPLE
Department of Geological Sciences and Geological Engineering, Queens University, Kingston, Ontario K7L 3N6, Canada
ELAINE K. BAKER
Australian ODP Secretariat, School of Geosciences, University of Sydney, New South Wales 2006, Australia
PETER T. HARRIS
Petroleum and Marine Division, Geoscience Australia, GPO Box 378, Canberra, ACT 2601, Australia
AND
MICHAEL G. HUGHES
School of Geosciences, Division of Geology and Geophysics, University of Sydney, New South Wales 2006, Australia

ABSTRACT: The delta of the Fly River is tide dominated, both morphologically and sedimentologically: the spring tidal range reaches 5 m, with near-bed tidal currents that commonly exceed 1 m/s, whereas wave influence is minimal, except during the monsoon period, when winds blow onshore. The delta plain displays a classic funnel-shaped geometry, with three main distributaries that flare seaward from a common bifurcation point 110 km inland. Of these, the southernmost is the main route for sediment export; the northernmost channel is effectively abandoned at present and is experiencing widespread erosion by tidal currents. Despite the net export of mud and sand, mutually evasive tidal-current patterns have created a series of elongate tidal bars within both active and abandoned distributary channels and in the distributary-mouth-bar area. The later is a zone of bedload convergence, with net seaward transport on its landward side and net landward transport on its seaward flank. This limits the offshore movement of sand.

The deposits of the Fly delta are dominated by mud, because the river flows along the low-gradient axis of the foreland basin before reaching the sea, causing deposition of most sediment coarser than fine to very fine sand farther inland. Distributary-channel bases are floored by a thin unit of cross-bedded and rippled sand and mud-pebble conglomerate. These are typically overlain abruptly by mud deposits formed by dense fluid-mud bodies that form in the channel bottoms during spring tides. The mud layers in this facies are anomalously thick (commonly > 1 cm), and channel-floor deposits are characterized by interbedding of the coarsest and finest sediments. Above these muds, the sediments are pervasively heterolithic and show a net upward coarsening to about the mid-depth level on the tidal bars because of the thinning of the mud layers. These bars may contain 50% (or more) mud and display lateral-accretion bedding. Bioturbation is scarce to absent. The sediments then fine upward into the intertidal zone. Clear indications of a tidal origin for the heterolithic stratification are relatively uncommon, although tidal rhythmites are present locally, including within active channels. The mouth-bar deposits are predominantly sand and also contain lateral-accretion bedding. The delta-front facies are heterolithic, with both millimeter- and decimeter-thick sand/mud alternations, all with a limited degree of bioturbation. The prodelta consists of biologically homogenized mud. The stratigraphic organization of the deposits reflects the fact that the delta plain aggraded vertically during the last part of the postglacial sea-level rise, producing a complex stack of channel deposits, followed by progradation of the mouth bars by as much as 40–50 km.

INTRODUCTION

Tide-dominated (as opposed to tide-influenced) deltaic systems are poorly known compared to other types of deltas. Indeed, some authors have questioned their very existence (Bhattacharya and Walker, 1992; Walker, 1992), because they generally occur in large embayments and therefore lack a discrete shoreline protuberance. However, if "delta" is defined in more fundamental terms as "a progradational sediment body at the mouth of a river, formed of sediment supplied by the river, and containing fluvially-influenced deposits" (Dalrymple, 1999, 2000; cf. Dalrymple et al., 1992), then tide-dominated deltas are arguably the most abundant delta type today. In fact, eight of the twelve largest deltas in the world, including four of the top five (the Ganges–Brahmaputra, Amazon, Chang Jiang, and Irawaddy) are either tide dominated or experience significant tidal influence (Middleton, 1991).

Previous work on tide-dominated deltas is patchy. The Ord River, which is commonly used as a type example (e.g., Wright, 1985), has been described only in a reconnaissance fashion (Wright et al., 1975) and may, in fact, be an estuary and not a delta at all (Dalrymple et al., 1992). The intertidal flats on the margins of the Colorado River delta have received considerable attention (Thompson, 1968), but the distributaries and mouth-bar area have not been examined in detail (Meckel, 1975). Work has also appeared recently describing deposits of the Amazon (Jaeger and Nittrouer, 1995; Kuehl et al., 1996), Ganges–Brahmaputra (Segall and Kuehl, 1994), and Rajang (Staub and Esterle, 1993) deltas, but these studies are generally localized in geographic or thematic scope and do not give a comprehensive picture of the entire deltaic system. As a result, attempts to produce a general facies model for tide-dominated deltas (e.g., Dalrymple, 1992) have relied heavily on facies information from tide-influenced (i.e., mixed-energy) deltas such as the Mahakam (Allen et al., 1979; Gastaldo et al., 1995) or the Niger (Allen, 1970). However, such systems show significant morphological differences from strongly tide-dominated deltas; hence, their applicability to tide-dominated systems can be questioned.

The purpose of this paper is to help fill this gap, by providing a comprehensive overview of the depositional processes, geomorphology, sedimentology, and stratigraphy of the Fly River delta, located in the southeastern portion of Papua New Guinea (Fig. 1). The Fly River is the 17th largest river in the world in terms of sediment discharge, and its delta is a classic tide-dominated system (Fisher et al., 1969; Galloway, 1975), with a funnel-shaped mouth and numerous distributaries separated by shore-normal tidal bars

Tropical Deltas of Southeast Asia—Sedimentology, Stratigraphy, and Petroleum Geology
SEPM Special Publication No. 76, Copyright © 2003
SEPM (Society for Sedimentary Geology), ISBN 1-56576-086-7, p. 147–173.

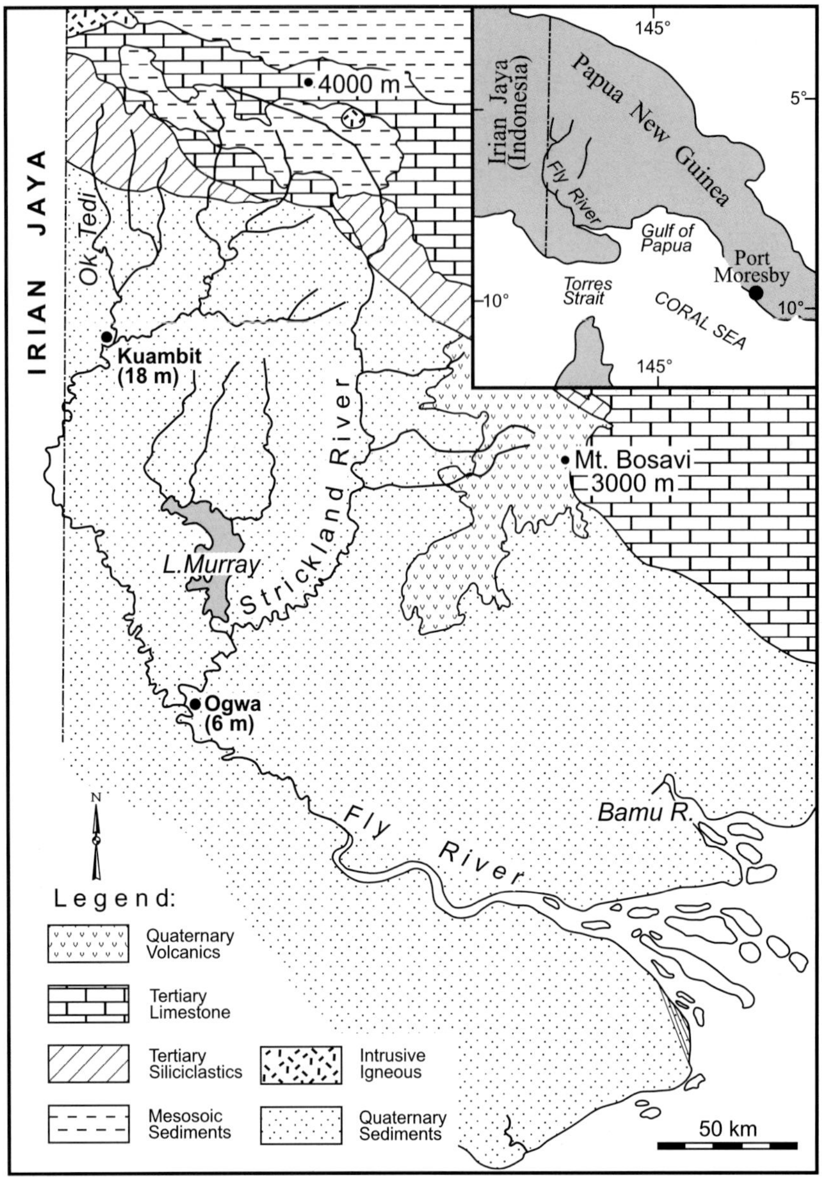

FIG. 1.—General location map of the Fly River drainage system showing bedrock geology (after Harris, 1994). Note the low river gradients. Beach ridges along the south side of the delta suggest that the river mouth transgressed to approximately the inner end of the beach-ridge plain during the early to mid-Holocene, before beginning to prograde seaward during the current highstand.

and islands (Fig. 2). It lies in a foreland basin, in an area of long-term net deposition, and so is influenced minimally by pre-existing bedrock topography. As a result, it is a potentially valuable analogue for similar deposits in ancient sedimentary basins.

Although the area is remote, there has been sufficient previous work to provide a reasonably complete picture of the process framework. Previously published work includes studies of the hydrodynamics (e.g., Alongi et al., 1992; Wolanski et al., 1990; Wolanski et al., 1995a; Wolanski et al., 1995b; Wolanski et al., 1998; Wolanski and Eagle, 1991; Wolanski and Gibbs, 1995), as well as reconnaissance-level sedimentological investigations by Harris et al. (1993) and Baker et al. (1995). Sediments of the adjacent Gulf of Papua have been described by Harris (1988, 1995) and Harris et al. (1996). The most comprehensive previous synthesis of the existing data is presented by Baker (1998).

METHODOLOGY AND DATA BASE

This study is based mainly on a three-week cruise in January, 1994, that investigated the distributary channels and mouth-bar region of the delta. In total, almost 500 km of high-resolution seismic data were obtained using an ORE Geopulse Boomer, together with approximately 170 km of 100 kHz sidescan sonograms. Bathymetric data were collected continuously throughout the cruise, because large parts of the delta were uncharted or the bathymetry had changed. Surveying was concentrated in four small, representative areas: (1) at the first large meander bend in the tidally influenced river channel upstream of the delta "Apex" (i.e., the region where the river splits into distributary channels; Fig. 2); (2) at the "Apex"; (3) in the outer part of the Far Northern Entrance, an "inactive" distributary; and

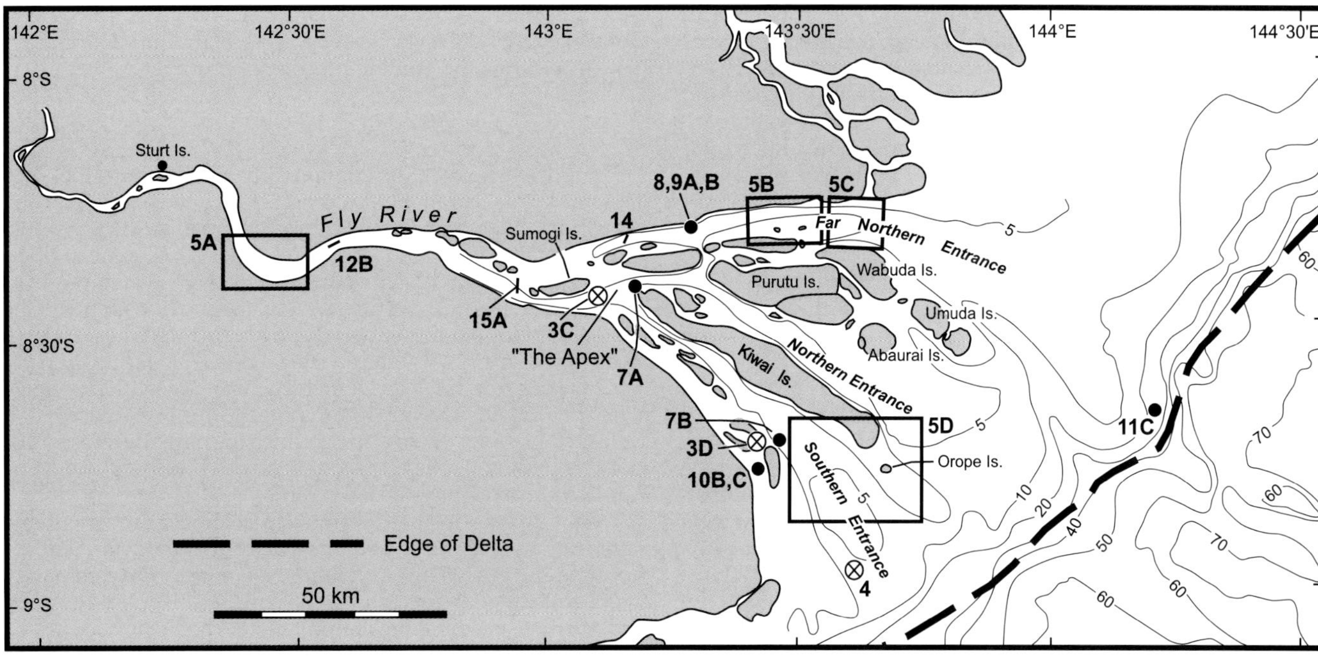

FIG. 2.—Map of Fly River delta and adjacent part of the Gulf of Papua continental shelf. Bathymetric contours in meters. The "Edge of Delta" line represents the toe of the prodelta slope (after Harris et al., 1996). Sediment on the shelf to the southeast of this line contains more than 60% carbonate. Boxes give locations of detailed morphological maps shown in Figure 5. Other numbers indicate the location of current-meter data (⊗), seismic sections (short lines), and cores (•) shown in the relevant figures.

(4) at the mouth of the Southern Entrance, an active distributary, and on the adjacent distributary-mouth bars.

One hundred and fifty-nine grab samples were obtained on detailed grid networks (0.5–3.5 km sample spacing) within the four detailed study areas, using a 0.25 m^2 Shipek grab sampler. Thirty vibracores (maximum length 5.3 m) were collected from subtidal and intertidal areas. They were subsequently split and logged at a centimeter scale. Five of them were X-radiographed after splitting. Textural analyses were performed on grab and core samples using a Malvern Laser Particle Size Analyser.

To monitor the tidal currents, Aanderaa RCM-7 self-recording current meters were deployed one meter above the bottom at six locations (one at the "Apex", area 2; three in the Far Northern Entrance, area 3; and two in the Southern Entrance, area 4) for periods of up to 14 days. In addition, boundary-layer measurements of current speed and suspended-sediment concentrations were recorded over one or two tidal cycles at three sites, one in or adjacent to each of areas 2, 3, and 4.

These observations and data are supplemented by additional grab samples and cores from the delta front, prodelta, and adjacent shelf obtained on previous cruises (see Harris et al. 1996). Long-term current-meter data are also available from one site on the prodelta (Harris, 1995). Readers are referred to Baker (1998) for a detailed tabulation of the data base.

ENVIRONMENTAL SETTING

Location and General Morphology

The headwaters of the Fly River and its main tributary, the Strickland River, lie in the highlands along the collisional northern margin of the Australian lithospheric plate. The river flows southward through the foreland fold and thrust belt, into the adjacent foreland basin (Fig. 1). There it turns eastward and runs along the basin axis for a distance of over 700 km before reaching the Gulf of Papua. Gradients are extremely low: the river is 6 m above sea level (asl) at Ogwa, 430 km from the sea (slope 1.4 x 10^{-5} degrees from the coast to Ogwa), and only 18 m asl at Kuambit, 850 km from the coast (slope 2.9 x 10^{-5} degrees between Ogwa and Kuambit). The thalweg of the river is continuously below mean sea level approximately 500 km inland.

The river ends in a funnel-shaped, tide-dominated delta that spans 100 km of the coastline (Fig. 2). It contains three main distributaries (the Southern, Northern, and Far Northern Entrances) that branch from a common point (the "Apex"), 110 km inland. These distributaries are separated by elongated, vegetated islands that pass seaward into an extensive (ca. 20 km wide) area with water depths generally less than 10 m. This distributary-mouth bar and delta-front area is occupied by shore-normal tidal sand bars. Seaward of this, the prodelta slopes relatively steeply seaward (slope 0.0025°), meeting the topographically irregular shelf surface at a depth of 40–50 m (Fig. 2). Map and airphoto records of coastline change indicate that the delta is prograding, at an average rate of about 6 m/a (Harris et al., 1993). Seismic data also show that prodelta muds downlap progradationally onto the modern shelf carbonates that mark the maximum flooding surface (Harris et al., 1996). Thus, the system is actively exporting significant amounts of river-supplied sediment, including sand, to the sea (Harris et al., 1993; Harris et al., 1996; Robertson and Alongi, 1995) and is a delta as defined above.

Water and Sediment Discharge

The climate of the source is tropical and ever wet, with rainfall exceeding 10 m/a! Discharge varies seasonally, but the variation is not great, typically reaching only ± 25% of the mean flow (Wolanski et al., 1997). Salinity measurements show that the limit of salt-water intrusion is located approximately 50 km inland from the coast (i.e., about halfway between the coast and the "Apex") in the Southern Entrance, whereas salt water penetrates nearly to the "Apex" (i.e., 100 km inland) in the Far Northern Entrance (Wolanski and Eagle, 1991; Wolanski et al., 1997). Such observations have been used to calculate that 60–80% of the river outflow reaches the Gulf of Papua by way of the Southern distributary, with less than 10% of the discharge exiting by way of the Far Northern Entrance (Wolanski et al., 1997). Thus, the Far Northern distributary is essentially inactive with respect to sediment export at present and is temporarily abandoned.

The combination of high rainfall, mountainous terrain, and abundance of easily weathered volcanic, sedimentary, and weakly metamorphosed bedrock leads to rapid rates of erosion (3–4 mm / a; Pickup, 1984) and very high sediment loads in the Fly River (estimated sediment discharge 85 million metric tons / yr; Harris et al., 1993). Mining in the mountainous headwaters of the Ok Tedi tributary (Fig. 1) is believed to have caused a 40% increase in the sediment discharge since 1985 (Eagle and Higgins, 1990), but this has not produced a recognizable change in the delta's morphology or deposits.

The sediment source provides a wide range of grain sizes, but the low river gradient in the foreland basin leads to rapid deposition of the coarser fractions. Gravel and the coarser sand sizes are dropped close to the mountain front, while deposition of the medium and fine sand occurs in inland parts of the foreland basin (Pickup, 1984; Parker, 1991; Dietrich et al., 1999). This causes aggradation of the channel and flanking levees, which in turn leads to blockage of tributaries and the development of lakes (e.g., Lake Murray; Fig. 1; Löffler, 1977). As a result of this selective deposition, only fine-grained sediment reaches the delta. Pickup (1984; see also Higgens et al., 1987) has estimated that perhaps as little as 10% of this sediment is of sand size, most of which is fine to very fine sand (Baker, 1998). Despite the tropical climate, silt-size material predominates in the mud fraction, with clay-size material constituting less than 20% of the suspended load (Wolanski and Gibbs, 1995; Baker, 1998).

The composition of the sediment reflects the lithologic heterogeneity and immaturity of the source area. Sands are dark colored and are composed of (in decreasing abundance) lithic fragments, quartz, plagioclase feldspar, pyroxene, and hornblende (Baker, 1998). Heavy minerals can be abundant and include ilmenite, zircon, rutile, and monazite. The clay-mineral fraction consists of weakly crystalline kaolinite, chlorite, illite, and montmorillonite (Baker, 1998). These compositions are uniform throughout the delta, indicating that river-supplied sediment is reaching the distributary-mouth bar and prodelta.

Wave Climate

The wave climate in the Gulf of Papua is strongly tied to the seasonal monsoon. Wave energy is highest from May to October, when winds blow onshore, but even then the waves are not large by global standards (significant wave height only 1.3 m; Thom and Wright, 1983). During the rest of the year, wave energy is minimal.

The open, funnel-shaped mouth of tide-dominated deltas raises the possibility that ocean waves might penetrate into the delta-plain distributaries and have a greater influence on sedimentation there than they do in other delta types. However, qualitative observations suggest that the shallow distributary-mouth bars, the tops of which at least locally reach into the intertidal zone, block all but small, short-period waves, most of which are generated locally. The scarcity of beaches at the high-tide level, except at the very seaward ends of the interdistributary islands, confirms the limited importance of waves.

Tides and Tidal Currents

The tides are mixed, predominantly semidiurnal (Fig. 3A). The semidiurnal component is most strongly developed during spring tides, but a pronounced diurnal inequality exists, producing alternately higher and lower high tides. During neap tides, the semidiurnal component nearly disappears, leaving only a small, essentially diurnal tide. The tidal range at the mouth of the delta is 3.8 m (upper mesotidal) during spring tides but less than 1 m during neap tides (Wolanski and Eagle 1991). Within the delta distributaries, the tides are hypersynchronous, with the range increasing inland. The maximum range (5.1 m; i.e., macrotidal) occurs at the "Apex", landward of which the range decreases. The tidal limit is situated near the junction with the Strickland River (Fig. 1), approximately 400 km inland.

Tidal currents in the distributary channels are essentially rectilinear, paralleling the channel margins. Maximum speeds vary with the tidal range and mimic the mixed, predominantly semidiurnal tidal character of the system (Fig. 3B). During spring tides, current speeds, as measured 1 meter above the bed, exceed 0.5–0.8 m / s (maximum recorded value 2.3 m / s) at all monitored sites in the distributary channels, including the "Apex" area (see also Wolanski and Eagle, 1991; Wolanski et al., 1995a). Slack-water periods are relatively brief, with the duration of speeds less than 0.1 m / s being only 10–20 minutes. By contrast, during neap tides, peak speeds are typically of the order of only 0.2–0.5 m / s and slack-water periods last from as little as 1–2 hours to as much as 24 hours.

On the delta front and prodelta, tidal currents weaken rapidly offshore. A current meter located on the delta front seaward of the Southern Entrance recorded maximum current speeds of approximately 0.4 m / s (Fig. 4), oriented in an onshore–offshore direction. Residual water movement is to the north, parallel to the coast (Wolanski et al., 1990; Harris, 1995; Wolanski et al., 1999).

Suspended-Sediment Distribution

Large amounts of fine-grained sediment are in suspension throughout the entire system, with suspended-sediment concentration (SSC) values rarely falling below 0.5 g / l (Wolanski and Eagle, 1991; Wolanski and Gibbs, 1995; Wolanski et al., 1995a). SSC values vary over individual tidal cycles as well as the neap–spring cycle (Figs. 3C, D, 4), with the lowest concentrations occurring when the current speeds are low (i.e., slack-water periods and neap tides), and the highest during or slightly after spring tides because of greater resuspension from the bed. During spring tides, a well developed turbidity maximum (the zone of elevated suspended-sediment concentrations) is formed in the region stretching from above the "Apex" to the distributary-mouth bar area (Wolanski et al., 1995a; Wolanski and Gibbs, 1995). SSC values in this area are very high, with depth-averaged values typically 1–3 g / l.

Beneath the turbidity maximum, fluid-mud bodies are widely developed (Wolanski and Eagle, 1991; M. Hughes, personal communication, 1994; Wolanski et al., 1995a). These dense, mobile, near-bed suspensions (SSC values > 10 g / l; maximum recorded 40 g / l; Figs. 3C, D) are typically up to 1 m thick (cf. Wolanski and Eagle, 1991) and occupy the deepest parts of the channels. They form during slack-water periods when the suspended sediment begins to settle and are dispersed to varying degrees when the tidal-current speeds increase. Fluid muds are

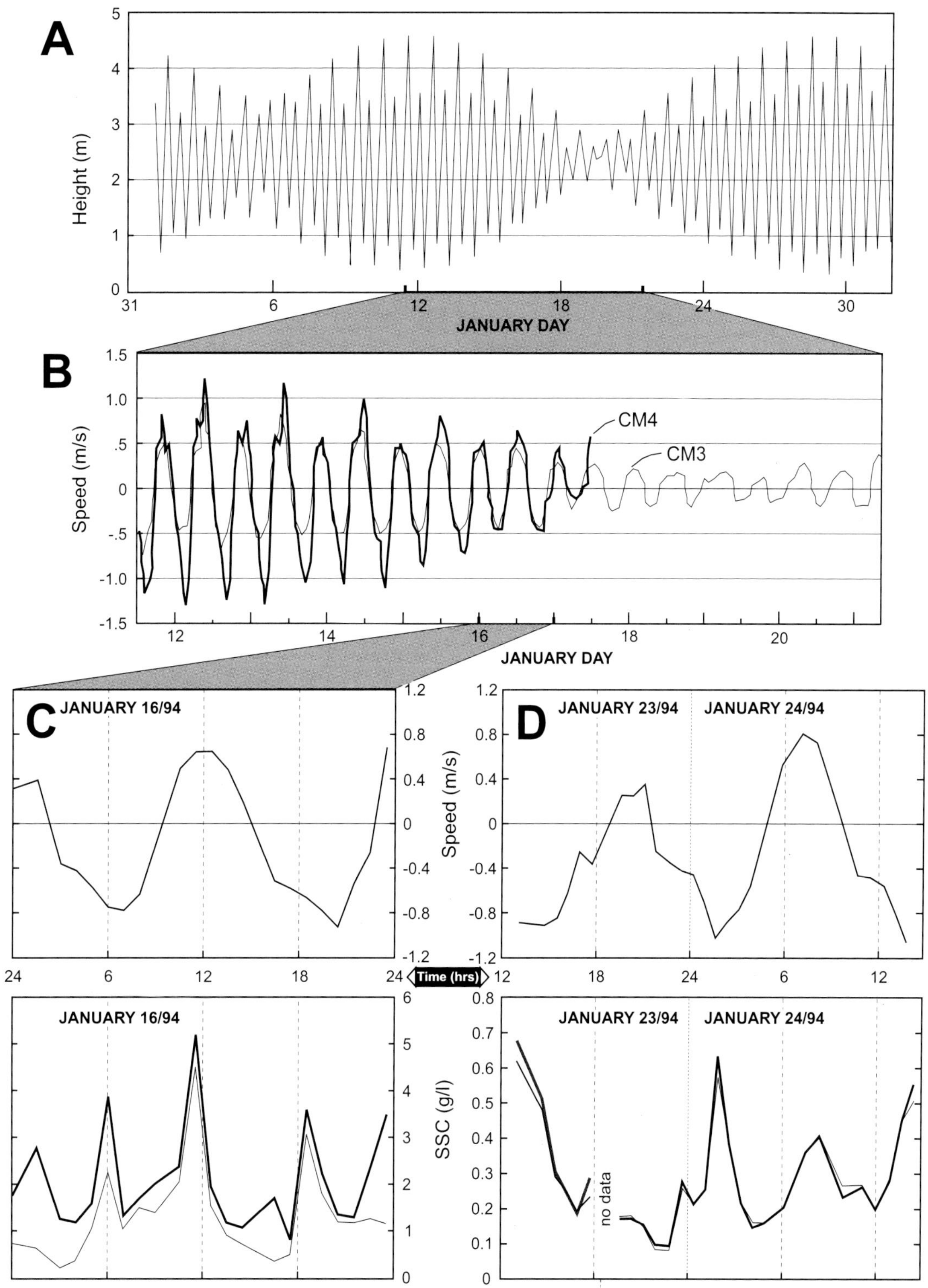

FIG. 3.—**A)** Predicted tidal curve for the mouth of the delta, based on data supplied by the National Tidal Facility, Adelaide, Australia (from Baker, 1998). **B)** Tidal-current speeds (positive = flood; negative = ebb), one meter above the bed, for two stations in the Far Northern Entrance (see Fig. 5C for locations). Station CM3 is flood dominated, and station CM4 is ebb dominated. **C, D)** Tidal current speeds (positive = flood; negative = ebb) and suspended-sediment concentrations (SSC) for two sites in the distributary channels (see Fig. 2 for locations). Speeds were measured one meter above the bed. The SSC values were measured 20 cm (heavy lines) and 95 cm (light lines) above the bed; note the different scales in Parts C and D. Both data sets were obtained when tidal ranges were intermediate between springs and neaps, C) following spring tides (see shaded link to part B), and D) following neap tides.

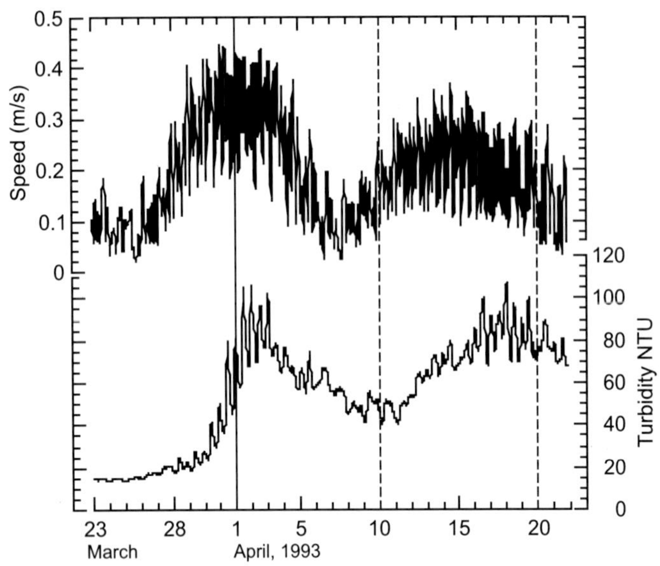

FIG. 4.—Tidal-current speeds and water-turbidity values (NTU, nephelometer transmissivity units) 1 m above the bed, measured at a water depth of 7 m on the delta front, seaward of the Southern Entrance (see Fig. 2 for location; after Harris, 1995, his fig. 5). The NTUs cannot be converted directly to SSC values (g/l) because of a lack of calibration data, but mean SSC values of a few tens of milligrams per liter have been reported from the delta-front area (Robertson et al., 1993). Variations in turbidity over individual semidiurnal tidal cycles (small oscillations) are superimposed on longer-term variations tied to the neap-spring cycle. Note the slight lag in the maximum NTU values relative to peak current speeds at spring tides (April 1 and 15).

best developed during spring tides when SSC values are highest (Wolanski and Gibbs, 1995); prolonged periods of weak currents lead to their depletion by deposition.

SSC values decrease rapidly seaward across the delta-front and prodelta areas (compare Figs. 3C, D and 4; Harris et al., 1989; Robertson et al., 1993; Wolanski and Alongi, 1995), and there is limited offshore movement of sediment despite the shore-normal tidal currents. The offshore limit of mud deposition occurs only 50 km from the coast, in water depths of 40–50 m (Fig. 2), because the sediment falls out of suspension and/or is advected to the north by a coast-hugging, residual current (Harris, 1995; Wolanski and Alongi, 1995; Wolanski et al., 1995b; Harris et al., 1996). Thus, the middle and outer shelf are free of terrigenous-clastic sediment, and carbonate sedimentation occurs (Harris et al., 1996).

Sediment Transport Patterns

The direction of net water movement is seaward through all of the distributaries, with long-term net speeds of only a few centimeters per second (Wolanski et al., 1995a). As discussed above, the geological evidence indicates that mud and sand are reaching the distributary-mouth bar and prodelta areas, so there must be net seaward transport of sediment for the delta as a whole. In detail, however, the patterns of sediment transport are more complex.

The transport behavior of the fine-grained suspended sediment has been studied at length by Wolanski and others (Wolanski and Gibbs, 1995; Wolanski et al., 1995a, 1997). They suggest that a residual landward flow of bottom water occurs, driven by a weak, salinity-driven estuarine (baroclinic) circulation. This, coupled with "tidal pumping" (i.e., a general flood dominance associated with deformation of the tidal wave in shallow water) and the behavior of the flocs (the strongest flocs are those containing the finest sediment), causes a selective trapping of the clay-size material in the turbidity maximum. Silt-size material presumably has a greater chance of escaping the system.

The available current-meter data (e.g., Fig. 3B) and an analysis of the channel-floor morphology (e.g., Fig. 5) indicate that most of the tidal–fluvial channel above the "Apex" is ebb (river) dominated, with flood-dominated areas restricted to sheltered areas on the downstream ends of point bars (Fig. 5A). Within the distributary channels, the sand-size (bedload) material moves in mutually evasive transport paths, with both seaward-directed and landward-directed net transport. As discussed further below, the elongate tidal bars within the distributary channels and on the distributary-mouth bar experience opposing directions of net sediment transport of either side of their crests (Figs. 5B–D). Net bedload transport on the delta front is directed onshore (Harris, 1995).

GEOMORPHOLOGY

The Fly River delta exhibits the classic morphology of a tide-dominated delta (cf. Fisher et al., 1969; Coleman and Wright, 1975; Wright, 1985) and can be divided into four geomorphic regions: the tidal-fluvial channel above the "Apex"; the subaerial delta plain (the distributary channels and islands); the distributary mouth bars; and the delta front and prodelta. Below we describe each of these in turn. The sediments accumulating in each area are discussed in the following section.

Tidal–Fluvial Channel

This region consists of a single channel, with only rare development of islands (e.g., Sturt Island; Figs. 2, 5A). The channel is gently meandering, with both the radius of curvature and the channel width increasing in a downstream direction (Fig. 6), because of the gradual increase in the tidal flux (cf. Langbein in Myrick and Leopold, 1963). Channel depth is variable, ranging from about 10–15 m at the outside of meander bends (Fig. 5A) and in local, bedrock-controlled constrictions, to 7–10 m in shallower, crossover sections. Prominent, headward-terminating (blind) channels called flood barbs (Robinson, 1960) are present on the downstream ends of point bars in the seaward part of this section (Fig. 5A). The shallow areas that separate the flood barbs from the main channel are the only features resembling elongate tidal bars in this zone. The point bars have relatively flat tops with well developed intertidal flats, whereas the outer edge dips relatively steeply (slopes < 1°) into the channel thalweg. Elsewhere in this zone, and especially on the outsides of meander bends, the channel banks are steeper and an erosional cliff commonly extends from the overbank surface to near the low-tide level. Below this, narrow, temporary mudflat accumulations occur. Undercutting by river and tidal currents causes episodic slumping of the cutbanks. In general, however, the tidal–fluvial channel is stable, with few significant changes detectable over the last 30 years.

By comparison, the channels and bars in the transition from the single tidal–fluvial channel to the distributary channels at the delta "Apex" (Fig. 2) experience rapid morphological change (Hughes and Baker, 1996). Here, the channel is subdivided into two or three branches by transient, elongate bars that can build into the intertidal zone and even become vegetated before being eroded. Diversions of flow from one channel to another, perhaps

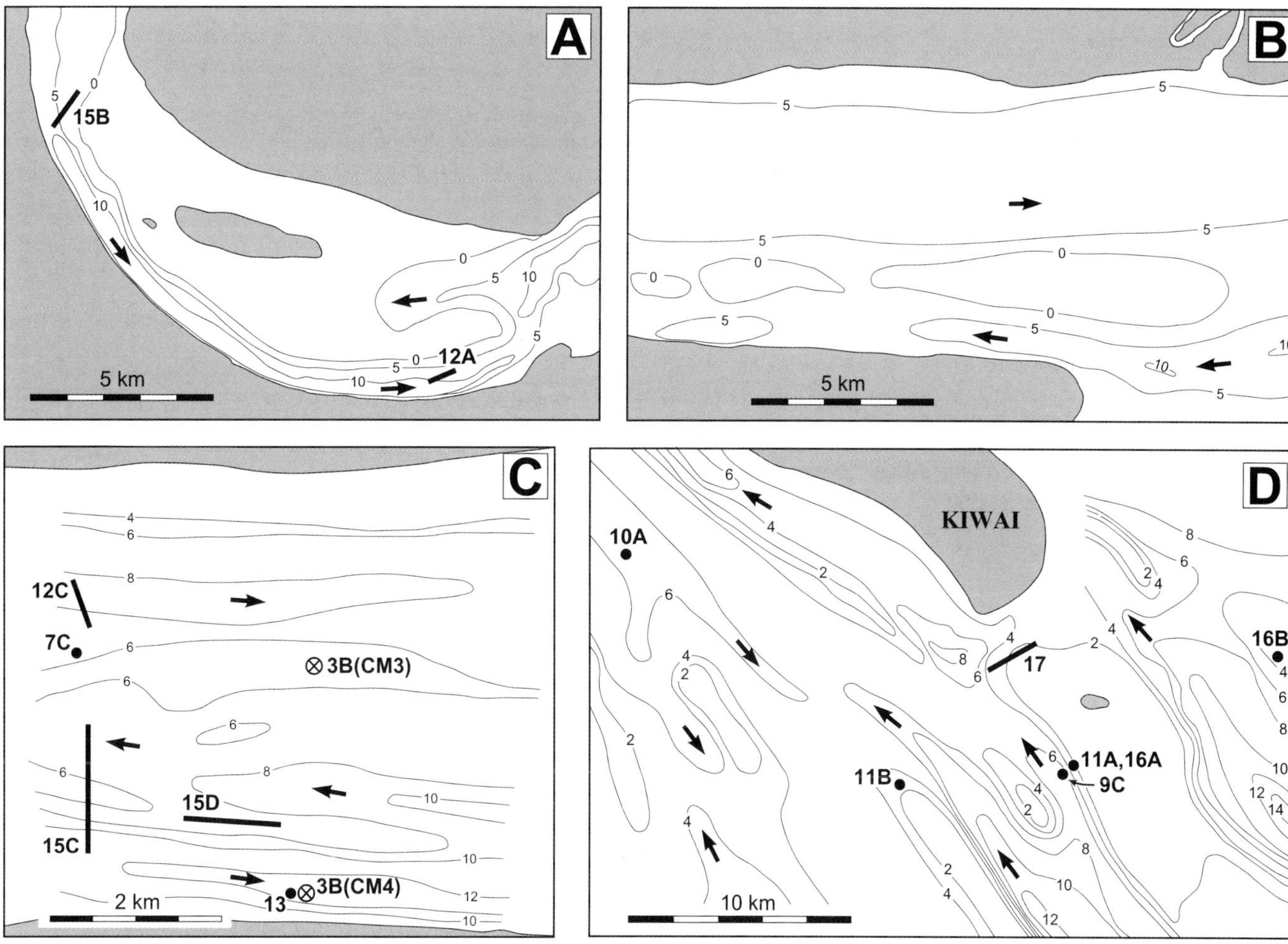

FIG. 5.—Morphology of selected parts of the Fly River delta (see Fig. 2 for locations), showing locations of current-meter data (⊗), seismic and sidescan images (lines), and cores (●) shown in the relevant figures. Bathymetric contours in meters. Datum mean low water in Parts A and B (from SMEC, 1981), mean high water in Parts C and D (this study, 1994). Arrows indicate inferred net transport direction. **A)** Large meander bend in the tidal–fluvial reach inland of the delta "Apex". A blind flood barb occurs at the seaward end of the point bar. **B)** Middle section of Far Northern Channel showing headward-terminating flood-dominant channel along the south shore, separated from the main ebb channel by an elongate tidal bar. The bar crest bends to the south and attaches to the shore just off the left edge of the map, producing a J-shaped bar. Low areas between emergent segments are swatchways. **C)** Outer part of Far Northern Entrance (detailed measurement area 3), showing part of a U-shaped tidal bar: the two elongate bars (depths < 6 m) join immediately west of map area, encircling a headward-closing, flood-dominant channel. **D)** Outer part of Southern Entrance and adjacent mouth-bar region. Interdigitation of alternately headward-terminating (flood-dominant) and seaward-terminating (ebb dominant) channels is evident. Note the broad, flat-topped bar lying seaward of Kiwai Island.

as a result of a minor morphological change, can have major repercussions throughout the rest of the delta. For instance, in 1981, the main channel lay to the north of Sumogi Island (Fig. 2; SMEC 1981), whereas since at least 1990 it has passed to the south of this island (Wolanski et al., 1997). This switch apparently led to the abandonment of the Far Northern Entrance and caused the Southern Entrance to become the active distributary.

Distributary Channels and Islands

The main distributary channels are straight to very gently sinuous and flare seaward: the widths of individual channels increase seaward from 2–5 km near the "Apex" to 25–35 km at the coast (Fig. 2). Depths reach a maximum of about 20 m at constrictions between islands but are more commonly 5–10 m. These main channels are subdivided into a series of mutually evasive, flood- and ebb-dominated (sub)channels by a series of elongate tidal bars (Fig. 5B–D). These bars have reliefs that reach 5–8 m. Their widths and lengths are 0.5–2 km and several hundred meters to several kilometers, respectively. Bathymetric data are scarce, but the available information, coupled with inferences drawn from other tidal systems (cf. Robinson, 1960; Dalrymple et al., 1992), indicate that the ebb channel is probably continuous with the river channel, whereas the flood channels terminate headward as flood barbs (Fig. 5B, C). The ebb-dominated thalweg in each distributary is more sinuous than the channel margins, with the sinuosity decreasing seaward, just as it does in the tidal–fluvial channel above the "Apex" (cf. Fig. 6). Erosion on the outsides of the meander bends is responsible for the common occurrence of concave segments along the margins of the islands

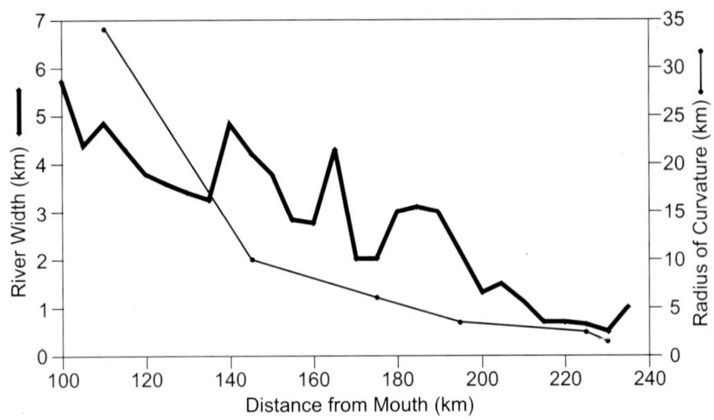

FIG. 6.—Longitudinal variation in river width (bank to bank at high-tide level; heavy line) and radius of curvature of meander bends (light line) in the tidally influenced reach above the delta "Apex". The anomalously wide sections at 140 km and 165 km correspond to the first two major point bars (the 135–155 km segment is shown in Fig. 5A). The narrows at 170 km correspond to an outcrop of Tertiary bedrock near Sturt Island (Fig. 2). The exponential seaward flaring is not clearly evident because of these local irregularities and the omission of the outer 100 km, where the widening occurs most rapidly.

(see particularly the southern coast of Kiwai Island; Fig. 2). The tidal bars that separate the ebb and flood channels are predominantly straight, but they almost always have a curved segment at their landward ends. Thus, they can have either a J-shape if they attach to a channel bank (Fig. 5B) or a U-shape if they are isolated in the center of a channel (Fig. 5C). In areas where active deposition is ongoing, elongate tidal bars aggrade into the intertidal zone (Fig. 5B) and are eventually colonized by vegetation. This is apparently the origin of the many small elongate islands that occur throughout the delta, including those along the southern side of the Southern Entrance (Fig. 2).

Some of the larger islands that lie between the Southern, Northern, and Far Northern Entrances (e.g., Kiwai Island; Fig. 2) have a nucleus of pre-Holocene sediment (Harris et al., 1993), whereas others are believed to be entirely depositional in origin. These depositional islands appear to have grown from emergent distributary-mouth bars, in the manner observable today at the seaward tip of Kiwai Island (Fig. 5D; see more below). Seaward progradation of these bars produces the elongate series of islands that separate the main distributaries.

Accretionary intertidal flats are relatively rare along the margins of the distributaries, being most extensive in the outer part of the subaerial delta plain, where coastline progradation is taking place. Farther inland, the margins of the channels are commonly bordered by erosional cliffs and / or nondepositional to erosional tidal flats. Channel banks in these areas are steep to nearly vertical. Depositional tidal flats do occur locally, however, mainly on the insides of meanders of the channel thalweg.

Distributary-Mouth Bars

The subaqueous portion of the delta plain occupies the relatively shallow (generally less than 10 m deep) area immediately seaward of the interdistributary islands (Fig. 2). This area is incompletely surveyed, so its character is not known in detail. The available data show that the morphology is an extension of the channel-bar topography that characterizes the distributary channels (Fig. 5D). Thus, the distributary-mouth bars consist of a series of coast-normal, elongate, tidal bars that may extend up to 15–20 km offshore. These bars are 1–6 km wide, the widest ones being in line with the interdistributary islands. The relief from the channel bottom to adjacent bar crest reaches 15 m at the mouths of the distributary channels (Fig. 5D), where the bar crests are highest and the channel bases are deepest. From there, relief decreases in an offshore direction. In most areas, bar crests are rounded to relatively sharp and remain shallowly submerged at low tide. By contrast, bars lying directly seaward of the interdistributary islands have a distinctly flat-topped profile and rise into the intertidal zone. This indicates that bar growth is greatest in line with the existing islands, thereby leading to their seaward extension. Orope Island, which lies seaward of Kiwai Island (Fig. 5D), has come into existence relatively recently, illustrating the continued progradation of this part of the delta.

The channels separating the tidal bars display an interdigitation of flood-dominated and ebb-dominated segments. Ebb-dominated (seaward-terminating) channels are more abundant on the landward sides of the distributary mouth bars, whereas flood-dominated (landward-terminating) channels predominate on the seaward sides (Fig. 5D). This presumably reflects river dominance in landward areas, whereas the flood dominance on the seaward side (cf. Harris, 1995) may be a result of deformation of the tidal wave as it propagates into shallow water. Thus, the mouth-bar zone is a bedload convergence (cf. Johnson et al., 1982), thereby limiting the offshore transport of sand. As a result, tide-dominated deltas may not to act as suppliers of sand to the shelf, as has been suggested by some workers (e.g., Levell, 1980).

Delta Front and Prodelta

The morphology of this area is not known in detail. In general, it slopes gently seaward and is relatively smooth and featureless; to date, prominent gullies and slump scars have not been noted. The only major feature is a broad (ca. 10 km wide), low-relief channel defined by the 10 m isobath that extends seaward from the Far Northern Entrance, terminating at a water depth of 30–40 m, 40 km offshore (Fig. 2). The direct association of this unique channel with this abandoned distributary suggests that it is an erosional feature that has been created since the river's discharge was diverted to the Southern Entrance.

SEDIMENTOLOGY

As might be expected because of the scarcity of sand reaching the delta, the deposits are dominated by mud. Nevertheless, most of the sediments are heterolithic, with interstratification of sand or coarse silt and mud (silt and / or clay) on scales ranging from millimeter-thick laminae to decimeter-thick beds (Figs. 7–11). Although the deposits appear monotonous on cursory examination, more detailed study reveals that these heterolithic sediments are highly diverse. Consideration of all available cores and grab samples indicates that the deposits can be subdivided into ten environmentally significant facies, based primarily on bulk grain size (which is mainly a function of the relative proportion of sand and mud layers), the thickness of the sand–mud layering, and the degree of bioturbation or rooting. The physical structures in the sands are of less importance here because of the narrow range of sand sizes and the resulting difficulty of clearly identifying structures in cores, but they would probably provide an extra criterion in a rock succession. Note that in the following facies descriptions all mud-layer thicknesses refer to the unconsolidated state of the muds. Because water content is commonly high, thicknesses are likely to be reduced considerably during compaction.

Delta-Plain Channels, Tidal Bars, and Tidal Flats

As described above, the distributary channels are characterized by strong tidal currents and high to very high suspended-sediment concentrations. In most circumstances, tidal-current speeds are highest in the deeper portions of channels and decrease upward and outward onto the tidal bars and marginal tidal flats. As a result, there is a tendency for the sands to fine upward. On the other hand, SSC values are generally highest in the deepest parts of the channels, because of the presence of bottom-hugging fluid muds. Thus, the rate of mud deposition decreases upward. Because of these opposing tendencies, channel-bottom deposits can be either sandy or very muddy, depending on the relative importance of strong tidal currents or fluid-mud deposition. The deposits at intermediate elevations tend to be heterolithic in nature, whereas those on the intertidal flats are generally muddy because of weak tidal currents.

Channel-Floor Sands and Mud-Pebble Conglomerates (Facies 1 and 2).—

Coarse-grained, channel-bottom facies 1 and 2 are of relatively limited extent; they form mainly, but not exclusively, close to areas of erosion, such as near the outside of meander bends (Fig. 12A, B) and throughout much of the abandoned Far Northern Entrance (Fig. 12C). These areas are also almost the only places where dunes are developed: dunes are presumably not present elsewhere because the sand size is below the lower limit of the dune stability field, but they occur in or adjacent to areas of erosion because tidal currents winnow the finer fractions, leaving a coarser lag. In general, these coarse sediments are typically less than 1–2 m thick (Fig. 7) and are largely restricted to the deeper parts of channels.

The sands constituting facies 1 are among the coarsest in the delta, consisting mainly of very fine sand (average mean grain size 100–110 μm), but with the coarse tail reaching medium sand. These sands are moderately clean, but they contain a significant silt component (15–20%) and some interstitial clay-size material (average 5%) (Baker 1998). Large-scale cross-beds (set thicknesses 5–15 cm) characterize the coarser variants of this facies (Fig. 7A, B), whereas ripple cross-lamination predominates elsewhere, especially in the transition to facies 3 or 4. Discrete mud layers are generally absent, but they occur rarely as drapes within or between cross-beds.

Mud-pebble conglomerates (facies 2) are common immediately above erosion surfaces (Figs. 7A, 13), the abundance of mud pebbles decreasing upward through an interval of 0.5–1.5 m. The pebbles are typically 0.5–1 cm in diameter (maximum length rarely exceeds 5 cm) and are generally rounded. Tabular forms are moderately common; their thickness suggests derivation from the underlying heterolithic deposits. Pieces of wood (Fig. 13), some reaching several centimeters in length, are locally abundant. Scattered and broken pieces of shell occur in this facies in the more seaward areas (Fig. 7B), but they are abundant only rarely (Fig. 13).

Channel-Floor Muds (Facies 3).—

Depositional areas in the deeper portions of the tidal–fluvial and distributary channels are underlain by muddy deposits. Grab samples and cores from such areas (e.g., the outer part of the Southern Entrance) reveal the presence of a surface layer, 20–100 cm thick, of "jelly mud" that has a high water content and minimal strength. Bulk grain-size analyses show that this sediment consists mainly of silt-size to clay-size particles (average mean grain size 10–15 μm with 10–20% clay-size material; Baker, 1998). Seismic records show the presence of continuous reflectors that broadly conform to the modern sediment surface, but the reflection from sediment surface is commonly weak to almost nonexistent (Fig. 14), presumably because of the high water content of the mud.

The individual mud layers in this facies are typically more than 0.5 cm thick and commonly thicker than 1 cm, with some being more than 2 cm (Figs. 7B, C, 8, 9B, 10A); these channel-bottom mud layers are thicker than those found at higher elevations in the channels (compare Fig. 10A with Figs. 8, 10B, C). They are texturally homogeneous with sharp bases (Fig. 9B). Organic material is generally absent, but some contain uniformly disseminated organic material, which may be accompanied by dispersed grains of coarser silt and/or very fine sand with no evidence of grading. Internal lamination is generally absent, although some X-radiographs reveal a faint horizontal streakiness (Fig. 9B) resulting from subtle segregation of grains of different sizes and/or variations in organic content. There may also be a sharp break between the main body of the layer and a thin (typically 1 mm thick) clay cap. Bioturbation is generally absent.

Sand, which constitutes less than 10% of the sediment, generally occurs as discrete laminae (Fig. 10A; see also Figs. 7B, 8), most of which are less than one millimeter thick, but reaching 0.5 cm in some instances. The sandier variants of this facies are gradational into either facies 1 or heterolithic facies 4 (see below). Internal structures are generally not visible in the sands, but many occurrences take the form of discontinuous, small lenses that represent starved-ripple formsets. Small load structures (spacing 1–3 cm; amplitude < 1–3 mm) are abundant at the bases of the thicker sands, as are erosional features (Fig. 9A). In most cases, sand-layer thicknesses do not vary in a regular way between layers, but symmetrical or asymmetrical upward-thickening and upward-thinning successions do occur (Fig. 9A, B).

Because of the thick, homogeneous character of the mud layers and the channel-bottom position of this facies, these muds are believed to represent fluid-mud deposits. The high SSC values within the fluid-mud bodies leads to hindered settling and gives the suspension a finite shear strength (Mehta et al., 1989; Wolanski and Eagle, 1991). These characteristics inhibit vertical segregation of materials with different settling velocities, leading to the development of the mud layers that lack grading and contain uniformly disseminated organic material (cf. Kirby and Parker, 1983). The pronounced decrease in SSC values that occurs as tidal-current speeds decrease from spring to neap tides (cf. Figs. 3C, D, 4) suggests that the thickest mud layers are deposited then rather than during neap tides (see Figs. 8, 9B).

The small amount of sand in facies 3 is somewhat surprising, given that near-bed current speeds reach 0.5–1 m/s (Fig. 3C). Therefore, the lack of sand is not due to low energy levels; indeed, the sand in the facies 3 muds is commonly texturally similar to the relatively coarse sands of facies 1, resulting in bimodal bulk compositions (Figs. 7C, 13). Instead, it appears that the fluid-mud deposits blanket the channel floor, preventing the remobilization of previously deposited sand and slowing the introduction of sand from adjacent areas.

Thin-Bedded Heterolithic Stratification Deposited on Tidal Bars and Point Bars (Facies 4).—

Deposition at intermediate depths in the channels occurs on within-channel elongate (tidal) bars or bank-margin point bars. Seismic data show that both bar types contain lateral-accretion deposits (Fig. 15). Stratal dips are generally less than 1–2°. In current-parallel sections, the stratification in point bars is gently

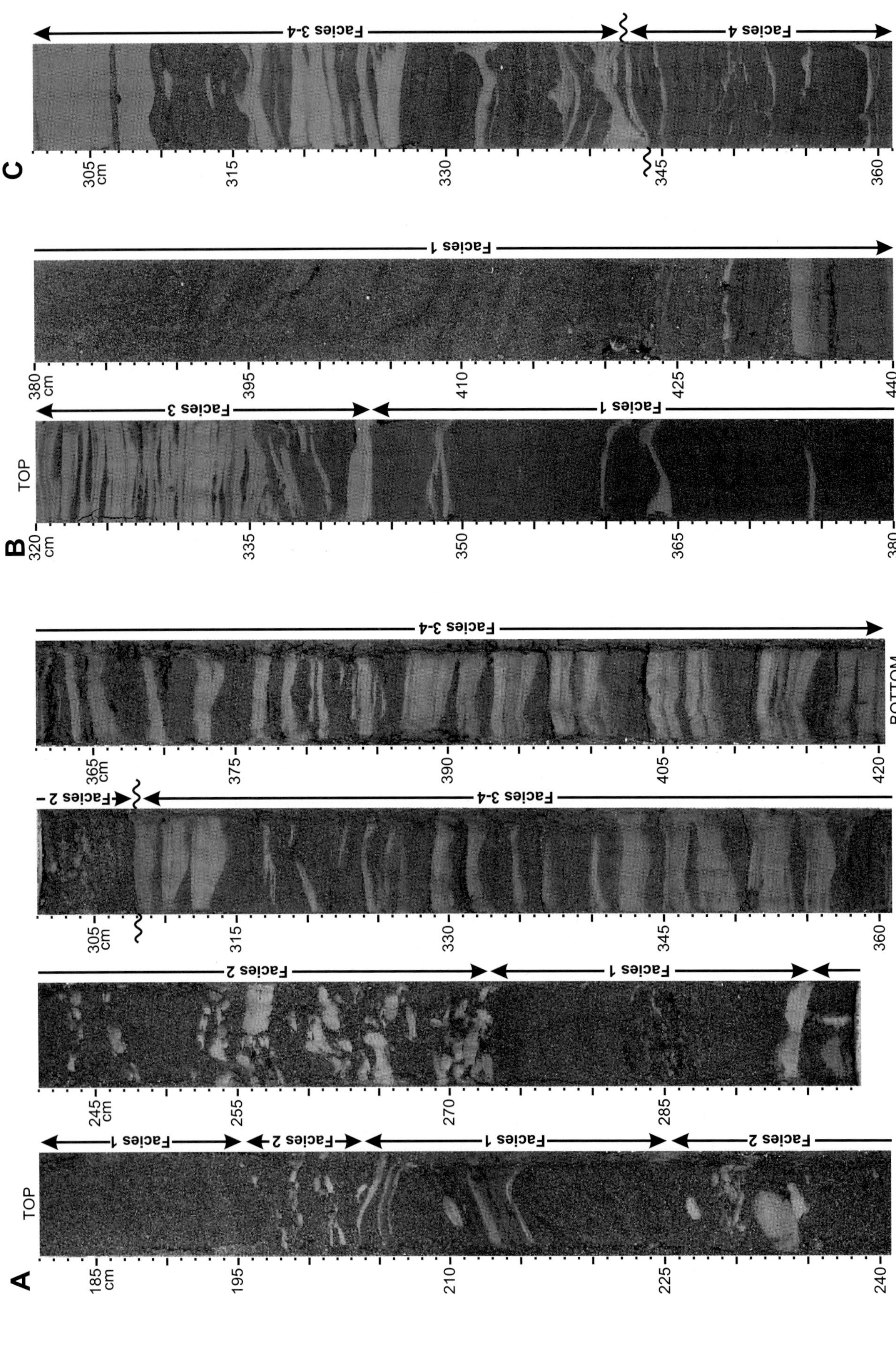
A
TOP
185 cm
195
210
225
240
Facies 1
Facies 2
Facies 1
Facies 2
245 cm
255
270
285
Facies 2
Facies 1
305 cm
315
330
345
360
Facies 2
Facies 3-4
365 cm
375
390
405
420
Facies 3-4
BOTTOM
B
TOP
320 cm
335
350
365
380
Facies 3
Facies 1
380 cm
395
410
425
440
Facies 1
BOTTOM
C
305 cm
315
330
345
360
Facies 3-4
Facies 4

domed (Fig. 15A), because of the arcuate nature of the point bar, whereas axis-parallel sections of tidal bars show horizontal bedding (Fig. 15C). The continuity of these reflectors suggests that individual beds extend for considerable distances.

Cores from both bar types show that these lateral-accretion deposits consist of thin bedded heterolithic stratification (facies 4; Figs. 7A- lower part, 8) that is characterized by a millimeter- to centimeter-scale alternation of sandy layers (fine sand to coarse silt) and muddy layers (finer silt and clay), although in some instances decimeter-scale beds of either sand or mud are also present. The proportion of sand ranges from more than 80% (transitional to facies 1) to less than 20% (transitional to facies 3 and 5). Most commonly, however, sand / silt layers constitute 40–70% of the deposit. Slump structures consisting of contorted bedding occur sporadically, but they are not abundant.

Mud layers in this facies are thinner, on average, than those in facies 3 (compare upper part of Fig. 8 with Figs. 7B, C, and 10A); indeed, muds more than about 1 cm thick are not common. Most mud layers thicker than this are composite, with silt and / or sand stringers less than 1 mm thick spaced 2–5 mm apart (Fig. 9A). Thicker mud layers that lack any internal subdivision are generally restricted to the topographically and stratigraphically lower occurrences of this facies and are indistinguishable from the fluid-mud layers that occur in facies 3 (Figs. 8, 9B, 10A). Some of the mud layers are internally graded but most appear to be texturally uniform. The contact with the underlying sand is most commonly abrupt, especially on the stoss sides of ripple foresets in the sand, but gradational contacts are present on the lee sides of these same ripples. Comminuted organic material may occur at the sand–mud contact. Most mud layers are continuous, and flaser bedding is rare (Fig. 7C). Bioturbation is generally scarce to absent (Figs. 8, 9A) but becomes more common in the seaward parts of the distributary channels, and especially in the outer part of the Far Northern Entrance, where the salinity is higher.

The sand layers in facies 4 are rarely more than 5 cm thick (Figs. 7A, 8). Formsets and cross lamination produced by current ripples are widespread. Current directions are generally unidirectional, but opposing orientations are present in rare instances. In mud-dominated intervals, the sand layers are typically < 5 mm thick and the ripples occur as disconnected lenses (lensen bedding). Sand layers thicker than about 5–10 cm commonly contain mud rip-up clasts (Fig. 8), which are interpreted to represent minor erosional events. Small load structures are moderately common on the bases of sand layers, especially where they overlie muds more than about 0.5–1 cm thick.

Many cores show a quasi-regular thickening and thinning of successive sand layers (e.g., Figs. 8, 9A), producing cycles 10–30 cm thick composed of 10–25 sand–mud couplets. Many successions also display alternately thicker and thinner sand layers (Figs. 7A, 8). It is tempting to interpret such features as the product of variations in tidal-current speeds as a result of the neap–spring cycle and the pronounced diurnal inequality, respectively (Fig. 3A, B). However, an unequivocal tidal cyclicity is obvious in only relatively few instances (Fig. 7A). Much more commonly, the number of layers in a cycle is significantly less than the expected number (28), the cycles are strongly asymmetric (upward-thickening or upward-thinning half cycles), and the thick–thin alternation persists for only a small number of repetitions (Fig. 8). As a result, analyses utilizing the techniques of de Boer et al. (1989) fail to detect a statistically significant tidal signal in most instances (Baker 1998). In addition, ^{210}Pb profiles in most cores yield sedimentation rates of less than 10 cm / yr (Baker 1998), which might be taken to imply that the decimeter-thick thickening and thinning trends are annual cycles, and that the individual sand–mud couplets represent neap–spring cycles, as suggested by Allen (1991) for somewhat similar deposits in the Gironde estuary. Such slower rates of sedimentation are consistent with the presence of bioturbation in some instances (Fig. 9A).

However, the absence of a clear tidal signal in most heterolithic successions and the slow net sedimentation rates do not preclude a tidal origin for many of the individual sand–mud couplets. Erosion during spring tides (perhaps producing the thin mud-pebble layers), amalgamation of mud layers during neap tides, the presence of anomalously thick fluid-mud deposits (Fig. 9B), and / or the sporadic presence of couplets produced during the subordinate tide (Fig. 7A) will obscure the pure tidal signal, producing statistically insignificant tests of rhythmicity. Furthermore, analyses of the vertical variation of copper content (Baker, 1998), which shows a dramatic increase above background levels beginning approximately in 1987 as a result of tailings discharge from the Ok Tedi mine in the headwaters of the Fly River (see also Alongi et al., 1991), indicate that sedimentation rates locally exceed 1m / yr and commonly exceed 20 cm / yr. Thus, the calculated sedimentation rate increases as the measurement time decreases, a feature that is common to many environments (Sadler, 1999; C. Nittrouer, personal communication, 2000). As Li et al. (2000) have shown, individual tidal cycles may produce discrete

←

FIG. 7.—Core photographs showing distributary-channel-bottom deposits. [NOTE: In all core photos, sand is dark because of the presence of lithic fragments, silty mud is a medium gray, and clay is light gray. Straight vertical indentations and horizontal striations are an artifact of cleaning the soft sediment surface.] **A)** Heterolithic deposits (transitional between facies 3 and 4), erosionally overlain at 308 cm by mud-pebble conglomerate (facies 2) and cross-bedded sand (facies 1). See Fig. 2 for location. The thickness of the mud layers below 308 cm indicates that deposition occurred in a channel-bottom setting, in the presence of fluid muds, as does the overall upward-coarsening trend. These heterolithic deposits show three orders of tidal cyclicity: (1) most of the muds below 390 cm are doublets, separated by a thin (~ 2 mm) sand; (2) the thicker sands (> 5 mm thick) show a prominent thick–thin alternation; and (3) a larger-scale variation in sand-layer and mud-layer thicknesses is present, with the thickest muds (which probably formed as fluid muds were deposited following spring tides) occurring at 308–315 cm, 340–360 cm, and 385–410 cm. There are 26 thick sands between the two lower muddy intervals. These observations indicate that the layering is a tidal rhythmite containing neap–spring cycles and evidence of the diurnal inequality (the thick–thin alternation; cf. Fig. 3A, B). The thinnest sands in the mud doublets were deposited by the subordinate tide. **B)** Fine (to medium) sands (facies 1) overlain with a rapid transition by channel-floor muds (facies 3; see Fig. 2 for location). The sands fine upward and contain cross bedding from 375–420 cm and ripple cross lamination above this. The basal sands contain shell fragments; thus, this lag may reflect an earlier period of abandonment of the Southern Entrance. **C)** Subtle erosional break at 344 cm separates very fine rippled sand with thin mud flasers (below) from fine–medium sand with thick mud drapes (above). Both units are assignable to facies 4, but the upper unit is interpreted as a channel-bottom deposit because of the association of relatively coarse sand and thick mud layers, whereas the lower unit formed higher within a channel, perhaps even in the lower intertidal zone. Sandy channel-floor facies 1 and 2 are not present above the channel-bottom erosion surface. See Fig. 5C for location.

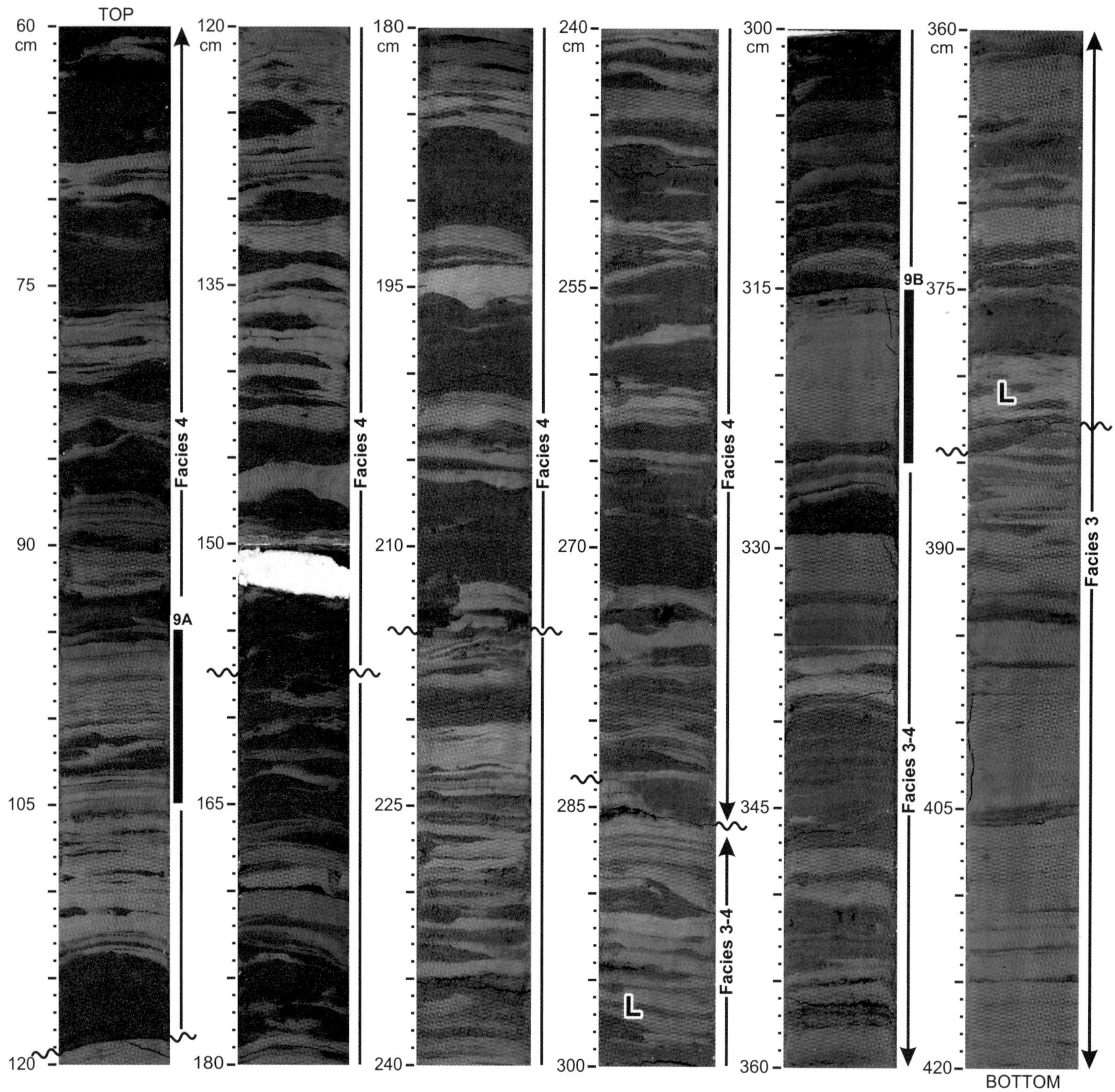

FIG. 8.—Upward-coarsening heterolithic succession (facies 4 gradationally overlying facies 3) from a bank-attached bar in the Far Northern Entrance (see Fig. 2 for location). Muddy deposits below about 285 cm contain many mud layers up to 1+ cm thick interpreted to be fluid-mud deposits that accumulated in a channel-bottom setting. Note the overall upward thinning of mud layers. The lenticular nature and the irregular tops of many sand layers (e.g., 84, 130, and 196 cm) reflect the presence of ripple formsets. Some of the sand–mud couplets are graded, as indicated by the gradual upward lightening (e.g., 305–310 cm depth); other mud layers have a sharp base, especially where they overlie ripples. Irregular variations in sand content occur throughout the core, but clear evidence of tidal cyclicity is weak to absent; for example, an alternation of thicker and thinner sand layers occurs for a short interval between 130–140 cm but is not pervasive. Subtle erosional breaks are inferred at 120 cm (shell fragments in sand), 157 cm (mud pebbles in sand), and 215, 285, and 385 cm (truncation of underlying strata), but the time gaps are minimal because there are no abrupt changes in sediment consolidation. Loading (L) is locally abundant. Burrowing is present, but is not abundant (see Fig. 9A). Vertical bars at 95–105 cm and 315–325 cm indicate detailed sections shown in Fig. 9A, B. The white material at 150–152 cm is styrofoam filling a gap between core segments.

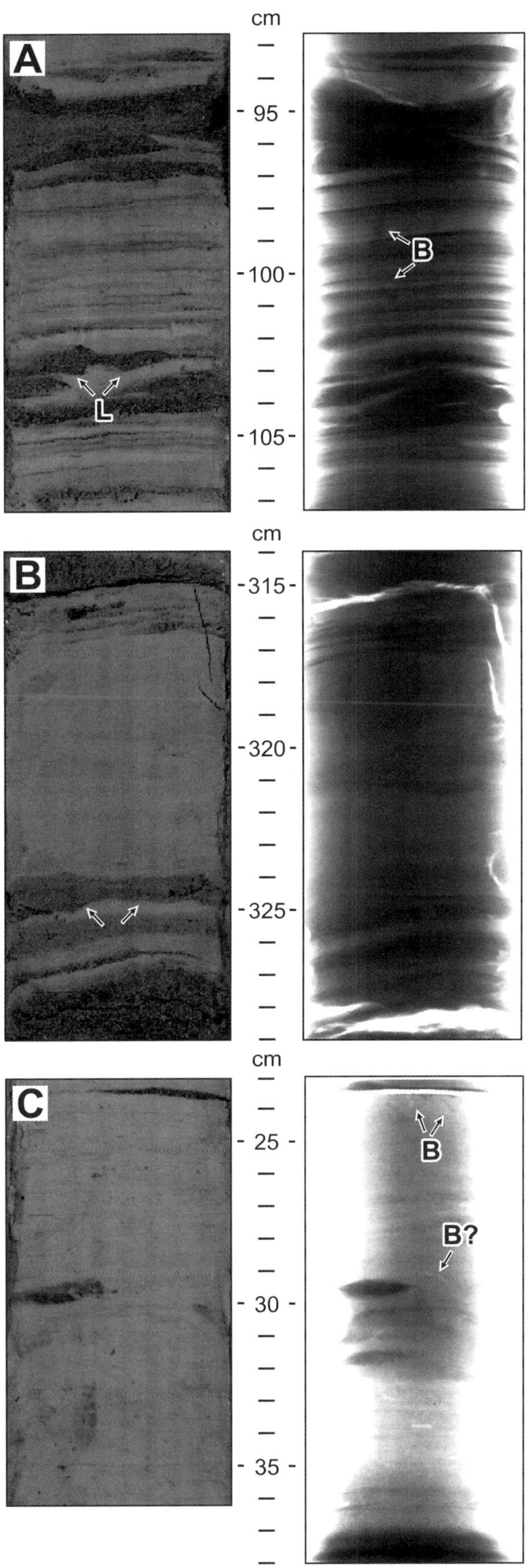

layers, but only a small fraction are ultimately preserved. Therefore, it is possible that many of the individual sand–mud alternations within the tidal bars and point bars of the Fly delta (facies 4) are the product of individual tidal cycles rather than neap–spring cycles.

Thinly Laminated Heterolithic and Rooted, Tidal-flat Deposits (Facies 5 and 6).—

Most of the intertidal flats are relatively muddy, although sandy flats are present locally, low in the intertidal zone. These sandy flats contain deposits like those of facies 1, grading upward into facies 4 heterolithic stratification. More typically, the intertidal flats contain thinly laminated heterolithic stratification (facies 5; Fig. 10B) in which the alternation of coarser and finer layers occurs on a millimeter scale; individual mud laminae rarely exceed 1–2 mm in thickness. Regular, "pinstripe lamination" is present in some instances, but the lamination is generally somewhat irregular, because of the presence of ripples. Lenticular bedding is moderately common. The coarser layers consist of silt and very fine sand. Slump structures are locally present.

The upper part of the intertidal zone is vegetated by grasses throughout the inner part of the delta plain, whereas mangroves colonize the higher parts of the flats in seaward areas where the water is brackish. As a result, root-turbation increases gradually upward through the intertidal zone (Fig. 10C), ultimately producing unstratified, mottled mud (facies 6) in the uppermost intertidal and supratidal zones. These muds are weakly to strongly oxidized, because of periodic exposure. Some desiccation cracking is present, but it is relatively rare.

←

FIG. 9.—Core-photo–X-radiograph pairs showing detailed structure of mud layers (lighter colored in both images). **A)** Thinly interlaminated mud and very fine sand and silt from facies 4. What appear in core to be single mud layers are shown in X-radiographs to consist of 2–5 subtle silt–clay laminae (2–3 mm thick). On a larger scale, sand-layer thicknesses show a progressive upward decrease (depth range 104–100 cm) and then an increase (depth range 100–95 cm), with ripple cross lamination occurring only in the thicker sands at the base and top. This might represent a spring–neap–spring cycle. However, the composite nature of many of the mud layers is not easily explained in this interpretation. The vertical burrow ("B") originates at the top of the muddy interval, implying that there is a break at that level. L = load structures. See Fig. 8 for context. **B)** Thick mud layer with diffuse siltier intervals from deposits transitional between facies 3 and 4. Discrete burrows are not present. Its position in the core (see Fig. 8) suggests that it formed in a channel-bottom setting. Its characteristics suggest rapid deposition, perhaps from a fluid-mud body. It may have formed during the passage from spring to neap tides as the fluid muds created by spring-tide resuspension dissipated by deposition; note that the mud and sand layers below the anomalous, massive mud are thicker than the corresponding layers above it, suggesting a decrease in current strength and SSC values. The irregular upper surface of the mud layer at the arrows may reflect partial erosion of an originally thicker mud layer: fluid muds began to consolidate, but the current speeds were still high enough to resuspend part of the deposit. **C)** Massive-appearing muds from the delta-front (facies 9; see Fig. 5D for location). A faint, streaky lamination and an indistinct mottled appearance are evident in the X-radiograph. Discrete burrows may be present ("B"). See text for discussion of origin.

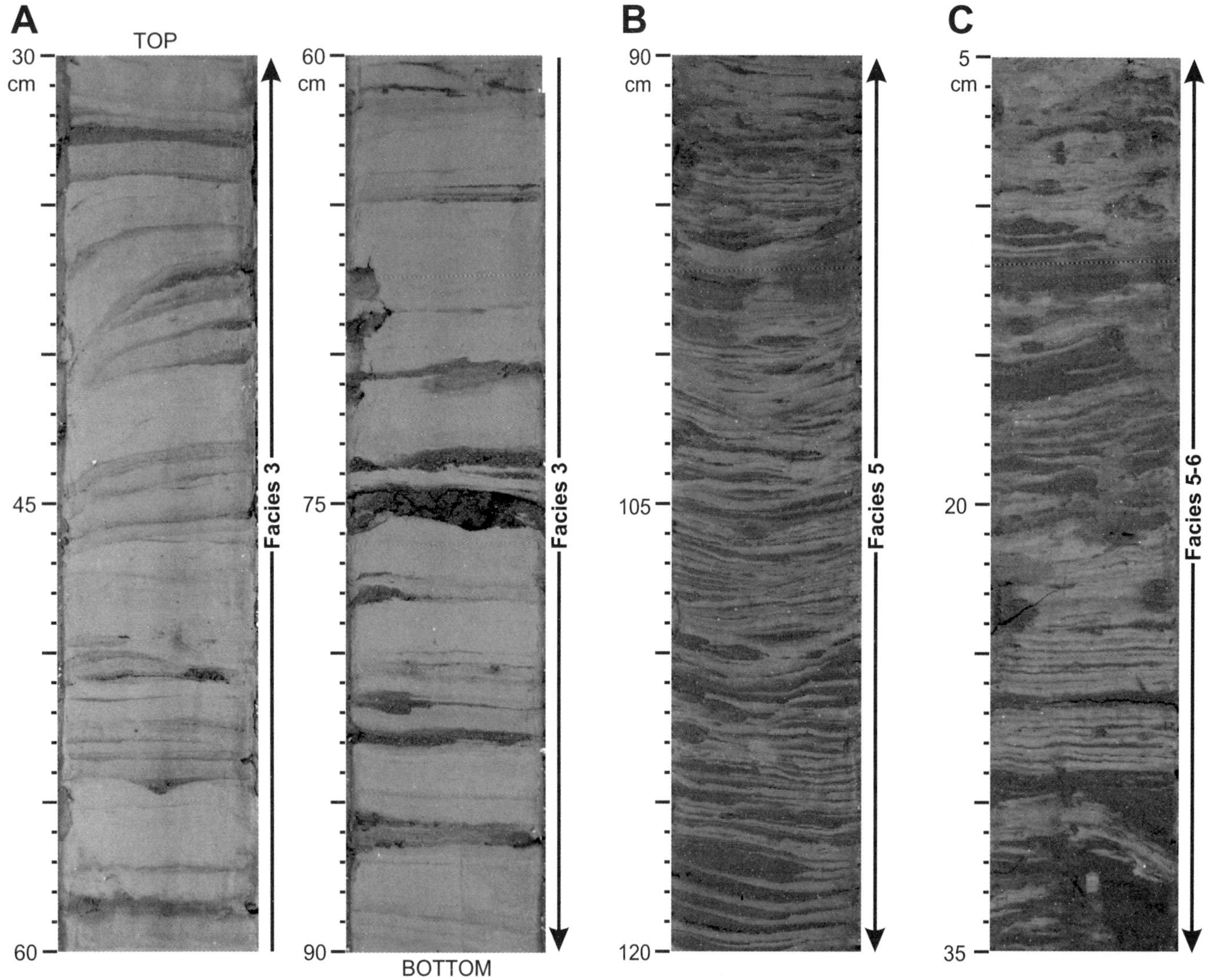

FIG. 10.—Comparison of channel-bottom (A) and intertidal (B, C) muds. **A)** Laminated, thick muds (facies 3) from the bottom of the Southern Entrance, the main active distributary (see Fig. 5D for location). Thicker interbedded sand layers (e.g., at 76 cm) are relatively coarse (i.e., fine–medium), as is typical of channel-bottom locations. **B, C)** Thinly laminated muds and silty, very fine sand from the lower (B; facies 5) and upper (C; transitional between facies 5 and 6) intertidal zone (see Fig. 2 for location). Rooting and invertebrate burrows are evident in Part C. Unlike the channel-bottom deposits, intertidal mud layers rarely exceed 2 mm thick.

The supratidal (overbank) zone farther from the channels was not sampled in this study. Observations from the air indicate that the near-channel regions are heavily treed, whereas areas farther from the channels tend to become more swampy, suggesting that low levees are developed. However, crevasse splays are not present. It is not known whether peat is accumulating in the swampy areas.

Mouth-Bar, Delta-Front, and Prodelta Region

Sedimentation in the area seaward of the distributary channels occurs under the influence of a rapid seaward decrease in tidal-current speeds and suspended-sediment concentrations (compare Figs. 3, 4), an increase in salinity, a net landward transport of bedload, and a northward advection of suspended sediment. Wave influence is important seasonally in shallow-water areas. All of these factors lead to a seaward decrease in grain size and sedimentation rate, coupled with an increase in biological activity. Because of the small number of cores from this large region, the following facies subdivisions must be considered preliminary.

Distributary-Mouth-Bar Sands (Facies 7).—

The available grab samples and cores demonstrate that the shore-normal distributary-mouth bars consist of clean, well-sorted, fine sand to coarse silt (average mean size 100 μm; average 5% clay-size material) that is texturally similar to facies 1 channel-floor sands. The structures observed include cross-bedding in the coarser sands (fine–medium sand; Fig. 16B) and low-angle lamination in the finer sands (Fig. 16A), which may represent plane

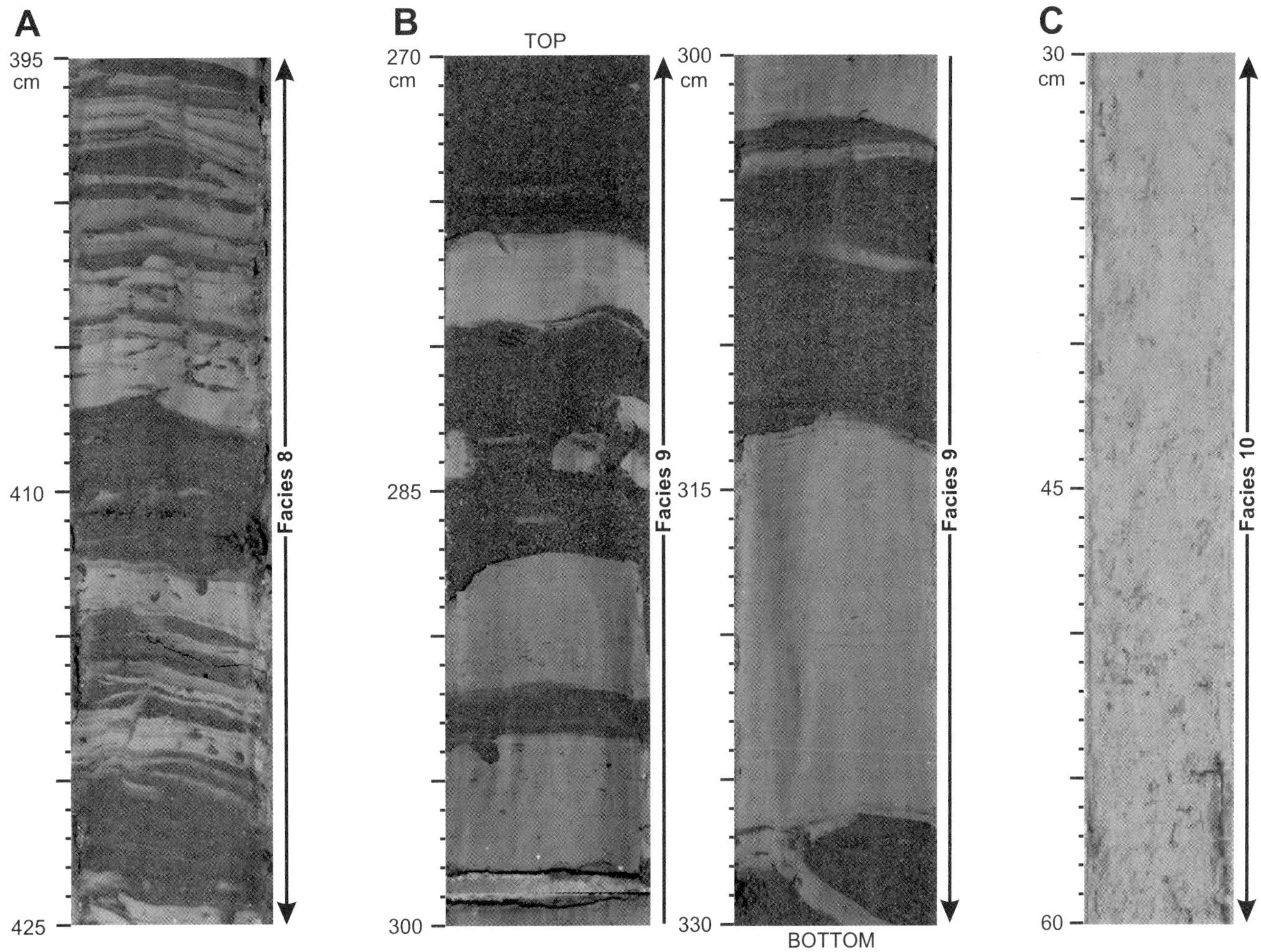

FIG. 11.—Muddy facies of the delta front (A, B) and prodelta (C). **A)** Thinly laminated to bedded mud and fine to very fine sand with discrete burrows (facies 8) that formed where sedimentation rates are low enough and salinities are sufficiently high to allow benthic colonization. See Fig. 5D for location. **B)** Medium-bedded alternation of massive to faintly laminated muds and fine to very fine sand (facies 9). Core comes from channel between the interdistributary bars (see Fig. 5D for location) and is interpreted to penetrate slightly older, delta-front sediments. See Fig. 9C for X-radiograph of correlative deposits. The black specks in the mud layers at 291 cm and 295 cm are finely comminuted, disseminated organic material. A burrow is present at the top of the mud bed at 294 cm. The mud pebbles at 285 cm suggest that a slightly longer break occurs at this level. **C)** Bioturbated sandy mud from the distal prodelta (facies 10; see Fig. 2 for location).

bed or hummocky cross stratification. On the higher parts of the bars, mud layers are generally rare and thin (< 2-3 mm thick), because the SSC values are low this high in the water column. Wave action also inhibits mud deposition and/or resuspends any slack-water deposits. Deeper in the channels between the bars, however, thicker mud layers are present (Fig. 16B), because of higher SSC values in low-lying areas. Mud-pebble conglomerates (facies 2) similar to those within the distributary channels are present in the bottoms of these interbar channels. Seismic records are of poor quality, with low resolution and limited penetration, perhaps because of the uniform texture and hard-packed nature (due to wave action) of the sands. Nevertheless, they suggest the presence of lateral-accretion bedding (Fig. 17) with dips of up to 2°. Together with the cores, these records suggest that sand thicknesses reach 5–10 m.

Bioturbated, Heterolithic Stratification of the Delta Front (Facies 8 and 9).—

Cores from the delta front, or that penetrate older delta-front deposits beneath the bases of mouth-bar channels, contain heterolithic stratification of two distinct types. The first (facies 8) consists of deposits that resemble the thin-bedded heterolithic stratification of facies 4 and 5, with the exception that bioturbation is much more abundant (Figs. 11A, 16B). Nowhere, however, is bioturbation sufficient to significantly blur the original stratification. The thickness of the mud layers varies from as little as 1–2 mm to as much as 1 cm. Most sand layers are of the same order of thickness, but some reach several centimeters thick. The sands are generally flat laminated, but ripples are present in some layers. Noncyclic variations in the relative proportions of sand

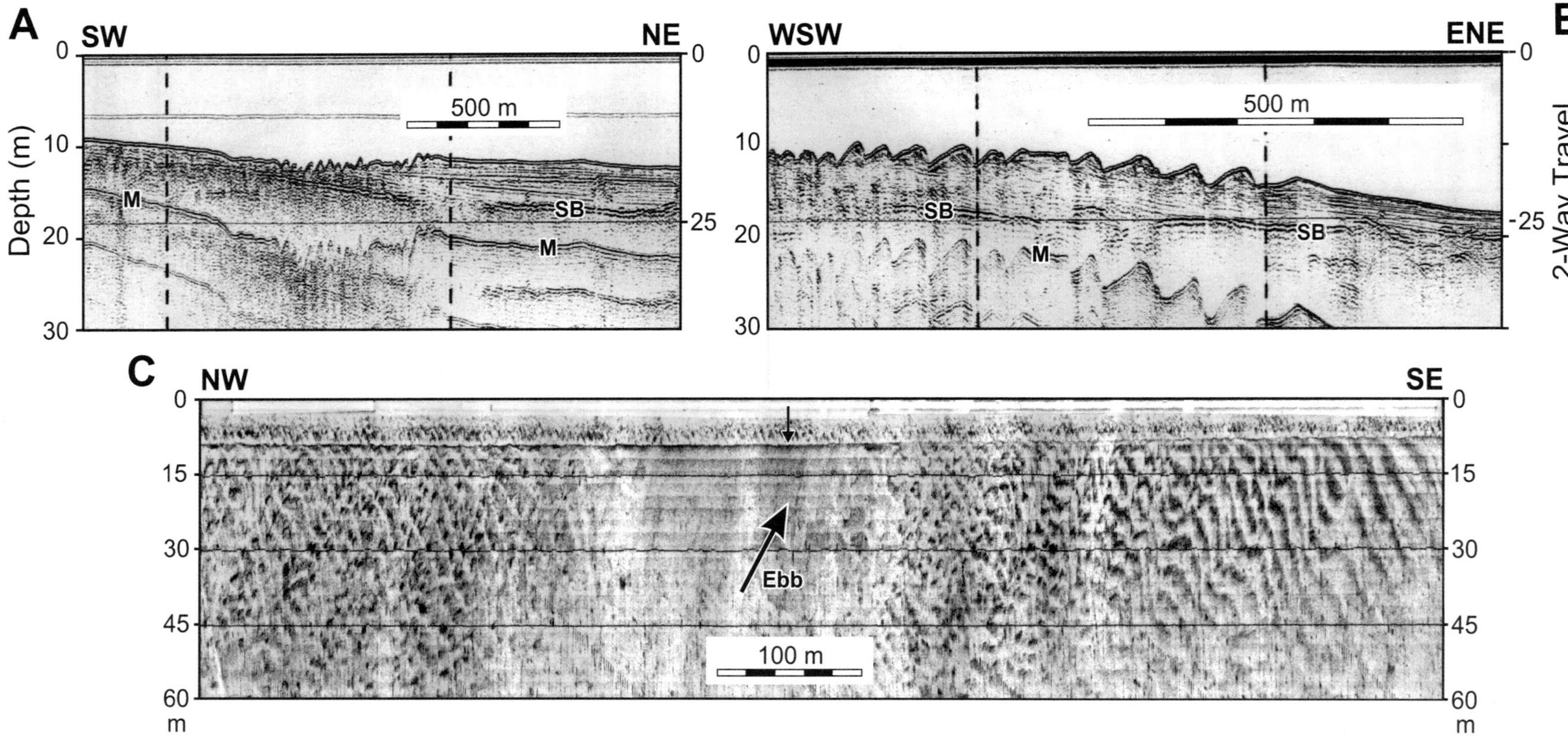

FIG. 12.—**A, B)** Seismic records showing ebb-oriented dunes occurring only in areas of erosion. (See Figs. 2 and 5A, respectively, for locations). Note that cross stratification is visible within the larger dunes in Part B. The inclined strata beneath the present-day erosion surface formed on the flank of an elongate tidal bar (A) and a tidal-fluvial point bar (B), and are interpreted to consist of heterolithic strata (facies 4). These stata downlap onto a prominent reflector interpreted as the late Pleistocene sequence boundary (SB). M = multiple. All two-way travel times have been converted to depths assuming a speed of sound of 1500 m/s. **C)** Sidescan sonar record from the outer part of the Far Northern Entrance (see Fig. 5C for location), showing parts of two sand ribbons consisting of medium 3D dunes. This region is generally erosional: three of the five cores in the area show sand with mud pebbles and/or shells at the surface, whereas the other two have a thin (< 85 cm) surface layer of very soft, recently deposited mud (facies 3) over an erosion surface.

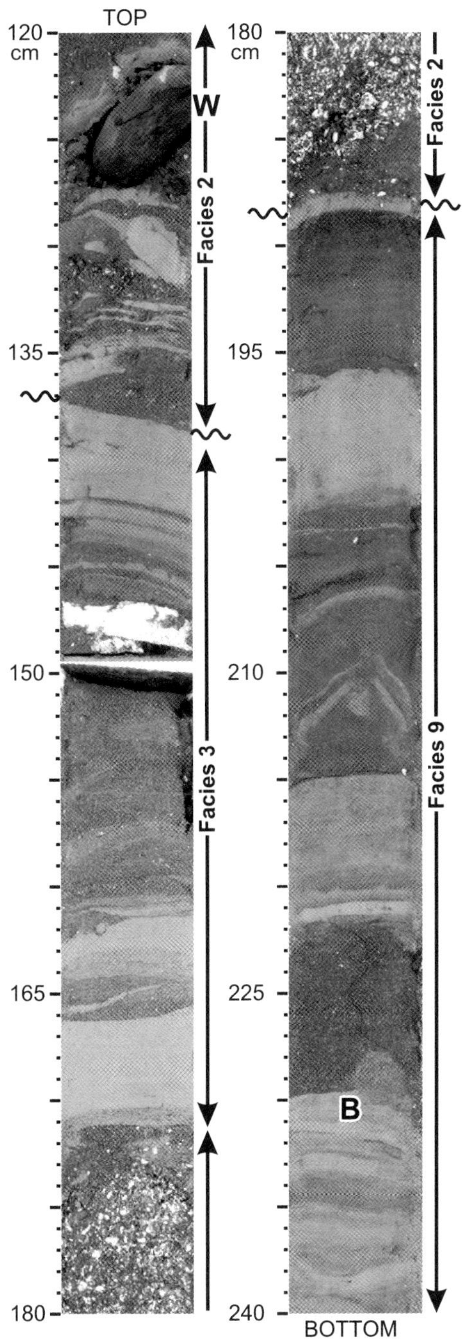

FIG. 13.—Burrowed ("B"), medium-bedded sands and muds (facies 9; cf. Fig. 11A) of inferred delta-front origin erosionally overlain at 188 cm by shell lag (facies 2) and thick, channel-bottom, fluid-mud beds (facies 3). See Fig. 5C for location. This shell concentration (dated 6260 ± 60 yr BP; Beta-93007) is anomalous in the Fly delta and is interpreted to represent a period of distributary abandonment and channel-floor erosion in a distal setting, where conditions (lower sedimentation rates; higher salinity) allowed bivalves to live nearby. Sands associated with the thick mud beds are fine–medium grained. A second channel-floor erosion surface at 137 cm is overlain by facies 2 (mud-pebble conglomerate) containing a rounded wood pebble (W). The white material at 147–150 cm is styrofoam filling a gap between core segments.

and mud are evident, but unequivocal evidence of a tidal origin is lacking. Apparently, this facies forms in an area with moderately fast sedimentation (i.e., too fast for significant bioturbation) but with lower rates of accumulation and higher salinities than exist within the distributary channels. It may, therefore, be a distributary-proximal, delta-front facies, but its spatial distribution is unknown. It is similar in appearance to the laminated sands and muds described by Jaeger and Nittrouer (1995; see also Kuehl et al., 1996) from areas immediately seaward of the mouth of the Amazon River. By analogy, therefore, the millimeter-scale sand–mud alternations in facies 8 may represent individual tidal cycles, whereas the thicker sand layers formed by amalgamation of sand layers during spring tides.

The second delta-front facies (facies 9) consists of interbedded sands and muds that are 5–15 cm thick (i.e., considerably thicker than most of the heterolithic stratification observed in facies 4, 5, or 8; Figs. 11B, 13). The sands appear massive because of their uniform, fine grain size (grading is not observed), but flat lamination and ripple cross-lamination is present in some cases. Mud pebbles occur in rare instances, as do mud layers a few millimeters thick. Most of the mud layers appear massive in both freshly cut cores and X-radiographs, whereas others contain a faint horizontal lamination (e.g., Fig. 9C). A more distinct layering is present in some occurrences (Fig. 13). Bioturbation is present in most examples, where it consists of a faint mottled appearance with scattered, discrete burrows that are commonly concentrated at the tops of mud layers (Figs. 9C, 11B, 13). However, bioturbation is never pervasive and is not present everywhere. The sand content of the muds is generally minimal (Fig. 11B) but is locally significant (Fig. 13). Organic material is disseminated throughout some mud layers (Fig. 11B). The contact between the mud and the underlying sand is typically sharp, with sand loaded into the mud in some instances.

A very similar style of stratification has been documented from the inner shelf, offshore of the Amazon River (Kuehl et al., 1996). According to their interpretation, the thick mud layers are deposited from fluid muds that persist for months or even longer. During times of increased wave activity (seasonal or during discrete storms), the soft upper part of this surface mud layer is eroded and sand, which may contain mud pebbles, is deposited. Colonization of the erosion surface is possible prior to sand deposition, producing the bioturbation at the tops of the mud layers. Sand stringers within the mud layer, or mud layers in the sand, presumably represent less intense variations in energy levels, perhaps associated with neap–spring cycles or smaller storms. Harris et al. (1993) had previously interpreted each sand–mud couplet as an annual deposit, on the basis of the correspondence between layer thickness and the annual sedimentation rate determined using ^{210}Pb profiles. Their inference that the sand was deposited during times of onshore winds and higher wave activity, whereas the muds accumulated during times of offshore winds and minimal sea-bed disturbance, is generally consistent with the Kuehl et al. (1996) interpretation.

Bioturbated Prodelta Sandy Muds (Facies 10).—

In more distal, prodelta areas, deposits consist of uniform sandy mud that is intensely burrowed (Fig. 11C). Presumably the mud accumulates slowly, largely from suspension, and the sand is introduced during storms, with biogenic homogenization during fair weather. Shell debris, including whole shells, is present in small amounts, with the carbonate content increasing seaward as the sedimentation rate of siliciclastic mud decreases (Harris et al., 1996). In seismic sections, these deposits can be seen to downlap onto and prograde over the *Halimeda*-rich

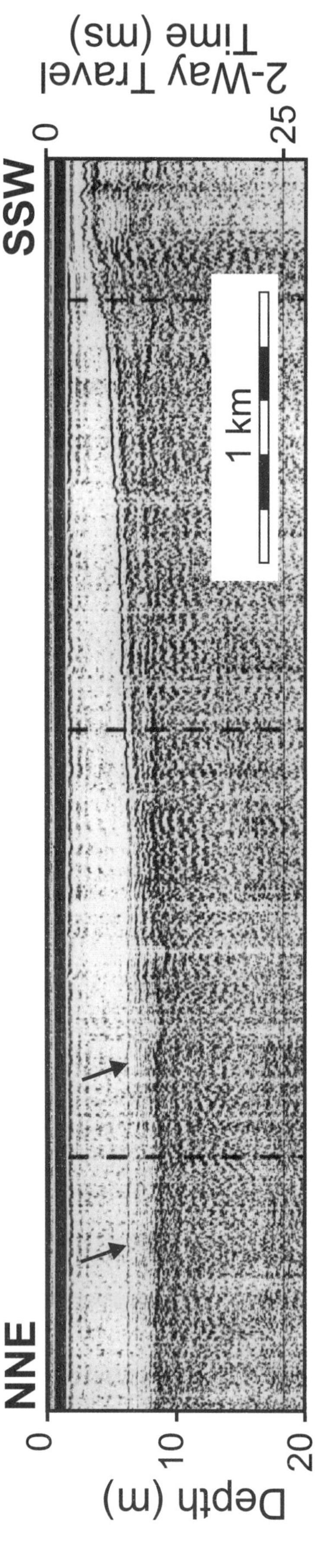

FIG. 14.—Seismic section from inner part of Far Northern Entrance (see Fig. 2 for location) showing gradual weakening of the channel-bottom reflector (indicated by arrows) as it passes laterally from a tidal bar with small dunes (SSW end) into the adjacent channel. Cores from other areas show that such weak, channel-bottom reflectors are caused by the presence of very soft ("jelly") muds deposited from fluid-mud bodies (Fig. 10A).

carbonate gravels that mark the maximum flooding surface on the shelf (Harris et al., 1996).

STRATIGRAPHIC ORGANIZATION

The stratigraphic organization of the deposits of the Fly delta (Fig. 18) must be considered at two scales: that of individual facies successions deposited within specific subenvironments, and the larger-scale, overall stacking of deposits caused by the response of the delta to the postglacial changes in sea level.

Facies Successions

Distributary Channels.—

The facies succession produced within the distributary channels (Fig. 18B) begins with an erosion surface (Fig. 7A). In most cases, this is immediately overlain by mud-pebble conglomerate (facies 2) that pass gradationally upward into cross-bedded sand (facies 1). The thickness of these two facies is typically 0.5–1 m but ranges between zero (Fig. 7C) and 2–3 m. If present, these deposits are the coarsest sediments in the succession, in terms of both absolute sand size and the low proportion of mud. Most commonly, these deposits are overlain abruptly but gradationally by 1–3 m of facies 3 channel-floor muds that are commonly the finest-grained deposits in the succession (Figs. 7B, 10A). They in turn pass gradually upward into heterolithic facies 4 by the addition of progressively thicker sands (Fig. 8). The mud layers also become thinner upward, on average, because of deposition at higher elevations, where the suspended-sediment concentration is less. The result is an upward-coarsening succession, with the coarsest heterolithic facies occurring near mid-channel depth on the elongate tidal bars and point bars (Fig. 18B). This upward-coarsening trend is confirmed by grab samples, which show a weak but consistent inverse correlation between mean grain size and water depth in all four of the areas of detailed sample (Baker 1998). At still higher elevations on the fringing tidal flats, the thickness and abundance of sand layers decrease more rapidly than the thickness of the mud layers, because of the upward decrease in tidal-current speed. This yields an upward-fining trend (Fig. 18B) in the upper part of the channel deposit (facies 4 → facies 5). The entire succession is capped by rooted upper intertidal and supratidal deposits (facies 6).

In some cores, the lower, upward-coarsening part of the succession is not clearly evident, perhaps because conditions were not favorable for the development or preservation of fluid-mud deposits (facies 3). Even in such situations, however, the mud layers in the heterolithic deposits (facies 4 and 5) show an upward decrease in thickness.

Mouth Bar–Delta Front.—

The prograding, prodelta–delta front–distributary mouth bar portion of the delta is expected to generate the anticipated upward-coarsening succession typical of all deltas (Fig. 18C). In the Fly delta, this succession begins with bioturbated sandy mud of the prodelta (facies 9), which passes upward into bioturbated heterolithic stratification (facies 8 and 9) of the delta front and then into the sands of the distributary-mouth bars (facies 7), which contain progressively fewer mud layers at higher elevations (Figs. 16, 18C). However, this gradational succession may not be as common as might be anticipated. Two of the cores intersecting the distributary-mouth elongate bars show that these sands erosionally overlie interbedded sands and muds of the delta front (Figs. 16B, 18D). Mud-pebble con-

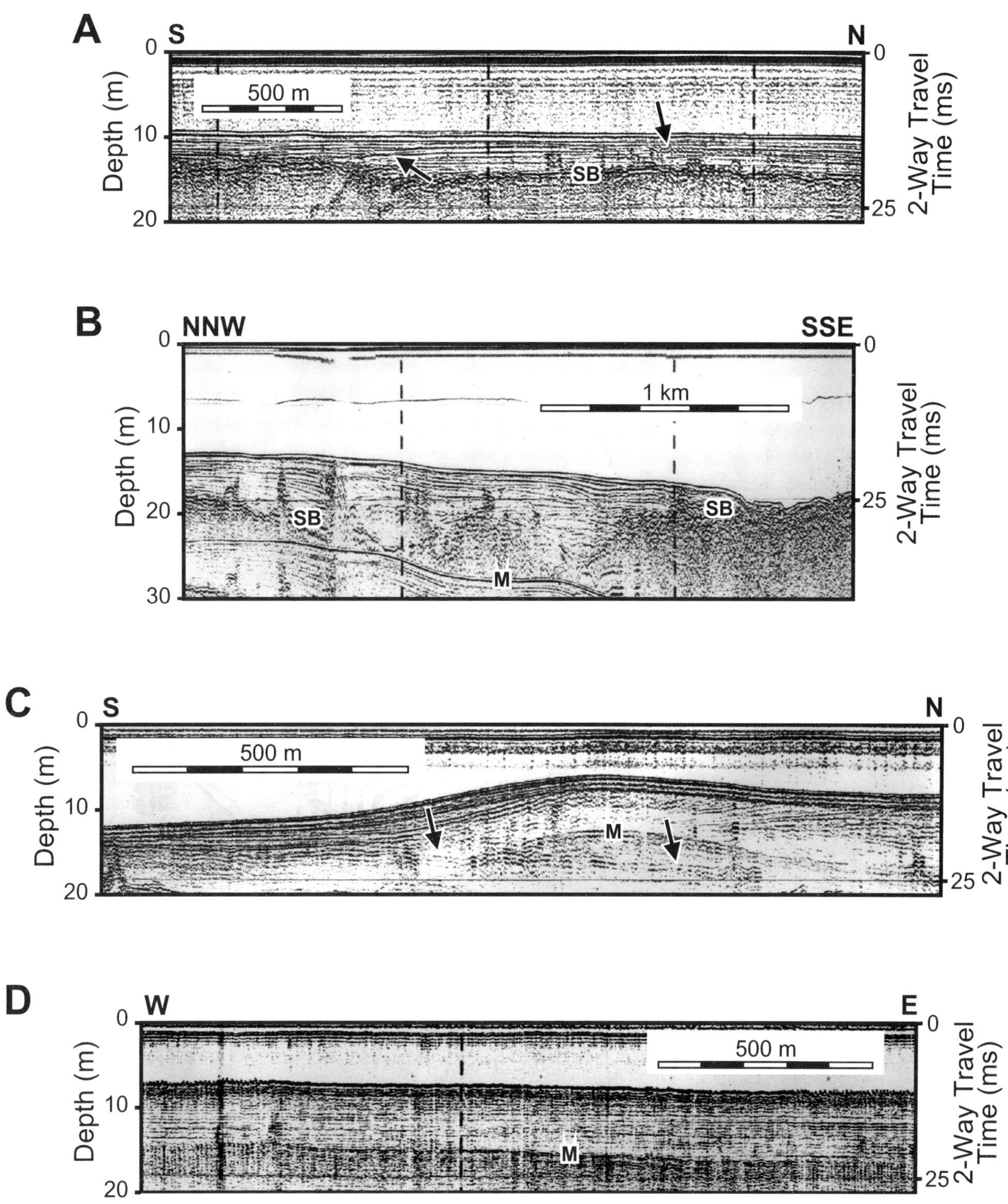

FIG. 15.—**A, B)** Flow-parallel (A) and flow-oblique (B) sections of a tidal point bar (see Figs. 2 and 5A, respectively, for locations). In Part A, note mounded geometry of toe of former point bar (right half of section) that is onlapped on the left by younger deposits (at left arrow). Vertical and lateral variations in reflector character are interpreted to represent variations between muddier, more heterolithic deposits (continuous, strong reflectors) and sandier sediment overlying erosion surfaces (broken reflectors; see especially at right arrow). In Part B, note conformable draping of stratification and downlap onto late Pleistocene sequence boundary (SB). **C, D)** Seismic sections transverse (C) and parallel (D) to axis of tidal bar in Far Northern Entrance (see Fig. 5C for location). The stratal dip in (C) reaches approximately 1.5°. The stratification clearly reflects lateral accretion to the south, with simultaneous erosion of the northern flank of the bar. Cores show that heterolithic strata (facies 4) underlie the modern erosion surface. Arrows point to possible location of sequence boundary.

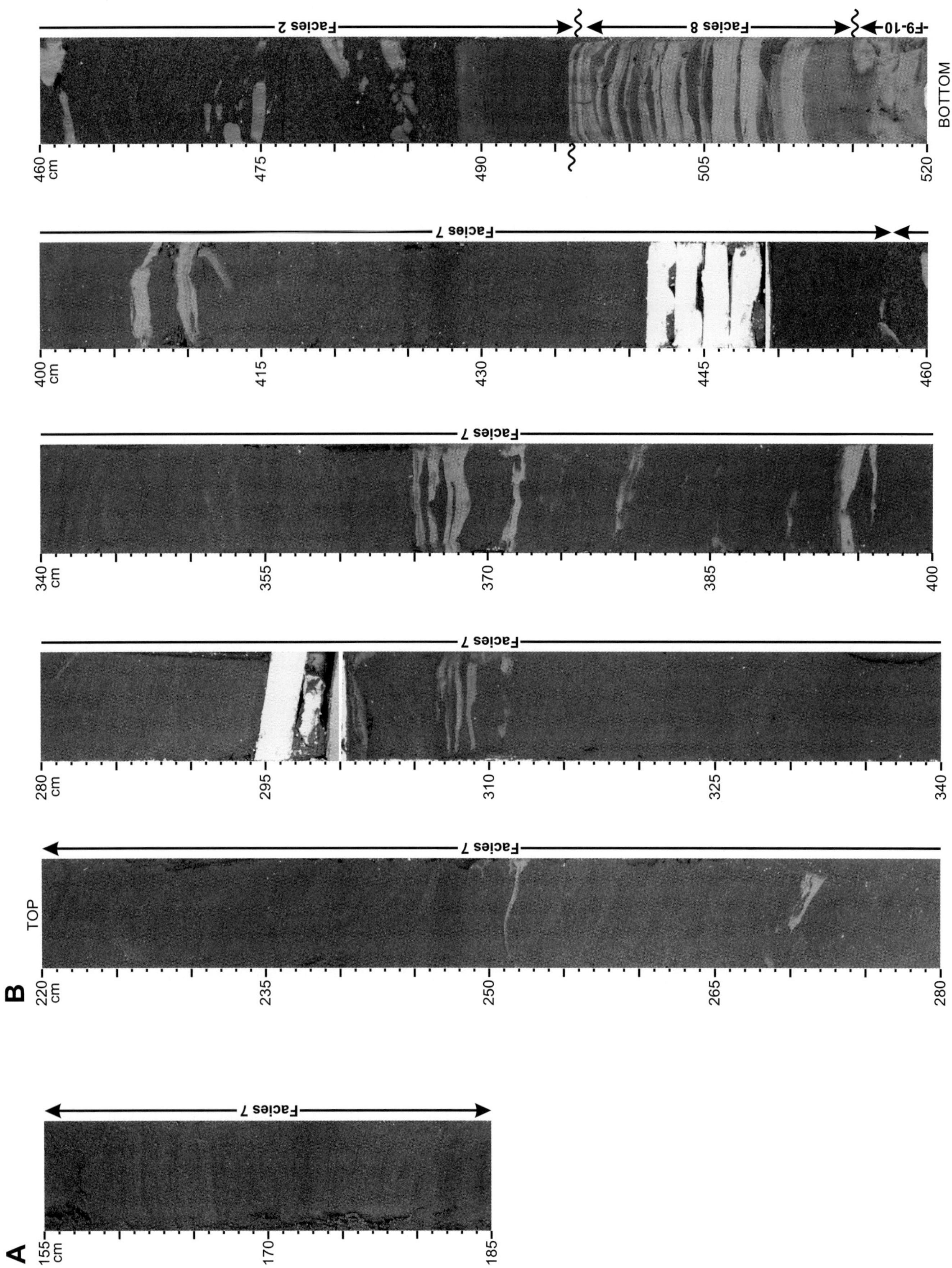
A
B
TOP
BOTTOM
Facies 2
Facies 8
F9-10
Facies 7
155 cm
170
185
220 cm
235
250
265
280
280 cm
295
310
325
340
340 cm
355
370
385
400
400 cm
415
430
445
460
460 cm
475
490
505
520

glomerates are well developed at the contact. This sharp-based sand presumably reflects lateral movement of the interbar channels, as indicated by the presence of lateral-accretion bedding within the bars (Fig. 17).

Delta switching in response to the diversion of discharge between distributaries at the delta "Apex" may also generate an erosion surface in the delta-front region (Fig. 18E). As suggested above, the large channel-form feature dissecting the delta front seaward of the Far Northern Entrance (Fig. 2) may be a result of erosion by tidal currents in the absence of significant sediment introduction. Observations in the outer part of the Far Northern Entrance show that a coarse-grained lag is being generated (Fig. 13), with widespread development of dunes (Fig. 12C). Because of the limited fresh-water influx and resulting relatively high salinity in this area, shell material is more abundant in this lag than in lags formed elsewhere in the delta.

Sequence-Stratigraphic Evolution

On a regional scale, the deposits in the modern delta complex and on the adjacent shelf show a stacking pattern that is consistent with the postglacial sea-level history. During oxygen-isotope stage 2 (ca. 18,000 yr B.P.), the Fly River flowed southeastward across the exposed Gulf of Papua shelf, incising a valley that extends to the shelf edge (Harris, 1994; Harris et al., 1996). The main axis of this valley underlies the Southern Entrance, on the basis of mapping of the depth of the late Pleistocene sequence boundary (Sanders, 1996), and continues inland beneath the present course of the Fly River. As postglacial sea level rose, the sediment load of the river was deposited within the valley, causing aggradation and valley filling (Harris, 1994). The widespread lack of transgressive sediment outside the valley on the shelf has been interpreted by Harris et al. (1996) to be the result of northward transport by a coastal current. However, there appears to have been at least one episode when the rate of the sea-level rise or the rate of northward removal slowed sufficiently that the river was able to prograde. The resulting lobate deposit is situated in 50–60 m of water, to the southeast of the modern delta margin. It overfills the valley and extends beyond it (Harris, 1994, his figs. 6, 7; see Fig. 18A, upper right), with internal clinoforms that downlap onto the underlying sequence boundary and/or flooding surface. This sediment body, which is about 40 km NW–SE by 50 km SW–NE and up to 20 m thick, represents a parasequence that is part of the transgressive systems tract (Harris et al., 1996).

←

FIG. 16.—Distributary-mouth-bar deposits. **A)** Relatively clean, flat-laminated sands (facies 7) from near the crest of an elongate bar. Note the absence of mud drapes. See Fig. 5D for location. **B)** Cross-bedded sands (facies 7) and mud-pebble conglomerates (facies 2) erosionally overlying delta-front deposits (facies 8-10; see Fig. 5D for location). The mouth-bar sands represent a channel-margin deposit, and they fine upward. The upward decrease in the abundance and thickness of mud drapes suggests that they were formed by deposition from fluid muds in the bottom of a channel between the bars. Many of the drapes are doublets. The basal 4 cm of the delta-front sediments consist of bioturbated mud (facies 10 or a mud bed of facies 9), which is sharply overlain by 20 cm of slightly bioturbated, thin-bedded muds and fine to very fine sands (496–516 cm; facies 8). The white material at 295 cm and 441–449 cm is styrofoam filling gaps between core segments.

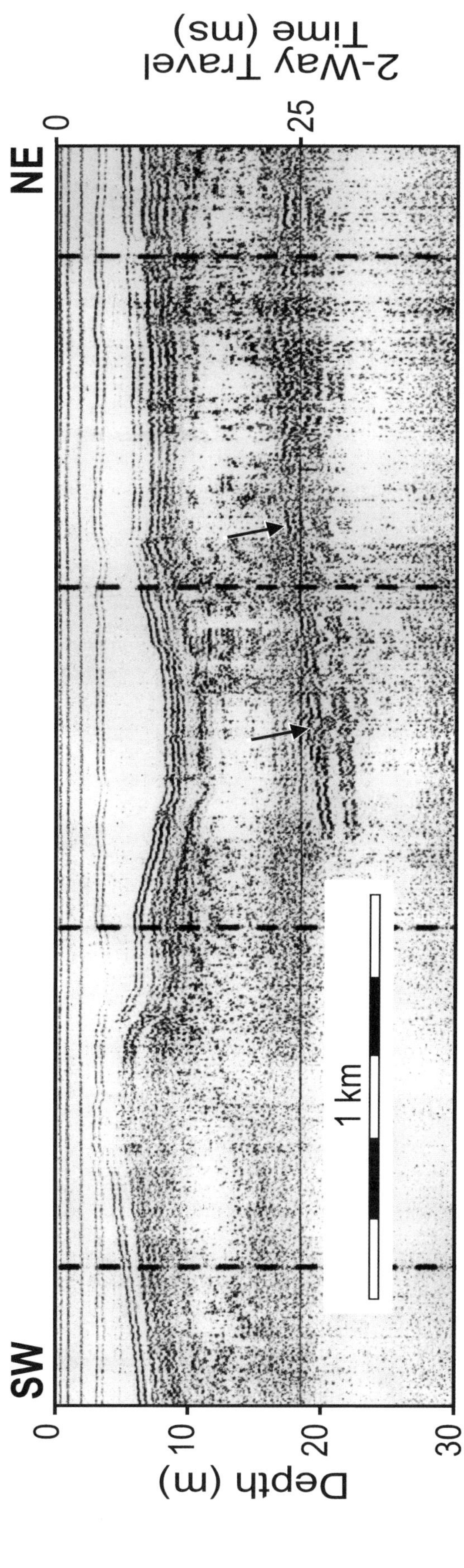

FIG. 17.—Seismic section across proximal end of distributary-mouth bar, north side of Southern Entrance. See Fig. 5D for location. Note inclined stratification within bar (maximum dips ~2°); poor definition of reflectors is likely a result of the homogeneous sandy nature of the sediments. The bar sands may overlie Holocene delta-front deposits and/or Pleistocene material at the right side of the record (at arrows).

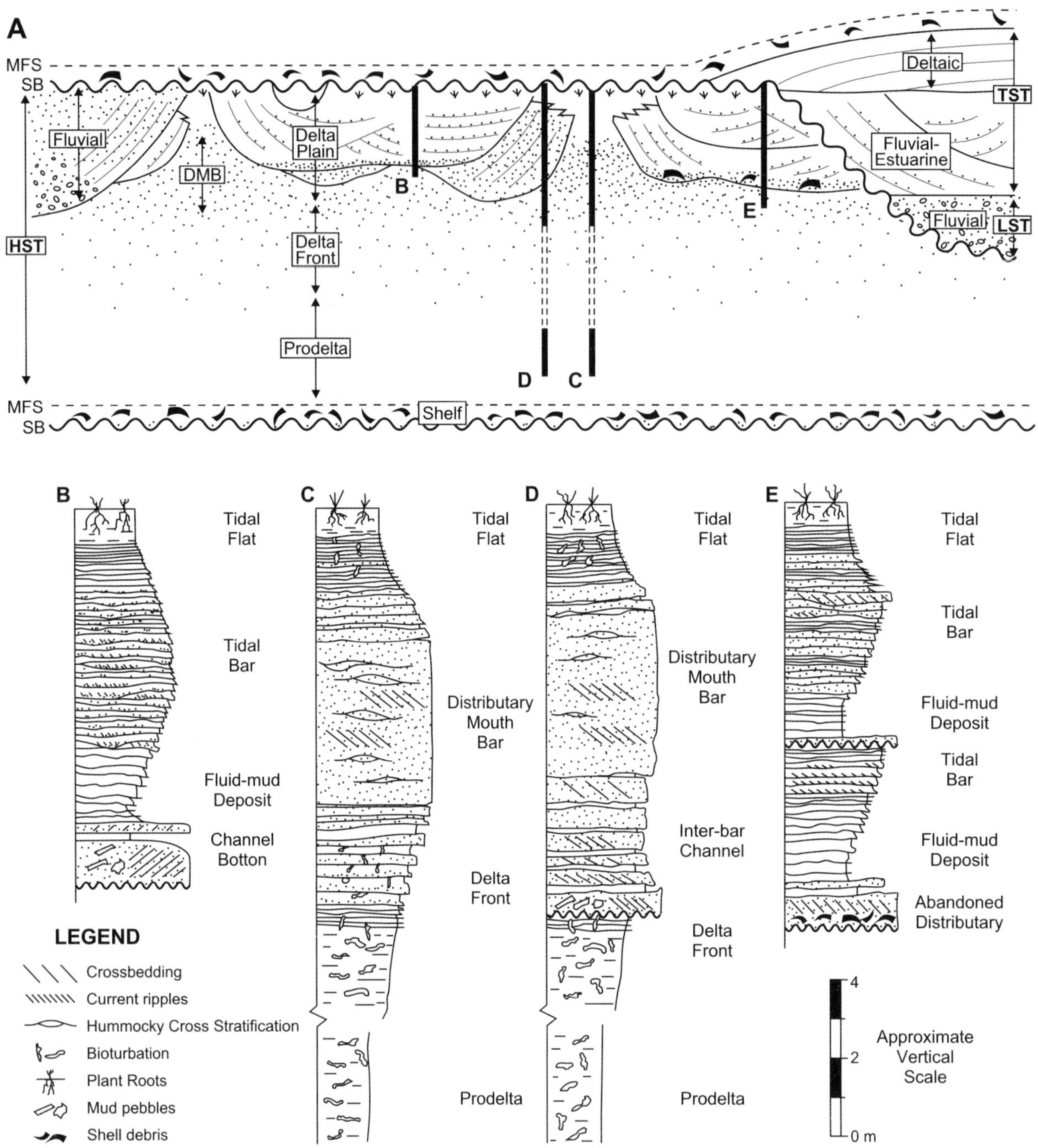

FIG. 18.—**A)** Schematic coast-parallel (N–S) section showing the stratigraphic organization of the complete deltaic succession in a sequence-stratigraphic context: transgression of a sequence boundary; highstand progradation of prodelta, delta-front, delta-plain, and fluvial (left side) environments; lowstand incision to create an incised valley (right side); and transgressive backfilling of the valley by fluvial and estuarine deposits. The deltaic body in the upper right corner formed during a stillstand in the transgression. The length of the section is 50–80 km. When constructing the section, it was assumed that the highstand progradation occurred during a time of very slightly rising to stable sea level, so that the thickness of the deltaic deposits more or less reflects the bathymetry of the system. Farther landward, near the point of turnaround from transgression to regression, the thickness of the delta-plain facies will be greater because of aggradation, with vertical stacking of channel-bottom deposits. By contrast, thicknesses will be reduced if deposition occurred during a relative sea-level fall. The shelly deposits overlying the combined sequence-boundary and flooding surfaces represent open-shelf deposits and will not be present in more inland areas. DMB = distributary mouth bar. **B-E)** Schematic vertical sections illustrating typical facies successions: (B) distributary channel (cf. Figs. 7, 8, 10); (C) prograding mouth bar, delta front, and prodelta; (D) mouth-bar–channel deposit (cf. Fig. 16B) overlying delta-front and prodelta facies; and (E) abandoned-distributary scour surface (cf. Fig. 13) overlain by two stacked, distributary-channel successions formed after channel reoccupation.

After this stillstand, transgression continued until sometime in the early Holocene, at which point the present-day progradation began. Seismic data (e.g., Figs. 12A, 15A, D) and cores (e.g., Fig. 7A, C) from the inner part of the delta reveal that the bases of the modern channels are generally separated from the late Pleistocene sequence boundary by more than 5 m of sediment containing facies identical to those of the present-day channels. ^{14}C dates on these deposits yield ages that are consistently less than 8000 yr BP. Numerous erosion surfaces marked by abrupt increases in sand content (facies 1 overlying facies 4), mud-pebble conglomerates (facies 2), and / or sudden increases in mud-layer thickness (facies 3 or 4 overlying facies 4 or 5) punctuate this succession (Fig. 7). Thus, vertical aggradation occurred during the turnaround from transgression to regression, presumably because the high sediment discharge initiated progradation before sea level stabilized at its present elevation approximately 6500 years ago (Thom and Roy, 1983). The point of maximum transgression of the distributary mouths is not known with certainty. However, beach ridges of presumed mid-Holocene age that are present to the south of the Fly delta (Fig. 1; Löffler, 1977; Baker, 1998, fig. 5.18) imply that the distributary-mouth region lay approximately halfway between the present coast and the "Apex". This is supported by the presence of delta-front heterolithic deposits (facies 8 and 9) with an age of 6260 yr BP in the bases of cores from the outer part of the Far Northern Entrance (Fig. 13) and also beneath the present-day mouth bars (Fig. 16B). This indicates that the Holocene regression has moved the coastline seaward by as much as 40–50 km, a distance that is consistent with the present-day rate of progradation (~ 6 m / a; Harris et al., 1993). Figure 18A shows the vertical succession of facies that would be expected if progradation continued for another 100–150 km, such that the single tidal–fluvial channel inland of the "Apex" came to overlie prodelta deposits.

DISCUSSION

The geomorphology and sedimentology of the Fly River delta show clearly that it is tide dominated. However, the tectonic and climatic settings (labile source rocks in a tropical, ever-wet climate, with deposition in a low-gradient, foreland basin) cause the deposits to be rich in mud. Therefore, two questions need to be examined: (1) To what extent are the deposits clearly tidal in character (i.e., would a muddy deposit such as this be readily identified as a tide-dominated delta in the ancient record)? (2) How representative is the Fly delta likely to be of tide-dominated deltaic systems? The hydrocarbon reservoir potential of the deposits is considered briefly at the end of this section.

In many of the cores, there is little direct evidence for the tidal nature for the deposits. Despite the inferred tidal origin of much of the layering in the more thinly layered facies (facies 3, 4, 5, and 8), statistically significant rhythmicity (cf. de Boer et al., 1989) and definitive tidal rhythmites (cf. Archer, 1998) are rarely developed. Nevertheless, some degree of cyclicity is moderately common, as is a thick–thin alternation of layer thicknesses that may reflect either the pronounced diurnal inequality of the system or the fortnightly variation in the intensity of spring-tidal currents. As in other tidal deposits, bipolar paleocurrent indicators in close stratigraphic proximity (e.g., herringbone cross lamination; superposition of current ripples with opposite facing directions) are not common because of the dominance of either the ebb or the flood at any given location. Perhaps the most pervasive, but indirect, tidal indicator is the heterolithic nature of the deposits (cf. Nio and Yang, 1991; Dalrymple, 1992).

The pervasive development of heterolithic stratification poses particular problems for the identification and interpretation of ancient tide-dominated delta deposits and their distinction from the deposits of other tide-dominated environments (e.g., macrotidal estuaries). For instance, the distinction between deltas and estuaries as defined in the Introduction (cf. Dalrymple et al., 1992; Dalrymple, 1999, 2000) lies in the identification of the facies stacking pattern: deltas are progradational, whereas estuaries are transgressive. This in turn requires that superficially similar heterolithic stratification formed in distinct environments (compare, for example, facies 4, 5, and 8; Figs. 10B, 11A) must be correctly distinguished and interpreted. Thus, as noted in the preceding facies descriptions, careful attention must be paid to all attributes (layer thickness, grain size, physical and biogenic structures, associated facies, etc.) of both the mud and sand components of the layering, in order to produce an environmentally significant subdivision of the heterolithic stratification (see Martinius et al., 2001, for an instructive example). In particular, this study has demonstrated that the thickness of the mud layers is very important in distinguishing between subenvironments, because of the significant vertical (within channels) and longitudinal (with respect to the turbidity maximum) variations in suspended-sediment concentration.

We believe that many of the features described here are widely applicable, despite the specific tectonic setting of the Fly system, which limits the size and amount of sand supplied to the delta. The general geomorphology of the tidal bar–channel systems in the distributaries (Fig. 5) appears fundamentally similar to those documented in other macrotidal environments (Robinson, 1960; Dalrymple et al., 1990; Dalrymple and Rhodes, 1995), apparently because the process responsible for bar formation (the so-called "Huthnance process"; cf. Dalrymple and Rhodes, 1995) is not dramatically affected by the high proportion of mud in the system (note that the amount of mud in the distributary-channel tidal bars approaches 50% or more in many instances). Similarly, the entire system of distributary channels and intervening islands appears likely to be largely insensitive to the sand / mud ratio because the geometry is mainly determined by processes operating in the distributary-mouth region (see previous discussion of the elongation of Kiwai Island), where sand predominates. Thus, paleogeographic reconstructions showing the pervasive development of a braided channel-bar pattern in sandy tidal systems (e.g., Van Wagoner, 1991) are unlikely to be correct.

In this regard, it is important to note that the tidal-bar model proposed by Mutti et al. (1985; see Dalrymple, 1992, fig. 30) and cited by many workers (e.g., Van Wagoner, 1991; Johnson and Baldwin, 1996) is not supported by the geomorphic and seismic observations reported here (Figs. 5, 15, 17). These data show that tidal bars generate lateral-accretion deposits, not the forward-accretion deposits (small- to medium-scale cross beds migrating down gently dipping master bedding planes) depicted in the Mutti et al. (1985) model. Instead, we suggest that deposits described by the Mutti et al. (1985) are compound dunes, not tidal bars.

The presence of more and coarser sand would cause the distributary channels and mouth-bar region to be sandier and to contain a greater abundance of cross-bedded facies than seen in the Fly delta. However, muddy channel-bottom deposits similar to facies 3 and the muddier variants of facies 4 (Figs. 7, 8) would probably still exist, because the processes responsible for trapping suspended sediment to create a turbidity maximum (see summaries in Allen, 1991; Hughes et al., 1998) operate in all systems. Thus, thick mud drapes should be typical of the deeper parts of delta-plain channels, although this may not be unique to deltas, as indicated by the presence of fluid muds in estuarine settings (e.g., Kirby and Parker, 1983; Allen 1991). Indeed, Allen (1991; see also Fenies and Tastet, 1998) has reported that channels in the Gironde estuary coarsen upward to near the low-tide level because of the formation of fluid muds in the channel bottoms, even though the Gironde is overall much coarser grained than the Fly. Furthermore, the abundance of heterolithic stratification, combined with the active migration of the channels, should cause mud-pebble conglomerates to be a particularly characteristic feature of this setting (cf. Fenies and Tastet, 1998). The apparent similarity of the delta-front facies of the Fly and Amazon deltas noted above is a further indication that the results of this study have broader application.

At the largest scale, the stratigraphic organization of the Fly delta (early Holocene transgression, followed by *in situ* aggradation and then by progradation) illustrates that tide-dominated systems obey the same fundamental sequence-stratigraphic rules as other coastal systems. The challenge is, however, to distinguish between the many erosion surfaces that will occur in tidal deposits: channel-base scours related to autocyclic channel-bar migration; channel-base scours related to distributary abandonment and the ensuing enlargement by tidal-ravinement processes; and channel-base scours caused by fluvial incision during sea-level falls (i.e., the sequence boundary). In all cases, the deposits immediately overlying the surface in an axial location hold the key to its origin and sequence-stratigraphic significance.

Previous attempts to develop a facies model for tide-dominated deltas (e.g., Wright, 1985; Dalrymple, 1992; Emery and Myers, 1996) were hindered by the lack of detailed facies descriptions, and changes are required in light of the results presented here. With regard to the models presented by Dalrymple (1992, fig. 29), the most notable differences found in the Fly system are: distributary channels and channels in abandoned parts of the delta commonly do not show a simple upward-fining trend because of the presence of channel-bottom fluid-mud deposits; thin-bedded heterolithic stratification is more prevalent in delta-plain and delta-front environments than shown; and upward-coarsening distributary-mouth-bar successions may not be common because of the lateral migration of the interbar channels (Fig. 18A). Furthermore, predictions that tidal rhythmites (*sensu stricto*) may be developed in delta-front and prodelta settings (Dalrymple, 1992) are not substantiated by this study.

Because of the high mud content, the hydrocarbon reservoir potential of a system like the Fly delta is mixed. The mouth-bar sands represent by far the best reservoir facies, because of their relatively great thickness (5–10 m), large lateral extent, both parallel and perpendicular to the coast (many tens of kilometers and perhaps > 10 km, respectively), and relative scarcity of mud baffles or barriers. Channel-floor lags are probably the next best reservoir facies, but they are generally thin in the Fly delta and may be of limited lateral extent because of their restriction to areas of local erosion. The lag produced during the abandonment phase of a distributary is potentially more widespread, but carbonate cementation may be a problem because of the incorporated shell material. The heterolithic deposits of the tidal bars and point bars within the distributary channels have poor reservoir potential, because of the abundance and lateral continuity of the mud layers. The channel facies of systems with a higher sand content have better reservoir potential, but mud drapes may still be widespread, producing permeability baffles.

SUMMARY AND CONCLUSIONS

The Fly River is the 17th largest river in the world in terms of sediment load. Its headwaters are in an active continental-arc setting, but it flows for more than 700 km parallel to the axis of the north-Australian foreland basin before emptying into the Gulf of Papua. As a result, almost all sediment coarser than fine to very fine sand is deposited far inland, and the sediment reaching the coast is predominantly silt, with about 10% sand-size material. Because of the large sediment input, the modern delta is prograding at a rate of about 6 m/a and has advanced 40–50 km in the 6000–8000 years since the local end of the postglacial transgression. The turnaround from transgression to progradation was accompanied by vertical aggradation and the stacking of distributary-channel deposits.

The delta is tide dominated because it is subjected to tidal currents that generally exceed 0.5–1 m/s at spring tides, when the tidal range reaches 5.1 m. Wave energy is generally minimal, because of its latitudinal position. Because of the low gradient, tidal influence is felt up to 400 km inland. The tidal currents are responsible for the formation of a funnel-shaped set of three main distributary channels (typically 5–10 m deep and up to 25–35 km wide at the coast), which are separated by a series of elongate, vegetated islands. Within these channels, tidal flow has produced a series of elongate tidal bars that separate subchannels with opposing directions of residual bedload transport. Coast-normal, elongate tidal bars also characterize the distributary-mouth-bar and delta-front region, which extends 15–20 km offshore. Seaward of this, the prodelta slopes relatively steeply seaward, meeting the shelf surface in 30–40 m of water, 40 km offshore.

Small morphological changes at the point where the distributaries join are responsible for diverting flow from one distributary to another (i.e., the process of "delta switching"). At present, 60–80% of the sediment discharge reaches the sea by way of the southernmost distributary, whereas the northernmost channel is largely abandoned. As a result, tidal currents are actively scouring this channel, generating sand-rich and shell-rich lags. Dunes are widely developed in this lag, but they are never large because of the general scarcity of sand.

The abundance of suspended sediment, combined with hydrodynamic trapping mechanisms and flocculation, lead to the development of fluid muds (SSC values > 10 g/l) in the bottoms of channels throughout the delta. These fluid muds generate anomalously muddy channel-bottom deposits with individual mud layers that commonly reach or exceed 1 cm in thickness. These muddy deposits overlie thin (generally < 1–2 m thick) units of cross-bedded and/or rippled sand and mud-pebble conglomerate that are produced as lags in localized areas of channel-floor erosion. The fluid-mud deposits are themselves overlain by heterolithic stratification that shows an overall upward coarsening to about the mid-channel elevation, where the mud content is 30–60%. These heterolithic deposits occur as gently inclined lateral-accretion bedding that forms on the elongate tidal bars and bank-margin point bars. (The Mutti et al., 1985, tidal-bar model is not supported by the observations obtained in this study.) At higher elevations on the tidal flats, the sediments become progressively finer and more thinly laminated. Throughout the channel succession, the mud layers decrease in thickness upward, because of the decreasing SSC values.

The distributary-mouth bars consist of fine to very fine sand containing flat lamination (hummocky cross stratification?) and cross bedding. Mud layers are not abundant in exposed positions, because of wave resuspension, but they do occur in interbar channels. Lateral migration of these channels causes many of the mouth-bar sands to be sharp based and produces lateral-accretion bedding within the bars. Gradational upward-coarsening successions from biologically homogenized prodelta muds, through heterolithic delta-front deposits, into mouth-bar sands may not be volumetrically dominant. However, more work is needed to understand the mouth-bar and delta-front region.

Correct environmental and sequence-stratigraphic interpretation of a tide-dominated system such as the Fly requires careful distinction between subtly different types of heterolithic stratification. The thickness and internal character of the mud layers is particularly important in this regard, because of the need to recognize the deposits formed by fluid muds. If good coast-parallel outcrop exists, changes in channel width might also be a useful proximal–distal indicator, because of the seaward increase in channel size.

ACKNOWLEDGMENTS

The field work on which this paper is based was supported by grants from the Australian Defence Science and Technology Organization and the Australian Research Council (Ref. No. A39131170) to PTH. Ok Tedi Mining Corp. is thanked for allowing use of the RV Western Venturer, the captain and crew of which assisted in every way possible. Work by RWD was funded by a grant from the Canadian Natural Sciences and Engineering Research Council (No. 7553). Ela Rusak and Thomas Ullrich (Queen's University) drafted the figures. We also thank Alan Archer, Howard Johnson, Erik Kvale, Terry Lukie, and Brian Willis for their constructive reviews.

REFERENCES

ALLEN, G.P., LAURIER, D., AND THOUVENIN, J., 1979, Etude sédimentologique du delta de la Mahakam: Compagnie Française des Pétroles, Notes et Mémoires 15, 156 p.

ALLEN, G.P., 1991, Sedimentary processes and facies in the Gironde estuary: a recent model for macrotidal estuarine systems, *in* Smith, D.G., Reinson, G.E., Zaitlin, B.A., and Rahmani, R.A., eds., Clastic Tidal Sedimentology: Canadian Society of Petroleum Geologists, Memoir 16, p. 29–40.

ALLEN, J.R.L., 1970, Sediments of the modern Niger delta, a summary and review, *in* Morgan, J.P., ed., Deltaic Sedimentation, Modern and Ancient: Society of Economic Paleontologists and Mineralogists, Special Publication 15, p. 138–151.

ALONGI, D.M., CHRISTOFFERSON, P., TIRENDI, F., AND ROBERTSON, A., 1992, The influence of freshwater and material exported on sedimentary facies and benthic processes within the Fly Delta and adjacent Gulf of Papua (Papua New Guinea): Continental Shelf Research, v. 12, p. 287–326.

ALONGI, D.M., TIRENDI, F., AND CHRISTOFFERSON, P., 1991, Vertical profiles of copper in sediments from the Fly Delta and Gulf of Papua: Marine Pollution Bulletin, v. 22, p. 253–255.

ARCHER, A.W., 1998, Hierarchy of controls on cyclic rhythmite deposition: Carboniferous basins of eastern and mid-continental U.S.A., *in* Alexander, C.R., Davis, R.A., and Henry, V.J., eds., Tidalites: Processes and Products: SEPM, Special Publication 61, p. 59–68.

BAKER, E.K., 1998, Sedimentation in a tidally dominated delta and the impact of upstream mining: Fly River, Papua New Guinea: unpublished Ph.D. thesis, University of Sydney, Sydney, Australia, 245 p.

BAKER, E.K., HARRIS, P.T., SHORT, S.A., AND KEENE, J.B., 1995, Patterns of sedimentation in the Fly River Delta, *in* Flemming, B.W., and Bartholomä, A., eds., Tidal Signatures in Modern and Ancient Sediments: International Association of Sedimentologists, Special Publication 24, p. 193–211.

BHATTACHARYA, J.P., AND WALKER, R.G., 1992, Deltas, *in* Walker, R.G., and James, N.P., eds., Facies Models; Response to Sea Level Change: St. John's, Newfoundland, Geological Association of Canada, p. 157–177.

COLEMAN, J.M., AND WRIGHT, L.D., 1975, Modern river deltas: variability of processes and sand bodies, *in* Broussard, M.L., ed., Deltas, Models for Exploration. Houston, Texas, Houston Geological Society, p. 99–149.

DALRYMPLE, R.W., 1992, Tidal depositional systems, *in* Walker, R.G., and James, N.P., eds., Facies Models; Response to Sea Level Change: St. John's, Newfoundland, Geological Association of Canada, p. 195–218.

DALRYMPLE, R.W., 1999, Tide-dominated deltas: Do they exist or are they all estuaries? (abstract): American Association of Petroleum Geologists, Annual Meeting, San Antonio, Official Program, p. A29–30.

DALRYMPLE, R.W., 2000, Tidal bars, estuaries versus deltas, and sequence boundaries: lessons and challenges from the Bay of Fundy and Fly River delta (abstract): Tidalite 2000, Seoul, Korea, Abstracts, p. 27–29.

DALRYMPLE, R.W., KNIGHT, R.J., ZAITLIN, B.A., AND MIDDLETON, G.V., 1990, Dynamics and facies model of a macrotidal sand bar complex: Sedimentology, v. 37, p. 577–612.

DALRYMPLE, R.W., AND RHODES, R.N., 1995, Estuarine dunes and barforms, *in* Perillo, G.M., ed. Geomorphology and Sedimentology of Estuaries: Amsterdam, Elsevier, Developments in Sedimentology, v. 53, p. 359–422.

DALRYMPLE, R.W., ZAITLIN, B.A., AND BOYD, R., 1992, Estuarine facies models: conceptual basis and stratigraphic implications: Journal of Sedimentary Petrology, v. 62, p. 1130–1146.

DE BOER, P.L., OOST, A.P., AND VISSER, M.J., 1989, The diurnal inequality of the tides as a parameter for recognizing tidal influences: Journal of Sedimentary Petrology, v. 59, p. 912–921.

DIETRICH, W.E., DAY, G., AND PARKER, G., 1999, The Fly River, Papua New Guinea: Inferences about river dynamics, floodplain sedimentation and fate of sediment, *in* Miller, A.J., and Gupta, A., eds., Varieties in Fluvial Form: New York, Wiley, p. 345–376.

EAGLE, A.M., AND HIGGINS, R.J., 1990, Environmental investigations on the effects of the Ok Tedi copper mine in the Fly River system, *in* Lawrence, D., and Cransfield-Smith, T., eds., Torres Strait Baseline Study. Workshop Series 16, Great Barrier Reef Marine Park Authority, 535 p.

EMERY, D., AND MYERS, K.J., 1996, Sequence Stratigraphy: Oxford, U.K., Blackwell Science, 297 p.

FENIES, H., AND TASTET, J.P., 1998, Facies and architecture of an estuarine tidal bar (the Trompeloup bar, Gironde Estuary, SW France): Marine Geology, v. 150, p. 149–169.

FISHER, W.L., BROWN, L.F., SCOTT, A.J., AND MCGOWEN, J.H., 1969, Delta Systems in the Exploration for Oil and Gas, A Research Colloquium: Austin, Texas, Texas Bureau of Economic Geology, 204 p.

GALLOWAY, W.E., 1975, Process framework for describing the morphologic and stratigraphic evolution of deltaic depositional systems, *in* Brossard, M.L., ed., Deltas: Models for Exploration: Houston, Texas, Houston Geological Society, p. 87–88.

GASTALDO, R.A., ALLEN, G.P., AND HUC, A.Y., 1995, The tidal character of fluvial sediments of the modern Mahakam River delta, Kalimantan, Indonesia, *in* Flemming, B.W., and Bartholomä, A., eds., Tidal Signatures in Modern and Ancient Sediments: International Association of Sedimentologists, Special Publication, 24, p. 171–181.

HARRIS, P.T., 1988, Sediments, bedforms and bedload transport pathways on the continental shelf adjacent to Torres Strait, Australia–Papua New Guinea: Continental Shelf Research, v. 8, p. 979–1003.

HARRIS, P.T., 1994, Incised valleys and backstepping deltaic deposits in a foreland-basin setting, Torres Strait and Gulf of Papua, Australia, *in* Dalrymple, R.W., Boyd, R., and Zaitlin, B.A., eds., Incised-Valley Systems: Origin and Sedimentary Sequences: SEPM, Special Publication 51, p. 97–108.

HARRIS, P.T., 1995, Muddy waters: The physical sedimentology of Torres Strait, *in* Bellwood, O., Choat, H., and Saxenta, N., eds., Recent Advances in Marine Science and Technology '94, Proceedings of the PACON Conference: Townsville, James Cook University of North Queensland, p. 149–159.

HARRIS, P.T., BAKER, E.K., COLE, A.R., AND SHORT, S.A., 1993, Preliminary study of sedimentation in the tidally dominated Fly River Delta, Gulf of Papua: Continental Shelf Research, v. 13, p. 441–472.

HARRIS, P.T., BAKER, E.K., AND SCHNEIDER, P.M., 1989, Sandwave movement, currents and sedimentation in Torres Strait: Summary of results obtained during a cruise of HMAS Flinders in February, 1989: University of Sydney, Ocean Sciences Institute, Report 40, 106 p.

HARRIS, P.T., PATTIARATCHI, C.B., KEENE, J.B., DALRYMPLE, R.W., GARDNER, J.V., BAKER, E.K., COLE, A.R., MITCHELL, D., GIBBS, P., AND SCHROEDER, W.W., 1996, Late Quaternary deltaic and carbonate sedimentation in the Gulf of Papua foreland basin: response to sea-level change: Journal of Sedimentary Research, v. 66, p. 801–819.

HIGGENS, R.J., PICKUP, G., AND CLOKE, P.S., 1987, Estimating the transport and deposition of mining waste at Ok Tedi, *in* Thorne, C.R., Bathurst, J.C., and Hey, R.D., eds., Sediment Transport in Gravel Bed Rivers: New York, John Wiley & Sons, p. 949–976.

HUGHES, M.G., AND BAKER, E.K., 1996, Using satellite remote sensing data to identify coastline changes in the Fly River delta, PNG: A Pilot study. University of Sydney, Ocean Sciences Institute, Report 68, 21 p.

HUGHES, M.G., HARRIS, P.T., AND HUBBLE, T.C.T., 1998, Dynamics of the turbidity maximum zone in a micro-tidal estuary: Hawkesbury River, Australia: Sedimentology, v. 45, p. 397–410.

JAEGER, J.M., AND NITTROUER, C.A., 1995, Tidal controls on the formation of fine-scale sedimentary strata near the Amazon river mouth: Marine Geology, v. 125, p. 259–281.

JOHNSON, H.D., AND BALDWIN, C.T., 1996, Shallow siliciclastic seas, *in* Reading, H.G., ed., Sedimentary Environments; Processes, Facies and Stratigraphy: Oxford, U.K., Blackwell Science, p. 232–280.

JOHNSON, M.A., KENYON, N.H., BELDERSON, R.H., AND STRIDE, A.H., 1982, Sand transport, *in* Stride, A.H., ed., Offshore Tidal Sands; Processes and Deposits: London, Chapman & Hall, p. 58–94.

KIRBY, R., AND PARKER, W.R., 1983, Distribution and behavior of fine sediment in the Severn Estuary and inner Bristol Channel, U.K.: Canadian Journal of Fisheries and Aquatic Sciences, v. 40, Supplement 1, p. 83–95.

KUEHL, S.A., NITTROUER, C.A., ALLISON, M.A., FARIA, L.E.C., DAKUT, D.A., MAEGER, J.M., PACIONI, T.D., FIGUEIREDO, A.G., AND UNDERKOFFLER, E.C., 1996, Sediment deposition, accumulation, and seabed dynamics in an energetic fine-grained coastal environment: Continental Shelf Research, v. 16, p. 787–815.

LEVELL, B.K., 1980, Late Precambrian tidal shelf deposit, the Lower Sandfjord Formation, Finnmark, Northern Norway: Sedimentology, v. 27, p. 539–557.

LI, C., WANG, P., DAIDU. F., BING, D., AND TIESONG, L., 2000, Open-coast intertidal deposits and the preservation potential of individual laminae: a case study from east-central China: Sedimentology, v. 47, p. 1039–1051.

LÖFFLER, E., 1977, Geomorphology of Papua New Guinea: Canberra, Australia, Australian National University Press, 258 p.

MARTINIUS, A.W., KAAS, I., NÆSS, A., HELGESEN, G., KJÆREFJORD, J.M., AND LEITH, D.A., 2001, Sedimentology of the heterolithic and tide-dominated Tilje Formation (Early Jurassic, Halten Terrace, offshore mid-Norway), *in* Martinsen, O.J., and Dreyer, T., eds., Sedimentary Environments Offshore Norway—Paleozoic to Recent: Norwegian Petroleum Society, Special Publication 10, p. 103–144.

MECKEL, L.D., 1975, Holocene sand bodies in the Colorado delta area, northern Gulf of California, *in* Broussard, M.L., ed., Deltas: Models for Exploration: Houston, Houston Geological Society, p. 239–265.

MEHTA, A.J., HAYTER, E.J., PARKER, W.R., KRONE, R.B., AND TEETER, A.M., 1989, Cohesive sediment transport. I: Process description: Journal of Hydraulic Engineering, v. 115, p. 1076–1093.

MIDDLETON, G.V., 1991, A short historical review of clastic tidal sedimentology, *in* Smith, D.G., Reinson, G.E., Zaitlin, B.A., and Rahmani, R.A., eds., Clastic Tidal Sedimentology: Canadian Society of Petroleum Geologists, Memoir 16, p. ix–xv.

MUTTI, E., ROSELL, J., ALLEN, G.P., FONNESU, F., AND SGAVETTI, M., 1985, The Eocene Baronia tide dominated delta-shelf system in the Ager Basin, *in* Mila, M.D., and Rosell, J., eds., International Association of Sedimentologists, 6th European Regional Meeting, Lleida, Spain, Excursion Guidebook, p. 579–600.

MYRICK, R.M., AND LEOPOLD, L.B., 1963, Hydraulic geometry of a small tidal estuary: U.S. Geological Survey, Professional Paper 422B, 18 p.

NIO, S.-D., AND YANG, C.S., 1991, Diagnostic attributes of clastic tidal deposits, *in* Smith, D.G., Reinson, G.E., Zaitlin, B.A., and Rahmani, R.A., eds., Clastic Tidal Sedimentology: Canadian Society of Petroleum Geologists, Memoir 16, p. 3–28.

PARKER, G., 1991, Selective sorting and abrasion of river gravel. I: Theory: Journal of Hydraulic Engineering, v. 117, p. 131–171.

PICKUP, G., 1984, Landforms, hydrology and sedimentation in the Fly and lower Purari, Papua New Guinea, *in* Schick A.P., ed., Channel Processes—Water, Sediment, Catchment Controls: Braunschweig, Catena Supplement, v. 5, p. 1–17.

ROBERTSON, A.I., AND ALONGI, D.M., 1995, Role of riverine mangrove forests in organic carbon export to the tropical coastal ocean: A preliminary mass balance for the Fly Delta (Papua New Guinea): Geo-Marine Letters, v. 15, p. 134–139.

ROBERTSON, A.I., DANIEL, P.A., DIXON, P., AND ALONGI, D.M., 1993, Pelagic biological processes along a salinity gradient in the Fly delta and adjacent river plume (Papua New Guinea): Continental Shelf Research, v. 13, p. 205–224.

ROBINSON, A.H.W., 1960, Ebb–flood channel systems in sandy bays and estuaries: Geography, v. 45, p. 183–199.

SADLER, P. M., 1999, The influence of hiatuses on sediment accumulation rates, *in* Bruns, P., and Hass, H.C., eds., On the Determination of Sediment Accumulation Rates: Zurich, Switzerland, Trans Tech Publishers Ltd., GeoResearch Forum 5, p. 15–40.

SANDERS, J., 1996, A high-resolution seismic study of the architecture and facies of the Holocene sediments in the Fly River Delta, Papua New Guinea: unpublished B.Sc. (Hons.) thesis, Queen's University, Kingston, Canada, 84 p.

SEGALL, M.P., AND KUEHL, S.A., 1994, Sedimentary structures on the Bengal shelf: a multi-scale approach to sedimentary fabric interpretation: Sedimentary Geology, v. 93, p. 165–180.

SMEC (Snowy Mountains Engineering Corporation), 1981, Fly River Chart Folio—Korimoro Point to Kiunga: Ok Tedi Mining Ltd., unpublished report, 44 p.

STAUB, J.R., AND ESTERLE, J.S., 1993, Provenance and sediment dispersal in the Rajang River delta / coastal plain system, Sarawak, East Malaysia: Sedimentary Geology, v. 85, p. 191–201.

THOM, B.G., AND ROY, P.S., 1983, Sea-level change in New South Wales over the past 15,000 years, *in* Hopley, D., ed., Australian Sea Levels in the Past 15,000 Years: A Review: Townsville, Australia, James Cook University, Department of Geography, p. 64–84.

THOM, B.G., AND WRIGHT, L.D., 1983, Geomorphology of the Purari Delta, *in* Petr, T., ed., The Purari—Tropical Environment of a High Rainfall River Basin: The Hague, Junk Publishers, p. 47–65.

THOMPSON, R.W., 1968, Tidal Flat Sedimentation on the Colorado River Delta, Northwestern Gulf of California: Geological Society of America, Memoir 107, 413 p.

VAN WAGONER, J.C., 1991, High-frequency sequence stratigraphy and facies architecture of the Sego Sandstone in the Book Cliffs of western Colorado and eastern Utah, *in* Van Wagoner, J.C., Nummedal, D., Jones, C.R., Taylor, D.R., Jennette, D.C., and Riley, G.W., eds., Sequence Stratigraphy—Applications to Shelf Sandstone Reservoirs: Outcrop to Subsurface Examples: American Association of Petroleum Geologists, p. 1–22.

WALKER, R.G., 1992, Facies, facies models and modern stratigraphic concepts, *in* Walker, R.G., and James, N.P., eds., Facies Models; Response to Sea Level Change: St. John's, Newfoundland, Geological Association of Canada, p. 1–14.

WOLANSKI, E., AND ALONGI, D.M., 1995, A hypothesis for the formation of a mud bank in the Gulf of Papua: Geo-Marine Letters, v. 15, p. 166–171.

WOLANSKI, E., AND EAGLE, M., 1991, Oceanography and sediment transport, Fly River estuary and Gulf of Papua, *in* 10th Australasian Conference on Coastal and Ocean Engineering, Proceedings: Auckland, New Zealand, p. 453–457.

WOLANSKI, E., AND GIBBS, R.J., 1995, Flocculation of suspended sediment in the Fly River estuary, Papua New Guinea: Journal of Coastal Research, v. 11, p. 754–762.

WOLANSKI, E., GIBBS, R.J., SPAGNOL, S., KING, B., AND BRUNSKILL, G.J., 1998, Inorganic sediment budget in the mangrove-fringed Fly River delta, Papua New Guinea: Mangroves and Salt Marshes, v. 2, p. 85–98.

WOLANSKI, E., KING, B., AND GALLOWAY, D., 1995a, Dynamics of the turbidity maximum in the Fly River estuary, Papua New Guinea: Estuarine, Coastal and Shelf Science, v. 40, p. 321–337.

WOLANSKI, E., KING, B., AND GALLOWAY, D., 1997, Salinity intrusion into the Fly River estuary, Papua New Guinea: Journal of Coastal Research, v. 13, p. 983–994.

WOLANSKI, E., NORRO, A., AND KING, B., 1995b, Water circulation in the Gulf of Papua: Continental Shelf Research, v. 15, p. 185–212.

WOLANSKI, E., SPAGNOL, S., KING, B., AND AYUKAI, T., 1999, Patchiness in the Fly River plume in Torres Strait: Journal of Marine Systems, v. 18, p. 369–381.

WOLANSKI, E., TRENORDEN, M., AND KING, B., 1990, Fly River estuary and Gulf of Papua: Oceanography and sediment dynamics data report: Report to Ok Tedi Mining Ltd., Australian Institute of Marine Science, Townsville, Australia, 32 p.

WRIGHT, L.D., 1985, River deltas, *in* Davis, R.A., Jr., ed., Coastal Sedimentary Environments, 2nd Edition: New York, Springer-Verlag, p. 1–76.

WRIGHT, L.D., COLEMAN, J.M., AND THOM, B.G., 1975, Sediment transport and deposition in a macrotidal river channel, Ord River, Western Australia, *in* Cronin, L.E., ed., Estuarine Research, v. II: New York, Academic Press, p. 309–322.

DELTA EVOLUTION MODEL INFERRED FROM THE HOLOCENE MEKONG DELTA, SOUTHERN VIETNAM

SUSUMU TANABE
Graduate School of Science and Technology, Niigata University, Ikarashi-2 8050, Niigata 950-2181, Japan
Present address: Sedimentary Geology Research Group, Geological Survey of Japan,
AIST. Central 7, Higashi 1-1-1, Tsukuba 305-8567, Japan
e-mail: s.tanabe@aist.go.jp
THI KIM OANH TA AND VAN LAP NGUYEN
Sub-Institute of Geography, National Research Center for Science and Technology,
Mac Dinh Chi 1, Dist. 1, Ho Chi Minh, Vietnam
MASAAKI TATEISHI AND IWAO KOBAYASHI
Department of Geology, Niigata University, Ikarashi-2 8050, Niigata 950-2181, Japan
AND
YOSHIKI SAITO
MRE, Geological Survey of Japan, AIST. Central 7, Higashi 1-1-1, Tsukuba 305-8567, Japan

ABSTRACT: The Holocene Mekong Delta is a good example of a delta that has evolved from a tide-dominated to a tide–wave mixed-type delta in its shape, sediment facies, and progradation rate during the regression after the maximum flooding during the Holocene. This evolution indicates that tide-dominated deltas developed well in the early phase of the highstand systems tract and that they changed into more wave-dominated deltas in response to subsequent regression.

The Mekong Delta, which is located at the southeast tip of the Indochina Peninsula, is one of the largest deltas in Southeast Asia, with a large delta plain ranked third largest in the world. The present Mekong Delta is classified as a tide-dominated delta, but close to a wave-dominated delta, in the triangular deltaic depositional classification system. Two rotary-drilled cores, VL1 and TV1, were obtained from the flood plain and strand plain, respectively, of the delta plain. Detailed core descriptions, high-resolution ^{14}C dating by accelerator mass spectrometry (AMS), and topographical analysis of the delta plain indicate that the Mekong Delta evolved from a tide-dominated delta from 6.5 to 2.5 ka to a tide–wave mixed delta from 2.5 ka to the present. The present delta is characterized by beach ridges on the strand plain, dominantly wave-influenced sediments, and a low progradation rate due to longshore sediment dispersal by waves.

As the sea level rises, marine flooding causes coastal inundation and the formation of embayments, which amplifies tides and shelters deltas from waves, so that they are more tide dominated. However, as a delta progrades and its shoreline migrates seaward, the resultant smoothed coastline, narrow shelf, and increased wave energy create a more wave-dominated coastal setting. This change will become more pronounced as the delta continues to prograde towards shelf margins.

INTRODUCTION

Deltas are important coastal depositional systems and display sedimentary strata that are typical of those formed during periods of regression. Presently, the rivers of Asia and Oceania play an important role in the global sediment budget. These rivers contribute about 70–80% of the world sediment flux from the land to the ocean (Milliman and Meade, 1983; Milliman and Syvitski, 1992), and these huge sediment deposits have formed large deltas during the Holocene (e.g., Ganges–Brahmaputra River, Goodbred and Kuehl, 2000; Southeast Asian rivers, Woodroffe, 2000; Mekong River, Nguyen et al., 2000; Yellow River, Saito et al., 2000; Yangtze River, Hori et al., 2002). Another reason for the wide distribution of large deltas in these regions is the stable or slightly falling sea level of the last 5–6 ka due to hydroisostasy (Clark et al., 1978). The highest sea levels in the Holocene were recorded at 2.5–4.5 m above the present sea level at 5–7 ka (e.g., Mekong Delta, Nguyen et al., 2000; Chaophraya Delta, Somboon and Thiramongkol, 1992). The long sea-level highstand during the Holocene affords us a good opportunity to analyze delta evolution both in a highstand systems tract and in a regressive system.

Deltas have been classified according to a triangular system, in which sediment input, wave energy flux, and tidal energy flux form the apices (Galloway, 1975; Orton and Reading, 1993) and using a ternary coastal classification system based on process and sediment source (Boyd et al., 1992). However, the evolution of coastal systems from transgressive estuaries to deltas, and an appropriate classification system for tide-dominated deltas, are still being debated (Boyd et al., 1992; Walker, 1992; Dalrymple et al., 1992; Reading and Collinson, 1996). Here we treat deltas as prograding, regressive systems (Reading and Collinson, 1996) formed mainly by river sediment supply. The wave–tide classification system for coastal environments by Davis and Hayes (1984) and sedimentary-facies data are used to evaluate the influence of waves and tides in this paper.

The purpose of this paper is to describe the sediment characteristics of the Mekong Delta and to construct a model for delta evolution based on detailed analyses of samples from two boreholes taken from the Mekong delta plain in 1998. The Mekong River is the tenth largest river in the world in terms of water discharge and the ninth largest with respect to sediment discharge (Milliman and Ren, 1995). It also has the third widest low-lying delta plain in the world. The evolution of the Mekong Delta during the Holocene and its modern sediments have been discussed previously, mainly on the basis of topographical and surface-sediment analyses (Gagliano and McIntire, 1968; Kolb and Dornbusch, 1975; Nguyen et al., 2000; Woodroffe, 2000). However, analyses on sediments from deep boreholes penetrating the Holocene successions with detailed core descriptions were not reported in these studies. Moreover, a detailed chronostratigraphic analysis of the sequence stratigraphy and delta evolution has not been done because of the limited number of ^{14}C

Tropical Deltas of Southeast Asia—Sedimentology, Stratigraphy, and Petroleum Geology
SEPM Special Publication No. 76, Copyright © 2003
SEPM (Society for Sedimentary Geology), ISBN 1-56576-086-7, p. 175–188.

ages available for the subsurface sediments. This paper is the first report with a detailed description of the deltaic sediments and high-resolution ^{14}C dates from the Mekong Delta.

GENERAL SETTING

The Mekong Delta, one of largest deltas in Southeast Asia, is located at the mouth of the Mekong River. The river has its headwaters in the mountains of Tibet, and it flows down the Indochina Peninsula and empties into the South China Sea in southern Vietnam (Fig. 1A). The river is 4350 km long and has a drainage area of 810,000 km^2 (National Astronomical Observatory, 1999). Its water discharge is 470 km^3/yr and its sediment discharge is 160 million t/yr (Milliman et al., 1995), or tenth and ninth largest in the world, respectively. The water discharge varies by season because most of the drainage area is under a monsoonal tropical regime. The flow at Phnom Penh, Cambodia, reaches a maximum in October (typically 39,000 m^3/s) and a minimum in May (about 1700 m^3/s) (Gagliano and McIntire, 1968; Wolanski et al., 1996; Wolanski et al., 1998). Tonle Sap Lake in Cambodia (Fig. 1B) plays an important role as a water reservoir during the rainy season (Kolb and Dornbusch, 1975) by helping to regulate the flow of the principal distributaries of the Mekong, known as the Mekong and Bassac rivers (Fig. 1C) (Woodroffe, 2000).

The coastal environment of the Mekong Delta is classified as mixed-energy (tide-dominated) according to Hayes (1979) and Davis and Hayes (1984). The mean tidal range is 2.5 ± 0.1 m (Gagliano and McIntire, 1968; Coleman, 1981; Wolanski et al., 1996; Nguyen et al., 2000), and the maximum tidal range is 3.2–3.8 m

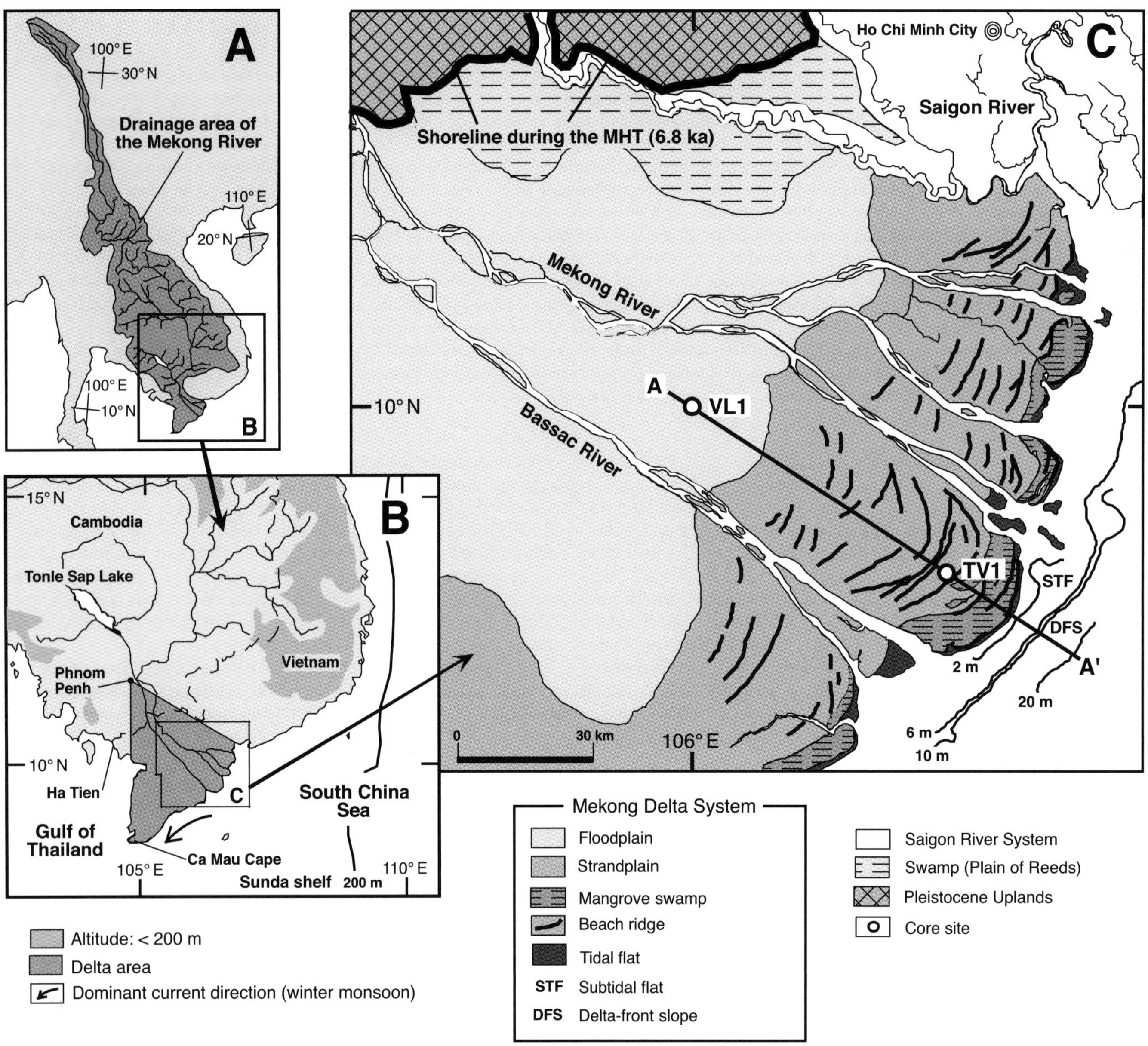

FIG. 1.—Index map and geographical map of the Mekong Delta. **A)** The drainage area of the Mekong River; **B)** the area of the Mekong Delta; **C)** map of the Mekong Delta with coring sites. The map is modified after Nguyen et al. (2000) with bathymetric data after National Geographical Directorate of Vietnam (1965). MHT: Middle Holocene transgression maximum.

(Wolanski et al., 1996; Nguyen et al., 2000). The mean wave height is 0.9 m (Tran Duc Thanh, personal communication). Southwestward coastal and longshore currents generated by winter monsoons are dominant (Gagliano and McIntire, 1968) (Fig. 1B).

The Mekong Delta is defined as the roughly triangular area between Phnom Penh, the Saigon River mouth, and Ca Mau Cape (Fig. 1B) (Gagliano and McIntire, 1968). The area of the delta is between 62,520 and 93,781 km^2 (Orton and Reading, 1993; Nguyen et al., 2000), and it is relatively wide for the size of the drainage area (Coleman and Roberts, 1989). The delta is bounded on the northeast by Pleistocene uplands, swamps (the Plain of Reeds), and the Saigon River system (Fig. 1C). The delta system can be classified into flood plain and strand plain, with mangrove swamp along the shore. The strand plain is characterized by rows of beach ridges, trending northeast to southwest and 3 to 10 m in height, and inter-ridge swamps. The beach ridges often have a diverging, bifurcating pattern formed by southwestward spit progradation (National Geographic Directorate of Vietnam, 1965; Nguyen et al., 2000). The modern active coastal environment is divided into mangroves, beach ridges, sand flats (foreshore) in the upper part of intertidal zone, intertidal to subtidal flat, delta front, and prodelta (Fig. 1C). The water depth of the boundary between the subtidal flat and the delta front is about 10 m.

The modern Mekong Delta system has several distributary channels known as the Bassac and Mekong rivers. The Mekong River consists of three major distributary channels. The positions of these distributary channels are estimated, from the distribution of the beach ridges, which indicate interdistributary plains, to have been relatively stable over the last 2–3 ka.

The mid-Holocene sea-level highstand was recorded at 2.5 or 4.5 m above the present sea level about 5–7 ka (calibrated from Nguyen et al., 2000). The shoreline at that time (nearly equal to the maximum extent of transgression) was located in Cambodia (Fig. 1C). Most of the present Mekong delta plain in Vietnam was shallow sea. Then the delta prograded southeast, toward the South China Sea (Gagliano and McIntire, 1968; Kolb and Dornbusch, 1975; Nguyen et al., 2000) with an average rate of 10–16 km^2/yr.

The Pleistocene uplands, which have heights from 5 to 6 m, consist of shallow marine successions with a tan-colored oxidized lithology (Nguyen et al., 2000). They are located at a depth of 20 to 55 m under the Holocene sediments with coarse-grained lag deposits at the boundary (Ta et al., 1999).

MATERIALS AND METHODS

The TV1 (Tra Vinh) and VL1 (Vinh Long) borehole samples were obtained from the Mekong delta plain (Fig. 1C) by rotary drilling in January and February 1998. Core recovery was more than 95%. The TV1 core was obtained from the back-swamp area between beach ridges in the strand plain (latitude 9° 41' N, longitude 106° 27' E, elevation 3 m). The core length was 27 m. The VL1 core was drilled on the floodplain (latitude 10° 02' N, longitude 105° 59' E, elevation 2 m). The core length was 42 m.

The cores were split and described. Slab samples (5 cm x 25 cm x 1 cm) were collected from whole cores. These slab samples were photographed, X-radiographed, and described in detail in terms of lithology, grain size, and sedimentary structures. The relative amount of shell fragments, microfossils (such as foraminifers, burrows, and plant fragments, were classified as none (N), few (F), some (S), and abundant (A) (Figs. 2, 3). Laminae in each slab sample of the VL1 core were counted to determine the frequency of the lamination (Fig. 3). Sand and mud content was measured every 10 or 20 cm for both cores using a 63-micron sieve according to the method of Kumon and Tateishi (1998). ^{14}C ages were obtained on fifteen samples of shell fragments and organic materials by accelerator mass spectrometry (AMS) through Beta Analytic Inc. Calendar years were calculated from conventional ^{14}C ages based on Stuiver et al. (1998). All of ages are discussed in units of calendar years in this paper except the age of a shell, dated 4,3420 ± 980 yr BP, in the TV1 core.

SEDIMENTARY UNITS, FACIES AND DELTAIC SUCCESSIONS

Sedimentary Units

The TV1 and VL1 cores were divided into sedimentary units I, IIa, and IIb in ascending order. Each unit was characterized by a combination of lithology, sedimentary structures, texture, contact character, succession character, fossils, and ^{14}C ages (Figs. 2, 3, Table 1). Unit I consisted of basement strata of late Pleistocene shallow marine sediments. The Holocene strata unconformably overlying Unit I were divided into units IIa and IIb, which were transgressive lag and deltaic sediments, respectively. These units and their facies are described below.

Unit I.
Depths in Cores: TV1, 27.00–21.56 m; VL1, 42.00–36.40 m.—

Unit I is composed of light yellow to bright brown clay (Fig. 4A). In the TV1 core, bauxite pebbles are scattered in the clay. Sand pipes with lengths of 10–30 mm and widths of 10–20 mm, filled by shell fragments and fine to medium sand, were recognized. In the VL1 core, lenticular bands of fine sand 3–8 mm thick with mud drapes 1–2 mm thick were found in the clay. The ^{14}C age of shells from this unit of the TV 1 core was 43,420 ± 980 yr BP.

Interpretation.—The ^{14}C age and sediment facies indicate that Unit I consists of late Pleistocene tide-influenced shallow marine sediments. Moreover, the colors of the clays suggest oxidation and weathering (Kolb and Dornbusch, 1975; Nguyen et al., 2000), and the bauxite pebbles, which would have formed under a lateritic weathering regime on the land surface (Parton et al., 1995), indicate that these marine sediments were exposed on land before the Holocene.

Unit IIa.
Depths in Cores: TV1, 21.56–21.50 m; VL1, 36.40–35.84 m.—

This unit is composed of yellowish, poorly sorted granules and pebbles with a yellowish fine-sand or sandy-silt matrix (Fig. 4B), unconformably overlying Unit I with a sharp, erosional contact. Bauxite pebbles were observed in this unit as well. The unit fines upward, from the lowest part, which has a sand content of more than 70%, to the upper part, with a sand content of less than 40%. The number of granules and pebbles decreases upward. Shell fragments and microfossils such as foraminifers are abundant. The unit grades into Unit IIb. The ^{14}C dates show an age of 6.1 ka in the VL1 core and 2.7 ka in the TV1 core for this unit.

Interpretation.—Unit IIa overlies Pleistocene Unit I and is covered by the deltaic-succession Unit IIb. The abundance of microfossil and molluscan-shell fragments indicates that this unit formed during marine flooding in the middle Holocene. The ^{14}C age in VL1 corresponds to almost the same time period as the maximum Holocene flooding. On the other hand, the younger ^{14}C age in TV1 suggests that the shelf was starved for sediments after the Holocene maximum. These results indicate that the erosional surface between units I and IIa can be interpreted as a

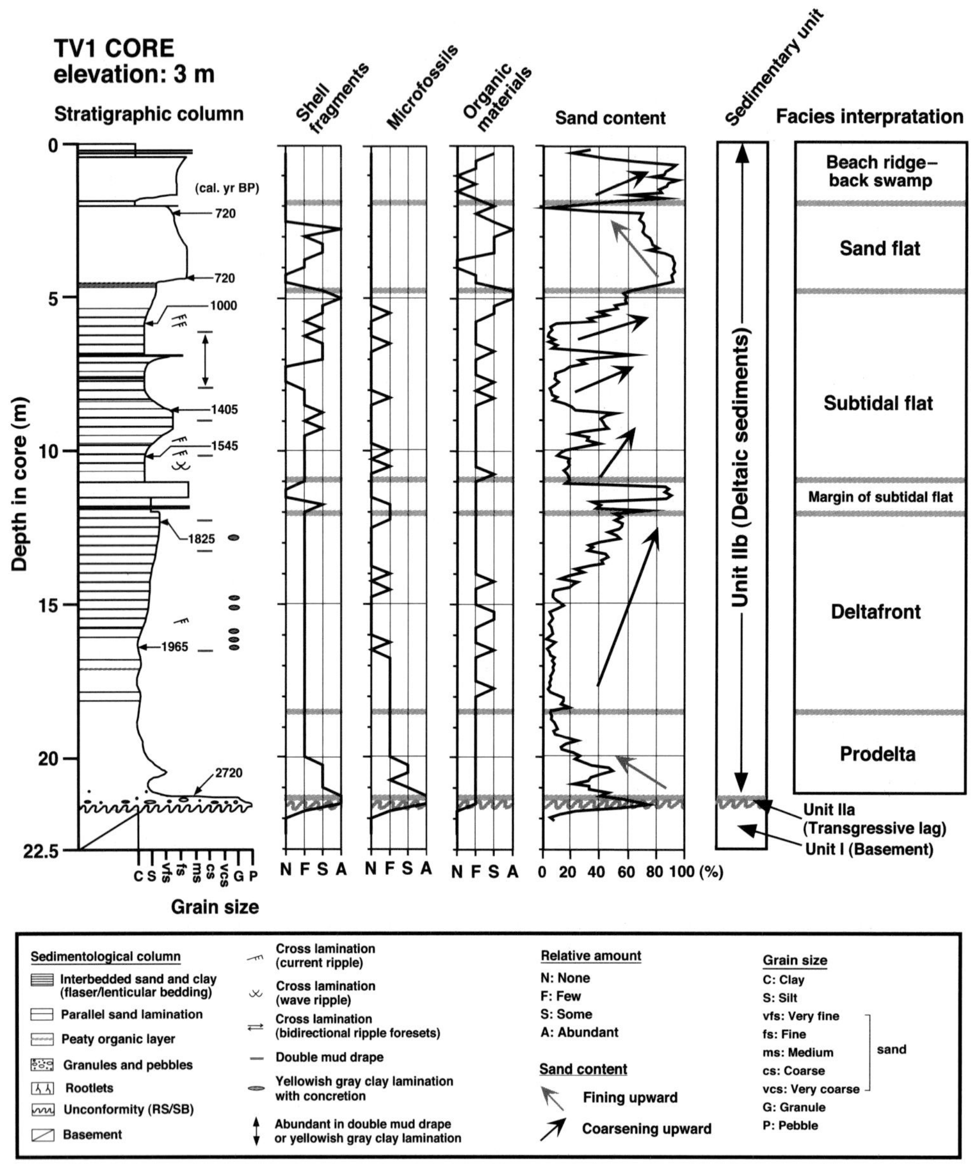

FIG. 2.—Lithology and sediment facies of the TV1 core.

sequence boundary (SB) and ravinement surface (RS) and that this unit is a transgressive lag or condensed section.

Unit IIb.
Depths in Cores: TV1, 21.50–0 m; VL1, 35.84–0 m.—

Unit IIb consists of brownish interlaminated clay with flaser and lenticular bedding and light brownish sand in the TV1 core, and consists of gray clay, brownish interlaminated clay, and lenticular bedding in the VL1 core. Sand content and the frequency of the lamination show a general upward-coarsening pattern in both cores. The abundance of microfossils in TV1 and of shell fragments and burrows in VL1 decreases upward. In contrast, the abundance of organic materials increases to the tops of the cores. These data indicate that this unit reflects an upward-shallowing succession. ^{14}C ages of both cores show that the VL1 and TV1 sediments were deposited mostly between 6.1 and 3.4 ka and between 2.7 and 0.7 ka, respectively.

The upward-coarsening and upward-shallowing successions of this unit reflect delta progradation after the maximum flooding at ca. 6 ka. Detailed sediment facies are described below.

Sedimentary Facies of the Deltaic Succession: TV1 Core

Unit IIb in the TV1 core consists of the following six sedimentary facies: prodelta, delta front, subtidal-flat margin, subtidal flat, intertidal sand flat, and beach ridge to back swamp.

Prodelta sediments.
Depth in Core: 21.50–18.70 m.—

These sediments consist of an upward-fining succession from sandy silt to bioturbated massive clay (Fig. 4C). Sand content decreases upward from 53% to 3% (Fig. 2). The sandy silt is very thinly laminated in parts (< 1 mm thick) in X-radiographs. Round-shaped burrows (3–30 mm in diameter) are scattered in the sandy

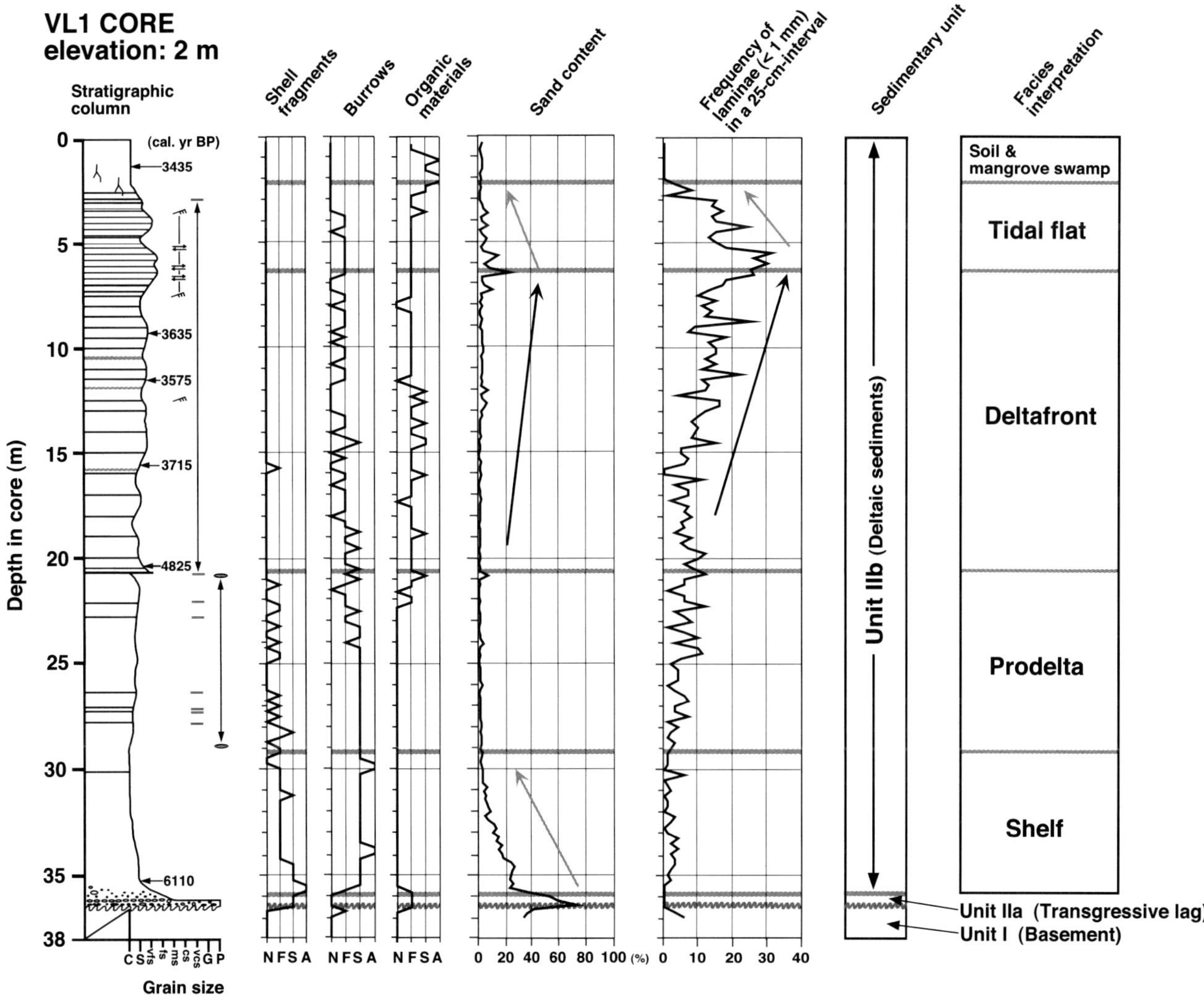

FIG. 3.—Lithology and sediment facies of the VL1 core.

silt. Very fine to fine sand, small shell fragments (smaller than 1 mm), and foraminifers fill these burrows. The amount of shell fragments and microfossils and the number of round-shaped burrows decreases upward. The boundary with the underlying Unit IIa is gradual. The accumulation rate in this interval was lower than in the overlying facies (Fig. 5).

Interpretation.—Abundant shell fragments and marine microfossils indicate a relatively strong open-marine influence, whereas the fine grain size with the low accumulation rate suggests a weak riverine influence. Moreover, because this facies was overlain by an upward-coarsening succession, it is interpreted as a prodelta environment on the shelf. An upward-fining succession with a ^{14}C age of 2.7 ka is interpreted as the result of sediment mixing by bioturbation between starved shelf sediments and increasing prodelta mud.

Delta-Front Sediments.
Depth in Core: 18.70–12.13 m.—

Delta-front sediments overlie the prodelta massive clay with a gradational contact and display an upward-coarsening succession to alternating layers of sand and mud with flaser bedding (Fig. 4D). The increase in the number of sand laminae and partings is responsible for this upward-coarsening pattern. Sand content increases from 6% to 55% (Fig. 2). The sand laminae and partings are composed of layers of coarse silt and very fine sand 3–10 mm thick. They are parallel or isolated lenticular in form. Some of them form couplets with double mud drapes, 1 mm to several millimeters thick, and current–ripple lamination. Yellowish-gray spots of clay and $CaCO_3$-rich concretions 5–10 mm in diameter are scattered in these sediments. The amount of microfossils decreases upward.

Interpretation.—The upward-coarsening succession of these sediments implies delta progradation in the delta-front area. The occurrence of cross-lamination with double mud drapes indicates that the tidal influence increased upward. Present oceanographic data from 50 km off the delta coast at more than 20 m water depth show that tidal effects prevail on the bottom sediments (Anikiyev et al., 1986). The present offshore profiles of the delta show that the delta front slopes to 20–30 m water depth (National Geographic Directorate of Vietnam, 1965; Gagliano and McIntire, 1968). Thus, these sediments are interpreted as delta-front sediments.

TABLE 1.—AMS ^{14}C dating results from the TV1 and VL1 cores.

TV1 core

Depth in core (m)	Sedimentary Unit	Material	$\delta^{13}C$ (‰)	Conventional ^{14}C age (yr BP)	Intercept of ^{14}C age with calibration curve (cal. yr BP)	Calibrated ^{14}C age (cal. yr BP)
2.62	II b	Shell	-0.8	1190 ± 40	720	795 to 655
4.56	II b	Shell	-1.4	1190 ± 40	720	795 to 655
6.52	II b	Shell	-0.6	1470 ± 40	1000	1105 to 930
8.65	II b	Shell	0.3	1870 ± 40	1405	1510 to 1320
10.38	II b	Shell	-0.3	2000 ± 40	1545	1665 to 1480
12.44	II b	Shell	-0.6	2230 ± 40	1825	1915 to 1725
16.56	II b	Shell	-0.3	2350 ± 40	1965	2060 to 1870
21.43	II b	Shell	0.8	2930 ± 40	2720	2760 to 2675
25.79	I	Shell	0.8	43,420 ± 980	———	———

VL1 core

Depth in core (m)	Sedimentary Unit	Material	$\delta^{13}C$ (‰)	Conventional ^{14}C age (yr BP)	Intercept of ^{14}C age with calibration curve (cal. yr BP)	Calibrated ^{14}C age (cal. yr BP)
1.55	II b	Plant material	-29.1	3210 ± 50	3435	3550 to 3350
9.94	II b	Plant material	-29.4	3390 ± 40	3635	3710 to 3550
11.96	II b	Plant material	-32.6	3340 ± 40	3575	3675 to 3470
15.8	II b	Plant material	-27.6	3480 ± 50	3715	3870 to 3630
20.61	II b	Plant material	-28.3	4210 ± 30	4825	4870 to 4805 / 4760 to 4640
35.66	II b	Shell	0.3	5700 ± 40	6110	6195 to 5985

^{14}C ages and $^{13}C/^{12}C$ ratios ($\delta^{13}C$) were measured by accelerator mass spectrometry (AMS) through Beta Analytic Inc. (Geo-Science Laboratory, Japan).

Subtidal-Flat-Margin Sediments. Depth in Core: 12.13–11.00 m.—

These sediments consist of alternating sand, 8–50 cm in thickness, and clay, 5–34 cm in thickness. The sands are well sorted and medium-grained. Sand content is more than 90% (Fig. 2). Sand beds contain no shell fragments or microfossils. This interval is the coarsest of the subtidal sediment facies penetrated.

Interpretation.—Well-sorted, medium sand implies that the sand beds formed under relatively high-energy conditions. Because this facies was overlain by a finer succession with wave ripples, it is estimated to be located near the boundary between the delta front and the subtidal flat, which generally receives the strongest wave energy. The water depth at the break in slope from the delta front to the subtidal flat is 5–10 m at present.

Subtidal-Flat Sediments. Depth in Core: 11.00–4.87 m.—

Most of the sediments of this facies are composed of alternating sand and mud with lenticular and flaser bedding 2–15 mm thick (Fig. 4E). Cross-lamination can be identified in some parts. Wave-ripple lamination is also visible in X-radiographs. Most bedding is accompanied by mud drapes or double mud drapes. The sand beds consist of very fine to fine sand. The association of lenticular and flaser beds create sets of gradually upward-coarsening successions abruptly bounded by upward-fining successions. Sand content varies from 2–17% to 50–60% in these sets (Fig. 2). Each upward-coarsening set was about 2 m thick.

Interpretation.—Flaser and lenticular bedding and mud drapes suggest that this facies was deposited under a tide-dominated environment (cf. Reineck and Singh, 1980). Moreover, wave-ripple lamination indicates the influence of waves. According to the present-day offshore profile of the delta and the sediment dynamics on the subtidal flat, this facies corresponds to a subtidal-flat environment. Currently monsoon-related ocean currents form megaripples (subaqueous dunes) with heights of nearly 1 m on the flat (Gagliano and McIntire, 1968). Gradually upward-coarsening sets in this facies are interpreted as the result of migration of these subaqueous dunes on the flat.

Intertidal-Sand-Flat Sediments. Depth in Core: 4.78–2.05 m.—

These sediments are composed of sand beds, up to 240 cm thick, which gradually fine upward from well-sorted medium sand to silty sand (Fig. 4F). Sand content decreases upward from 90% to 70% (Fig. 2). Organic materials are abundant, in contrast to scarce shell fragments. A 20-cm-thick brownish black clay bed forms the boundary between these sediments and the underlying subtidal-flat sediments.

Interpretation.—Upward-fining, well-sorted, thick sand beds are interpreted as tidal sand flats by comparison with modern data (Nguyen et al., 2000). The flats could be divided into lower, middle, and upper sandy flats by the lithology. They changed

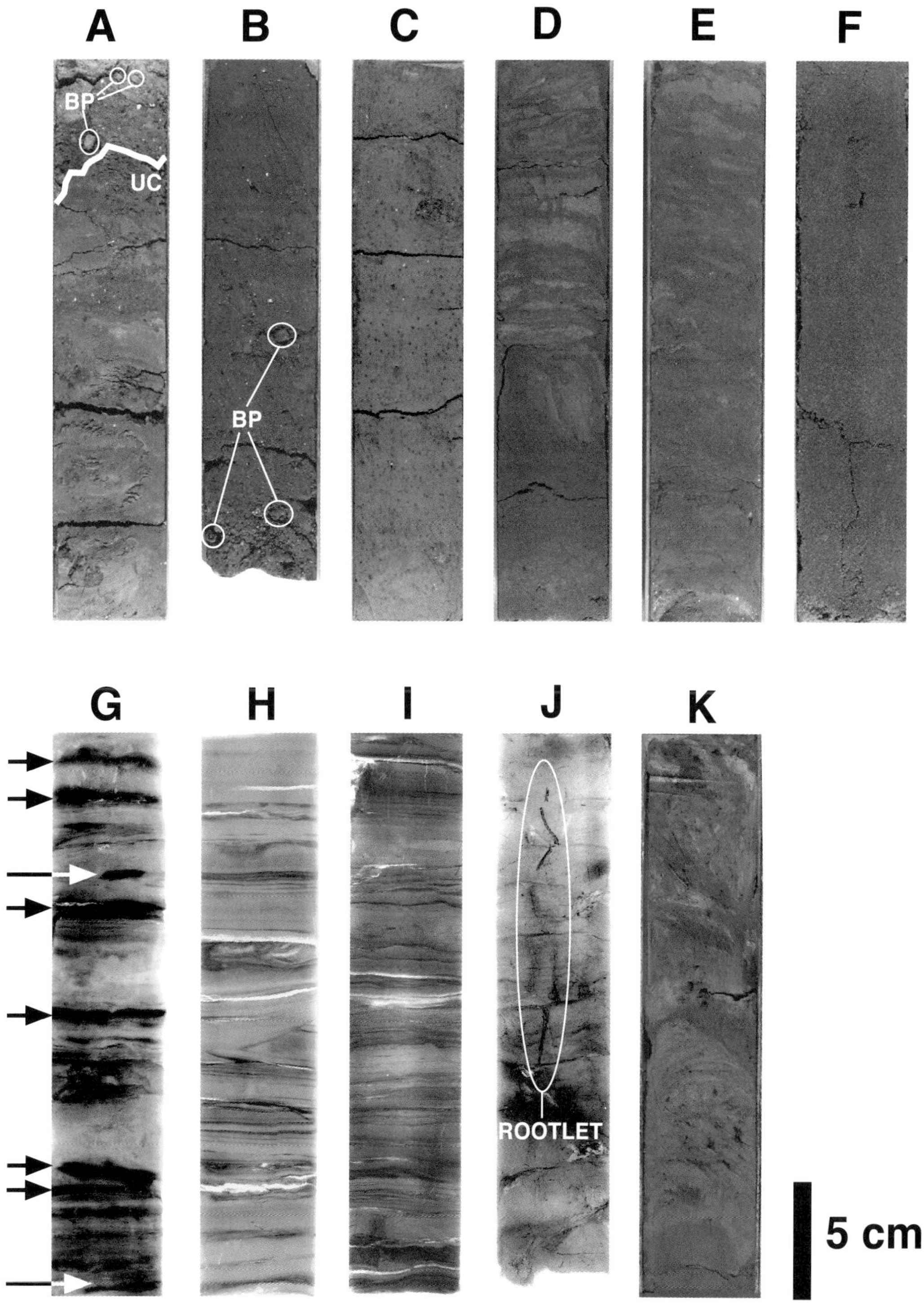

FIG. 4.—Selected photographs and X-ray radiographs of the slab samples obtained from TV1 and LV1 cores. Parts G, H, I, and J show the X-ray radiographs, the others are photographs. **A)** (TV1 core: 21.75–21.50 m): Poorly sorted granules and pebbles (Unit Ia) unconformably overlie lateritic clay (Unit I). The UC means unconformity. The BP means bauxite pebble. **B)** (TV1 core: 21.49–21.25 m): Poorly sorted granules and pebbles (Unit IIa). The BP means bauxite pebble. **C)** (TV1 core: 19.75–19.50 m): Massive clay (prodelta, Unit IIb). **D)** (TV1 core: 12.75–12.50 m): Flaser bedding (delta front, Unit IIb). **E)** (TV1 core: 9.25–9.00 m): Flaser bedding (subtidal flat, Unit IIb). **F)** (TV1 core: 3.75–3.50 m): Well-sorted medium sand (sand flat, Unit IIb). **G)** (VL1 core: 26.75–26.50 m): Color alternation of gray and yellowish-gray clay. The black and white arrows show the yellowish-gray calcareous clay (prodelta, Unit IIb). **H)** (VL1 core: 20.75–20.50 m): Poorly laminated clay (delta front, Unit IIb). **I)** (VL1 core: 5.50–5.25 m): Lenticular bedding (tidal flat, Unit IIb). **J)** (VL1 core: 1.49–1.25 m): Rooted clay (mangrove swamp, Unit IIb). **K)** (VL1 core: 0.25–0 m): Lateritic clay (soil, Unit IIb).

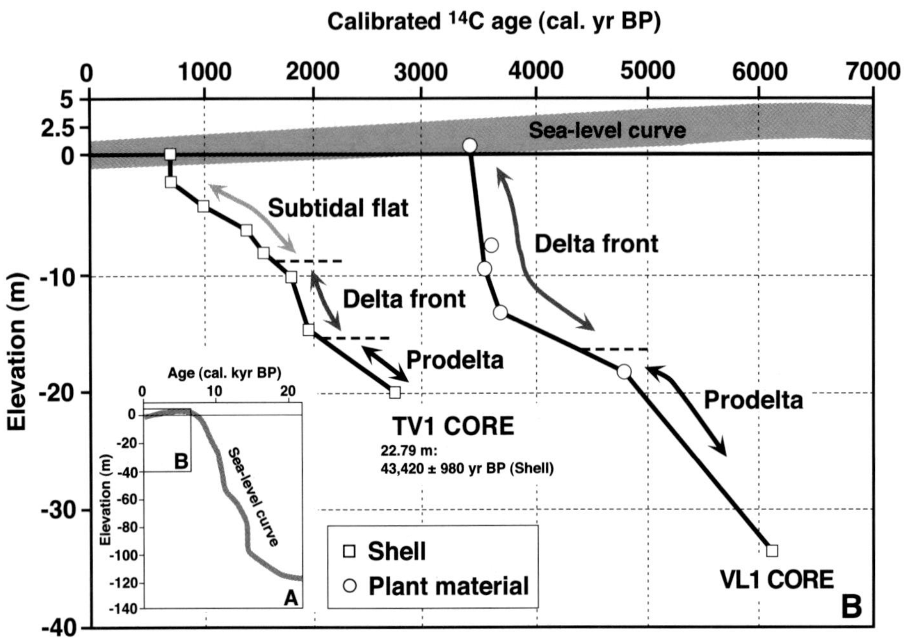

FIG. 5.—Accumulation curves for the TV1 and VL1 cores and the related sedimentary facies. The sea-level curves for the last 20 kyr are illustrated based on Hanebuth et al. (2000) and Nguyen et al. (2000). The age of shell fragments, dated 43,420 ± 980 yr BP, is not calibrated.

from 90% medium sand in the lower flat, to 60–70% medium to fine sand in the middle flat, to muddy sand in the upper flat (Fig. 2). These changes suggest landward fining laterally or upward fining vertically.

Beach-Ridge to Back-Swamp Sediments.
Depth in Core: 2.05–0 m.—

This facies consists of alternating beds of sand and clay. The sand beds of the 1.84–0.46 m interval coarsen upward slightly from fine sand to well-sorted medium sand. Sand content increases from 80% to 90% (Fig. 2). Some wood fragments are observed in the clay. The top 7 cm of this interval is composed of dark reddish brown clay with abundant organic matter.

Interpretation.—This facies is interpreted as beach-ridge to back-swamp deposits. Because the depth of this facies in the core was from 2.05 m to 0 m, these deposits are located at an altitude just above mean sea level of the recent delta and are part of a small rise in the strand plain.

Sedimentary Facies of the Deltaic Succession: VL1 Core

Unit IIb in the VL1 core consisted of the following six sedimentary facies: shelf, prodelta, delta front, tidal flat, mangrove swamp, and soil.

Shelf Sediments.
Depth in Core: 35.84–29.00 m.—

The sediments consist of an upward-fining succession from scattered granules in clayey silt to massive clay. The boundary with the underlying Unit IIa, the transgressive lag, is sharp. The sand content decreased upward from 36% to 2% (Fig. 3). Although the clay is extensively bioturbated, few very thin (up to 3 mm in thickness) parallel laminae composed of silt could be rarely observed in X-radiographs. The horizon at about 29 m depth is the finest part of the core with respect to grain size. Small shell fragments (up to 2 mm) and foraminifers are relatively richly scattered in this facies. Diatoms indicated a neritic to open-sea environment (Van Lap Nguyen, unpublished data).

Interpretation.—Upward-fining clay lithology and intense bioturbation with abundant shell fragments represent an environment under low-energy conditions and weak delta influence. Because the diatom assemblage suggests a neritic or more open-sea environment, this facies is interpreted as shelf sediments.

Prodelta Sediments.
Depth in Core: 29.00–20.81 m.—

These sediments are composed of gray, partly laminated clay (Fig. 4G). The sand content is less than a few percent. Very thin parallel laminates, 1 mm-thick, composed of silt to very fine sand laminates and covered with mud drapes or double mud drapes are observed. These thin sand layers increased upward. Yellowish-gray calcareous clay layers, a few millimeters to 2 cm thick, are abundant throughout this interval (Fig. 4G). Calcareous concretions were also observed.

Interpretation.—The lithologic character of this facies is very similar to that of prodelta sediments described by Coleman (1981). The decrease in burrows and shell fragments and the increase in laminates and sand content may represent a retreating offshore environment and the approach of the delta front by progradation. Laminates with double mud drapes suggest that the site was under a tide-dominated environment.

Delta-Front Sediments.
Depth in Core: 20.81–6.40 m.—

These sediments are characterized by a continuous upward-coarsening succession consisting of brownish laminated clays with thin lenticular coarse silts and sands (Fig. 4H). This interval is bounded by a sharp contact with the underlying prodelta facies, accompanied by a 2-cm-thick sand layer. Sand content ranges from less than 1% to 23% and increases slightly upward from less than 1% in the lower 9 m to around 3% in the upper 6 m (Fig. 3). The coarsest part of this core is at 6.40 m depth at the boundary with the overlying facies. The frequency of laminae (number of laminae greater than 1 mm thick in a 25 cm-interval) clearly shows an upward-coarsening succession. The lenticular and parallel silt to sand laminae and partings were 1 to 10 mm thick. Some of the sand layers showed ripple cross-laminae and are accompanied by mud drapes. Bidirectional cross-lamination is found in the upper part of this interval. Organic materials such as plant fragments are condensed into thin layers (5–10 mm in thickness) of brownish-black clay. A few shell fragments were present.

Interpretation.—This facies is interpreted as a tide-dominated delta front by the upward coarsening, the increase in organic materials, the bidirectional cross-lamination, and the mud drapes. Upward-coarsening lithology is common in delta-front deposits (Reineck and Singh, 1980). Abundant double mud drapes, lenticular bedding, and bidirectional cross-lamination indicate that the delta front was a tide-dominated environment (Reineck and Singh, 1980; de Boer, 1998).

Tidal-Flat Sediments.
Depth in Core: 6.40–2.60 m.—

This interval is characterized by an upward-fining succession consisting of alternating silt or sand and clay (Fig. 4I). The frequency and thickness of the silt/sand layers decreases upward. Sand content ranges from 1% to 10–20% (Fig. 3). Ripple cross lamination, bidirectional ripple foreset bedding, and double mud drapes are commonly observed. The relative amount of burrows and bioturbation decreases upward. On the other hand, the relative amount of organic materials such as plant fragments and rootlets increases upward.

Interpretation.—The characteristics are very similar to those of regressive upward-fining tidal-flat sediments (Dalrymple, 1992). Thus, this unit is interpreted as having been deposited in a tidal flat.

Mangrove-Swamp Sediments.
Depth in Core: 2.6–0.4 m.—

This interval is characterized by brownish to grayish-brown clay containing abundant rootlets and organic materials (Fig. 4J). The color changes from brownish gray to grayish brown at a depth of 2.0 m. Sets of very thin parallel laminae are scarcely interbedded with layers of mica or carbonaceous material 10–20 mm thick. Gypsum and sulfur crystallized on the surface of dried samples.

Interpretation.—Clay sediments with abundant rootlets and plant materials overlying tidal-flat sediments are interpreted as mangrove-swamp sediments. At one-meter depth under the floodplain area of the VL1 site, the gray clay is rich in mangrove pollen (Woodroffe, 2000). The gypsum and sulfur crystallization also support the interpretation as a salt-influenced mangrove-swamp environment.

Soil.
Depth in Core: 0.4–0 m.—

This lithology consists of brownish-gray or dull yellowish-brown lateritic clay soil (Fig. 4K). Gypsum and sulfur crystallized on the surface of dried samples.

SEQUENCE STRATIGRAPHY

Late Quaternary sequences in the Mekong Delta, southern Vietnam, are not well understood. Unit I, with a ^{14}C age of 43,420 ± 980 yr BP, indicates the presence of late Pleistocene marine successions below the delta plain, although further investigation is needed. These Pleistocene sequences were eroded during the sea-level lowstand at the time of the last glacial maximum (LGM). A lowering of sea level of about 120 m has been estimated on the Sunda Shelf (Hanebuth et al., 2000) (Fig. 5A). At that time, at least one deep incised valley was formed in the Mekong delta area. Ta et al. (1999) reported an incised valley more than 55 m deep about 50 km north of the TV1 site that had been mostly filled during the transgression by fluvial and estuarine sediments. The spatial location of the valley and the topography of the erosional surface are still not clear. The borehole sites TV1 and VL1 are located outside of the main incised valley, which may have been formed by the paleo–Mekong River during the LGM. Therefore, transgressive successions from the incised valley are missing in the cores. The basal erosional surface bounding Units I and Unit II is considered to be the result of a wave ravinement process during the Holocene rise in sea level. However, the lag deposits overlying the erosional surface and the upward-fining successions above them are not consistent with the period of the rise of sea level in the early Holocene. The age of these lag sediments represents the regressive period after the maximum transgression. Moreover, this erosion does not represent a marine erosion surface of regression, because the relative fall of sea level after the maximum transgression was only 2.5–4.5 m here, and the erosional effects of wave energy do not influence such deep shelf environments. It is possible that the erosional surface formed during the transgression, and some residual lag and sandy sediments remained. Because of the landward shift of the depositional zone of the Mekong Delta around the time of the maximum transgression, the borehole sites were sediment-starved environments, and microfossils and shell fragments were relatively common. As the delta prograded, the depositional zone also migrated seaward. The increasing mud supply and continuous bioturbation of the underlying sands created an upward-fining succession that ended as the main depocenter migrated farther seaward. These successions show that an upward-fining succession does not necessarily imply transgression and that the maximum flooding surface and the horizon of finest grain size are discordant in a sequence, and that the horizon with finest grain size is diachronous.

ACCUMULATION CURVES AND DELTA PROGRADATION

The accumulation curves at the two borehole sites give important information about the Holocene Mekong Delta, such as (1) the relationship between the accumulation rate and the sediment facies, (2) when the delta front and prodelta passed these sites, and (3) changes in the delta progradation rate.

Relationship with Sediment Facies

The accumulation rate and the sedimentary facies are closely related. The relationship between them is shown in Figure 5B. Both curves indicate a low accumulation rate in the prodelta facies and a high accumulation rate in the delta-front facies. Approximate rates were less than 0.7 cm / yr in prodelta facies, 3.0 cm / yr in the delta-front facies, and 0.7 cm / yr in the subtidal-flat facies at the TV1 site, and less than 0.4 cm / yr in the upper part of the prodelta facies, less than 1.2 cm / yr in the lower part of the prodelta facies, and 5.1 cm / yr in the delta-front facies at the VL1 site. These facies and accumulation rates are closely related to subaqueous topography, as will be discussed later.

Delta Progradation and Progradation-Rate Change

Delta Progradation Rate.—

Accumulation curves at the borehole sites also indicate when each sedimentary environment (facies), such as the prodelta, delta front, tidal flat, and shoreline, passed the site. If the effects of compaction on the Holocene muddy sediments are ignored, progradation rates of the delta system can be calculated on the basis of age–depth data at the borehole sites and present bathymetry data. The average annual progradation rates between the VL1 and TV1 sites were approximately 23 m at the shoreline, 25 m at 5 m depth, 35 m at 10 m depth, and 30 m at 15 m depth, which occurred between 3.7 and 0.7 ka. Moreover, the rates between the TV1 site and the present sea floor were approximately 20 m at the shoreline, 19 m at 5 m depth, 13 m at 10 m depth, and 11 m at 15 m depth, which occurred during the last 2 ka. These data show a recent decrease in progradation rate for the Mekong Delta (Fig. 6). The shoreline migration rate between the maximum shoreline position (6.8 ka; calibrated from Nguyen et al., 2000) and the VL1 site was ca. 40 m / yr. This rate is higher than the rates at the 10 m and 15 m depth between the borehole sites.

These rates are linked with sediment facies and surface topography in the delta plain from the flood plain to the strand plain. The progradation rate of the Mekong Delta decreased for the last 2–3 ka, and beach ridges developed.

With respect to the progradation rates between borehole sites, the rates at 5 m depth and the shoreline are small in comparison with those at 10 m and 15 m depth and the shoreline migration rate for 6.8–3.4 ka. A possible explanation is that this rate was influenced by the slow progradation of the strand plain between the borehole sites. Chronostratigraphic lines on the profile clearly show that the proportion of strand-plain-type sediments in the delta between the two sites increased upward (Fig. 6). If a delta system progrades seaward while keeping its subaqueous morphology, progradation rates at all depths should be the same. So if the delta system migrated from the VL1 site toward the TV1 site while keeping its shape with a rate of 32.5 m / yr, which is the average rate at 10–15 m depth, the estimated progradation rate of the strand-plain-type delta between the two sites is 18 m / yr at the shoreline and 21 m / yr at 5 m depth when the strand-plain-type delta started at 2.5 ka. These estimated values are similar to rates between the TV1 site and the present sea floor. Therefore, we conclude that after the maximum transgression, the delta progradation rate was about 40 m / yr between 6.8 and ca. 4 ka, and 30–35 m / yr between ca. 4 and 2.5 ka. This rate changed to 19–20 m / yr for the shoreline to

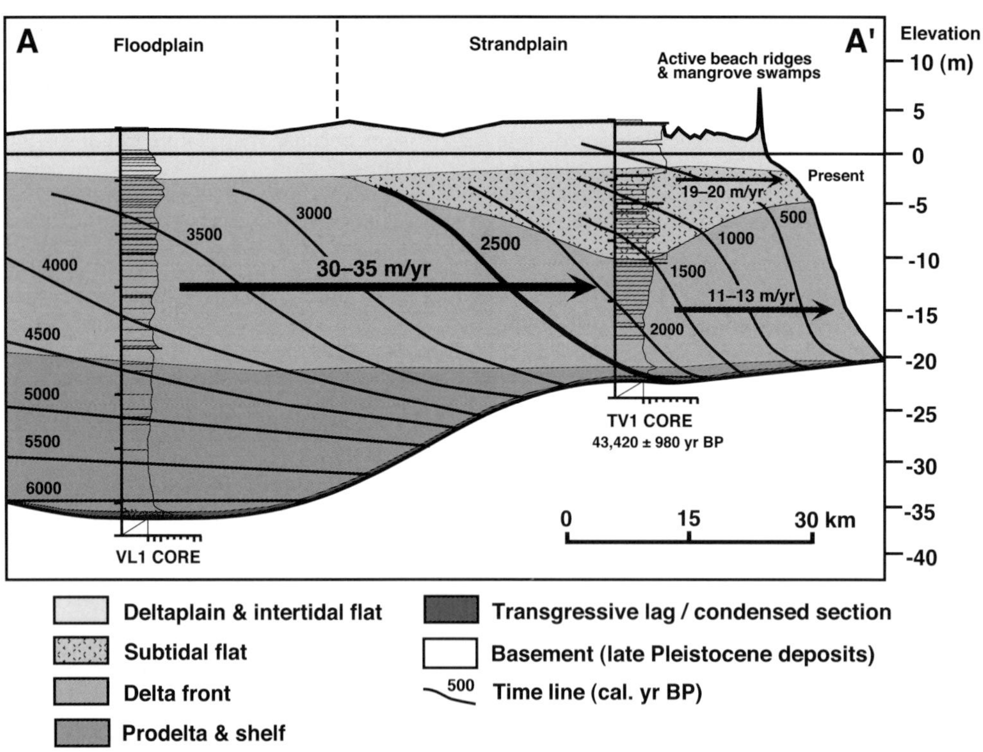

FIG. 6.—Delta profile of section A–A' with sediment facies and estimated time lines. The location is shown in Figure 1. The present subaqueous delta profile is modified after National Geographical Directorate of Vietnam (1965). The age of shell fragments, dated 43420 ± 980 yr BP, is not calibrated.

5 m depth and to about 11–13 m/yr for 10–15 m depth as the beach ridges developed (Fig. 6).

Changes in Subaqueous Delta Topography.—

The accumulation rate for each sedimentary facies and the progradation rate of the delta system give topographical information about the subaqueous delta. If the slope of the delta front is stable in relation to delta progradation, the gradient can be estimated from data both on accumulation rates and on progradation rates of the delta front. Because the accumulation rate was 5.0 cm/yr and the progradation rate was about 50 m/yr at the VL1 site, the estimated paleo-gradient of the delta front is 0.001. Inasmuch as the development of the subtidal-flat facies was poor at the VL1 site, a gentle slope from the subtidal zone to the delta front is expected. This gradient and the muddy sediment facies are very similar to those of the present delta front of the Chaophraya Delta, Thailand. For this estimate, the compaction effect was neglected. By the same reasoning, the estimated paleo-gradients of the delta front and subtidal flat at the TV1 site were 0.0025 and 0.0003, respectively. The present gradients for the delta front and subtidal flat are 0.0015–0.003 and 0.0005–0.0007, respectively. There is a break in slope between the subtidal-flat and the delta-front environments.

The sediment facies of the TV1 core shows a wave-influenced sandy subtidal facies. Gagliano and McIntire (1968) described subaqueous dunes on a subtidal flat formed by southwestward longshore currents. These waves and related longshore currents affect the sediment facies: for subaqueous topography as subtidal flats and for subaerial topography as beach ridges. The slope of the shallow marine parts of the delta has changed. The average gradient became steeper from the VL1 site to the present location because of different progradation rates at each water depth. Relatively high progradation rates of the shoreline and subtidal flats yielded these steep gradients

However, the averaged progradation rates of the shallow marine part of the delta have decreased about 50%. Possible reasons are sediment dispersal toward the southwest by longshore currents (Gagliano and McIntire, 1968) and the broadening of the delta front. Even present-day Ca Mau Cape, which is the southwest tip of the delta, prograded at a rate of 122 ha/yr from 1885 to 1985 (Nguyen et al., 1999). The abrupt topographical change from flood plain to strand plain supports wave-induced sediment dispersal, as predicted by the presence of more wave-influenced facies at the TV1 site.

EVOLUTION OF THE MEKONG DELTA

Morphological and Sedimentological Evolution

The modern Mekong Delta has been classified as a tide-dominated delta, but it is somewhat similar to a wave-dominated delta because of the distribution of the series of beach ridges (Galloway, 1975; Orton and Reading, 1993). Moreover, in the coastal classification system using mean wave height and mean tidal range (Hayes, 1979; Davis and Hayes, 1984; Orton and Reading, 1993), the Mekong Delta is classified as a tide-dominated to mixed-energy (tide) environment. Detailed sediment-facies analysis of the TV1 and VL1 cores indicate that the sediment facies of the Holocene Mekong Delta were tide-dominated. For the last 2.5 ka, however, the subtidal-flat and sand-flat sediments have been influenced by waves, and beach ridges have developed on the delta plain. Before 2.5 ka, the delta was in a more tide-dominated environment, and it is not easy to find wave-influenced features in the sediments and surface topography. These changes to the delta during the middle to late Holocene indicate that the Mekong Delta evolved from a tide-dominated environment into a wave-influenced tide-dominated environment. This delta evolution is seen in the changes in the sediment facies, facies succession, subaqueous delta morphology, delta progradation rate, and subaerial delta morphology mentioned above. Schematic and idealized delta evolution models, typical successions, and coastal morphodynamics are shown in Fig. 7. Two stages are shown, the VL1 stage and the TV1 stage.

During the VL1 stage, the Mekong River had two main channels on the delta plain. These channels were mostly stable, and lateral channel migration may not have occurred because there are no obvious abundant channels on the delta plain and a channel-fill succession has not been reported from the interchannel plains. The basement in the Ca Mau Cape area (about 10 m deep) and the headland east of Ho Chi Minh City sheltered the area from waves on the northeast. The Mekong Delta therefore prograded into a calm, sheltered embayment.

After delta progradation proceeded beyond the headland, the Mekong Delta formed a convex coastal feature projecting into the South China Sea, and it felt the effects of waves from the northeast. The sediment facies, subaqueous delta morphology with subtidal flats, delta progradation rate, and subaerial delta morphology characterized by beach ridges form the present as well as the TV1-stage delta.

Delta Evolution According to a Sequence Stratigraphic Scheme

The tide-dominated delta does not fit particularly well into the sequence-stratigraphic scheme. The evolution of coastal systems from transgressive estuaries to deltas, and the classification of tide-dominated deltas using ternary plots (Galloway, 1975) are still debatable (Boyd et al., 1992; Walker, 1992; Dalrymple et al., 1992; Reading and Collinson, 1996). However, tide-dominated deltas surely exist (Orton and Reading, 1993). Moreover, deltas as prograding, regressive systems (Reading and Collinson, 1996) should be discriminated from transgressive estuaries (Dalrymple, this volume). As sea level rose, drowned valleys and embayments developed on the shelves. These embayments often enlarged the tidal influence, and they were sheltered from waves. These circumstances supported the development of tide-dominated deltas or estuary development just after the maximum transgression. Therefore, deltas in the early phase of the highstand systems tract (HST) were susceptible to strong tidal influences. A similar example is seen in the Red River (Song Hong) Delta in northern Vietnam (Mathers et al., 1996; Mathers and Zalasiewicz, 1999). The delta plain of the Red River Delta is classified as alluvial-dominated, tidal-dominated, or wave-dominated. Wave-dominated systems are characterized by series of sandy beach ridges, as in the southern, outer part of the Red River delta system. The northern, sheltered part of the Red River Delta and the western part are tidal-dominated and alluvial-dominated systems, respectively.

As a delta progrades across a shelf, the delta moves to a more exposed open coast and is more influenced by waves and storms. Figure 8 is a schematic diagram of delta evolution on a wave–tide diagram for coastal classification after Davis and Hayes (1984). Waves and tide conditions of the early stage of the Mekong Delta after the maximum transgression and of a future stage are estimated on the basis of present data from the South China Sea region. Other tide-dominated deltas such the Fly delta and the Mahakam delta located in the marginal seas of Southeast Asia (Baker et al., 1995; Allen and Chambers, 1998) should also show

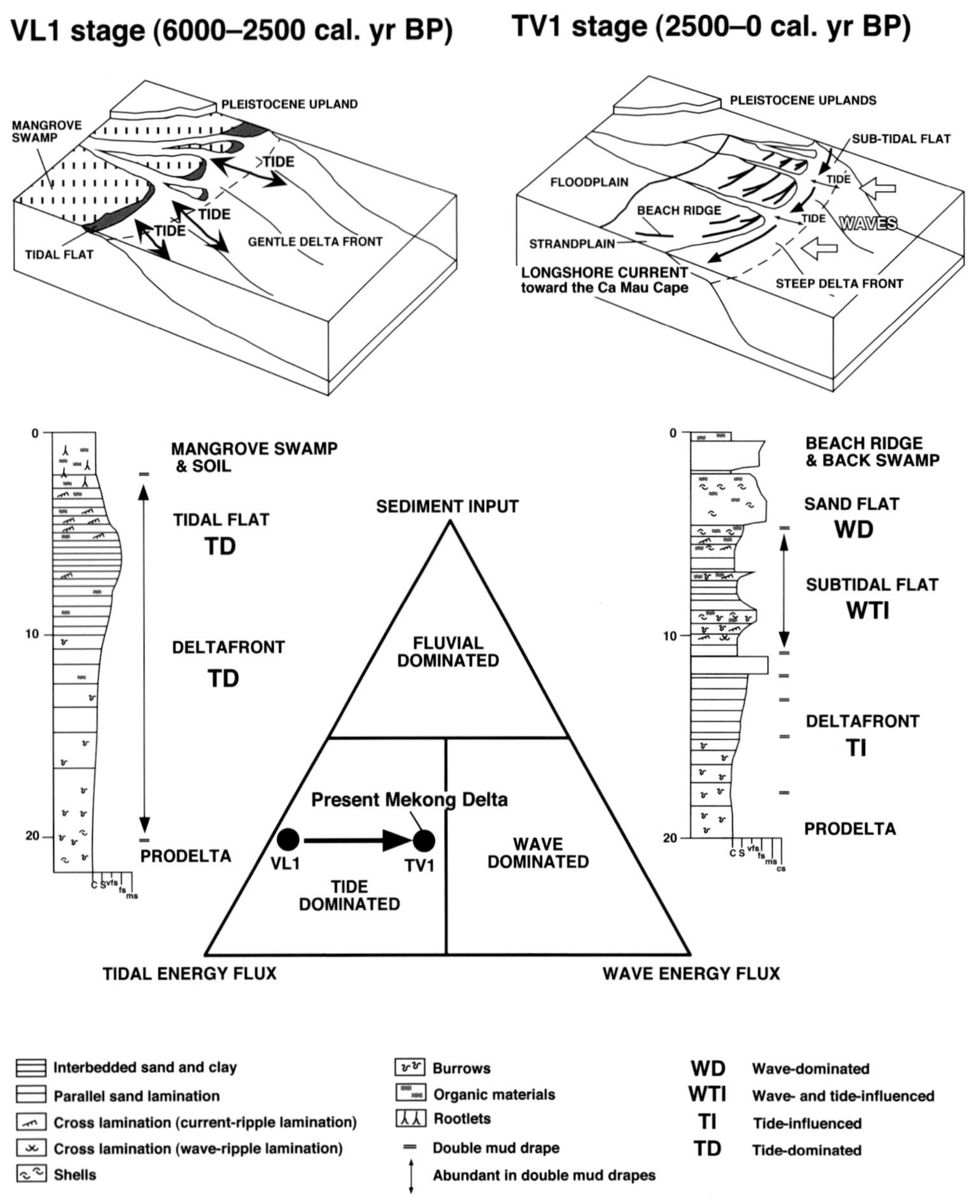

FIG. 7.—Evolution of the Mekong Delta from the VL1 stage to the TV1 stage. Triangular classification of deltaic depositional systems is modified after Galloway (1975). The plot of the present Mekong Delta is after Galloway (1975).

a similar evolution from tide-dominated to wave-dominated as delta progradation proceeds.

CONCLUSIONS

1. Two rotary-drilled cores, TV1 and VL1, were obtained from the strand plain and flood plain, respectively, of the Mekong Delta. These cores consisted of three Units: Unit I, late Pleistocene shallow marine; Unit IIa; Middle to late Holocene upward-fining succession accompanied by lags at its base and unconformably overlying Unit I; Unit IIb, Middle to late Holocene progradational deltaic succession.

2. The deltaic succession included prodelta, delta-front, subtidal-flat margin, subtidal-flat, sand-flat, and beach-ridge to back-swamp facies in the TV1 core, and shelf, prodelta, delta-front, tidal-flat, and mangrove-swamp facies in the VL1 core, in ascending order. The TV1 core taken from the strand plain showed more wave-influenced sediment facies.

3. Fifteen AMS^{14}C ages from these two cores indicate that the accumulation rates were closely related to the sedimentary facies: low rates for prodelta facies, and high rates for delta-front facies. The accumulation rates in the VL1 core were higher than in the TV1 core.

4. The estimated progradation rates of the Mekong Delta were about 30–40 m / yr before 2.5 ka and 11–20 m / yr for the last 2.5 ka. Subaqueous delta morphology, sediment facies, subaerial delta topography, and progradation rates changed at about 2.5 ka from a tide-dominated setting to a wave-influenced tide-dominated setting.

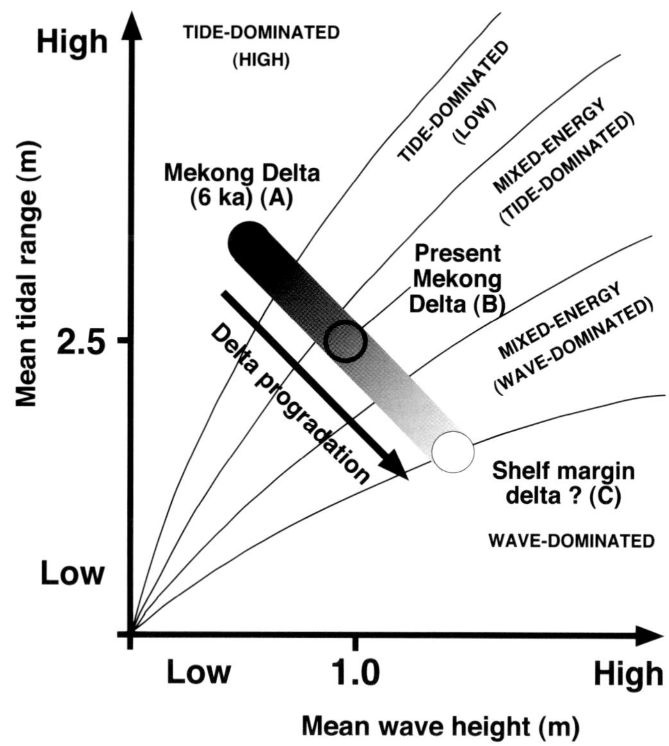

FIG. 8.—Changes in coastal conditions of the Mekong Delta. **A)** Early stage of highstand deltas; **B)** present delta; **C)** estimated future stage. The directions of the arrow show the change and evolution of coastal conditions by delta progradation.

5. The evolution of the Mekong Delta after the maximum marine transgression during the middle to late Holocene is a good example of delta evolution in marginal seas in tropical regions. Deltas evolved from a tide-dominated setting to a more wave-dominated setting as delta progradation encountered more exposed coastal environments.

ACKNOWLEDGMENTS

We are indebted to Dr. F. Hasan Sidi of Conoco Indonesia Inc. Ltd. for encouraging us to submit our paper for this special publication. Most of this study was carried out for the master's thesis of the first author (ST) in Shinshu University under the kindly supervision of Prof. Fujio Kumon of Shinshu University. The authors are also grateful to an anonymous reviewer whose critical review has greatly improved the manuscript. Core drilling and AMS ^{14}C dating were conducted through a research program on "Studies on Comprehensive Assessment of Impacts Sea-Level Rise" of the Global Environmental Research Fund of the Ministry of the Environment in Japan.

REFERENCES

ALLEN, G.P., AND CHAMBERS, J.L.C., 1998, Sedimentation in the Modern and Miocene Mahakam Delta: Indonesian Petroleum Association, Jakarta, 236 p.

ANIKIYEV, V.V., ZAYTSEV, O.V., HIEU, T.T., SAVIL'YEVA, I.I., STARODUBTSEV, Y., AND SHUMILIN, Y., 1986, Variation in space–time distribution of suspended matter in the coastal zone of the Mekong River: Oceanology, v. 26, p. 725–729.

BAKER, E.K., HARRIS, P.T., KEENE, J.B., AND SHORT, S.A., 1995, Patterns of sedimentation in the macrotidal Fly River delta, Papua New Guinea, *in* Flemming, B.W., and Bartholomä, A., eds., Tidal Signatures in Modern and Ancient Sediment: International Association for Sedimentologists, Special Publication 24, p. 193–211.

BOYD, R., DALRYMPLE, R.W., AND ZAITLIN, B.A., 1992, Classification of clastic coastal depositional environments: Sedimentary Geology, v. 80, p. 139–150.

CLARK, J.A., FARRELL, W.E., AND PELTIER, W.R., 1978, Global changes in post-glacial sea level: a numerical calculation: Quaternary Research, v. 9, p. 265–287.

COLEMAN, J.M., 1981, Deltas: Processes of Deposition and Models for Exploration: Boston, International Human Resources Development Corporation, 124 p.

COLEMAN, J.M., AND ROBERTS, H.H., 1989, Deltaic coastal wetlands, *in* van der Linden, W.J.M., Cloetingh, S.A.P.L., Kaasschieter, J.P.K., van de Graaff, W.J.E., Vandenberghe, J., and van der Gun, J.A.M., eds., Coastal Lowlands: Geology and Geotechnology: Dordrecht, The Netherlands, Kluwer Academic Publishers, p. 1–24.

DAVIS, R.A., AND HAYES, M.O., 1984, What is a wave-dominated coast?: Marine Geology, v. 60, p. 313–329.

DALRYMPLE, R.W., 1992, Tidal depositional systems, *in* Walker, R.G., and James, N.P., eds., Facies Models: Response to Sea-Level Change: Waterloo, Ontario, Geological Association of Canada, p. 157–177.

DALRYMPLE, R.W., ZAITLIN, B.A., AND BOYD, R.A., 1992, Estuarine facies models: Conceptual basis and stratigraphic implications: Journal of Sedimentary Petrology, v. 62, p. 1130–1146.

DE BOER, P.L., 1998, Intertidal sediments: composition and structure, *in* Eisma, D., ed., Intertidal Deposits: River Mouths, Tidal Flats, and Coastal Lagoons: Boca Raton, Florida, CRC Press, p. 345–362.

GALLOWAY, W.E., 1975, Process framework for describing the morphologic and stratigraphic evolution of deltaic depositional systems, *in* Broussard, M.L., ed., Deltas: Models for Exploration: Houston, Texas, Houston Geological Society, p. 87–98.

Gagliano, S.M., and McIntire, W.G., 1968, Reports on the Mekong River Delta: Louisiana State University, Coastal Studies Institute, Technical Report # 57, 143 p.

Goodbred, S.L., Jr., and Kuehl, S.A., 2000, The significance of large sediment supply, active tectonism, and eustasy on margin sequence development: Late Quaternary stratigraphy and evolution of the Ganges–Brahmaputra delta: Sedimentary Geology, v. 133, p. 227–248.

Hanebuth, T., Stattegger, K., and Grootes, P.M., 2000, Rapid flooding of the Sunda Shelf: a late-glacial sea-level: Science, v. 288, p. 1033–1035.

Hayes, M.O., 1979, Barrier island morphology as a function of tidal and wave regime, *in* Leatherman, S.P., ed., Barrier Islands—From the Gulf of St. Lawrence to the Gulf of Mexico: New York, Academic Press, p. 1–27.

Hori, K., Saito, Y., Zhao, Q., and Wang, P., 2002, Architecture and evolution of the tide-dominated Changjiang (Yangtze) River delta, China: Sedimentary Geology, v. 146, p. 249–264.

Kolb, C.R., and Dornbusch, W.K., 1975, The Mississippi and Mekong deltas—a comparison, *in* Broussard, M.L., ed., Deltas: Models for Exploration: Houston Geological Society, p. 193–207.

Kumon, F., and Tateishi, M., eds., 1998, Research Manual on Clastic Sediments: Association of Geologic Collaboration Japan, Special Publication 29, 399 p. (in Japanese).

Mathers, S.J., Davies, J., McDonald, A., Zalasiewicz, J.A., and Marsh, S., 1996, The Red River Delta of Vietnam: British Geological Survey, Technical Report, WC/96/02, 41 p.

Mathers, S.J., and Zalasiewicz, J.A., 1999, Holocene sedimentary architecture of the Red River Delta, Vietnam: Journal of Coastal Research, v. 15, p. 314–325.

Milliman, J.D., and Meade, R.H., 1983, World-wide delivery of river sediments to the ocean: Journal of Geology, v. 91, p. 1–21.

Milliman, J.D., and Syvitski, J.P.M., 1992, Geomorphologic/tectonic control of sediment discharge to the ocean: the importance of small mountainous rivers: Journal of Geology, v. 100, p. 525–544.

Milliman, J.D., and Ren, M.-E., 1995, River flux to the sea: Impact of human intervention on river systems and adjacent coastal areas, *in* Eisma, D., ed., Climate Change: Impact on Coastal Habitation: Boca Raton, Florida, Lewis Publications, p. 57–83.

Milliman, J.D., Rutkowski, C., and Meybeck, M., 1995, River Discharge to the Sea: A Global River Index: Land–Ocean Interactions in the Coastal Zone (LOICZ) Core Project Office, 125 p.

National Astronomical Observatory, ed., 1999, Chronological Scientific Tables (Rika-nenpyo): Maruzen, Tokyo, 1058 p. (in Japanese).

National Geographic Directorate of Vietnam, 1965, Morphological map, sheet # 6228, 1:100,000 scale: National Geographic Directorate of Vietnam.

Nguyen, V.L., Ta, T.K.O., Tateishi, M., and Kobayashi, I.,1999. Coastal variation and saltwater intrusion on the coastal lowlands of the Mekong River Delta, Southern Vietnam, *in* Saito, Y., Ikehara, K., and Katayama, H., eds., Land–Sea Link In Asia: Geological Survey of Japan and JISTEC, p. 212–217.

Nguyen, V.L., Ta, T.K.O., and Tateishi, M., 2000, Late Holocene depositional environments and coastal evolution of the Mekong River Delta, Southern Vietnam: Journal of Asian Earth Sciences, v. 18, p. 427–439.

Orton, G.J., and Reading, H.G., 1993, Variability of deltaic process in terms of sediment supply, with particular emphasis on grain size: Sedimentology, v. 40, p. 475–512.

Parton, T.R., Humphreys, G.S., and Mitchell, P.B., 1995, Soils—A New Global View: London, UCL Press, 213 p.

Reading, H.G., and Collinson, J.D., 1996, Clastic coasts, *in* Reading, H.G., ed., Sedimentary Environments: Processes, Facies and Stratigraphy: Oxford, U.K., Blackwell Scientific Publications, p. 154–231.

Reineck, H.E., and Singh, I.B., 1980, Depositional Sedimentary Environments: Berlin, Springer-Verlag, 551 p.

Saito, Y., Wei, H., Zhou, Y., Nishimura, A., Sato, Y., and Yokota, S., 2000, Delta progradation and chenier formation in the Huanghe (Yellow River) Delta, China: Journal of Asian Earth Sciences, v. 18, p. 489–497.

Somboon, J.R.P., and Thiramongkol, N., 1992, Holocene highstand shoreline of the Chao Phraya delta, Thailand: Journal of Southeast Asian Earth Sciences, v. 7, p. 53–60.

Stuiver, M., Reimer, P.J., Bard, E., Beck, J.W., Burr, G.S., Hughen, K.A., Kromer, B., McCormac, G., van der Plicht, J., and Spurk, M., 1998, INCAL98 radiocarbon age calibration, 24,000–0 cal BP: Radiocarbon, v. 40, p. 1041–1083.

Ta, T.K.O., Nguyen, V.L., Tateishi, M., Kobayashi, I., and Ho, V.C., 1999, Some new results on Late Pleistocene–Holocene deposits of BenTre area in the Mekong River Delta, Southern Vietnam, *in* Saito, Y., Ikehara, K., and Katayama, H., eds., Land–Sea Link in Asia: Geological Survey of Japan and JISTEC, p. 218–223.

Walker, R.G., 1992, Facies, facies models and modern stratigraphic concepts, *in* Walker, R.G., and James, N.P., eds., Facies Models; Response to Sea Level Change: Waterloo, Ontario, Geological Association of Canada, p. 1–14.

Wolanski, E., Nguyen, N.H., Le, T.D., Nguyen, H.N., and Ngoc, T., 1996, Fine-sediment dynamics in the Mekong River Estuary, Vietnam: Estuarine, Coastal and Shelf Science, v. 43, p. 565–582.

Wolanski, E., Nguyen, H.N. and Spagnol, S., 1998, Sediment dynamics during low flow conditions in the Mekong River Estuary, Vietnam: Journal of Coastal Research, v. 14, p. 472–482.

Woodroffe, C.D., 2000, Deltaic and estuarine environments and their Late Quaternary dynamics on the Sunda and Sahul shelves: Journal of Asian Earth Sciences, v. 18, p. 393–413.

THE STRATIGRAPHIC EVOLUTION OF THE SUNDA SHELF DURING THE PAST FIFTY THOUSAND YEARS

TILL J. J. HANEBUTH

Department of Sedimentology/Paleoceanography, Faculty of Geosciences, University of Bremen, Klagenfurter Strasse, 28334 Bremen, Germany

e-mail: thanebuth@uni-bremen.de

AND

KARL STATTEGGER

Institute for Geosciences, Geology, Christian-Albrechts-University, Olshausenstrasse 40, 24 118 Kiel, Germany

ABSTRACT: The Sunda Shelf was extensively exposed during the last glaciation. As a result of the very low shelf gradient, this area is ideal for detailed studies of the sedimentary system and variations of coastal environments in response to sea-level changes. Thirty-six sediment cores from a transect inside the main paleo-valley of the North Sunda River on the middle Sunda Shelf were analyzed by sedimentological and geochemical methods to determine the facies associations. AMS ^{14}C dating of plant material and benthic foraminifera allowed detailed stratigraphic correlation offshore Malaysia for the first time. The drastic environmental transformations from terrestrial to marine conditions and the fate of the paleo–river system on the shelf are also reflected in deep-marine sedimentary records of the southern South China Sea.

Large foresets of a delta system extended basinward following sea-level lowering before the Last Glacial Maximum. A widespread marshy soil formed immediately after exposure, and sediment bypassing dominated during the lowstand. The subsequent sea-level rise caused a stepwise submergence, controlled predominantly by the local morphological conditions. Drowning was restricted mainly to the North Sunda River valley up to 13.5 thousand calibrated years (ka). After 13.5 ka, flooding of the extended Sunda plain (≤ -70 m below modern sea surface) caused a loss of the upper course of the river system combined with a sudden interruption of the terrigenous supply. The transgressive deposits, which were formed during the sea-level rise, were closely related to the rapidly migrating paleo-shoreline. A fully marine regime of condensed muds was established in Holocene times, reflecting the completely new hydrographical conditions due to the final dissection of the remaining fluvial structures, and the opening of the Karimata Strait in the South.

Three stratigraphic units, which are reliably dated offshore for the first time, are connected to regressive (-lowstand), transgressive, and stable phases. They are separated by stratigraphically important discontinuities. These units are correlated to the established Malaysian stratigraphy onshore referring to the *transitional unit*, the *alluvial complex*, and to the *younger sedimentary cover*.

INTRODUCTION

During the sea-level lowstand of the Last Glacial Maximum (LGM), current continental-shelf areas were exposed worldwide. The subsequent rise of sea level, by approximately 120 m (Fairbanks, 1989; Hanebuth et al., 2000), flooded these areas rapidly. The accompanying transformation from terrestrial to marine conditions caused significant environmental and climatic changes. The land ecosystems and fluvial depositional structures were replaced by those of the marine system, deposition centers shifted from the shelf break and slope to nearshore shelf regions, flooding of shallow straits altered the interconnected exchange of marine basins, new ocean currents formed, and the evaporation ratio increased (cf. Wang et al., 1999).

The Sunda Shelf is one of the largest tropical shelf areas. It is situated at the southwestern margin of the South China Sea, the largest of the southeast Asian marginal seas. The shelf is connected to the western Pacific and to the eastern Indian Ocean by narrow and shallow straits (less than -40 m). Modern water depth is -70 m on average. The shelf is extremely broad (up to 800 km) with a very low gradient (1: 9,000 in the transect area; the "old peneplain" of Van Bemmelen, 1949) down to the shelf break at -180 to -220 m. Large parts of it were exposed during the LGM, and catchment areas were drained by a few river systems into the South China Sea (Molengraaff and Weber, 1921).

The modern oceanographic conditions are microtidal; the modern tidal range is less than 2 m and was probably less during glacial times. The area is directly influenced by the monsoon, which exerts a distinct climatic seasonality (wind direction and strength, precipitation, temperature) and the biannual inversion of the surface currents in the South China Sea (Wyrtki, 1957; Wang et al., 1995).

The Sunda Shelf, with the Singapore Platform in its center, seems to have been tectonically stable since the Pliocene–Pleistocene (Tjia and Liew, 1996). A fault system extends from western Sarawak in a NNW direction running east of Natuna Island, but this has been inactive during the Late Pleistocene. Young vertical faults with a penetration of some tens of meters situated near the shelf break may have their origin in gravity mass movement.

Recent sea-floor topography in the proximal and middle part of the Sunda Shelf (Fig. 1) is characterized by overdeepened valleys and depressions of an ancient river system (Haile, 1969; Tjia, 1980). The central submarine valley of the North Sunda ("Molengraaff") River is up to 40 m deeper than the adjacent plain, is 10 to 50 km in width, and has some local depressions in its middle part (Dash et al., 1972). Assuming that the distal part belongs to the youngest river system, its lower course turned northward seaward of Natuna Island (Wong et al., 1999). An associated network of incised-channel fills is also developed in this broad valley and on the adjacent areas on the middle shelf at -100 to -70 m modern water depth.

The transect followed by the research cruise of R.V. *Sonne/115* "SUNDAFLUT" is positioned along the paleo-valley of the North Sunda River on the middle shelf (Stattegger et al., 1997; Fig. 1; Table 1). The current study is a detailed reconstruction of the lateral and vertical facies shifts that reflect changing conditions connected with the rapid postglacial sea-level rise.

A high-resolution reconstruction of the postglacial sea-level rise using organic-rich deposits from the paleo-coastline is the

Tropical Deltas of Southeast Asia—Sedimentology, Stratigraphy, and Petroleum Geology
SEPM Special Publication No. 76, Copyright © 2003
SEPM (Society for Sedimentary Geology), ISBN 1-56576-086-7, p. 189–200.

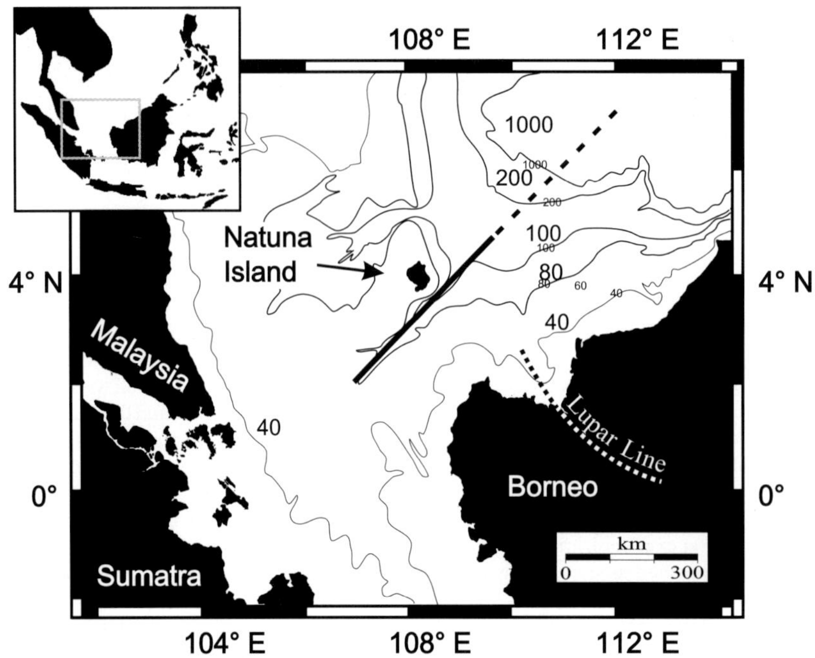

FIG. 1.—Map of the Sunda Shelf. Black line: transect location of cruise Sonne-115. Dotted line: outer part of the transect, not discussed here. The transect follows the paleo-valley of the North Sunda River. Water depth in meters.

first major result of the project (Hanebuth et al., 2000). The maximum lowstand of sea level during the LGM hereby clearly assigned to about -120 ± -3 m below modern sea surface on the Sunda Shelf is in agreement with the Barbados curve (Fairbanks, 1989, 1990). The sea-level rise between 21 and 11 cal. kyr BP in combination with the low shelf gradient caused an average lateral transgression rate of about 50 m yr^{-1}. During Meltwater Pulse 1A (14.6–14.3 ka) this rate even exceeded 450 m yr^{-1}.

MATERIALS AND METHODS

Thirty-six gravity and piston cores were taken from -126 to -70 m modern water depth along the transect. A number of sedimentological and geochemical methods were applied to identify the different types of facies contained in the cores.

The qualitatively and quantitatively analyzed content of the sieved coarse fraction (>63 μm) points to the sources of the material and the environment in which the deposition took place. X-radiography makes internal sedimentary structures visible, allowing analysis and interpretation of the energy conditions during the deposition as well as grain and material sorting. Further standard sedimentological parameters, such as grain size and magnetic susceptibility, lead to a first characterization. This was followed by specific analyses to determine the facies in detail. The organic carbon and carbonate contents (TOC, TIC) relate mainly to the depositional environment; $\delta^{13}C_{org}$ highlights the contrast and transition from terrestrial to fully marine conditions. The ash content of isolated plant material reflects the potential degree of decay as a result of the mixing by repeated reworking processes (Chapman, 1976; Gastaldo et al., 1993). Stable isotopes measured on planktonic and benthic foraminifera indicate general conditions on the shelf (e.g., freshwater lenses) but cannot contribute to the stratigraphy directly because of the very high sedimentation rates. Pollen analysis reveals the distance of a facies from the paleo-shoreline and from the river mouth during the time of deposition—on the one hand, the mixture of the pollen spectrum increases with increasing distance from the shore, and a river supplies pollen material from all parts of the hinterland; on the other hand, the content of mangrove pollen, which is not widely distributed in general, increases with proximity to the paleo-coastline.

The AMS ^{14}C ages were measured with a 3 MV HVEE Tandetron 4130 AMS system to a precision (counting statistics and machine statistics) of 0.3% for modern samples (Nadeau et al., 1997) at the Leibniz-Labor in Kiel. All ^{14}C ages presented here have been calibrated to calendar years BP with the latest calibration program CALIB 4.0 (Stuiver et al., 1998).

RESULTS

Description of Facies Types

More than 100 facies sections were distinguished in the 36 cores involved in this study and determined (Fig. 2) as other authors have previously demonstrated in other regions (e.g., Woodroffe and Mulrennan, 1993; Dalrymple et al., 1992; Dalrymple et al., 1994). The deposits were arranged into genetically related facies types. Three of the ten resulting types originated in a terrestrial environment and six are more or less directly coupled with the shoreline and nearshore (Fig. 3A). The ten different types are placed in stratigraphic order in the following (data partly from Schimanski, 1999; M. Kienast has kindly measured $\delta^{13}C_{bulk}$).

- *Modern Shelf* facies: The most seaward facies comprises a muddy foraminiferal sand which was deposited under fully marine conditions but which contains subordinately reworked material of terrestrial origin. The brownish-gray sediment is unconsolidated and without internal structures owing to bioturbation. The carbonate content ($C_{carb,\ bulk}$) exceeds 10%, and $\delta^{13}C_{org,\ bulk}$ of -22‰ also indicates a marine environment.

TABLE 1.—Sediment Cores and Coordinates used in this study (Cruise SO-115).

core number	coordinates latitude	coordinates longitude	water depth	core recovery
18269-2	4° 46.013' N	109° 26.321' E	113 (m)	881.5 (cm)
18270-4	4° 43.516' N	109° 28.630' E	106 (m)	140 (cm)
18271-3	4° 38.314' N	109° 32.971' E	116 (m)	576 (cm)
18272-2	4° 37.597' N	109° 33.632' E	118 (m)	338 (cm)
18273-2	4° 37.289' N	109° 33.931' E	126 (m)	348 (cm)
18274-3	4° 36.313' N	109° 34.818' E	117 (m)	755 (cm)
18275-3	4° 35.652' N	109° 35.539' E	109 (m)	597 (cm)
18276-2	4° 44.897' N	109° 44.837' E	116 (m)	721 (cm)
18295-2	4° 55.587' N	109° 17.865' E	119 (m)	823 (cm)
18296-2	4° 59.754' N	109° 14.446' E	118 (m)	244 (cm)
18297-3	4° 44.347' N	109° 01.916' E	111 (m)	396 (cm)
18298-2	4° 31.987' N	108° 49.508' E	102 (m)	587 (cm)
18299-1	4° 32.004' N	108° 49.537' E	102 (m)	580 (cm)
18300-2	4° 21.778' N	108° 39.215' E	91 (m)	885 (cm)
18301-2	4° 21.308' N	108° 38.811' E	93 (m)	582 (cm)
18302-2	4° 09.585' N	108° 34.535' E	83 (m)	598 (cm)
18303-2	4° 26.425' N	108° 55.491' E	107 (m)	736 (cm)
18304-2	4° 21.790' N	109° 00.157' E	104 (m)	228 (cm)
18305-2	4° 17.318' N	109° 04.599' E	109 (m)	514 (cm)
18306-4	3° 35.184' N	108° 26.422' E	89 (m)	165 (cm)
18307-2	3° 37.626' N	108° 31.648' E	100 (m)	943 (cm)
18308-2	3° 17.380' N	108° 47.143' E	80 (m)	105 (cm)
18309-2	3° 27.959' N	108° 41.174' E	83 (m)	597 (cm)
18310-2	3° 32.131' N	108° 32.131' E	100 (m)	568 (cm)
18311-2	3° 41.191' N	108° 27.093' E	60 (m)	468 (cm)
18312-2	3° 42.351' N	108° 42.380' E	101 (m)	667 (cm)
18313-2	3° 52.194' N	108° 52.226' E	98 (m)	620 (cm)
18314-2	3° 59.469' N	108° 59.473' E	100 (m)	370 (cm)
18315-2	2° 01.700' N	107° 02.000' E	69 (m)	583 (cm)
18316-2	2° 29.300' N	107° 27.500' E	71 (m)	597 (cm)
18317-3	2° 36.600' N	107° 22.500' E	95 (m)	197 (cm)
18318-3	2° 36.600' N	107° 22.500' E	87 (m)	406 (cm)
19319-2	2° 36.620' N	107° 22.502' E	81 (m)	empty
18320-2	2° 36.700' N	107° 22.500' E	76 (m)	492 (cm)
18321-2	2° 18.453' N	107° 25.326' E	109 (m)	569 (cm)
18322-2	2° 18.400' N	107° 37.900' E	70 (m)	493 (cm)
18323-2	2° 47.000' N	107° 53.200' E	92 (m)	540 (cm)

• *Inner Shelf* facies: This facies consists of highly bioturbated foraminiferal mud deposited under marine conditions ($C_{carb, bulk}$ of 2%). It contains ≤ 5% (exceptionally up to 20%) sand and consists partly of reworked material. This facies type can be distinguished from the *modern shelf* facies by a stiffer consistency of the material, by the color (light brown), and by an erosional surface at the top associated with large burrows. The terrestrial influence is also slightly increased, as shown by $\delta^{13}C_{org, bulk}$ of -22.5 to -24.5‰.

• *Lagoon* facies: In proximity to the shoreline, this type of facies was deposited under marine conditions, as indicated by fresh planktonic foraminifera, but comprising predominantly terrigenous material. The sand fraction (> 63 μm) contains 60–80% lithoclasts. This facies is completely bioturbated and frequently interspersed by organic patches. $\delta^{13}C_{org, bulk}$ also implies the decreasing distance to the coastline, with values of -23.5 to -27‰ in comparison to the former facies types.

• *Delta Front* facies: The facies shows monotonously fine (< 5% sand) and rhythmic lamination (tidal) with thin interbeds of pure sand or organic material (peaty layers). The few fresh foraminifera and $\delta^{13}C_{org, bulk}$ values of –25.5 to -26‰ indicate marine conditions during deposition. The lamination is partly bioturbated, and micro-crossbedding is common.

• *Delta Mouth* facies: The previous facies type is differentiated from the *delta mouth* facies on the basis of the thickness of the intercalations and the presence of ripple-bedded structures: thick layers of terrigenous sands and organic material occur frequently in the *delta mouth* facies, and flaser bedding and crossbedding are common—both indicating a hydrographic regime of higher energy with frequent changes and a strong river supply of terrigenous material.

• *Tidal Flat* facies: This *mud to sand flat* facies consists of fine rhythmic lamination (tidal) with flaser bedding, lenticular structures, and crossbedding. These deposits are composed of mud and sand and are rich in organic material (peaty layers) of terrigenous origin, although $\delta^{13}C_{org, bulk}$ values of -23.5 to -27‰ suggest an ultimate marine influence. Thick single layers of large benthic foraminifera and sand probably represent storm-event deposits.

• Other *shoreline* facies: Several facies sections also represent obvious shoreline deposits. They cannot be related to a certain zone within the coastline, but a continuous, very minor marine influence as well as an alternation with thick oxidized sand layers suggests their origin. Flaser bedding, crossbedding, and micro-channel fills indicate water flow and temporary exposure. Peaty intercalations are also abundant. Another thin relict horizon is very similar to the following *coastal swamp* facies, but it is too thin for an exact determination and has been destroyed by large burrows.

• *Coastal Swamp* facies: This conspicuous facies type is completely bioturbated and penetrated by abundantly occurring root remnants, which are preserved in their *in situ* vertical position. The pollen record proves that this horizon formed within a mangrove swamp. The organic content of these dark gray to blackish gray deposits is high ($C_{org, bulk}$ > 4%) and peaty patches are dispersed. $\delta^{13}C_{org, bulk}$ yields the terrestrial end member with values of -26.5 to -27.5‰.

• *Coastal* (including *Delta*) *Plain* facies: The most landward facies is composed of terrestrial deposits with a rhythmic lamination (overbank) which show partly lenticular and ripple bedding. The sand content is generally less than 5% but single layers of light or oxidized sands and peaty material are frequently intercalated.

• *Marshy Soil* facies: Also of great importance is a widespread soil formation. It is pale gray with numerous orange flames and mottles (lateritic wash), a fact that indicates alternating conditions of stagnation and ventilation. Sideritic spherulites occur frequently concentrated on layers. This soil horizon obviously incorporated underlying deposits during its formation and was strongly consolidated due to exposure. It is poor in organic matter ($C_{org, bulk}$ of 0.2%). The $\delta^{13}C_{org, bulk}$ values of -22 to -23‰ (once -16) potentially suggest a grassland vegetation, as should be expected in a marshy environment. The age control showed that this soil horizon formed directly after emergence during the sea-level lowering before the LGM.

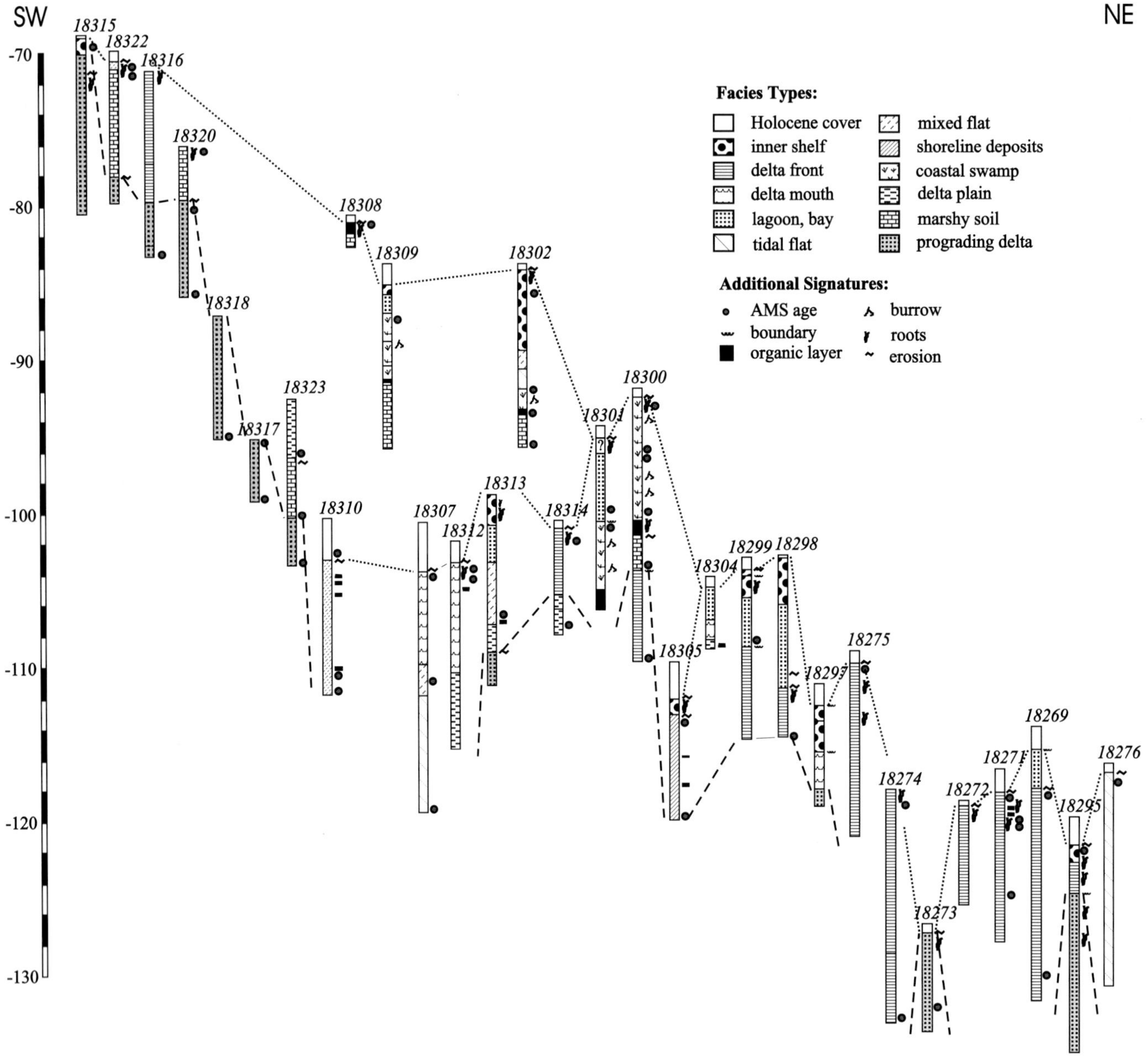

FIG. 2.—Lateral and vertical distribution of the facies of 28 cores (for core data see Table 1) along the transect. The dashed line separates the underlying regressive unit from the following transgressive unit. This long-lasting hiatus is marked by erosion and soil development. The dotted line indicates the condensed Holocene horizon covering the marine-erosional top of the transgressive deposits. Scale: depth of core top in meters below modern sea surface. Core columns have double length to this scale. Nine cores of Table 1 are not presented here because of insufficient age and facies control.

AMS ^{14}C dating

Control by high-precision AMS radiocarbon dating of terrigenous organic material in several cores enabled us to reconstruct the lateral and vertical facies shift over the past 50 ka (Table 2).

In a siliciclastic nearshore environment detailed AMS ^{14}C determination is necessary before precise statements can be made concerning the true age of deposition. During river transportation, fresh plant material is mixed with material that was eroded from older accumulation sources in the hinterland. After deposition, this process causes a prominent bias of the true age of deposition towards ages that are too old. We tried to estimate this mixing effect by measuring isolated organic fractions. The acid-alkali-acid extraction split the plant macrofossil material into insoluble residues and soluble humic acids.

The results are summarized briefly here. The insoluble residue material from peaty layers (nearshore) and from *in situ* preserved root remnants (mangrove) yield the most reliable ages. The coarser the organic material, the more exact is the determination of the depositional age. A dense sediment, which prevents ventilation of the sediment and migration of soluble acids, also

TABLE 2.—AMS ^{14}C ages. This list contains only ages relevant to major facies shifts

KIA †	core	depth	depth	sampled	age	deviation	
number	number	(m) #	(cm) +	material	(^{14}C yr)	+	–
former delta system							
5618	18315-2	69	40–42	bulk sediment *	47,670	2,090	1,660
3541	18320-2	76	200–202	peaty sediment *	44,800	1,850	1,500
3529	18300-2	91	879–881	bulk sediment *	> 41,510		
5977	18273-2	126	273	bulk sediment *	39,870	1,070	950
soil formation							
3531	18302-2	83	590	bulk sediment *	20,160	330	310
3528	18300-2	91	590–592	bulk sediment *	21,490	330	320
3544	18323-2	92	380–382	bulk sediment *	23,460	160	150
transgressive system							
5980	18276-2	116	60	bulk sediment *	17,680	80	80
5620	18322-2	70	67–69	bulk sediment *	11,025	55	50
Holocene cover							
6549	18295-2	119	163	*Rotalia* sp.	10,330	55	55
5978	18275-2	112	50	*Rotalia* sp.	9,910	60	60
5612	18312-2	101	80–81	*Rotalia* sp.	8,920	55	55
5615	18314-2	100	53–54	*Rotalia* sp.	6,435	50	50
5607	18308-2	80	23–24	*Rotalia* sp.	4,125	35	35
4646	18320-2	76	10–12	*Rotalia* sp.	6,420	50	50
5619	18322-2	70	49–51	*Rotalia* sp.	4,495	40	40
5617	18315-2	69	38–40	*Rotalia* sp.	8,630	55	55

† measured at the Leibniz–Labor, Kiel;
# below modern sea-surface;
+ below sea-bottom;
* residue of acid–alkali-acid extraction of organic material.

favors reliable ages. On the other hand, humic acids bear information about the proportion of the reworked material from older sources and about groundwater movement (seeps within the shoreline). The relative percentage of "^{14}C-dead" old material increases with increasing distance from the paleo-shoreline, which causes a bias in the age of up to 3.5 ka, as is the case in the delta-front facies. This is caused by the extension of the mixing effect with distance and by the fact that finer material is distributed preferentially over longer distances.

Benthic foraminifera (*Rotalia* sp.) were dated in fully marine environments, where terrigenous material was absent.

Grouping of Facies Types and Local Distribution

In general, long periods of nondeposition of several millennia are recorded, combined with a strong erosion activity. Deposition took place in episodic, rapid and locally restricted periods, and the facies record is interrupted in almost all cores. The spatial and temporal correlation of all facies sections led to a classification of cores into facies groups containing specific vertical facies successions. The five groups recognized are characterized by a predominant type of facies or by a typical facies succession. The environmental relations of these facies groups are schematically shown in Fig. 3B. On the Sunda Shelf, the local distribution of these groups fits fairly well with the modern sea-floor topography.

The "*Marine Front*" group is situated directly in front of the middle course of the North Sunda River valley in -126 to -109 m modern water depth. (The North Sunda River is incorrectly located at the -200 m isobath in most previous publications, which is the shelf break and not the paleo-coastline during the LGM.) Delta-front deposits predominate, although a tidal-flat facies is present in places. Nine cores contain these deposits.

The "*Surrounding Area*" group appears on the northern side of the central valley in a water depth of -111 to -102 m. This group comprises delta-front deposits (partly complemented by delta-mouth and mixed-flat facies) and is overlain by the deposits of lagoon and inner-shelf environments. It appears in six cores.

The "*Central River*" group is restricted between the flanks of the middle course of the paleo–river valley southeast of Natuna Island in -101 to -98 m water depth. It consists of nearshore facies such as delta-plain and tidal-flat deposits and is overlain by delta-front and delta-mouth facies. It occurs in six cores.

Adjacent to this group, the "*Swampy Plain*" group covers the area flanking the central valley in depths of -93 to -80 m. Here, the mangrove-swamp facies (on the Pleistocene surface) prevails, overlain by lagoon (partly delta-mouth) and shelf sediments. Five cores contain these facies.

Finally, the "*Uncovered Pleistocene*" group is the predominant group in the upper part of our transect. Its deposits consist of various facies of an older delta system which is exposed on the modern sea floor. It appears in seven cores.

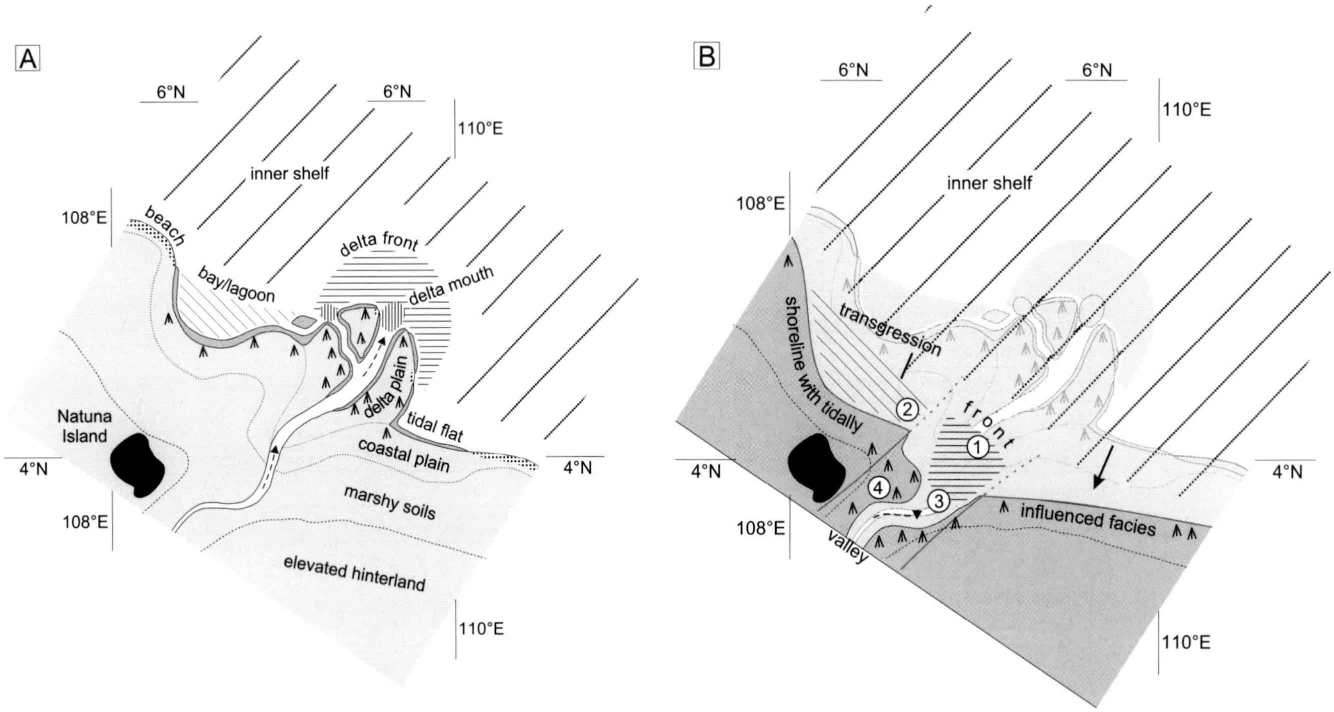

FIG. 3—**A)** Schematic distribution of facies types recognized in the Sunda Shelf cores around 17 cal. kyr BP. Paleo-environmental reconstruction according to modern analogues of lateral facies successions from the tropics. Triangles: mangrove swamps. **B)** Two thousand years later, during the initial phase of deglacial sea-level rise. Retreat of the river-related depositional system within the old valley depression. Because of the landward shift, younger distal facies are stacked on older facies of more landward origin. The margins of the valley are marked by gray lines. For tidal facies types see Part A. Numbers refer to facies groups in the text: (1) *Marine Front* group, (2) *Surrounding Area* group, (3) *Central River* group, (4) *Swampy Plain* group; the *Uncovered Pleistocene* group follows in shallower parts of the shelf (not shown here).

DISCUSSION

The Sedimentary History on the Sunda Shelf Over the Past 50 ka

The history of sedimentation on the Sunda Shelf during the past 50 ka is directly related to the influence of sea-level changes on variations of sediment supply and accommodation space.

Large delta foresets built up basinward in latest Pleistocene times (Fig. 4). These structures extend beyond the shelf break, thinning with increasing distance from the catchment area. The sea-level drop before the LGM resulted in elevation and the development of wide lagoon-like basins on the top of this foreset body, on which alluviation processes subsequently developed. Directly after this exposure, a widespread marshy pedogenesis of high-glacial age took place (e.g., Biswas, 1973; Kögler et al., 1982 [Sunda Shelf]; Umitsu, 1993 [Ganges delta]; Richter, 1998 [Mekong delta]) but was preserved only on the higher hinterland (above -100 m), possibly because of the short period of subaerial exposure and subsequent strong erosion.

During the sea-level lowstand (Fig. 5A) the coastline was stationary at about -120 m and the inner and middle parts of the shelf were exposed. The large central river was stable in its deep valley. Sediment bypass through this valley subsequently was the predominant process on the Sunda Shelf. Because the lower course of the North Sunda River left our transect to the northward (Wong et al., 1999), our investigations did not reveal extensive fluvial deposits associated with the lowermost part of the river system: if any, thin shoreline sediments were found. Previous authors found high sedimentation rates in front of the Sunda Shelf in the southern South China Sea (SCS) during the last glaciation (Broecker et al., 1988; Kienast, 1996). This clearly corresponds to the lack of sediments on the exposed plains beside the valley, indicating transport of material offshore into deeper water.

The initial slow sea-level rise after the LGM caused an immediate loss of the lower river course. Therefore, thick delta-front deposits were funneled in front of the newly formed river mouth. Basal shoreline-related deposits are absent, which is probably a result of reworking processes due to the migration of a narrow shore zone.

In the following period of sea-level rise, the middle river course gradually became submerged while the surrounding plains, which were still elevated some tens of meters, were modified by some associated channels. Delta-plain to delta-front facies, several meters thick, developed in this valley depression, shifting gradually upstream. The river system lost its deltaic character at this time and changed to an estuarine environment as a consequence of an increasing tidal-transgressive influence. Deposits from calm-water conditions (bay and lagoon) appear here for the first time.

Sea level rose dramatically from 14.6 to 14.3 ka (Fig. 5B) by about 16 m (Hanebuth et al., 2000), corresponding to Meltwater Pulse 1A observed in the open ocean (Bard et al., 1996). A complex network of ephemeral channels, probably belonging to a distributary river system, replaced the drowning middle course. These channels truncated the Pleistocene surface, including the marshy soil horizon, which proves the maximum age of the incision process. Mangroves dominated these protected shorelines. This plant community followed the rapid sea-level rise by trapping larger quantities of material delivered from the hinterland in their complex subaerial root networks, as is visible in the cores (proven by the comparison of the ages of *in situ* preserved root remnants and plant fragments in the fine-grained clastics, and by analysis of the freshness of the organic material by means of the ash content). The shoreline deposits were probably well preserved during the phase of accelerated sea-level rise because the interval of potential erosion activity was short. The gradual retreat of the depocenter in front of the landward-migrating paleo-coastline reduced the extent of the catchment area and caused a first fragmentation of the downstream river system, which is reflected in the general decrease of fluvial input into the southern South China Sea (Wang, 1999).

The surrounding plains were widely flooded when the sea surface reached the critical level of -70 to -60 m (13.5 to 12.5 ka; Fig. 5B, 5C). This subsequently caused a general loss of the established riverine system and a splitting into several individual small rivers, which were separated from each other and followed the retreating coastline. These smaller and shorter rivers brought less sediment down on the shelf, and only thin shallow-shelf deposits are preserved. The Pleistocene surface is exposed on the sea floor in the proximal parts of the shelf. A drastic reduction of the terrigenous input into the southern part of the South China Sea directly after 13.5 ka has been reported from sites at different distances to the shelf break (Broecker et al., 1988; Schönfeld and Kudrass, 1993; Wang et al., 1994). This decrease occurred simultaneously with the breakdown of the North Sunda River system and the subsequent reduction of the sediment accumulation observed in the area. This further indicates that the North Sunda River was the predominant factor in supplying terrigenous clastics into the southern South China Sea.

The second acceleration of sea-level rise after 11.5 ka (Meltwater Pulse 1B; Fairbanks, 1990) finally destroyed the uppermost fluvial structures of the North Sunda River. It flooded the land bridge between Borneo and the Peninsula Malaysia and started to open the Karimata Strait in the south (Fig. 5C). As a result, the general ocean circulation system produced strong water currents, which were forced by the monsoonal winds (Wyrtki, 1957, measured up to 40 cm / s in August), passing through the nozzle-like strait. In addition to nondeposition and reworking during the transgression, these currents may also have caused erosion leading to the recent exposure of the Pleistocene surface on the sea bottom. At this time, the last siliciclastic influence disappeared in front of the shelf margin, leaving purely hemipelagic conditions (Pelejero et al., 1999; Wang et al., 1999).

Thin foraminiferal muds of Holocene age cover the modern shelf. This new sedimentary regime was established at about 9 ka, after locally varying and depth-related depositional interruptions (erosion?) of several millennia. It apparently became consolidated in connection with the opening of the Karimata Strait. Hydrographic changes in hemipelagic deposition on the shelf slope at about 8 ka confirm this idea (S. Steinke, personal communication).

The Stratigraphic Framework

Three stratigraphic units resulted from sea-level changes over the past 50 ka (Table 3).

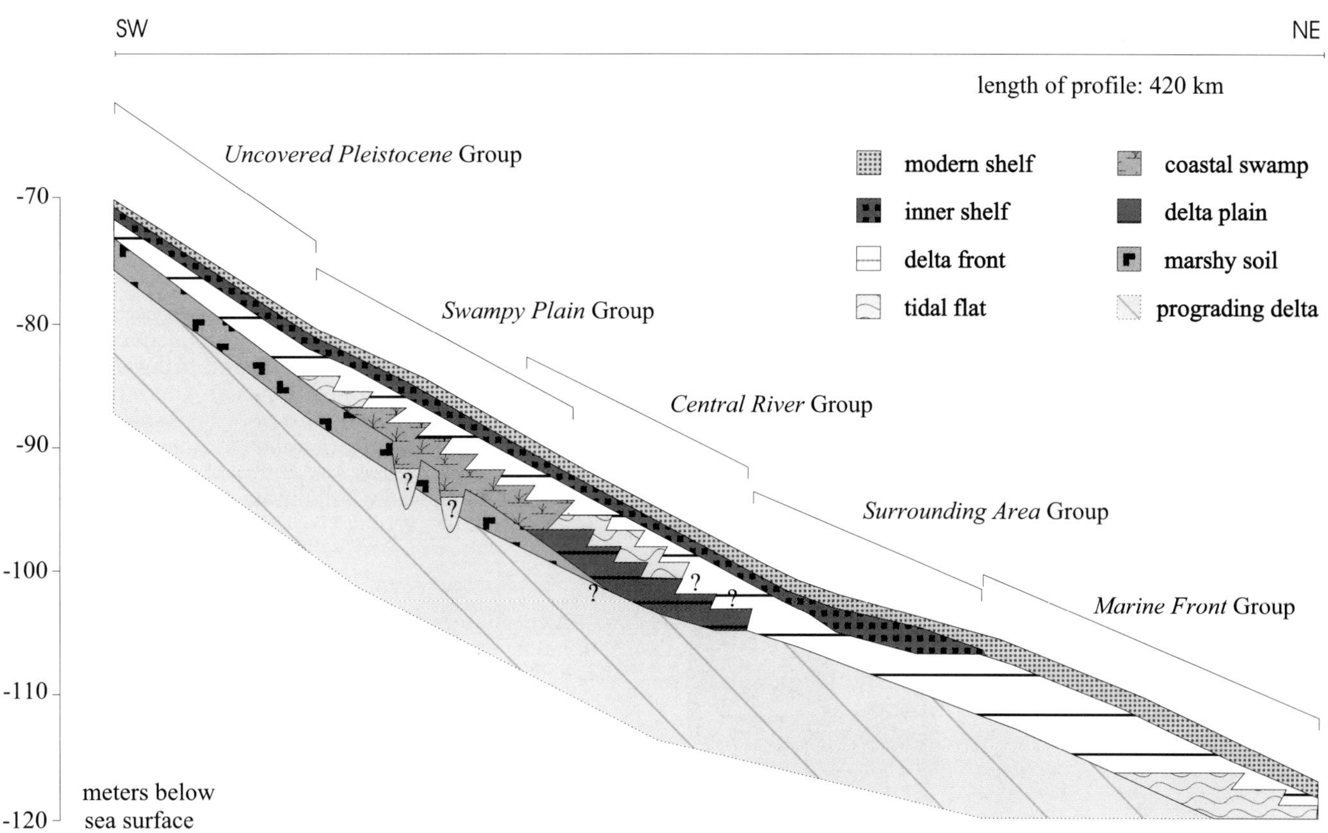

FIG. 4.—Schematic facies profile through the middle Sunda Shelf.

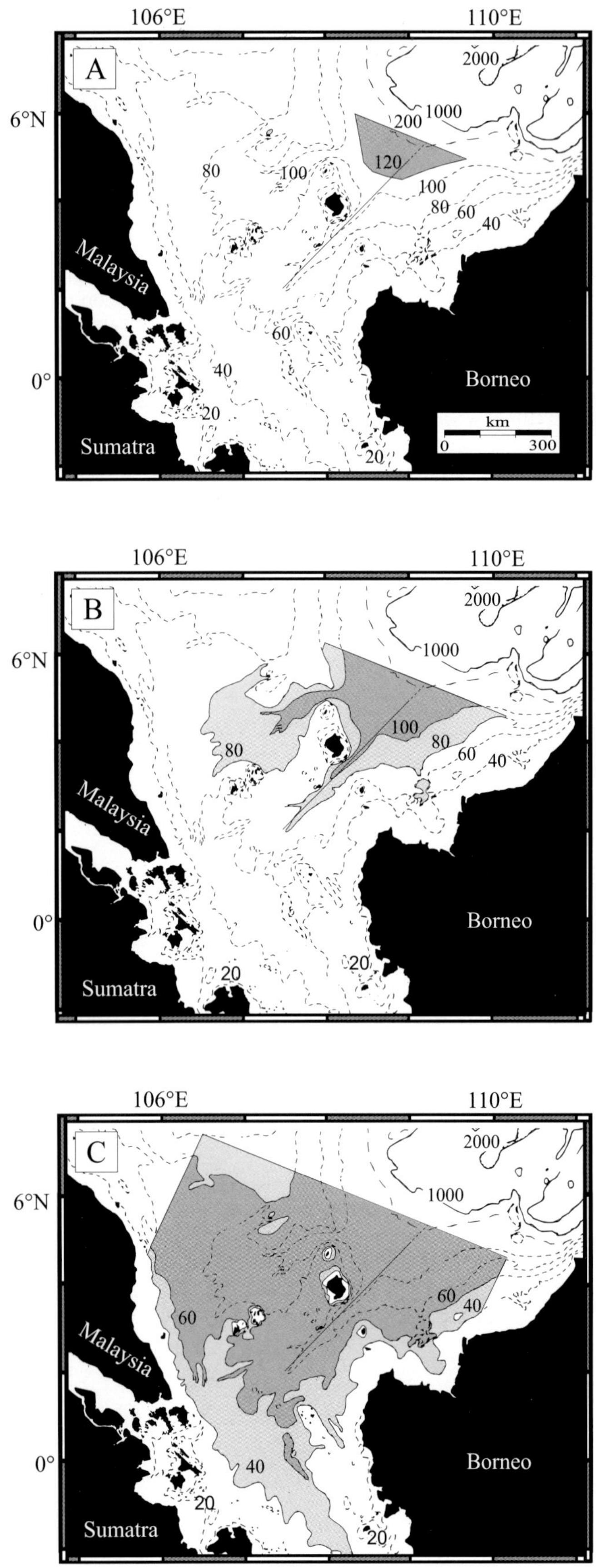

1. Regressive and Lowstand Unit: A basal thick delta foreset complex is documented in two thirds of all cores extending from the shallowest parts, near the present coastline, down to the outer shelf (cf. Wong et al., this volume). It comprises a variety of marine to terrestrial facies types which belong to the large basinward-directed foresets. Its age ranges between 50 ka (or older) and 23 ka (LGM), considering the limitation of the AMS ^{14}C method, and its thickness has been shown by shallow seismic surveys to be more than 40 m on average (Stattegger et al., 1997; Haft, 1999). A widely distributed soil was formed on the top of this unit during the early phase of subsequent exposure. These delta deposits and the soil formation are also described from areas closer to the Recent shoreline (Biswas, 1973; Emmel and Curray, 1982; Köhler, 1985).

2. Transgressive Unit: A succession of terrestrial, shoreline-related and subsequently marine facies overlies a widespread major discontinuity, which represents the Pleistocene surface, subaerial as well as submarine. The thickness of this unit increases with modern water depth. Several internal discontinuities accompany the jump-like facies shifts. The age of deposition ranges from 20.5 ka (LGM) to 13 ka.

3. Stable Unit: A second major discontinuity separates the uppermost fully marine unit of sub-Recent to Recent age (Holocene) from the underlying deposits. This blanket-like cover of smooth foraminiferal muds is starved of terrigenous material and is still subject to reworking, yielding ages between 10 and 4 ka. Its thickness is restricted to a few decimeters, in agreement with observations made in adjacent areas.

Using foraminifera that indicate paleo-depth, Biswas (1973) described various facies types of similar characteristics from nearshore and coastal environments, which are preserved in cores taken off East Malaysia. The Pleistocene regression surface is covered by lateritic soils overlain by deposits of lagoon, swamp, and marsh topped by younger thin muds. Sedimentary structures like these are known from adjacent regions as well, for example from the Gulf of Thailand (Pramonajee et al., 1986) and the Strait of Malacca (Emmel and Curray, 1982).

The deposits belong to two systems tracts according to a sequence-stratigraphical interpretation (Posamentier and Vail, 1988). The basal delta unit embodies deposits of the late sea-level highstand (forced regression) to lowstand systems tracts of the last glacial cycle. A top lowstand surface is developed on the top, marked by an extended soil formation (= sequence boundary). Transgressive nearshore deposits represent the thin transgressive systems tract—these are without stacking patterns, because

←

FIG. 5.—Successive phases of flooding constructed from modern bathymetry (in meters). The ancient fluvial structures are clearly visible in the morphology, because the younger sediment cover is of minor thickness. Single line: transect. Sea-level position **A)** gray—during lowstand (~ 21 ka, -120 m); **B)** dark gray—at ~ 15 ka, -100 m (moderate rate of sea-level rise); light gray—at ~ 14 ka, -80 m (after first phase of rapid rise); **C)** dark gray—at ca. 12.5 ka (phase of decelerating rate), -60 m; light gray—at ~ 10 ka (after second acceleration phase), -40 m (first connection to the Pacific Ocean in the south).

of the rapidity of the sea-level rise and the extreme extent of the shelf. The following maximum flooding surface with intense bioturbation and erosion patterns is overlain by sediments of the condensed section.

The Riverine System

The Importance of Incised-Valley Fills.—

Incision by two generations of fluvial drainage structures took place during shelf emergence. The central valley formed at the time of sea-level drop, when the large delta foresets crossed the shelf basinward. Therefore, the main drainage system was already established in an early stage of exposure (eventually following an older fluvial morphology). Because sea level did not reach the shelf break at about -180 m (-220 m) during the high-glacial lowstand, a separate valley incision is not evident there. The secondarily incised channel network developed during a phase of rapid sea-level rise and replaced the main river.

The deposits of the *transgressive systems tract* (TST) subsequently filled the valley and channel depressions. The main valley received sediments of partly more than 10 m thick, but not enough to fully compensate for its overdeepened relief, as can be seen at the modern sea floor. The incised channels, however, were completely filled (with rare exceptions). The existence of valleys and channels is of great importance: only there, shoreline and nearshore deposits were continuously trapped over longer periods and preserved. This fact contrasts with the depositional patterns on the surrounding plains, which mostly were affected by erosional activities. Nevertheless, the single depositional units of the TST appear scattered and without *stacking patterns* because of the rapidity of the sea-level rise and the very low shelf gradient.

The Mouth of the North Sunda River Through Time.—

Several environmental factors directed the symmetry of the North Sunda River mouth: high sediment supply, a microtidal ocean, a low shelf gradient, and biannual reversing longshore currents. But the predominant factor was the variation of sea level, which was probably responsible for the fact that a defined and expanded delta did not really build up at any time. We assume that a true delta may have developed during Holocene sea-level stability (lowstand), as observed worldwide since the last 6 ka of relative sea-level stability (Stanley and Warne, 1994), but its position and its deposits are still unknown. During the transgression, the system finally lost its deltaic character. The new estuarine conditions together with rapid backstepping, which was restricted mainly to the valley system, excluded delta progradation. This time was excellent for a massive extension of the mangrove swamp, covering numerous still-water areas.

The Major Factors Controlling Sedimentation.—

Three major factors dominated the depositional processes and the preservation potential, and were, therefore, responsible for the stratigraphic architecture of the Sunda Shelf.

1. Sea-level changes prepared the conditions that the deposystem had to respond to and prevented development

TABLE 3.—Response of the depositional system to sea-level changes

DEPOSYSTEM:	SEA LEVEL:	TIME INTERVAL
Regressive + lowstand phase:		
(a) large prograding delta foresets migrating basinwards, marine to terrestrial deposits	fall step by step	50 (or older) to younger than 23 cal. kyr BP
(b) stabilization of coastline, sediment bypassing through large river valley, deposition beyond shelf break, widespread soil (lateritic) formation	lowstand	ca. 21 ka (LGM)
Transgressive phase:		
(c) loss of lower river course by flooding, thick delta-front deposits in front of river mouth	moderate rise	20 to 15.5 ka
(d) depocenter retreats into middle course, deposits of delta plain followed by delta front, narrow coastal zone, first multichannel fills, change from deltaic to estuarine character	moderate rise	15.5 to 14.6 ka
(e) complete drowning of middle course, extensive mangrove distribution, short-lived channel network	rapid rise	14.6 to 14.3 ka
(f) final drowning of middle course and flooding of surrounding area, fragmentation of North Sunda river, extreme decrease of sediment supply	decelerating rise	14.3 to 11 ka
Stable phase:		
(g) thin fully marine muddy sediment, condensed horizon containing a high content of reworked material, covering the entire shelf area	stabilization	Holocene

of a mature coastline. Sea level determined the general timing and, therefore, permitted but also limited the main shelf construction.

2. The extreme shelf extension caused lateral shifts of the front of the depocenter over long distances, probably preventing the beginning of major autocyclic processes, of local sediment aggradation during lowering, and of stacking patterns during the transgression. On the other hand, the secondary channels that were incised in a complex network into the plains during the transgressive phase were refilled soon, because the riverine current velocity was low, owing to the low relief, and sea level gradually reached the level of the plains.

3. The huge catchment area in combination with the tropical precipitation intensity led to a continuous and high sediment input through time. As long as these sediments were supplied, the river complex was acting as one connected system and could continue to exist.

Stratigraphy off The Malaysia Peninsula

Three major disconformable surfaces of regional extent separate the latest Quaternary units offshore the Malaysia Peninsula and Borneo. The lower surface is marked by valley incision and soil formation that developed during exposure. The middle disconformity was formed by reworking during the subsequent transgression and contains incised channels. The upper surface represents the open-marine erosional base after the transgression crossed the area. The duration of the time gap between a surface (often showing a temporal shift) and the subsequent deposition depends on the local position and elevation and often exceeds several thousand years. A correlation of the Pleistocene sequences from localities around Malaysia and the Natuna Sea (Fig. 6) is presented in Figure 7.

The late Pleistocene basal *delta complex* (DC) on the Sunda Shelf is likely to correlate with the delta foresets in the adjacent Strait of Malacca described by Emmel and Curray (1982). Evans et al. (1995) claimed that the unit beneath the channel fills on the Sunda Shelf belongs to the upper part of the *older sedimentary cover* (OSC), although Aleva (1973) classified this unit into Miocene to Pliocene times. Batchelor (1979) placed it in the early Pleistocene. The *delta complex* probably corresponds to the *older marine unit* (OMU of Batchelor, 1979) representing the upper part of the *transitional unit*. Unfortunately, the consistency of this unit is poorly defined. Kamaludin et al. (1993) also determined the *older alluvium* (AO) as latest Pleistocene (67 to 28 ka).

The erosional "*exposed*" surface between the DC and the overlying deposits is present throughout the region. It is clearly time-regressive and is associated with the construction of the main valley. The attributes "abrasion" and "erosional", which have been used in the literature, reveal the quality of this discontinuity.

The *alluvial complex* (AC) of previous authors correlates with our *transgressive system* (TS) deposits. Aleva (1973) concluded that the reddish soil horizon developed on his marine *abrasion surface*. However, as we have shown, this soil formation is truncated by incised valleys of the *transgressive system*. Therefore, Aleva deduced an age for the AC (he carefully called this attempt "tentative connections") that is too old. Evans et al. (1995) adapted these estimates. A good correlation with our results arises if the *abrasion surface*, which Aleva has presented in his sonograms, is split into a lower incising surface at the base of the AC and an upper erosional but flat surface on the top, consistent with our seismic observations.

The uppermost depositional unit (*Holocene cover*, HC; *younger sedimentary cover*, YSC) has been described by all authors in a similar way. Aleva (1973) noted that alluvial fans occur in this sequence close to the island of Billiton in addition to the fully marine deposits.

The erosional surface at the base of the HC also appears continuously in the Sunda Shelf region and is always accompanied by bioturbation and reworking.

CONCLUSIONS

On the basis of 36 sediment cores, we have identified shoreline-related facies and their local distribution in the siliciclastic environment on the Sunda Shelf. This has allowed us to make a detailed reconstruction of the depositional succession as a response to sea-level changes over the past 50 kyr. In general, long gaps of nondeposition dominated the deposits, which were interrupted by ephemeral episodes of extremely rapid sedimentation. Five local groups of genetically related facies successions have been characterized. These groups belong to distinct phases defined by sea-level dynamics, morphological conditions, and sedimentation processes.

Large delta structures built up basinward as a consequence of the sea-level lowering before the Last Glacial Maximum (≥ 50 to 23 ka).

During the sea-level lowstand a widespread marshy soil horizon was formed. Sediment bypassed the North Sunda River valley and was carried out to the South China Sea. At that time there was little deposition on the shelf and high sedimentation rates in the southern South China Sea.

The initial rise of sea level caused a loss of the lowermost course of the drainage system. Thereby, sediments were funneled in front of the new river mouth.

The large river valley submerged subsequently, and an associated complex network of incised valleys appeared. The latter was also subsequently filled.

Sea level reached a critical elevation of -70 to -60 m after 13.5 ka, inasmuch as this is the general depth of the central Sunda plain. This resulted in a rapid submergence of the riverine system, which was dissected into local subordinate drainage structures. The sediment input into deeper parts of the South China Sea decreased continuously during this sea-level rise.

A completely new hydrographic system, in combination with a condensed sediment cover, was established after the drowning of the innermost parts of the shelf. This additionally caused the opening of a southern passage.

The incised valley and channels were essential for the trapping and preservation of transgressive deposits. Three major factors controlled the sedimentation process and, therefore, the depositional architecture of the shelf: sea-level changes gave the timing, the low shelf gradient was responsible for the sediment distribution, and the high riverine sediment supply ensured the continuation of the deposystem.

Offshore units have been dated in detail in the Malaysian region for the first time. Our stratigraphic results extend and provide a chronology for the offshore stratigraphy from adjacent areas.

ACKNOWLEDGMENTS

We sincerely acknowledge P.M. Grootes and his team of the Leibniz-Labor for Radiometric Dating and Isotope Research in Kiel for the professional procedure and the numerous constructive discussions. H. Florianova and many other diligent helpers are gratefully thanked. K.-J. Meyer (BGR) and H. Erlenkeuser (Leibniz-

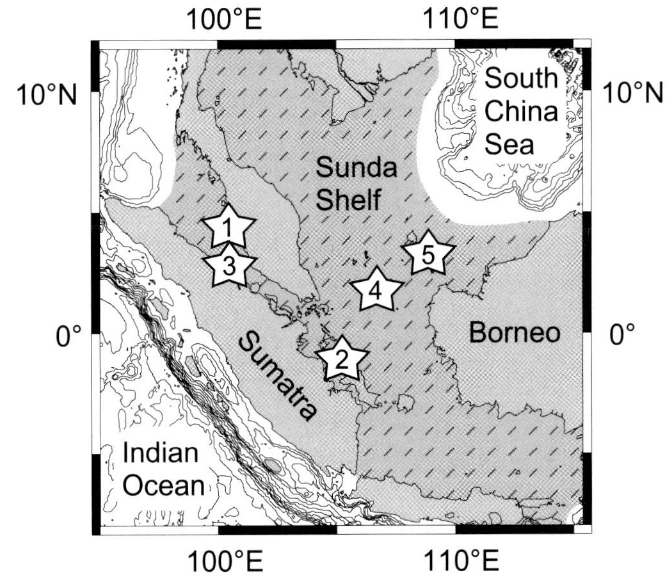

FIG. 6.—Overview map of the Sunda Shelf. Numbers in stars refer locations mentioned in Figure 7: (1) Batchelor (1979); (2) Aleva (1973); (3) Emmel and Curray (1982); (4) Evans et al. (1995); (5) this study.

Labor) have kindly analyzed pollen samples and stable isotopes. We acknowledge C.D. Woodroffe, H.D. Tjia, and the anonymous reviewers for their constructive comments on the manuscript and Y. Saito for his support. We also thank captain H. Andresen and his crew of R.V. *Sonne* for the hard work on cruise SO-115.

The project 03G0115A was financially supported by the German Ministry of Education, Science and Research, and Technology (Bundesministerium für Bildung und Forschung). TH acknowledges a stipend of the graduate school funded by the Deutsche Forschungsgemeinschaft (DFG).

REFERENCES

ALEVA, G.J.J., 1973, Aspects of the historical and physical geology of the Sunda Shelf essential to the exploration of submarine tin placers: Geologie en Mijnbouw, v. 52 (2), p. 79–91.

BARD, E., HAMELIN, B., ARNOLD, M., MONTAGGIONI, L., CABIACH, G., FAURE, G., AND ROUGERIE, F., 1996, Deglacial sea-level record from Tahiti corals and the timing of global meltwater discharge: Nature, v. 382, p. 241–244.

BATCHELOR, B.C., 1979, Discontinuously rising Late Cainozoic eustatic sea-levels, with special reference to Sundaland, Southeast Asia: Geologie en Mijnbouw, v. 58 (1), p. 1–20.

BISWAS, B., 1973, Quaternary changes in sea-level in the South China Sea: Geological Society of Malaysia, Bulletin, v. 6, p. 229–256.

BROECKER, W.S., ANDREE, M., KLAS, M., BONANI, G., WOLFLI, W., AND OESCHGER, H., 1988, New evidence from the South China Sea for an abrupt termination of the last glacial period: Nature, v. 333, p. 156–158.

CHAPMAN, V.J., 1976, Mangrove Vegetation: Vaduz, Liechtenstein, Cramer, 441 p.

DALRYMPLE, R.W., ZAITLIN, B.A., AND BOYD, R., 1992, Estuarine facies models: conceptual basis and stratigraphic implications: Journal of Sedimentary Petrology, v. 62, p. 1130–1146.

DALRYMPLE, R.W., BOYD. R., AND ZAITLIN, B.A., eds., 1994, Incised-Valley Systems: Origin and Sedimentary Sequences: SEPM, Special Publication 51, 391 p.

DASH, B.P., SHEPSTONE, C.M., DAYAL, S., GURU, S., HAINS, B.L.A., KING, G.A., AND RICKETTS, G.A., 1972, Seismic Investigation on the Northern Part of the Sunda Shelf South and East of Great Natuna Island: UN Economic Commission for Asia and the Far East, Committee for Coordination of Joint Prospecting for Mineral Sources in Asian Offshore Areas, Technical Bulletin, v. 6 (9), p. 179–196.

EMMEL, F.J., AND CURRAY, J.R., 1982, A Submerged Late Pleistocene delta and other features related to sea level changes in the Malacca Strait: Marine Geology, v. 47, p. 192–216.

	Lumut Dindings, offshore, -110 to 0 m Batchelor (1979)		**Pulau Tudjuh, offshore, -30 to -20 m** Aleva et al. (1973)		**Malacca Strait offshore, -100 to -80 m** Emmel and Curray (1982)		**cable route 2° N offshore, -86 m** Evans et al. (1995)		**North Sunda River valley offshore, -126 to -70 m** this work	
	unit	features	unit	features	unit	features	unit	features	unit	features
Holocene	YSC	neritic mud, beach sands	YSC	soft mud, blanketlike alluvial fans	Holocene	flat-lying, thin reworked material	YSC	cover	HC	marine mud, thin starved cover
(11 ka)	*disconf.* YA				*plane erosional surf.*		*ravinement surf.*		*max. flooding surface*	
Late Pleistocene	AC	floodplain, meander, peats, valleys	(*soil form.*)	soil, mottled, small creeks		peats, valleys, bypassing	AC	channel fills	TS	marine to terrestrial
LGM (21 ka)	*disconformity*	(*soil form.*)	*abrasion surf.*	truncation	delta foresets	prograding	?		*exposed surface*	soil formation, bypassing
Eem	OMU TU		AC	peat layers, stratification	*transgr. plane = OS*	truncation, valleys			DC ?	prograding
Middle Pleistocene	TU	braided stream, laterised, peat, erosional	(*basal disconf.*)	valleys,60 to 100 m below sea bottem			? ?			
	disconformity	(*soil form.*)								
Early Pleistocene	OSC	alluvial plain, gravelly, peats								
Tertiary	Sundaland Regolith	red beds	OSC	bedding planes			OSC			

FIG. 7.—Stratigraphic comparison for the Malaysian region. For locations see Figure 6. AC, alluvial complex; AS, abrasion surface; DC, delta complex; HC, Holocene cover; OMU, older marine unit; OS, old sea-floor erosional surface; OSC, older sedimentary cover; TS, transgressive system; TU, transitional unit; YA, younger alluvium; YSC, younger sedimentary cover. Hatched: problematic stratigraphic interpretation.

EVANS, C.D.R., BRETT, C.P., JAMES, J.W.C., AND HOLMES, R., 1995, Shallow seismic reflection profiles from the waters of East and Southeast Asia: an interpretation manual and atlas: British Geological Survey, Technical Report WC/94/60, p. 66–67.

FAIRBANKS, R.G., 1989, A 17,000-year glacio-eustatic sea level record: influence of glacial melting rates on the Younger Dryas event and deep-ocean circulation: Nature, v. 342, p. 637–642.

FAIRBANKS, R.G., 1990, The age and origin of the "Younger Dryas Climate Event" in Greenland ice cores: Paleoceanography, v. 5, p. 937–948.

GASTALDO, R.A., ALLEN, G.P., AND HUC, A.Y., 1993, Detrital peat formation in the Tropical Mahakam River Delta, Kalimantan, eastern Borneo: Sedimentation, Plant Composition, and Geochemistry, *in* Cobb, J.A., and Cecil, C.B., eds., Modern and Ancient Coal-Forming Environments: Geological Society of America, Special Paper 286. p. 107–118.

HAILE, N.S., 1969, Quaternary deposits and geomorphology of the Sunda Shelf off Malaysian shores: Études sur le Quaternaire dans le Monde. VIII. Congrès INQUA, v. I, p. 161–164.

HANEBUTH, T., STATTEGGER, K., AND GROOTES, P.M., 2000, Rapid flooding of the Sunda Shelf—A Late-Glacial Sea-Level Record: Science, v. 288, p. 1033–1035.

HAFT, C., 1999, Spätquartäre Sedimentation und Meeresspiegelschwankungen im südlichen Südchinesischen Meer abgeleitet aus air gun-Daten: unpublished diploma thesis, University of Hamburg, 64 p.

KAMALUDIN, B.H., NAKAMURA, T., PRICE, D.M., WOODROFFE, AND C.D., FUJII, S., 1993, Radiocarbon and thermoluminescence dating of the Old Alluvium from a coastal site in Perak, Malaysia: Sedimentary Geology, v. 83, p. 199–210.

KIENAST, M., 1996, Geschichte des Flußeintrags vom Sunda-Shelf: Abbild in hemipelagischen Sedimenten aus dem Südchinesichen Meer: unpublished diploma thesis, University of Kiel, 53 p.

KÖGLER, F.C., ANDRESEN, H., AND LINCKE, C., 1982, Final report SO 24, RF Reedereigemeinschaft Forschungsschiffahrt GmbH.

KÖHLER, B., 1985, Sedimentologie und Mineralogie der Oberflächensedimente im mittleren Sunda-Schelf (ein W–E-Profil zwischen Malaysia und Borneo): unpublished diploma thesis, University of Kiel, 75 p.

MOLENGRAAFF, G.A.F., AND WEBER, M., 1921, On the relation between the Pleistocene glacial period and the origin of the Sunda Sea (Java and South China Sea), and its influence on the distribution of coral reefs and on the land- and freshwater fauna: Proceedings of the Royal Academy, Amsterdam, v. 23 (1), p. 395–439.

NADEAU, M.-J., SCHLEICHER, M., GROOTES, P.M., ERLENKEUSER, H., GOTTDANG, A., MOUS, D.J.W., SARNTHEIN, J.M., AND WILLKOMM, H., 1997, The Leibniz-Labor AMS facility at the Christian-Albrechts University, Kiel, Germany: Nuclear Instruments and Methods B, v. 123, p. 22–30.

PELEJERO, C., KIENAST, M., WANG, L., AND GRIMALT, J.O., 1999, The flooding of Sundaland during the last deglaciation: Imprints in hemipelagic sediments from the southern South China Sea: Earth and Planetary Science Letters, v. 171, p. 661–671.

POSAMENTIER, H.W., AND VAIL, P.R., 1988, Eustatic controls on clastic deposition II—sequence and systems tract models, *in* Wilgus, C.K., Hastings, B.S., Kendall, C.G.St.C., Posamentier, H.W., Ross, C.A., and Van Wagoner, J.C., eds., Sea-Level Changes: An Integrated Approach: SEPM, Special Publication 42, p. 125–154.

PRAMOJANEE, P., HASTINGS, P., LIENGSAKUL, M., AND ENGAKUL, V., 1986, The Holocene transgression in Peninsula Thailand: GEOSEA V Proceedings v. 1, Geological Society of Malaysia, Bulletin, v. 19, p. 551–564.

RICHTER, A., 1998, Die Auswirkungen des spätpleistozänen Meeresspiegelanstiegs auf Sedimentationsprozesse im Paläo-Mekong Delta: unpublished diploma thesis, University of Kiel, 133 p.

SCHIMANSKI, A., 1999, Diversität und Entwicklung von Faziesräumen auf dem Sundaschelf im Verlauf der postpleistozänen Transgression: unpublished diploma thesis, University of Kiel, 65 p.

SCHÖNFELD, J., AND KUDRASS, H.R., 1993, Hemipelagic sediment accumulation rates in the South China Sea related to Late Quaternary sea-level changes: Quaternary Research, v. 40, p. 368–379.

STANLEY, D. J., AND WARNE, A. G., 1994, Worldwide initiation of Holocene marine deltas by deceleration of sea-level rise: Science, v. 265, p. 223–225.

STATTEGGER, K., KUHNT, W., WONG, H.K., AND OTHERS, 1997, Cruise Report SONNE 115 SUNDAFLUT: University of Kiel, Geologisch-Paläontologisches Institut, Berichte–Reports, v. 86, 211 p.

STUIVER, M., REIMER, P.J., BARD, E., BECK, J.W., BURR, G.S., HUGHEN, K.A., KROMER, B., MCCORMAC, G., VAN DER PLICHT, J., AND SPURK, M., 1998, INTCAL98 radiocarbon age calibration, 24,000–0 cal BP: Radiocarbon, v. 40, p. 1041–1084.

TJIA, H.D., 1980, The Sunda Shelf, SE Asia: Zeitschrift für Geomorphologie, N.F., v. 24, p. 405–427.

TJIA, H.D., AND LIEW, K.K., 1996, Changes in tectonic stress field in northern Sunda Shelf basin, *in* Hall, R., and Blundell, D., eds., Tectonic Evolution of Southeast Asia: Geological Society of London, Special Publication 106, p. 291–306.

UMITSU, M., 1993, Late Quaternary sedimentary environments and landforms in the Ganges Delta: Sedimentary Geology, v. 83, p. 177–186.

VAN BEMMELEN, R.W., 1949, The Geology of Indonesia, v. 1A: The Hague, Martinus Nijhoff, 723 p.

WANG, L., SARNTHEIN, M., ERLENKEUSER, H., GRIMALT, J., GROOTES, P., HEILIG, S., IVANOVA, E., KIENAST, M., PELEJERO, C., AND PFLAUMANN, U., 1999, East Asian monsoon climate during the Late Pleistocene: high-resolution sediment records from the South China Sea: Marine Geology, v. 156, p. 245–284.

WANG, P., 1999, Response of Western Pacific marginal seas to glacial cycles: paleoceanographic and sedimentological features: Marine Geology, v. 156, p. 5–40.

WANG, P., WANG, I., AND BIAN, Y., 1994, The Last Deglaciation in the South China Sea: 4th International Conference on Paleoceanography (ICP IV), Kiel, p. 294.

WANG, P., WANG, L., BIAN, Y., AND JIAN, Z., 1995, Late Quarternary Paleoceanography of the South China Sea: Surface circulation and carbonate cycles: Marine Geology, v. 127, p. 145–165.

WONG, H.K., HAFT, C., PAULSEN, A.-M., LÜDMANN, T., HÜBSCHER, C., AND GENG, J.G., 1999, Late Quaternary Sedimentation and Sea Level Fluctuations on the Sunda Shelf, southern South China Sea, *in* Saito, Y., Ikehara, K., and Katayama, H., eds., Land–Sea Link: Proceedings on an international workshop on sediment transport and storage in coastal sea-ocean system, Tsukuba, Japan, p. 205–211.

WOODROFFE, C.D., AND MULRENNAN, M.E., 1993, Geomorphology of the Lower Mary Plains, Northern Territory: Australian National University, North Australia Research Unit, Darwin, 152 p.

WYRTKI, K., 1957, Die Zirkulation an der Oberfläche der südostasiatischen Gewässer: Deutsche Hydrographische Zeitschrift, v. 1, p. 1–13.

QUATERNARY SEDIMENTATION IN THE MOLENGRAAFF PALEO-DELTA, NORTHERN SUNDA SHELF (SOUTHERN SOUTH CHINA SEA)

HOW KIN WONG, THOMAS LÜDMANN, CHRISTIANE HAFT, AND ALKE-MARIT PAULSEN
Institute of Biogeochemistry and Marine Chemistry, Center of Marine and Climate Research, Hamburg University, Bundesstrasse 55, D-20146 Hamburg, Germany
CHRISTIAN HÜBSCHER
Institute of Geophysics, Center of Marine and Climate Research, Hamburg University, Bundesstrasse 55, D-20146 Hamburg, Germany
AND
JIANHUA GENG
Department of Marine Geology and Geophysics, Tongji University, Shanghai, Peoples Republic of China

ABSTRACT: The seven seismic units recognized in the Molengraaff paleo-delta on the Sunda Shelf underlying the postglacial unit can be interpreted as prograding shelf-margin lowstand wedges deposited during forced regressions. Speculative age estimates of these units were made by correlating the interpreted lowstand wedges with the SPECMAP $\delta^{18}O$ curve. This correlation suggests that the oldest unit reaches back to 570 ka, that the mean sedimentation rates during glacial times are comparable to modern prodelta rates, and that the tectonic subsidence rate of the outermost shelf depressions during the latest Pleistocene was of the order of 27 cm/kyr. Subsidence (which was differential) also occurred in postglacial times. The rate at the shelf edge is of the order of 2.5 m/kyr, and the differential displacements are taken up by growth-fault activity.

The outer Sunda Shelf was the delta plain of the Molengraaff River system during the last glacial, during which major fluvial entrenchment did not take place because of the low gradient. Sand ridges formed by transgressive reworking of Late Pleistocene material mark the modern water-depth range of 135–145 m. Mass wasting and fan deposition are important processes on the continental slope during regressions and lowstands.

INTRODUCTION

With an area of 1.85 million km^2 the Sunda shelf is one of the largest continental shelves on earth (Tjia, 1980). During Pleistocene sea-level lowstands, it formed an extensive subaerial coastal plain dissected by river systems that formed large delta-fan complexes along the southwestern margin of the South China Sea. The most important of these river systems were the Molengraaff (or North Sunda) River and the paleo–Mekong River (Molengraaff and Weber, 1920; Molengraaff, 1922; see Fig. 1). During interglacial times and the postglacial transgression, the Sunda shelf was progressively flooded.

The geotectonic and structural framework of the South China Sea in general and of the Sunda Shelf in particular is relatively well known (Ben-Avraham and Emery, 1973; Lee and Lawver, 1994; Parke et al., 1971; Ru and Pigott, 1986; Schlüter et al., 1996; Taylor and Hayes, 1980, 1983; Xia and Zhou, 1993; Xia et al., 1994). The crustal structure has also been studied in detail (Hamilton, 1979; Houtz and Hayes, 1984; Ludwig et al., 1979). In contrast, systematic investigations of the architecture and characteristics of the Quaternary sediments and the processes controlling their distribution are limited. They include nearshore studies like that of Batchelor (1979a) and Emmel and Curray (1982), regional overviews exemplified by Evans et al. (1995), and reconnaissance surveys like that of Yuan et al. (1993), in which several phases of Quaternary delta formation accompanied by type 1 sequences and parasequences were described. Studies on sea-level fluctuations in this area are also not plentiful (Batchelor, 1979b; Geyl et al., 1979; Hantoro, 1997; Tija, 1996).

In 1997, we obtained new seismic data from the northern Sunda Shelf, namely from the Molengraaff paleo-delta, during cruise 115 of the R/V SONNE. Our study area is situated within the western Zengmu (or Sarawak) Basin between the Natuna Islands on the middle shelf and the South China Sea Basin beyond the continental slope (Fig. 1). Our data set consists of approximately 3,000 km of high-resolution air-gun profiles and about 4,000 km of very high-resolution PARASOUND profiles. In this paper, we report on these new data and their implications for Late Pleistocene to Holocene sedimentation and its controlling factors in this area.

GEOLOGICAL BACKGROUND

Structural Units and Quaternary Tectonics of the Zengmu Basin

The Zengmu Basin is characterized by three fault systems striking NW, NE, and N, respectively (Jiang et al., 1994; Zhong et al., 1996; Xia et al., 1994). Among these systems, that which trends NW is by far the best developed (Fig. 2). Except for the Shelf Margin Fault and the Zengmu II South Boundary Fault, which continued their activity into the Quaternary (Jiang et al., 1994 and our data), all faults of the NW–SE system were active from the Mesozoic to the Miocene (Asian Council on Petroleum (ASCOPE), 1981; Coordinating Committee for Coastal and Offshore Geoscience Programmes in East and Southeast Asia (CCOP)–Intergovernmental Oceanographic Commission (IOC), 1980). These faults divide the study area into a number of depressions and uplifts and controlled the structural and sedimentary evolution. Thus, while the Central Uplift and the structural units farther to the southwest were tectonically stable during the Quaternary, the Zengmu I and II Depressions have undergone a significant subsidence that was accommodated by vertical displacements along the boundary fault zones (Fig. 2). From the data given in Zhong et al. (1996), the subsidence rate averaged over the Quaternary can be estimated at 21 and 13 cm/kyr for the Zengmu I and II Depressions, respectively.

Tropical Deltas of Southeast Asia—Sedimentology, Stratigraphy, and Petroleum Geology
SEPM Special Publication No. 76, Copyright © 2003
SEPM (Society for Sedimentary Geology), ISBN 1-56576-086-7, p. 201–216.

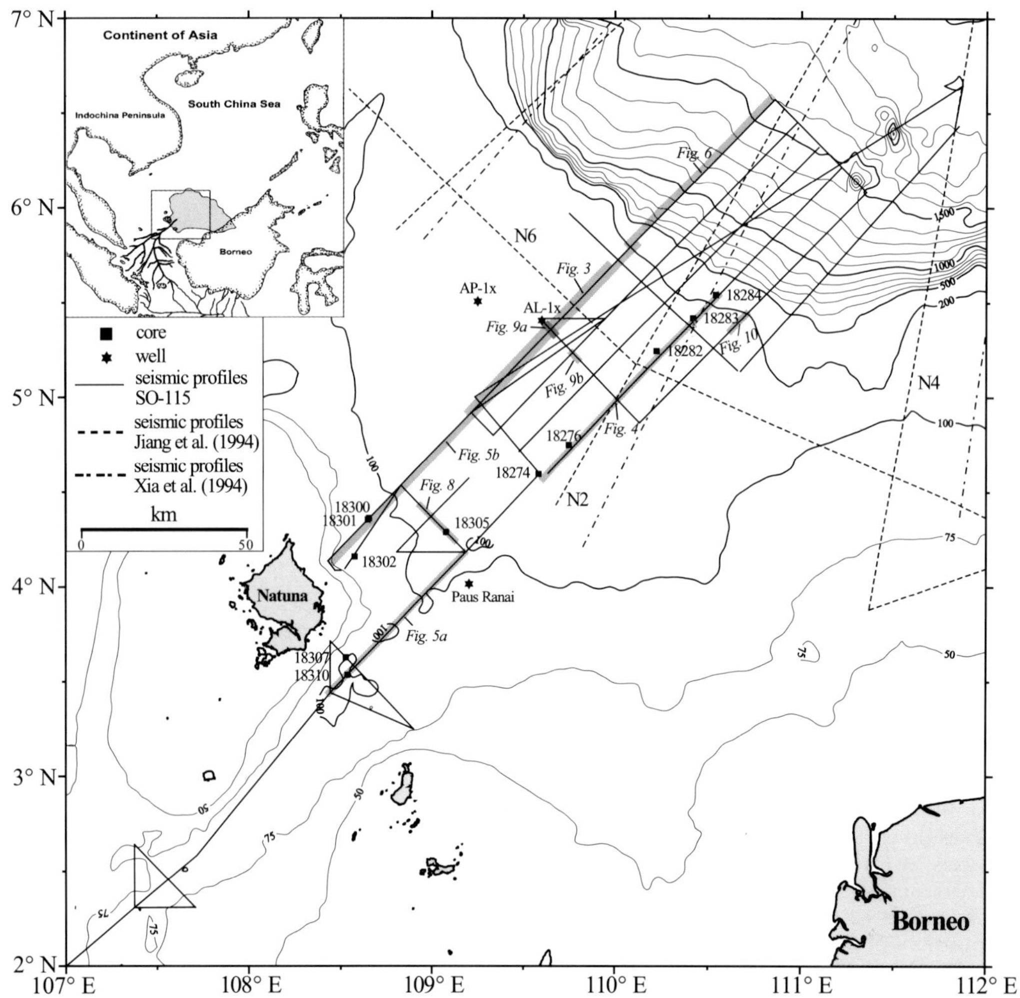

FIG. 1.—Locations of air-gun and PARASOUND profiles on the northern Sunda Shelf, cruise 115 of the R/V SONNE. Thick lines with numbers give the profile sections that are presented in later figures. The bathymetry is based on Hydrosweep data from our cruise, on the GEODAS data set, and on a depth chart of the South China Sea Institute of Oceanology (SCSIO, Academia Sinica, 1984). The area shown in this figure is marked in the insert, in which the glacial Molengraaff River and the Zengmu Basin (in gray) are shown.

UPPER PLEISTOCENE TO HOLOCENE SEISMIC STRATIGRAPHY OF THE MOLENGRAAFF PALEO-DELTA AREA

Upper Pleistocene to Holocene High-Frequency Seismic Units

Our air-gun and PARASOUND profiles show that sediments in the Molengraaff paleo-delta area can be divided into 8 seismic units, designated 1 to 8 from old to young (Figs. 3–5). Units 1 to 7 are generally wedge-shaped, although unit 4 is lenticular. Cores indicate that they consist of organic-rich, sandy, silty clays interpretable as deltaic and shoreline deposits. They commonly increase in thickness basinward, and terminate by toplap at the upper boundary and by downlap or are conformal at the lower boundary. Because of these terminations and their clinoformal internal reflection pattern, these units are considered to be progradational deltaic wedges of the Molengraaff delta deposited under the control of fourth-order sea-level cycles. Onlaps seen occasionally within a unit are interpreted to be due to autocyclic delta-lobe switching or minor sea-level fluctuations. Unit 8, in contrast, is a thin, discontinuous layer with a complex chaotic reflection pattern. Cores show that it comprises largely clayey and silty sands (locally in the form of NNE–SSW oriented sand ridges).

The shelf-margin progradational lowstand wedges (units 1–7) are attributed to forced regressions (Posamentier et al., 1992). They formed as part of the Molengraaff delta during sea-level lowstands in positions detached from their previous highstand shorelines. They are characterized in our profiles by basal unconformities and upper boundaries in the form of ravinement surfaces cut by erosional shoreface retreat. Lateral stacking of these lowstand wedges results from repeated high-frequency sea-level cycles (Piper and Aksu, 1992; Tesson et al., 1993).

Because the gradient of the outer Sunda Shelf is exceedingly low (1:2,000 or 0.03°), a small sea-level rise leads to a rapid and extensive transgression, so that the retrogradational transgressive systems tract (TST) is generally thin. During a sea-level highstand, the shoreline is at a position far removed from the shelf edge. The sedimentation rate on the outer shelf is therefore very low, and a condensed section results. Thus, the TST and the HST (highstand systems tract) are only occasionally resolved

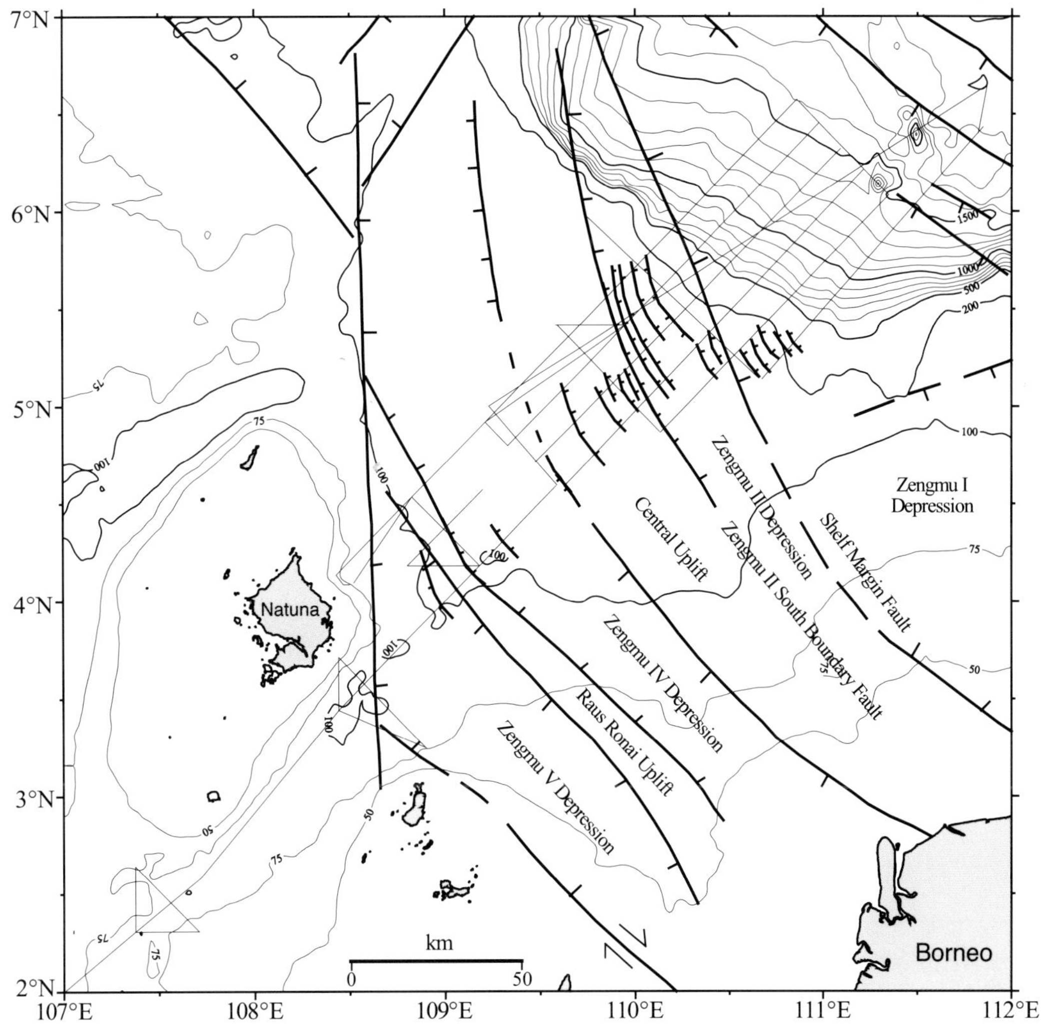

FIG. 2.—Map showing fault systems in the study area. These faults divide the area into a number of depressions and uplifts. The large faults are modified after Jiang et al. (1994). They have been verified by our data where possible. The smaller faults are from our profiles.

seismically in our very high-resolution PARASOUND profiles. The SPECMAP curve suggests that in general the rate of a eustatic sea-level rise is much higher than that of a fall. This implies that the lowstand systems tract (LST), which encompasses a sea-level fall, and the subsequent sea-level lowstand, can form over a considerably longer time interval in comparison with the HST or TST and hence is better developed. In addition, our profiles, although already up to 550 km in length, cover only about the distal half of the shelf. Deposits of the TST and HST not seen or not resolved in our data could have developed landwards of our profiles.

AMS ^{14}C Ages and Sediment Facies from Cores

Six cores were retrieved on or near the profiles 32 and 16 from water depths ranging from 109 to 226 m (Fig. 1). They penetrate a thin Holocene layer of olive gray silty or clayey sands with shells and shell fragments. Except for core 18305, closest to Natuna Island, this layer increases from about 14 cm to just under 1 m in thickness in the basinward direction (Fig. 4; Stattegger et al., 1997). Core 18284, on the upper slope at 226 m water depth, sampled a continuous Late Pleistocene–Holocene sequence (Stephan Steinke, personal communication, 2000). In contrast, AMS ^{14}C ages show that all the shelf stations are characterized by a hiatus below the mid-Holocene (Figs. 4, 5). At station 18282, on the outer shelf, the age of the sediments directly below the hiatus is 32.6 ka (Fig. 4). From the proximal outer shelf to the middle shelf, the corresponding ages are 17.7 ka at station 18276, mixed ages within the range of 14.6–16.7 ka at station 18274, and 15.3 ka at station 18305. These sediments were therefore laid down during the early transgression after the last glacial maximum (LGM), and they show a decreasing albeit indistinct age trend landwards. We suggest that the hiatus on the distal outer shelf is due to erosion on a transgressing shoreface, which resulted in a ravinement surface barely covered by Holocene sediments today. The proximal outer shelf and the middle shelf, however, are characterized not only by an erosional surface produced when they were subaerially exposed during the last glacial period (the interface between wedge 7 and sequence 8; Fig. 4) but also by post-LGM transgressive sediments. Because these areas were progressively flooded landwards, a landward-increasing age trend is expected.

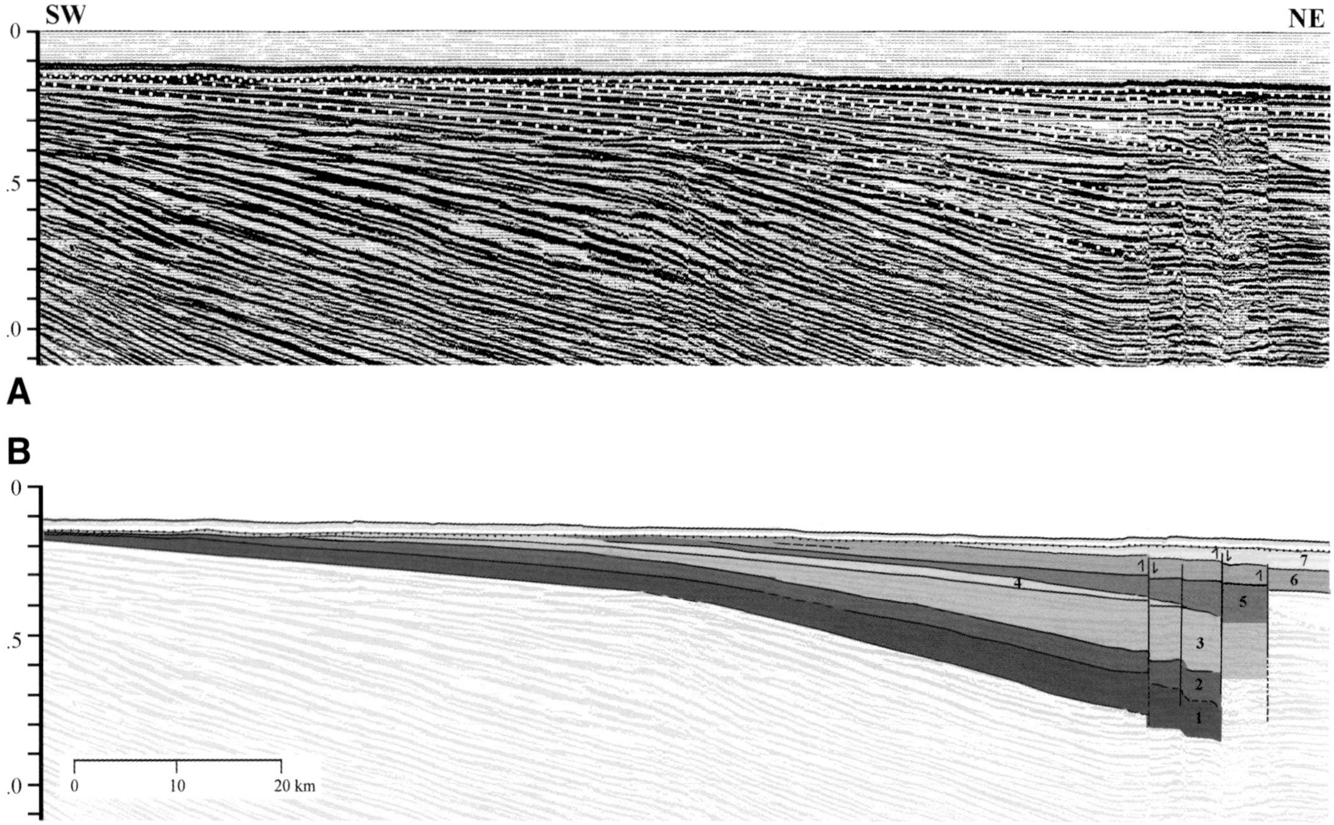

FIG. 3.—Part of air-gun profile 23, **A)** original and **B)** interpreted, showing the eight seismic units identified. See Fig. 1 for profile location.

Pedogenesis observed in cores retrieved from the middle Sunda Shelf suggests that this area was a land surface during the sea-level stillstand of the LGM (Hanebuth et al., 2000; Fig. 5). The depositional environment ranged from delta front, bay lagoon, delta plain, mudflat and mangrove swamp within the modern depth range of 126 to 70 m (Fig. 5; *Science* Supplementary Material to Hanebuth et al., 2000). Sedimentation within the period 21–13 ka was characterized by rapid lateral and vertical facies changes as well as sedimentation pulses and hiatuses. The uppermost reworked and mixed marine sands have highly varying AMS ^{14}C dates (Hanebuth et al., 2000). The age of 39.2 ka some 3 m below the erosional truncation of the LGM and the sediment facies of core 18300 suggest the existence of a pre-LGM delta at this location (Fig. 5). This age is older than that of 32.6 ka for the corresponding sediments of core 18282 on the outer shelf and is consistent with the fact that they are at the top of a truncated progradational wedge, so that their age should decrease basinward.

SEDIMENTATION PROCESSES AND TECTONIC SUBSIDENCE

Sedimentation Processes and Subsidence from the LGM to the Holocene

Subsidence.—

The observed seaward progradation of successively younger lowstand deltas and the overall progradation of the shelf edge suggest that during Pleistocene glacial times the rate of eustatic sea-level fall must have been higher than the subsidence rate at the shelf edge. This led to regression of the coastline to a location beyond the shelf edge, subaerial exposure of the Sunda Shelf, and formation of a type 1 sequence boundary. The following three pieces of evidence support this scenario. (1) The generally undulating erosional truncation marking Late Pleistocene subaerial erosion and the postglacial shoreface ravinement that extends to the shelf edge (see profile 16, Fig. 4). Although it is conceivable that part of this surface, namely that which was below sea level during the LGM (–125 m; Fairbanks, 1989; Schönfeld and Kudrass, 1993), was formed as a result of bottom-current erosion, its large extent (a strip of about 100 km width parallel to the shelf edge) makes this interpretation unlikely. (2) Mass-transport sediments and in particular lowstand fans with downlap terminations occur on the continental slope (Fig. 6). These fans attest to downcutting of rivers to a position beyond the shelf edge during a sea-level fall, when most of the fluvial sediment load as well as the material derived from the incised valleys were delivered to river mouths at or near the shelf edge, giving rise to a fan complex on the continental slope (Posamentier et al., 1988; Posamentier et al., 1991). (3) Sediment progradation and formation of shelf-margin sedimentary wedges characterized by a divergent reflection pattern during a sea-level lowstand.

The present-day shelf edge lies at a water depth of around 175 m (Figs. 4, 5). Assuming that the basinward limit of the postglacial ravinement surface (i.e., the shelf edge) marks the last-glacial sea level (-125 m), the shelf edge must have subsided about 50 m in the past 20 kyr. This yields an average subsidence rate for the shelf

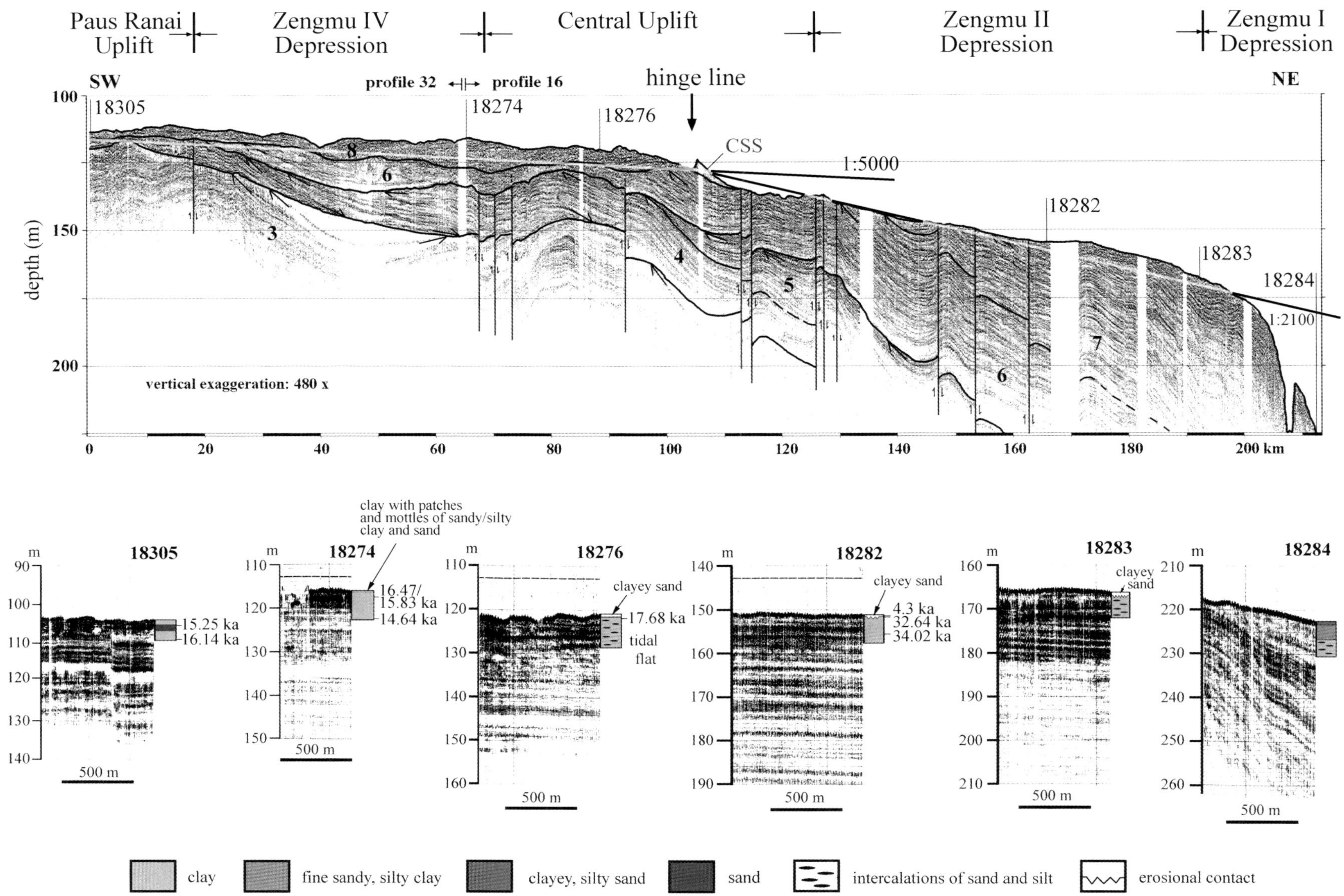

FIG. 4.—Part of PARASOUND profiles 32/16 showing prograding lowstand wedges on the outer shelf (seismic units 3–7) covered locally by post-LGM transgressive and highstand sediments (unit 8). The simplified lithology, AMS ^{14}C ages (after *Science* Supplementary Material to Hanebuth et al., 2000) and the upper part of PARASOUND profiles at coring stations on or near profiles 32/16 are also given. See Fig. 1 for profile and core station locations.

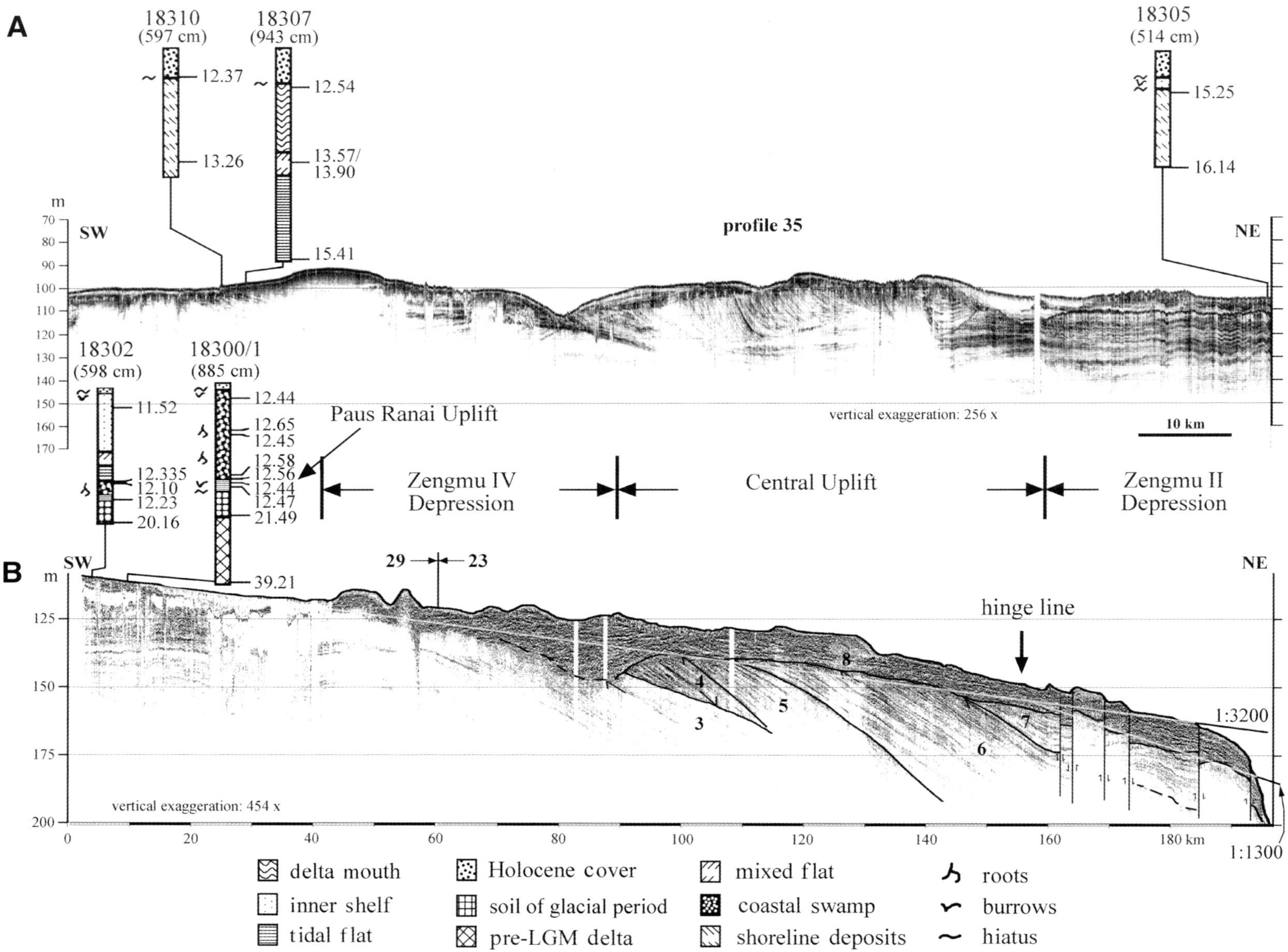

FIG. 5.—Part of **A)** PARASOUND profile 35 showing valleys incised into regressive deposits of the last glacial, **B)** and profiles 29/23 with prograding lowstand wedges on the outer shelf (seismic units 3–7) covered by post-LGM transgressive and highstand sediments (unit 8). AMS ^{14}C ages and the interpreted depositional environments of cores from stations on or near these profiles are also given (after *Science* Supplementary Material to Hanebuth et al., 2000). See Fig. 1 for profile and core station locations.

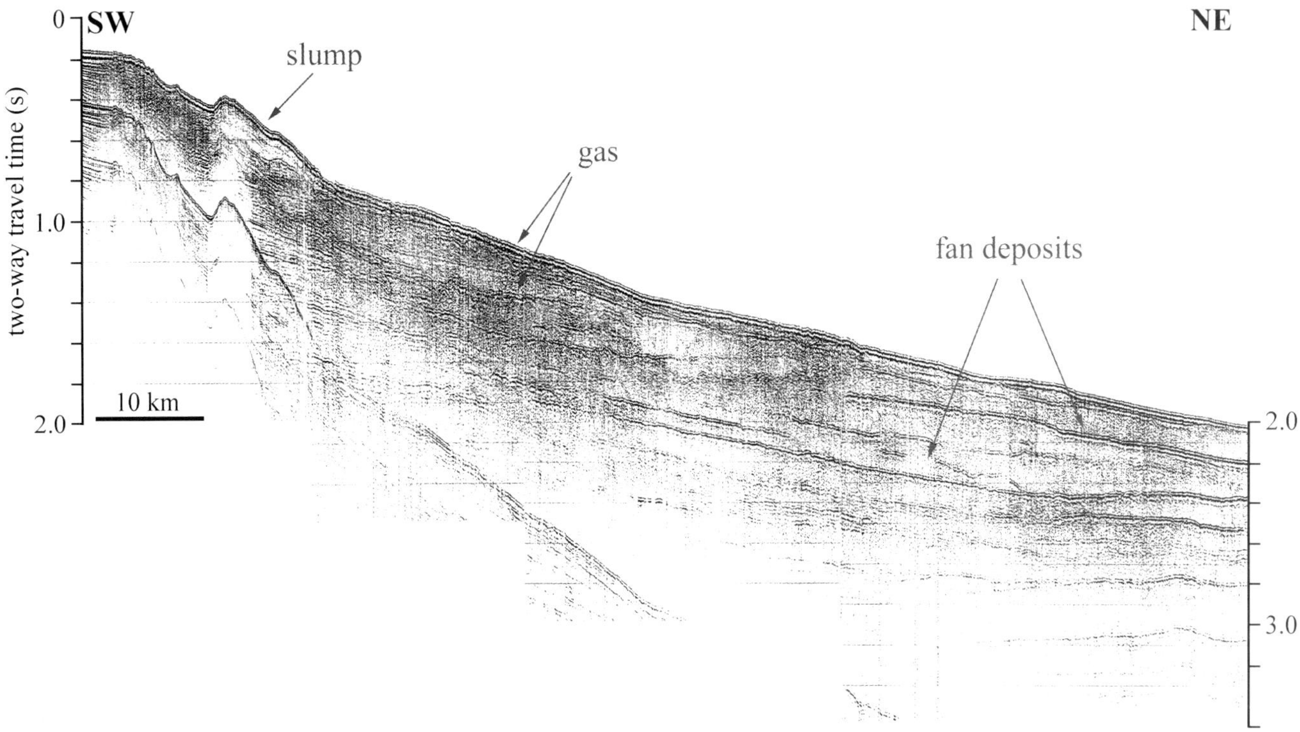

FIG. 6.—Part of air-gun profile 23 across the continental slope showing slump deposits and sediment fans. The acoustic masking is due to gas in the sediments.

edge of the order of 2.5 m / kyr, a rate that is an order of magnitude higher than the Quaternary average for the Zengmu I and II depressions but is applicable only to the shelf edge for the past 20 kyr. Despite the fact that this rate is comparable to the rates of subsidence attributed to sediment compaction in the modern Mississippi delta (30–60 cm / 100 yr; Reading and Collinson, 1996), the numerous steep, normal faults seen on our PARASOUND and air gun profiles demonstrate clearly that Holocene subsidence at the margin of the Molengraaff paleo-delta has an important tectonic component.

The landward boundaries of the Zengmu I and II Depressions mapped by Jiang et al. (1994) are plotted in Figure 2. These boundaries are marked in our profiles by (NW–SE-trending; Figs. 4 and 5) fault zones. The faults found beneath the Central Uplift do not reach the seafloor and are no longer active, in contrast to those within the Zengmu II Depression. The hinge line separating the tectonically stable area from the area of subsidence does not coincide exactly with the boundary between the Zengmu II Depression and the Central Uplift. On profile 16, it lies ca. 20 km landwards at approximately the position of the interpreted cemented sand structure (CSS). This position is marked by a sharp change in gradient of the erosional unconformity between units 7 and 8 (see lines labeled with gradient, Fig. 4; note vertical exaggeration of 480). The CSS lies at a modern water depth of about 125 m. During the LGM, the stretch from this structure to the modern shelf edge must have been horizontal at or very near sea level. The lower delta plain and its truncated surface is considerably less undulating than its subaerial landward continuation (Fig. 4). Since the LGM, subsidence with a progressively higher rate from the hinge line basinward produced the seafloor gradient of today, with the shelf edge lying some 50 m deeper than the CSS. For the profiles of Fig. 4, Holocene transgressive deposits occur only landward of the CSS, namely in the area that was mostly subaerial during the last glacial. In contrast, for the northwesternmost dip profile of Fig. 5B, postglacial sediments are found all the way to the shelf edge, suggesting that this profile lies closer to the center of the Pleistocene Molengraaff delta. The hinge point on this profile is about 8 km landward of the fault zone that separates the Zengmu II Depression from the Central Uplift, and the gradient of the erosional unconformity both landward (1:5000 vs. 1:3200) and basinward of the hinge line (1:2100 vs. 1:1300) is here steeper than on profile 32/16 (Fig. 4). We attribute this steepening to a probable increase in the Holocene subsidence rate in the direction of the center of the paleo-delta.

Sedimentation Processes.—

The study area can be divided into a number of sedimentary provinces (Fig. 7). Subprovince 1a on the shelf was just landward of the Molengraaff paleo-delta system during the LGM. It is characterized by a complex of channels of various ages and their associated floodplains. Subprovince 1b is interpreted to lie within the delta plain of the last glacial. Most prominent is the large (up to 8 km wide), slightly meandering channel of the last active delta (Figs. 7, 8). It has a number of distributaries that rapidly lose their identity in the basinward direction, suggesting that major fluvial entrenchment did not occur on the emergent outer shelf during the last lowstand because of the low shelf gradient (Miall, 1991). The age of the channel fill is postglacial, as suggested by all but one of the radiometric ages of the fill samples (Stattegger et al., 1998). In the immediate vicinity of this last active channel system is an older system that flowed northward out of our study area (Figs. 7, 8).

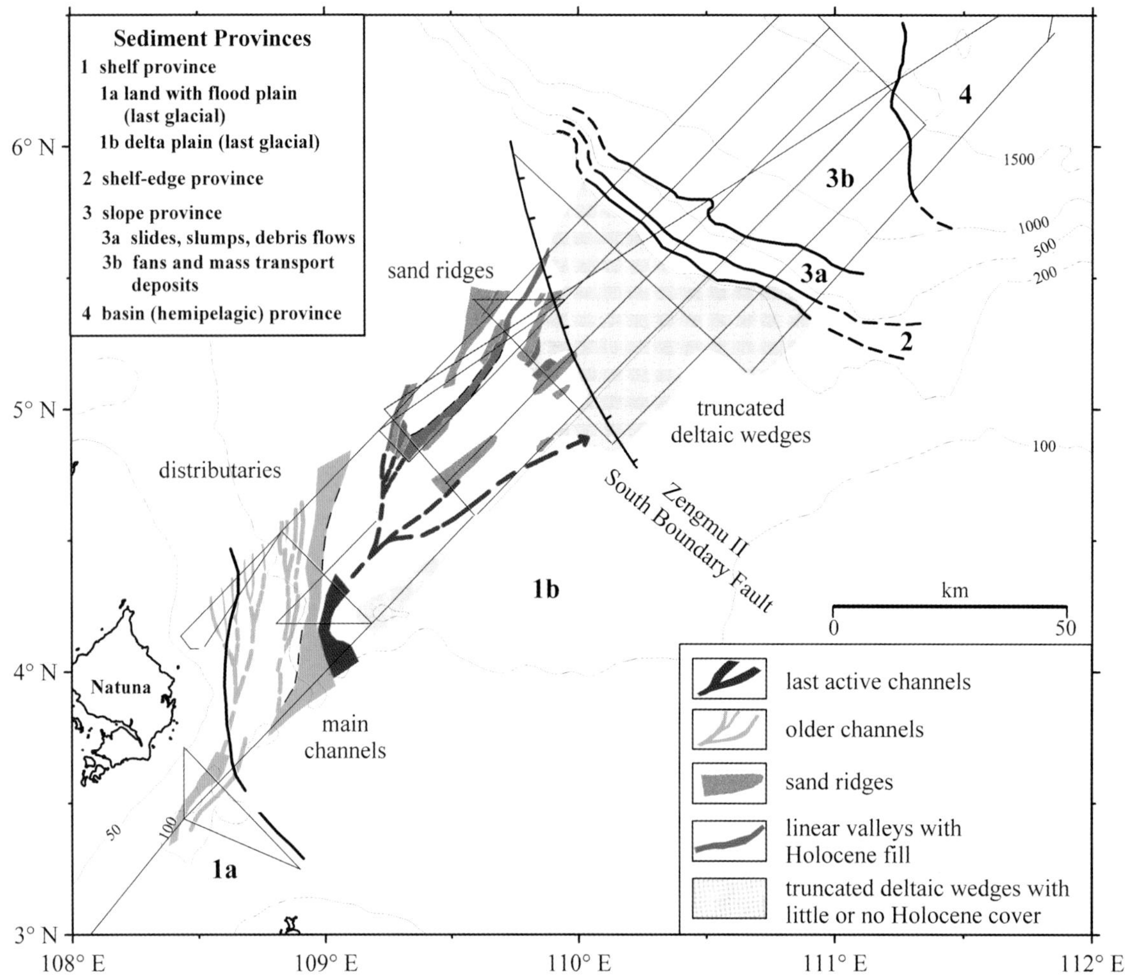

FIG. 7.—Sedimentary provinces (1 to 4) on the northern Sunda Shelf, slope and basin, as well as the positions of channels and sand ridges mapped in the Molengraaff paleo-delta. See text for explanation.

Within the modern depth range of 135–145 m and striking at a high angle to the shelf edge are surficial sand ridges, about 15 m high, up to 4 km wide, and ca. 50 km long (Fig. 7). They are highly asymmetric, with a "steep" (2.6°) slope to the WNW and a gentle gradient (0.2°) to the ESE. Some of these sand ridges consist of low-angle clinoforms downlapping a flat surface (Fig. 9a). They can be interpreted as transgressive deposits formed by upbuilding upon a postglacial transgressive surface during the Holocene sea-level rise (Stride et al., 1982). In contrast, other elongated, asymmetric sand bodies are capped by an irregular surface and appear to be erosional in origin (Fig. 9B). They consist of Late Pleistocene regressive sands reworked by transgressive processes after the LGM (Berné et al., 1994; Berné et al., 1998; Yang and Sun, 1988). A third variant is developed over a channel incision (not illustrated). Here, initial channel incision is succeeded by infill and sedimentary upbuilding. The geometry of these sand ridges indicates that they were formed under the influence of tidal currents. The water depth d at the time of formation can therefore be estimated from their average separation (λ, in m, here ca. 10,000) using the relationship $\lambda = 250d$ (Huthnance, 1982a, 1982b). This leads to $d = 40$ m. Thus, these sand bodies, which lie today at an average depth of 145 m, were probably formed or reworked at a sea level of about -105 m (ca. 15 ka; Hanebuth et al., 2000). Whether they represent reworked barrier islands or not we cannot tell; barrier-island remnants have not been observed on any of our seismic profiles.

The slope and basin provinces (2–4) are characterized by mass wasting and fan deposition during sea-level lowstands and by hemipelagic sedimentation during highstands.

Sedimentation Processes and Tectonic Subsidence Before the Last Glacial Maximum

Paleo-Shoreline Indicators and Their Corrected Modern Depths.—

We identified two types of features near the modern shelf edge that we interpret as approximate paleo-shoreline indicators. The first are aggradational shelf-margin carbonate deposits (Fig. 10; Sarg, 1988). They are characterized by high-amplitude reflectors, and a slope front with significant erosion, as well as an onlapping debris apron. These banks or margins could have formed when sea level fell, in which case they should have grown basinward while the older landward part was progressively subaerially exposed. Alternatively, they could have grown as sea level rose, to become progressively drowned under waters too deep for further growth. Although it is not clear which of these cases apply, the distal limit of the bank or margin would in both cases

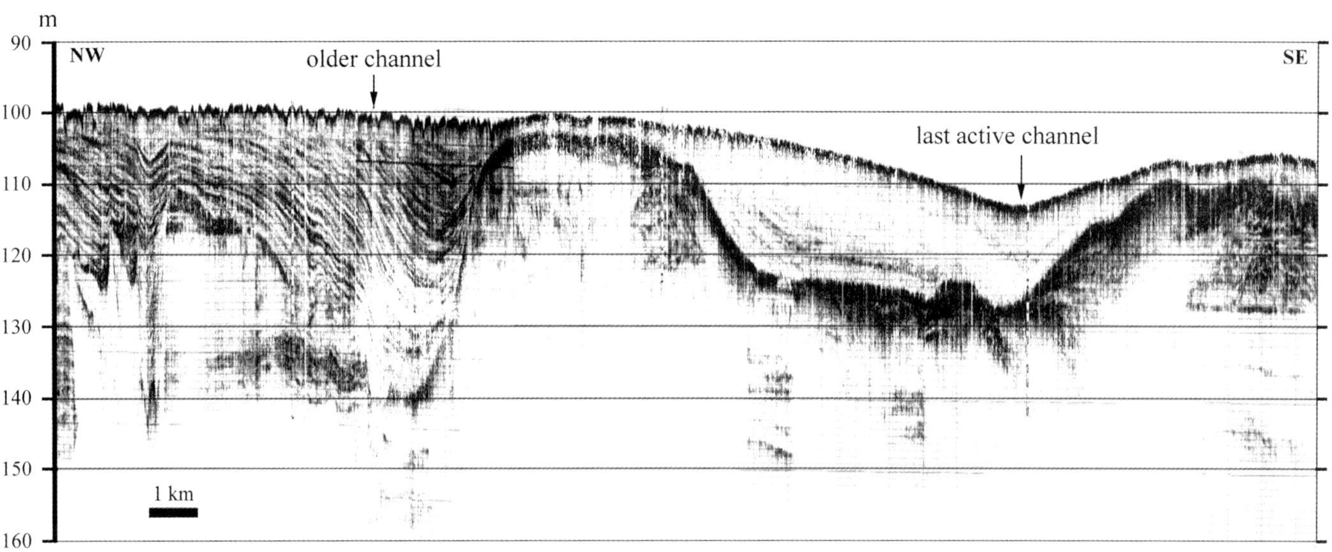

FIG. 8.—A ca. 7-km-wide, largely filled, recently active incised valley of the meandering Molengraaff River system southeast of an older channel. See Fig. 1 for profile location.

mark the approximate position of the sea-level lowstand. The second paleo-shoreline indicator is the topset-to-foreset transition (offlap break) of older buried delta lobes (Aksu et al., 1999). These two types of indicators were found in four of our six air-gun regional dip profiles, and their present-day depths both below the seafloor and below sea level were measured. The seismic sequences with which they are associated were also identified (Table 1).

We corrected the measured depths of the paleo-shoreline indicators for sediment compaction and (pointwise) isostatic compensation due to sediment loading. Thermal subsidence correction was neglected because it is < 2 m for any reasonable age assignments to the seismic units. The details of these correction computations can be found in the literature (Sclater and Christie, 1980; Bond and Kominz, 1984: Bond et al., 1989). It suffices to say that we used the backstripping method, which requires stratigraphic information to the basement to correct for the change in the depth of a given reflector as the result of increase in sediment compaction with burial depth. We assumed the average lithology compiled from drill holes in the Zengmu Basin and a basement depth of 7260 m (Ministry of Geology and Mineral Resources, PR China, 1992), as well as an exponential decrease in sediment porosity with depth (Fig. 11; Sclater and Christie, 1980). Because the initial layer thickness has to be known to compute the initial porosity, but a knowledge of the initial porosity is necessary to compute the layer thickness, an iterative approach was adopted. The most important source of error is uncertainties in lithology that results in uncertainties in porosity. The resulting errors (about 5 m) are an order of magnitude larger than those caused by uncertainties in basement age or in the total thickness of the lithologic column.

The difference between the corrected modern depth of a paleo-shoreline indicator and the eustatic sea level gives the tectonic subsidence since the indicator was formed (Table 1).

Speculative Estimates of Ages of the Seismic Units, Sedimentation Rates, and Tectonic Subsidence Rates.—

Rough age estimates can be interpolated from average sedimentation rates. The average lithostratigraphic column of the Zengmu Basin gives a Plio-Pleistocene to Holocene sediment thickness of 750 m with an age of 5.3 Ma at its base (Ministry of Geology and Mineral Resources, PR China, 1992). This implies an average sedimentation rate of 14 cm/kyr, which in turn yields a computed age of 2.15 Ma for the base of unit 1 inasmuch as the average thickness of units 1–8 is 305 m.

Although this average sedimentation rate is of the same order of magnitude as the measured rates in deep waters of the southern South China Sea (Table 2), it is inconsistent with the delta environment on the northern Sunda Shelf. Holocene sedimentation rates from modern prodeltas are commonly much higher and also much more spatially and temporally variable (83–550 cm/kyr; Díaz et al., 1990; Gensous et al., 1993; Li et al., 1981; Nelson and Maldonado, 1990; Wilkinson and Byrne, 1977). AMS ^{14}C ages for the uppermost sediments of our study area (Stattegger et al., 1997; Stattegger et al., 1998; Hanebuth et al., 1999; Hanebuth et al., 2000) yield rates of ca. 10–56 cm/kyr for the Holocene and 71–553 cm/kyr for the last glacial. The high glacial rates are consistent with the high sediment transport rates of Indonesian rivers today (Milliman and Meade, 1983) and the terrestrial nature of the Sunda Shelf during Pleistocene sea-level lowstands. Thus, ages extrapolated from the computed sedimentation rate of 14 cm/kyr must be rejected.

An alternative method is to correlate the interpreted lowstand wedges with the stacked SPECMAP $\delta^{18}O$ curve of Imbrie et al. (1993), because $\delta^{18}O$ can be looked upon as a proxy for sea level (Fig. 12). Because the $\delta^{18}O$ curve generally indicates slow sea-level falls and rapid sea-level rises, the sea-level lowstands commonly occur at or near the upper boundary of a lowstand wedge (Fig. 12). Thus, the age of the lowstand wedges and with them that of the associated paleo-shoreline indicators can be estimated.

Seismic unit 8 consists of postglacial transgressive and highstand deposits. By analogy with areas where the sedimentation is normal (Aksu et al., 1999; Piper and Aksu, 1992; Skene et al., 1998), unit 7 is considered to cover isotope stages 2–4. Despite the high sediment supply, unit 7 has only a limited thickness, which implies that a significant part of the sediment input was transported beyond the shelf edge onto the continental slope. Thus, unit 7 is correlated with the sea-level fall after stage 5 to the early rise after stage 2 (Fig. 12). In turn, units 6 and 5 encompass stages 6 and 8,

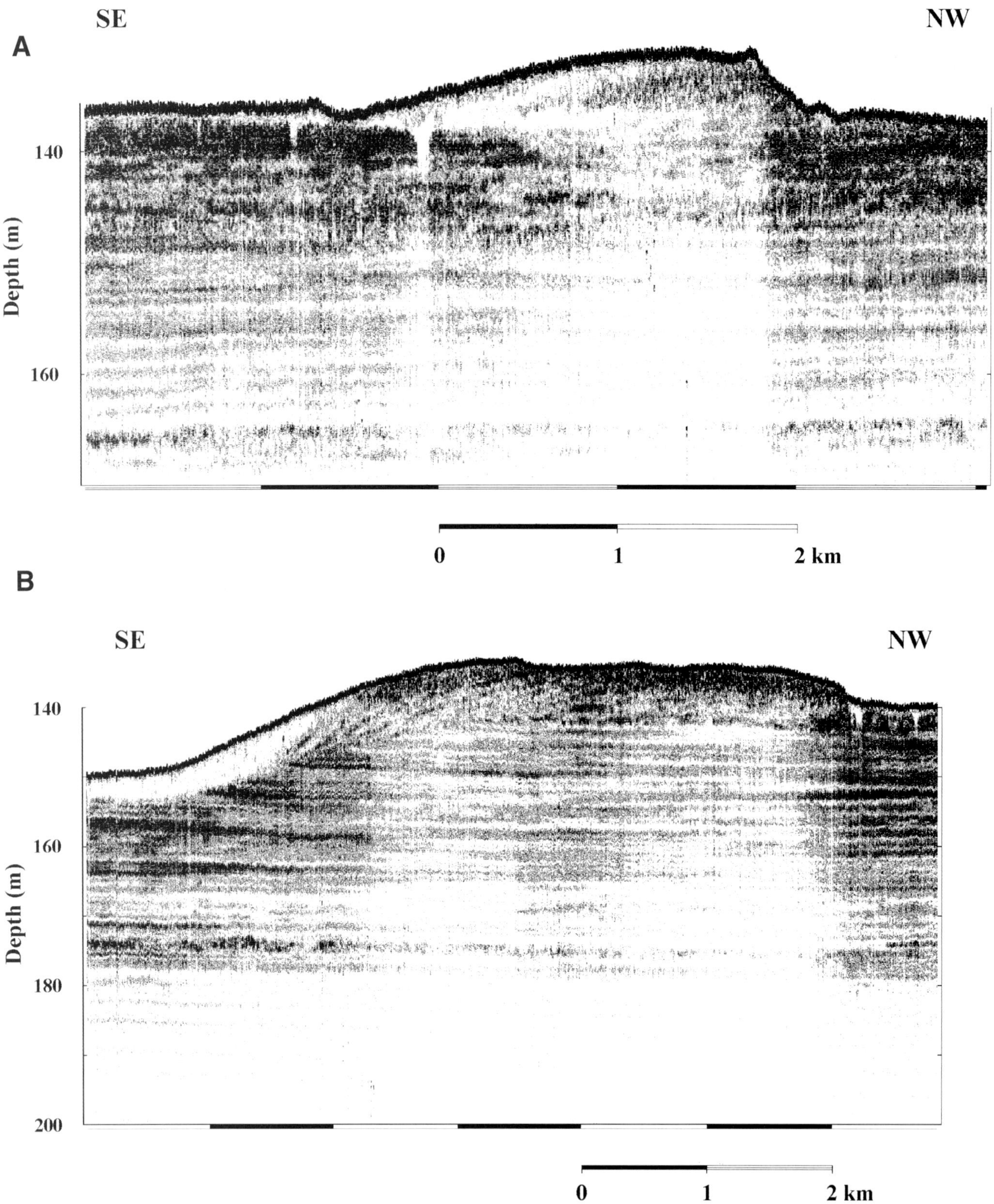

FIG. 9.—Cross section of a sand ridge formed by **A)** upbuilding and **B)** erosion. See text for explanations and Fig. 1 for profile locations.

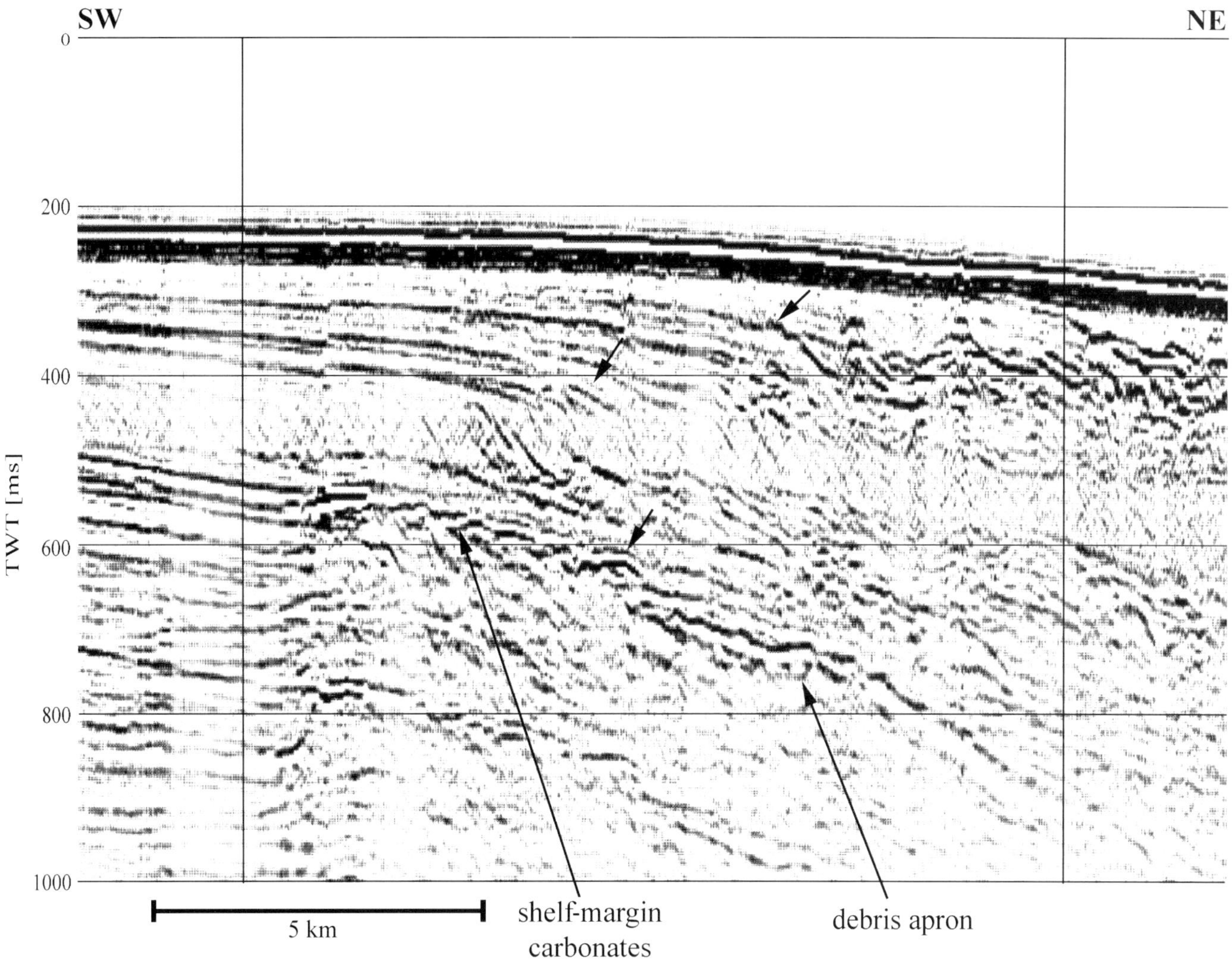

FIG. 10.—Part of air-gun profile 26 showing the two types of paleo-shoreline indicators (arrows). See text for explanation and Fig. 1 for profile location.

respectively. Unit 4 is interpreted to be the regressive unit deposited during the minor fifth-order sea-level cycle within the sea-level fall between stages 9 and 8 (Fig. 12), when the sea level was higher than the lowstand of unit 5. The downlap terminations and more landward position of unit 4 compared to unit 5 support our contention. Lastly, the lowstands of units 3, 2, and 1 are interpreted to include the isotope stages 10, 12, and 14, respectively. The speculative age assignments of the lowstand wedges 7 to 1 that result from this correlation are given in Fig. 12 and Table 3.

We are aware of the highly speculative nature of these age assignments, which will remain until data from proprietary wells are released. However, there is some circumstantial support for the assigned ages:

Sediment Thickness.—Jiang et al. (1994) and Jiang et al. (1996) recognized six regional unconformities in an area that includes the northern half of our study area. Although the details were not given, they assigned their unconformity T1 to the base of the Quaternary using correlations with results from drill holes. On their profile N4 (see Fig. 1 for profile location), the Quaternary thickness is 1.6 s twt at a water depth of 200 m (about 1280 m using a mean sediment velocity of 1.6 km/s; see Jiang et al., 1996, for sonobuoy velocity structure). On our profile 23, the base of unit 1 (presumably with an age of 570 ka) lies at 530 m subbottom at 200 m water depth. Linear extrapolation would give a Quaternary thickness of 1673 m at this location. Because the Quaternary section shows a basinal structure (the Molengraaff delta) centered at about 5.6° N, 110.4° E (Jiang et al., 1996, and our data) and profile 23 lies closer to the center of the basin than N4, the thickness values quoted here are plausible. Other comparisons give generally thicknesses for the 570–0 ka section from our data that are around 45% of the Quaternary thicknesses from Chinese data. In view of sediment compaction with time and the existence of the basin, our assigned age of unit 1 seems reasonable.

Sedimentation Rates.—The decompacted thickness of the seismic units averaged over the shelf areas where they occur are given in Table 3. Using these thickness values and the assigned ages of the unit boundaries, the true average sedimentation rates for the units can be calculated (Table 3). The values vary between 45 and 140 cm/kyr, with the lowest rates for the seismic units 7 and 4. These rates are compatible with Holocene rates of modern prodeltas (83–550 cm/kyr) and glacial rates deduced from AMS ^{14}C ages of cores from the study area (71–553 cm/kyr), suggesting

TABLE 1.—Corrected depth of paleo-coastline indicaors (averaged over four dip profiles), Northern Sunda Shelf.

Shoreline indicator at upper boundary of seismic unit	Present-day depth of shoreline indicator (m)		Modern water depth at location of indicator (m)	Decom-paction corr. (m)	Isostatic correction for sediment loading (m)	Assumed age of shoreline indicator (ka)	Corrected modern depth of paleo-shoreline (mbmsl)	Eustatic sea level (mbmsl)	Total tect. Subsid. (m) = (corrected modern depth of paleo-shoreline – eustatic sea level)
	Below seafloor	Below sea level							
7	0	175	175	0	0	18	175	125	50
6	100	266	166	57	28	135	181	118	63
5	141	315	174	80	41	267	194	100	92
3	275	458	183	157	79	345	222	113	109

mbmsl: meters below modern sea level

that the age assignments are plausible. As discussed above, the low rate for unit 7 is not surprising. Because unit 4 was laid down during a minor sea-level cycle, a low sedimentation rate also appears reasonable.

Tectonic Subsidence Rates.—The corrected modern depths of the paleo-shoreline indicators are shown in Table 2. When the eustatic sea-level depths at the assigned ages of the associated lowstands (relative to the modern sea level) are subtracted from these depths, the total amount of tectonic subsidence from the assigned age to today can be calculated (Fig. 12). By summing the tectonic subsidence curve deduced in this way and the eustatic curve, a relative regional sea-level curve can be constructed (Fig. 12). From the latest Pleistocene to the LGM (345–18 ka), we obtained an average tectonic subsidence rate of 27 cm/kyr for the combined Zengmu I and II depressions. Thereafter, the rate

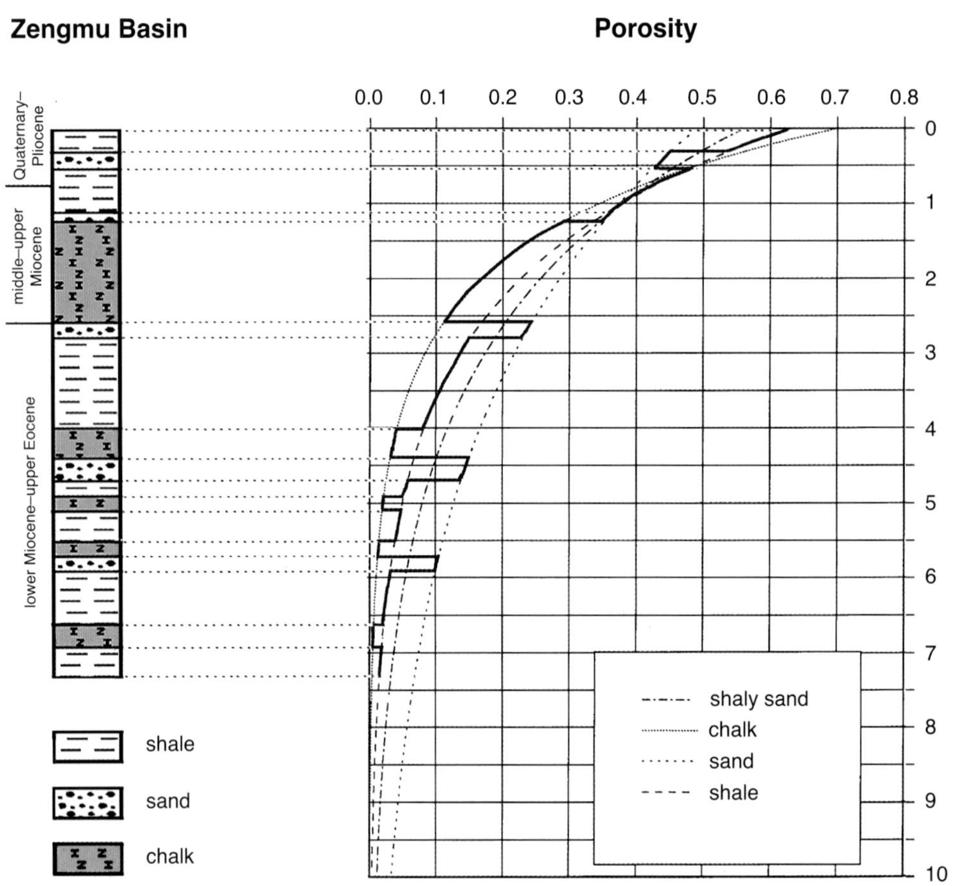

FIG. 11.—Average lithological column of the Zengmu Basin (after Ministry of Geology and Mineral Resources, PR China, 1992) as well as the porosity–depth distribution estimated using this lithology and the exponential decrease in sediment porosity with depth of Sclater and Christie (1980).

TABLE 2.—Average sedimentation rates, southern South China Sea.

Core	Area	Water Depth (m)	Time Period	Average Sedimentation Rate (cm/kyr)	Reference
V35-5	NE of our study area	1953	Holocene	15	Broecker et al. (1988)
			LG	33	
SO49-137SL	off Palawan Shelf	220	9.8–0 ka	0.66	Schönfeld and Kudrass (1993)
			18.3–9.8 ka	7.14	
various	off Sunda Shelf	deep waters	Holocene	2.5–7.1	Wang et al. (1995)
			LG	>10 (1.3–31)	
18284-3	off Sunda Shelf	226	Holocene	12 (10.3–56.4)	Stattegger (personal communication, 1999)
			LG	71.2–553	

LG = last glacial

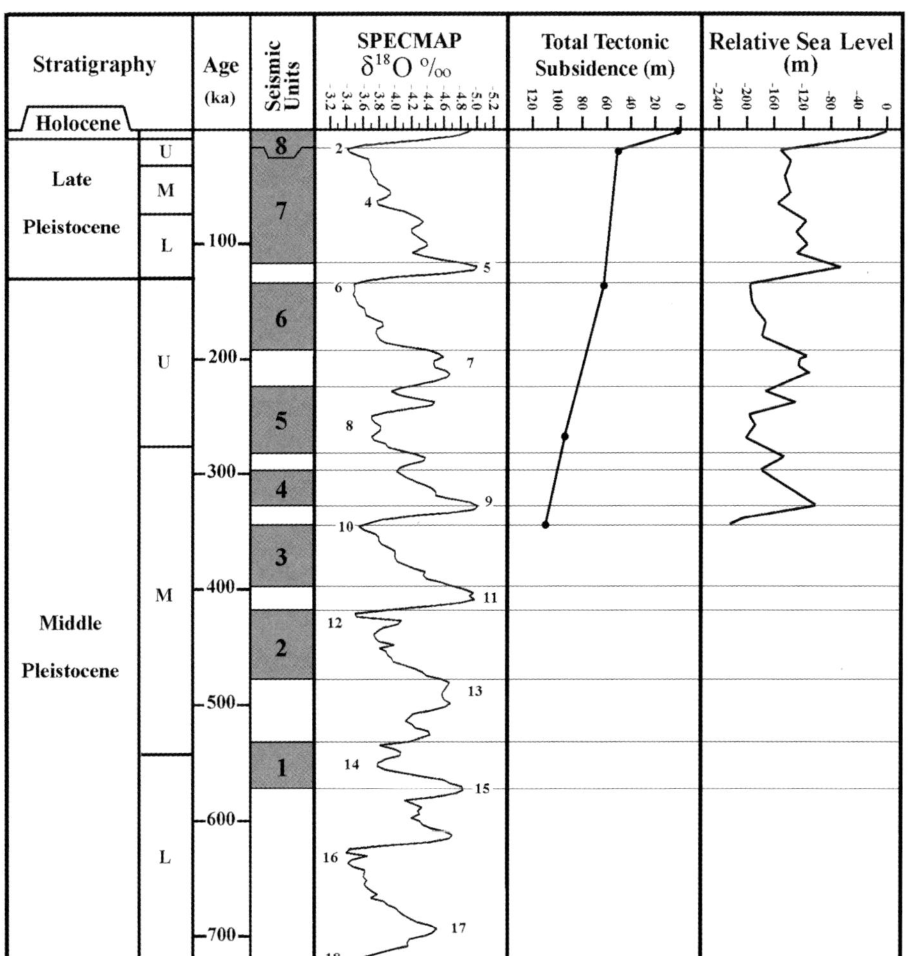

FIG. 12.—Correlation of the seismic units with the stacked SPECMAP $\delta^{18}O$ curve of Imbrie et al. (1993). Numbers near the $\delta^{18}O$ curves denote oxygen isotope stages. The total tectonic subsidence curve and a relative regional sea-level curve are also shown. See text for explanations.

TABLE 3.—Speculative sequence age estimates based on correlation with the SPECMAP $\delta^{18}O$ curve, and average sedimentation rates computed using these ages.

Seismic unit	Approximate unit age (ka)	Average corrected thickness (m)	Sedimentation duration (kyr)	Average sedimentation rate (cm/kyr)
8	18–0		18	
7	115–18	44	97	45
6	190–132	51	58	88
5	280–222?	46	58?	97
4	325–294	17	31	55
3	395–342	55	53	104
2	475–415	36	60	60
1	570–530	56	40	140

apparently increased sharply for reasons that are still unknown. Although this rate is somewhat higher than those computed for the two depressions for the entire Quaternary (21 and 13 cm/kyr; Zhong et al., 1996), it is still fairly comparable. Thus, again the age assignments seem reasonable. It should be noted that our subsidence results do not contradict the conclusions of Evans et al. (1995) and Tjia (1996) that the Sunda Shelf can be considered tectonically stable. Our results suggest that, at a local level, individual tectonic units may have undergone subsidence and differential subsidence, accommodated by growth-fault activity.

CONCLUSIONS

1. Eight seismic units were recognized in the Molengraaff paleodelta on the Sunda Shelf. Of these, seven are wedge-shaped and are interpreted as prograding shelf-margin lowstand wedges deposited during forced regressions of fourth-order or fifth-order sea-level cycles. In contrast, the highstand systems tracts are present largely only as condensed sections on the middle and outer shelves because deposition took place mostly on the inner shelf during sea-level highstands.

2. Postglacial differential subsidence occurred on the outermost Sunda Shelf. The hinge line is situated just landward of the fault zone between the Zengmu II Depression and the Central Uplift. The subsidence rate at the shelf edge was of the order of 2.5 m/kyr, the differential displacements being taken up by growth-fault activity.

3. A major part of the study area was taken up by the delta plain of the Molengraaff River system during the last glacial. During this time fluvial entrenchment did not take place on the emergent outer shelf because of the low gradient. Sand ridges formed by transgressive reworking of Late Pleistocene material mark the modern water-depth range of 135–145 m. Mass wasting and fan deposition were important processes on the continental slope during regressions and lowstands.

4. Speculative age estimates of the seismic units were made by correlating the interpreted lowstand wedges with the SPECMAP $\delta^{18}O$ curve. This correlation suggests that unit 1 reaches back to 570 ka, that the mean sedimentation rates during glacial times are comparable to modern prodelta rates, and that the tectonic subsidence rate of the Zengmu I and II depressions during the latest Pleistocene was of the order of 27 cm/kyr.

ACKNOWLEDGMENTS

We gratefully acknowledge the unfailing help of the captain, officers, and crew of the R/V SONNE during cruise SO-115 in the southern South China Sea and, in particular, financial support of this project (BMBF 03G0115B) by the German Federal Ministry of Education, Science, Research and Technology as well as able project management by the Projektträger BEO. Our thanks go to Drs. Till Hanebuth and Stephan Steinke, who provided us with AMS ^{14}C ages of relevant cores, and to Drs. Serge Berné and Richard Hiscott for their valuable constructive reviews, which led to a significant improvement of the manuscript.

REFERENCES

ALLEN, J.R.L., 1968, The nature and origin of bed-form hierarchies: Sedimentology, v. 10, p. 161–182.

AKSU, A.E., HISCOTT, R.N., AND YAŞAR, D., 1999, Oscillating water levels of the Marmara Sea and vigorous outflow into the Aegean Sea from Marmara Sea–Black Sea drainage corridor: Marine Geology, v. 153, p. 275–302.

Asian Council on Petroleum (ASCOPE), 1981, Tertiary Sedimentary Basins of the Gulf of Thailand and South China Sea: Stratigraphy, Structure and Hydrocarbon Occurrences: ASCOPE Secretariat, Jakarta, Indonesia.

BATCHELOR, B.C., 1979a, Geological characteristics of certain coastal and offshore placers as essential guides for tin exploration in Sundaland, Southeast Asia: Geological Society of Malaysia, Bulletin, v. 11, p. 283–313.

BATCHELOR, B.C., 1979b, Discontinuously rising Late Cenozoic eustatic sea-levels, with special reference to Sundaland, Southeast Asia: Geologie en Mijnbouw, v. 58, p. 1–20.

BEN AVRAHAM, Z., AND EMERY, K.O., 1973, Structural framework of Sunda Shelf: American Association of Petroleum Geologists, Bulletin, v. 57, p. 2323–2366.

BERNÉ, S., TRENTESAUX, A., STOLK, A., MISSIAEN, T., AND DE BATIST, M., 1994, Architecture and long term evolution of a tidal sandbank: The Middlekerke Bank (southern North Sea): Marine Geology, v. 121, p. 57–72.

BERNÉ, S., LERICOLAIS, G., MARSSET, T., BOURILLET, J. F., AND DE BATIST, M., 1998, Erosional shelf sand ridges and lowstand shorefaces: examples from tide and wave dominated environments of France: Journal of Sedimentary Research, v. 68, p. 540–555.

BOND, G.C., AND KOMINZ, M.A., 1984, Construction of tectonic subsidence curves for the early Paleozoic miogeocline, southern Rocky Mountains: Implication for subsidence mechanisms, age of breakup, and crustal thinning: Geological Society of America, Bulletin, v. 95, p. 155–173.

Bond, G.C., Kominz, M.A., Steckler, M.S., and Grotzinger, J.P., 1989, Role of thermal subsidence, flexure and eustasy in the evolution of early Paleozoic passive margin carbonate platforms, *in* Crevello, P.D., Wilson, J.L., Sarg, J.F. and Read, J.F., eds., Controls on Carbonate Platform and Basin Development: SEPM, Special Publication 44, p. 39–61.

Broecker, W.S., Andree, M., Klas, M., Bonani, G., Wolfli, W., Mix, A., and Oescheger, H., 1988, New evidence from the South China Sea for an abrupt termination of the last glacial period: Nature, v. 333, p. 156–158.

Coordinating Committee for Coastal and Offshore Geoscience Programmes in East and Southeast Asia (CCOP)– Intergovernmental Oceanographic Commission (IOC), 1980, Studies in East Asian Tectonics and Resources. Technical Support for Regional Offshore Prospecting in East Asia: RAS/80/003, CCOP Project Office, 141–151.

Díaz, J.I., Nelson, H., Barber, J. H., Jr., and Giró, S., 1990, Late Pleistocene and Holocene sedimentary facies on the Ebro continental shelf: Marine Geology, v. 95, p. 333–352.

Emmel, F.J., and Curray, J.R., 1982, A submerged Late Pleistocene delta and other features related to sea level changes in the Malacca Strait: Marine Geology, v. 47, p. 197–216.

Evans, C.D.R., Brett, C.P., James, J.W.C., and Holmes, R., 1995, Shallow Seismic Reflection Profiles from the Waters of East and Southeast Asia—An Interpretation Manual and Atlas: London, British Geological Survey, Technical Report WC/94/60, Overseas Geology Series, 94 p.

Fairbanks, R.G., 1989, A 17,000-year glacio-eustatic sea level record, influence of glacial melting rates on the Younger Dryas event and deep-ocean circulation: Nature, v. 342, p. 637–642.

Gensous, B., Williamson, D., and Tesson, M., 1993, Late Quaternary transgressive and highstand deposits of a deltaic shelf (Rhône Delta, France), *in* Posamentier, H.W., Summerhayes, C.P., Haq, B.U., and Allen, G.P., eds., Sequence Stratigraphy and Facies Associations: International Association of Sedimentologists, Special Publication 18, p. 197–211.

Geyl, M., Kudrass, H.-R., and Streif, H., 1979, Sea-level changes during the late Pleistocene and Holocene in the Strait of Malacca: Nature, v. 278, p. 441–443.

Hamilton, W., 1979, Tectonics of the Indonesian Region: U. S. Geological Survey, Professional Paper 1028, 245 p.

Hantoro, W.S., 1997, Quaternary sea level variations in the Pacific–Indian Ocean gateways: response and impact: Quaternary International, v. 37, p. 73–80.

Hanebuth, T., Stattegger, K., Grootes, P.M., and Nadeau, M., 1999, AMS-Untersuchungen am faziell komplexen Schelfsedimenten auf dem Sunda-Schelf, *in* Tagungsband zum Statusseminar 1999 "Meeresforschung mit FS Sonne", 10.-12.3.1999 in Freiberg, Forschungszentrum Jülich GmbH und Projektträger BEO des BMBF, p. 285–287.

Hanebuth, T., Stattegger, K., and Grootes, P.M., 2000, Rapid flooding of the Sunda Shelf: a late-glacial sea-level record: Science, v. 288, p. 1033–1035.

Houtz, R.E., and Hayes, D.E., 1984, Seismic refraction data from Sunda Shelf: American Association of Petroleum Geologists, Bulletin, v. 68, p. 1870–1878.

Huthnance, J.M., 1982a, On one mechanism forming linear sand banks: Estuarine, Coastal and Shelf Science, v. 14, p. 279–299.

Huthnance, J.M., 1982b, On the formation of sand banks of finite extent: Estuarine, Coastal and Shelf Science, v. 15, p. 277–299.

Imbrie, J., Berger, A., Boyle, E.A., Clemens, S.C., Duffy, A., Howard, W.R., Kukla, G., Kutzbach, J., Martinson, D.G., McIntyre, A., Mix, A.C., Molfino, B., Morley, J.J., Peterson, L.C., Pisias, N.G., Prell, W.L., Raymo, M.E., Shackleton, N.J., and Toggweiler, J.R., 1993, On the structure and origin of major glaciation cycles. 2: the 100,000-year cycle: Paleoceanography, v. 8, p. 99–735.

Jiang, S., Zhou, X., Huang, C., and Xia, K., 1994, The stratigraphy, structure and evolution of Zengmu Basin: Tectonophysics, v. 235, p. 51–62.

Jiang, S., Zhou, X., Jiang, J., Ye., X., Huang, Z., and Liu, D., 1996. Structural pattern and hydrocarbon prospects of the Zengmu Basin, *in* Xia, K., ed., Geology, Geophysics and Hydrocarbon Resources of the Nansha Islands and their Neighboring Areas: Beijing, Science Press, p. 141–164 (in Chinese).

Lee, T.-Y., and Lawver, L.A., 1994, Cenozoic plate reconstruction of the South China Sea region: Tectonophysics, v. 235, p. 149–180.

Li, C.X., Wan, T.J.T., Xue, S.Y., Guo, X.M., and Li, P., 1981, Holocene transgressive–regressive sequences in Yangtze Delta area: Collected Oceanic Works, Tianjin, Chinese Society of Oceanography, v. 4, no. 1, p. 89–98.

Ludwig, W.J., Kumar, N., and Houtz, R.E., 1979, Profiler-sonobuoy measurements in the South China Sea basin: Journal of Geophysical Research, v. 84, p. 3505–3518.

Miall, A.D., 1991, Stratigraphic sequences and their chronostratigraphic correlation: Journal of Sedimentary Petrology, v. 61, p. 497–505.

Milliman, J.D., and Meade, R.H., 1983, World-wide delivery of river sediment to the oceans: Journal of Geology, v. 91, p. 1–22.

Ministry of Geology and Mineral Resources, PR China, 1992, Sedimentary Basin Map, *in* Map Series on the Geology and Geophysics of Chinese Seas and the Adjacent Regions: Beijing, Geological Publishing House, 1:5,000,000.

Molengraaff, G.A.F., and Weber, M., 1920, On the relation between the Pleistocene glacial period and the origin of the Sunda Sea (Java and South China Sea), and its influence on the distribution of coral reefs and on the land- and freshwater fauna: Amsterdam, Koninklijk Akademie van Wetenschappen, Proceedings of the Section of Sciences, v. 23, 1st part, nos. 2 and 3, p. 395–439.

Molengraaff, G.A.F., 1922, De Geologie der Zeeen van Nederlandsch Oost Indiee, *in* Koninklijk Nederlandsch Aadrijkskundig Genootschap, De Zeeen van Nederlandsch Oost Indiee, Leiden, p. 272–357.

Nelson, C.H., and Maldonado, A., 1990, Factors controlling Late Cenozoic continental margin growth from the Ebro delta to the western Mediterranean deep sea: Marine Geology, v. 95, p. 419–440.

Parke, M.L., Jr., Emery, K.O., Szymankiewicz, R., and Reynolds, L.M., 1971, Structural framework of continental margin in South China Sea: American Association of Petroleum Geologists, Bulletin, v. 55, p. 723–751.

Piper, D.J.W., and Aksu, A.E., 1992, Architecture of stacked Quaternary deltas correlated with global oxygen isotope curve: Geology, v. 20, p. 415–418.

Posamentier, H.W., Hervey, M.T., and Vail, P.R., 1988, Eustatic control on clastic deposition I—conceptual framework, *in* Wilgus, C.K., Hastings, B.S., Posamentier H.W., Van Wagoner, J., Ross, C.A., and Kendall, C.G.St.C., eds., Sea-Level Changes: An Integrated Approach: SEPM, Special Publication 42, p. 109–124.

Posamentier, H.W., Erskine, R.D., and Mitchum, R.M., Jr., 1991, Models for submarine fan deposition within a sequence-stratigraphic framework, *in* Weimer, P., and Link, M.H., eds., Seismic Facies and Sedimentary Fan Processes of Submarine Fans and Turbidite Systems: New York, Springer-Verlag, p. 127–136.

Posamentier, H.W., Allen, G.P., James, D.P., and Tesson, M., 1992, Forced regressions in a sequence stratigraphic framework: Concepts, examples, and exploration significance: American Association of Petroleum Geologists, Bulletin, v. 76, p. 1687–1709.

Reading, H.G., and Collinson, J.D., 1996, Clastic coasts, in Reading, H.G., ed., Sedimentary Environments: Processes, Facies and Stratigraphy, 3rd Edition: Oxford, London, Blackwell Science, p. 154–231, see p. 195.

RU, K., AND PIGOTT, J.D., 1986, Episodic rifting and subsidence in the South China Sea: American Association of Petroleum Geologists, Bulletin, v. 70, p. 1136–1155.

SARG, J.F., 1988, Carbonate sequence stratigraphy, *in* Wilgus, C.K., Hastings, B.S., Posamentier H.W., Van Wagoner, J., Ross, C.A., and Kendall, C.G.St.C., eds., Sea-Level Changes: An Integrated Approach: SEPM, Special Publication 42, p. 155–181.

SCHLÜTER, H.U., HINZ, K., AND BLOCK, M., 1996, Tectono-stratigraphic terranes and detachment faulting of the South China Sea and Sulu Sea: Marine Geology, v. 130, p. 39–78.

SCHÖNFELD, J., AND KUDRASS, H.-R., 1993, Hemipelagic sediment accumulation rates in the South China Sea related to Late Quaternary sea level changes: Quaternary Research, v. 40, p. 368–379.

SCLATER, J.G., AND CHRISTIE, P.A.F., 1980, Continental stretching: An explanation of the post-Mid-Cretaceous subsidence of the central North Sea basin: Journal of Geophysical Research, v. 85, p. 3711–3739.

SCSIO (SOUTH CHINA SEA INSTITUTE OF OCEANOLOGY), ACADEMIA SINICA, 1984, Physiographic Map of the South China Sea, scale: 1,4 000 000: Tianjin, Map Publishing House.

SKENE, K.I., PIPER, D.J.W., AKSU, A.E., AND SYVITSKI, J.P.M., 1998, Evaluation of the global oxygen isotope curve as a proxy for Quaternary sea level by modeling of delta progradation: Journal of Sedimentary Research, v. 68, p. 1077–1092.

STATTEGGER, K., KUHNT, W., WONG, H.K., BÜHRING, C., HAFT, C., HANEBUTH, T., KAWAMURA, H., KIENAST, M., LORENC, S., LOTZ, B., LÜDMANN, T., LURATI, M., MÜHLHAN, N., PAULSEN, A.-M., PAULSEN, J., PRACHT, J., PUTAR-ROBERTS, A., HUNG, N.Q., RICHTER, A., SALOMON, B., SCHIMANSKI, A., STEINKE, S., SZAREK, R., NHAN, N.V., WEINELT, M., AND WINGUTH, C., 1997, Sequence Stratigraphy, Late Pleistocene–Holocene Sea Level Fluctuations and High Resolution Record of the Post-Pleistocene Transgression on the Sunda Shelf. Cruise Report Sonne 115, Sundaflut: Universität Kiel, Geologisch-Paläontologisches Institut, Berichte–Reports, no. 86, 211 p.

STATTEGGER, K., KUHNT, W., HANEBUTH, T., KAWAMURA, H., PAULSEN, J., RICHTER, A., SCHIMANSKI, A., STEINKE, S., SZAREK, R., AND WEINELT, M., 1998, SO-115A-Sundaflut: Stofflüsse, postglaziale Überflutung des Sunda- und Vietnamesischen Schelf: Auswirkungen auf Sediment- und Nährstoff-Flüsse: Christian-Albrechts-Universität zu Kiel, Zwischenbericht, 11 p. (unpublished).

STRIDE, A.H., BELDERSON, R.H., KENYON, N.H., AND JOHNSON, M.A., 1982, Offshore tidal deposits, sand sheet and sand bank facies, *in* Stride, A.H., ed., Offshore Tidal Sands; Processes and Deposits: London, Chapman & Hall, p. 95–125.

TAYLOR, B., AND HAYES, D.E., 1980, The tectonic evolution of the South China basin, *in* Hayes, D.E., ed., The Tectonic and Geologic Evolution of Southeast Asian Seas and Islands, Part I: Washington, D.C., American Geophysical Union, Geophysical Monograph 22, p. 89–104.

TAYLOR, B., AND HAYES, D.E., 1983, The South China Sea basin, *in* Hayes, D.E., ed., The Tectonic and Geologic Evolution of Southeast Asian Seas and Islands, Part II: Washington, D.C., American Geophysical Union, Geophysical Monograph 23, p. 24–56.

TESSON, M., ALLEN, G.P., AND RAVENNE, C., 1993, Late Pleistocene shelf-perched lowstand wedges on the Rhône continental shelf, *in* Posamentier, H. W., Summerhayes, C.P., Haq, B.U., and Allen, G.P., eds., Sequence Stratigraphy and Facies Associations: International Association of Sedimentologists, Special Publication 18, p. 183–193.

TJIA, H.D., 1980, The Sunda Shelf, Southeast Asia: Zeitschrift für Geomorphologie, v. 24, p. 405–427.

TJIA, H.D., 1996, Sea level changes in the tectonically stable Malay–Thai Peninsula: Quaternary International, v. 31, p. 95–101.

WANG, P., WANG, L., BIAN, Y., AND JIAN, Z., 1995, Late Quaternary paleoceanography of the South China Sea: surface circulation and carbonate cycles: Marine Geology, v. 27, p. 145–165.

WILKINSON, B.H., AND BYRNE, J.R., 1977, Lavaca Bay—transgressive deltaic sedimentation in Central Texas Estuary: American Association of Petroleum Geologists, Bulletin, v. 61, p. 527–545.

XIA, K.-Y., AND ZHOU, D., 1993, The geophysical characteristics and evolution of northern and southern margins of the South China Sea: Geological Society of Malaysia, Bulletin, v. 33, p. 223–240.

XIA, K.-Y., AND HUANG, C.-L., 1996, Hydrocarbon resources and a comprehensive analysis of the geology and geophysics of the area around the Nansha Islands, *in* Xia, K., ed., Geology, Geophysics and Hydrocarbon Resources of the Nansha Islands and their Neighboring Areas: Beijing, Science Press, p. 211–247 (in Chinese).

XIA, K.-Y., HUANG, C.-L., JIANG, S.-R., ZHANG, Y.-X., SU, D.-Q., XIA, S.-G., AND CHEN, Z.-R., 1994, Comparison of the tectonics and geophysics of the major structural belts between the northern and southern continental margins of the South China Sea: Tectonophysics, v. 235, p. 99–116.

YANG, C.-S., AND SUN, J.-S., 1988, Tidal sand ridges on the East China Sea shelf, *in* de Boer, P.L., van Gelder, A., and Nio, S.D., eds., Tide-Influenced Sedimentary Environments and Facies: Dordrecht, The Netherlands, Reidel Publishing Company, p. 23–38.

YUAN, Y., GE, Y., LIU, C., TANG, X., WANG, B., AND HAU, H., 1993, Quaternary shallow seismic stratigraphy of the area adjacent to the Nansha Archipelago, *in* The Multidisciplinary Oceanographic Expedition Team of Academia Sinica to the Nansha Islands, ed., Quaternary Sedimentary Geology of the Nansha Islands and the Adjacent Sea Area: Hubei Science and Technology Press, p. 300–320 (in Chinese).

ZHONG, J., HUANG, C., AND JEN, W., 1996, Analyses of subsidence of the northern Zengmu Basin since the Eocene, *in* Xia, K., ed., Geology, Geophysics and Hydrocarbon Resources of the Nansha Islands and their Neighboring Areas: Beijing, Science Press, p. 120–125 (in Chinese).

Tertiary Systems

SEQUENCE STRATIGRAPHY AND SYNDEPOSITIONAL TECTONICS OF UPPER MIOCENE AND PLIOCENE DELTAIC SEDIMENTS, OFFSHORE BRUNEI DARUSSALAM

ARTHUR SALLER AND GREGG BLAKE

Unocal Corporation, 14141 Southwest Freeway, Sugar Land, Texas 77478, U.S.A.

e-mail: asaller@unocal.com; ghblake@unocal.com

ABSTRACT: Two main deltaic systems, Champion and Baram, filled in much of the Brunei shelf during the Late Miocene, Pliocene, and Pleistocene. "Fourth-order" deltaic sequences, commonly 50–100 m thick on the outer shelf, have been identified in seismic data, wireline well logs, and biostratigraphic data. Average duration of each Pliocene sequence is between 100,000 and 200,000 years. Tide- and river-dominated deltas were common in embayments formed during late stages of transgression and early highstand, but wave-dominated deltas were prevalent in middle to late highstand parts of sequences. During highstands of sea level, shales and siltstones dominated at shelf margins, whereas reservoir-quality sandstones occur 1–3 km or more landward of the contemporaneous shelf margin.

From the Late Miocene through the Pleistocene, the Champion and Baram deltas prograded over thick mobile shales. The Champion delta (to the northeast) prograded approximately 40 km during the Late Miocene and 12 km in the Pliocene. The Baram delta (to the southwest) prograded approximately 40 km during the Pliocene. Delta progradation initiated movement on counter-regional growth faults on the upper slope that continued after the shelf margin prograded across those faults. The Frigate counter-regional fault system had 1–2 km of throw during the Pliocene and Pleistocene along the eastern part of the outer shelf. At the same time, much slower subsidence and even uplift occurred around diapiric shale ridges on inner shelf. Down-to-the-basin normal faults with 1–2 km of throw formed on the western part of the outer shelf (Baram delta) during the Pleistocene. Anticlinal folding and toe thrusts occurred on the middle and lower slope during times of major growth faulting (down-to-the-basin and counter-regional) on the outer shelf.

INTRODUCTION

The Champion and Baram deltas have been the location of major clastic sedimentation since the Miocene. These deltaic systems in Brunei contain giant oil and gas fields with total estimated ultimate recoveries in excess of 3 billion barrels of oil and 20 TCF of gas (summary of data in Sandal, 1996). Deltaic strata of the Brunei shelf (Figs. 1 and 2) show distinct patterns in sedimentation related to fluctuations in relative sea level and syndepositional tectonics. The purpose of this paper is to define and discuss a sequence stratigraphic framework for Upper Miocene, Pliocene, and Pleistocene strata on the Brunei shelf and relate that stratigraphic framework to syndepositional tectonics.

SETTING

Northwest Borneo is generally considered to be part of the southeastern Eurasian plate, but its Cenozoic tectonics have also involved the Indian and Pacific plates (Rangin et al., 1990). During the Cretaceous and early Tertiary, the oceanic crust at the southeastern margin of the Eurasian plate was subducted under Sundaland, resulting in the Crocker–Rajang accretionary belt, which now crops out in the Crocker–Rajang mountains along the Malaysia–Kalimantan border on Borneo (Fig. 1; Tan and Lamy, 1990). Collision of the South China Sea with Sundaland occurred along an oblique convergence front from the early Tertiary in the south to Middle Miocene in the north (Sandal, 1996). Collision and subduction of continental crust resulted in periods of uplift in the northwest Borneo area during the Late Miocene (Tan and Lamy, 1990).

Clastic deltaic systems have been prograding to the northwest away from the orogenic front since the Early Miocene. Areas that are currently offshore Brunei were dominated by deposition of the deep-water Setap shale during the Early and Middle Miocene (Wilson and Moss, 1999). Since the Middle Miocene, transtensional tectonics have dominated on the shelf to the west, transpressional tectonics on the shelf to the east, and a zone of toe thrusting has dominated on the middle to lower slope (Fig. 2; Sandal, 1996). Late Miocene to Pleistocene structures on the shelf include diapiric shale ridges, down-to-the-basin normal faults, and counter-regional growth faults which dip landward. Rapid Late Miocene progradation of shelf margins in Sabah and Sarawak resulted in a depositional embayment on the Brunei shelf that was filled by progradation of the Baram delta during the Pliocene and Pleistocene.

The data base for this study included well and seismic data. Biostratigraphic and wireline log data from approximately 23 wells on the Brunei shelf were used. Regional mapping was done on approximately 12,000 km of 2D seismic data at variable grid spacings ranging from less 1 km to approximately 5 km.

BIOSTRATIGRAPHY

Microfossils collected from ditch cuttings and sidewall cores have been analyzed to correlate deltaic sequences and systems tracts. Whenever possible, samples were worked at 5 and 10 m intervals in wells for both foraminiferal and calcareous nannofossils because larger sample intervals greatly reduce the ability to define sequence boundaries and systems tracts. Unlike the wells drilled on the modern inner shelf, where calcareous microfossils were poorly preserved and had poor recovery, wells drilled in the middle shelf to upper slope had moderate to good recovery and preservation of both foraminifera and calcareous nannofossils (Fig. 3). For the wells drilled in the inner-shelf area, palynology is usually the only fossil group that can provide any indication of age and paleoenvironment for the late Neogene, and hence, interpretations usually have more limited resolution. For this study, correlations between wells were made using the worldwide biochronologic zonations of Berggren et al. (1995) (Fig. 3). Figure 3 is a typical outer-shelf well that penetrated a progradational Upper Miocene to Pleistocene section in offshore Brunei. Paleoenvironmental interpretations are based on key benthic foraminiferal species and assemblages. Owing to the high depositional rates, biostratigraphic data are critical to correlation of seismic markers across major faults.

Tropical Deltas of Southeast Asia—Sedimentology, Stratigraphy, and Petroleum Geology
SEPM Special Publication No. 76, Copyright © 2003
SEPM (Society for Sedimentary Geology), ISBN 1-56576-086-7, p. 219–234.

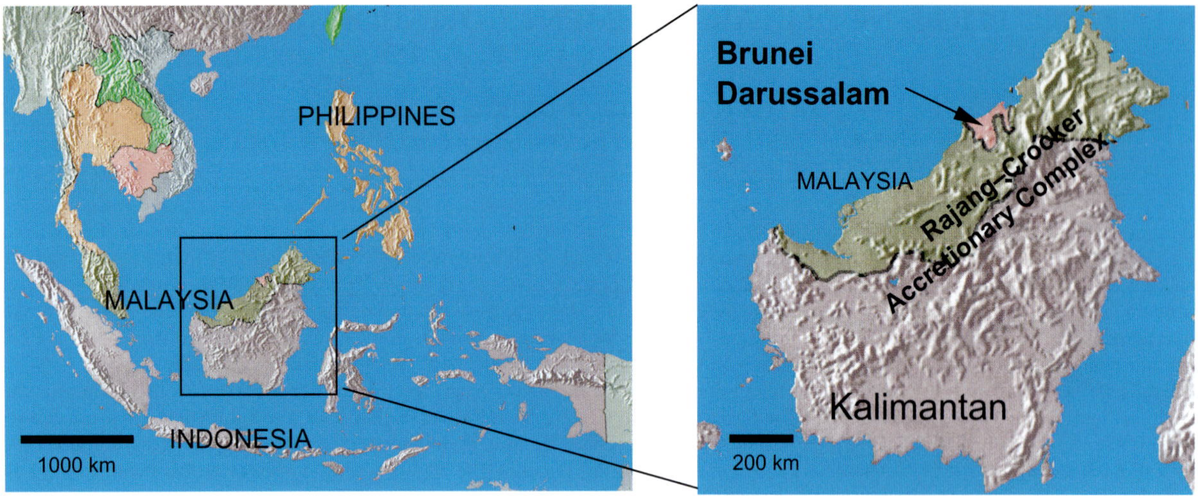

FIG. 1.—Maps showing the location of Brunei Darussalam on the northwest side of Borneo.

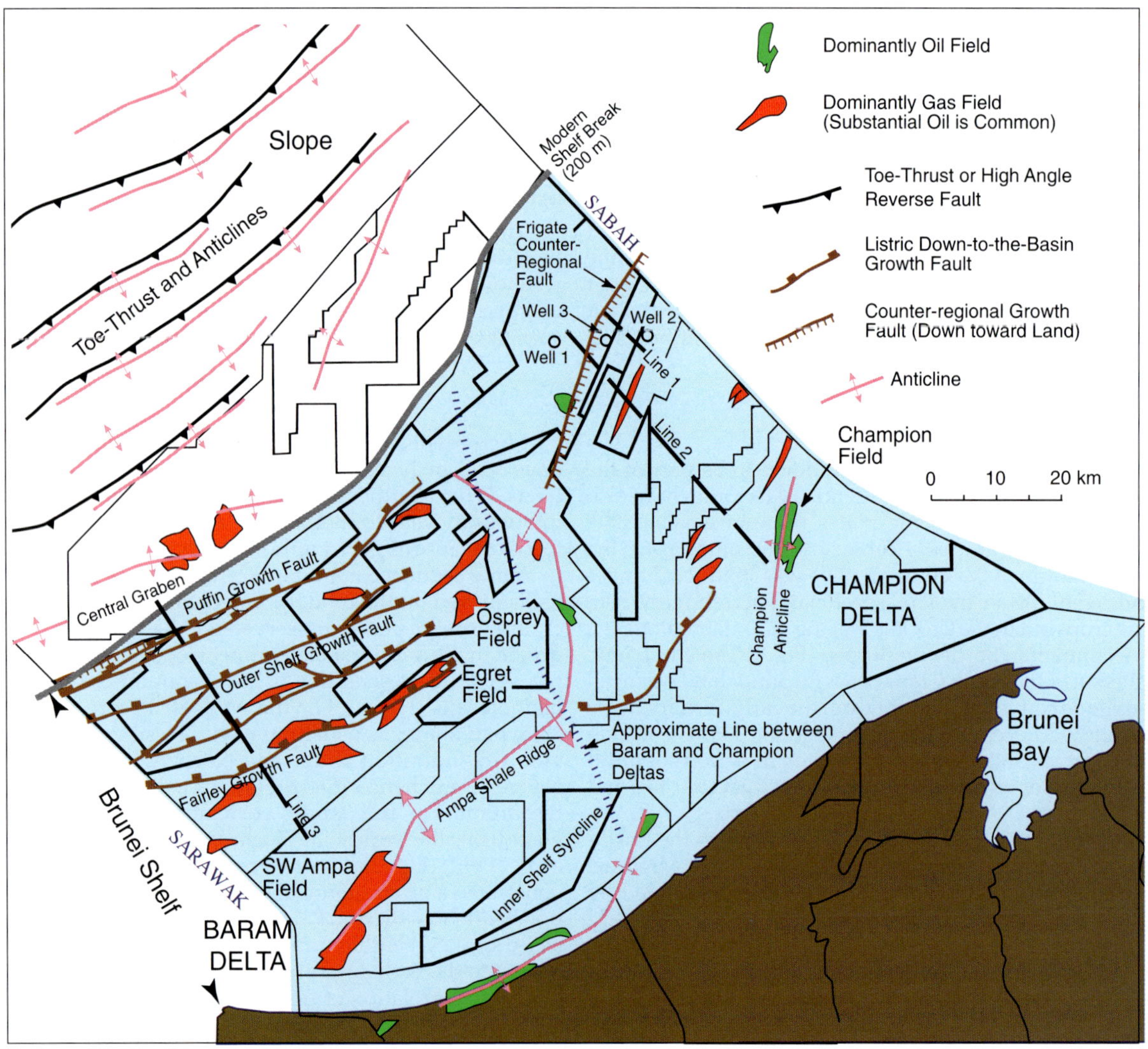

FIG. 2.—Map of the Brunei shelf. Locations of seismic lines and wells are approximate.

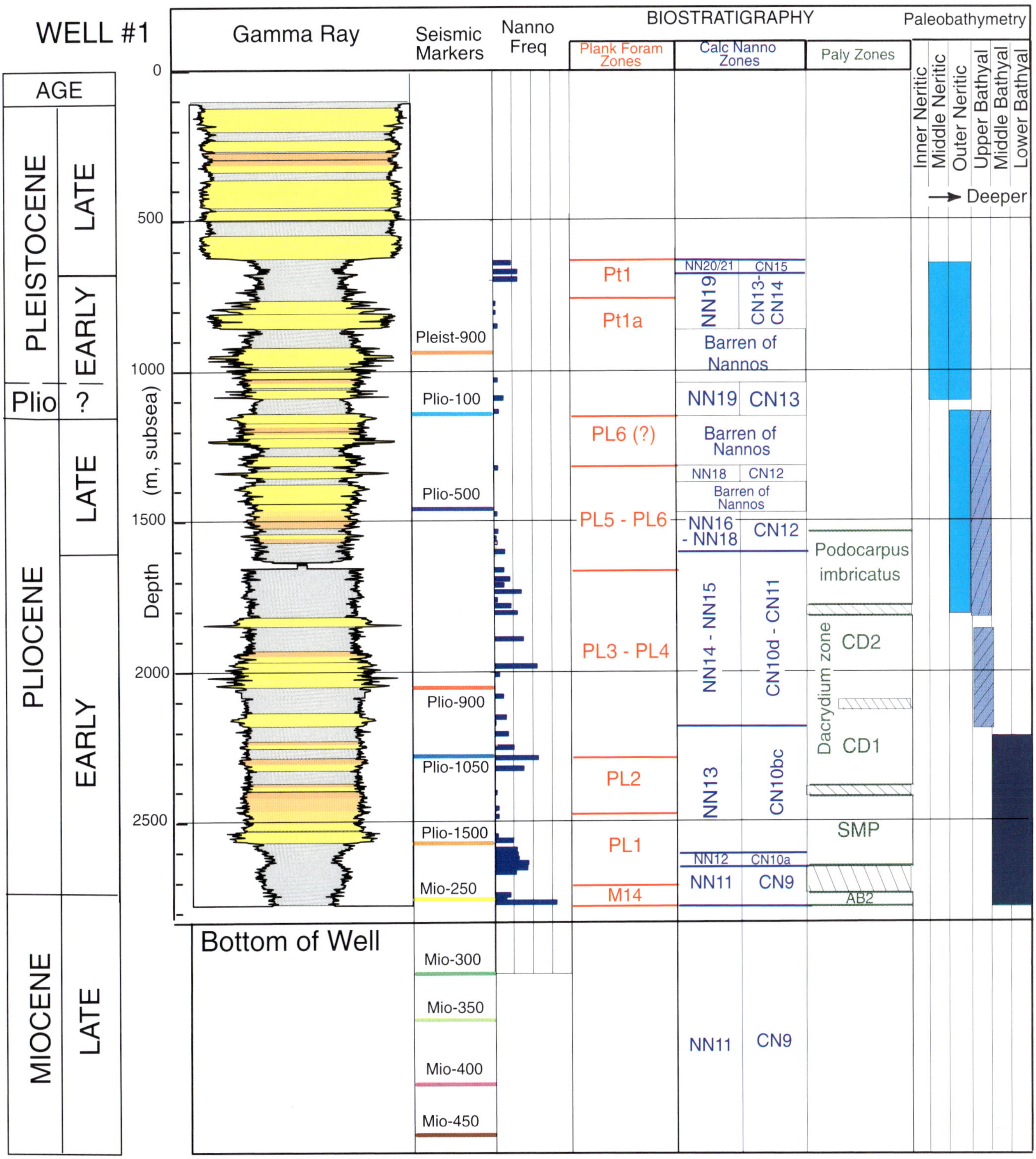

FIG. 3.—Well #1 with "butterflied" gamma ray, ages, seismic horizons, biostratigraphic zones, and paleobathymetry interpreted from microfossils. This is a typical well with progradation of middle- and outer-shelf strata over middle- and upper-slope strata.

HOLOCENE COASTAL ENVIRONMENTS

Although Brunei has a noted wave-dominated delta (Sandal, 1996), it also contains a variety of other coastal systems. The modern coastline of Brunei contains beaches and estuaries with river and tidal deltas in addition to the wave-dominated Baram delta (Lambiase et al., this volume). Broad open-ocean swells from the South China Sea have caused the modern Baram River to be a wave-dominated delta (Sandal, 1996: Figure 4, left) where sands are kept relatively close to the shoreline, generally within 10–15 km of the mouth of the Baram delta. Deposition of silt and clay dominate most of the middle and outer part of the modern Brunei shelf (Sandal, 1996). The Champion delta area now has a long sandy beach seaward of Brunei Bay, a large estuarine environment with tide-dominated deltas (Figure 4, right; Lambiase et al., this volume). Other minor rivers with associated spits also enter the South China Sea between the Baram delta and Brunei Bay.

A study of the Baram delta by Caline and Huong (1992) suggests that early, tide-dominated deltas pass upward to progradational wave-dominated deltas. Using satellite imagery to study the Holocene, Caline and Huong (1992) show that the Baram delta area was an embayment during the latest Holocene transgression and earliest highstand (5,400–5,000 yr BP). Shoals in that embayment were probably reworked by tidal currents. Between 5,000 and 2,000 yr BP, the embayment was converted into a wave-dominated delta protruding into the South China Sea (Caline and Huong, 1992). Deposition by tidal currents is also common in Miocene deltaic strata outcropping on the Brunei mainland (Lambiase et al., this volume; Sandal, 1996). These outcrops may include deltas which developed in protected bays and locally within wave-dominated deltas.

DELTAIC SEQUENCES

Seismic reflector geometries, log profiles, and biota were used to recognize sequences and systems tracts. Figure 5 shows the facies in systems tracts in a typical Brunei sequence. The facies and systems tracts are similar to models of Posamentier et al. (1988). By reasoning of Plint et al. (1992), the main sequences observed in seismic data and well logs are probably fourth-order sequences. Deltaic sequences in offshore Brunei vary considerably in thickness but are typically between 50 and 100 m on the middle to outer shelf. Larger variations in thickness are common in response to variations in accommodation caused by local tectonics. For example, sequences thicken across syndepositional faults and at the shelf–slope break but thin over shale ridges. These Brunei sequences are similar to those identified in Upper Miocene deltaic strata in Betty Field, on the middle shelf of adjacent Sarawak. Those sequences in Betty field consist of fining-up (transgressive) bases (6–15 m thick) that typically pass up to coarsening upward (progradational) packages also typically 50 m thick (Johnson et al., 1989). Johnson et al. (1989) interpreted most of that deposition as occurring in a shelf system dominated by wave energy.

Seismic Patterns

Pliocene deltaic sequences are most distinctive on seismic over outer-shelf and upper-slope environments (Fig. 6). Shelf margins are identified where topset reflectors increase abruptly in basinward dip into the foresets of the upper slope. Sequence boundaries are identified by one of three seismic geometries: (1) lowstand reflectors onlapping the slope, (2) toplap of prograding reflectors, or (3) landward shift (backstepping) of the shelf margin in transgressive system tracts at the bases of sequences (Fig. 6). Rarely are all three features together in a single sequence; usually only one or two are clearly present. The upper parts of sequences (highstand systems tract) commonly have prograding sigmoidal reflectors with downlap and local toplap. The systems-tract model in Figure 5 is consistent with geometries observed in seismic (Fig. 6) and lithologies observed in wells (Fig. 7) offshore Brunei.

Structural downwarping of the shelf and growth faulting have obscured many shelf-wide depositional geometries. As a result, some of the more basinward shelf margins may actually

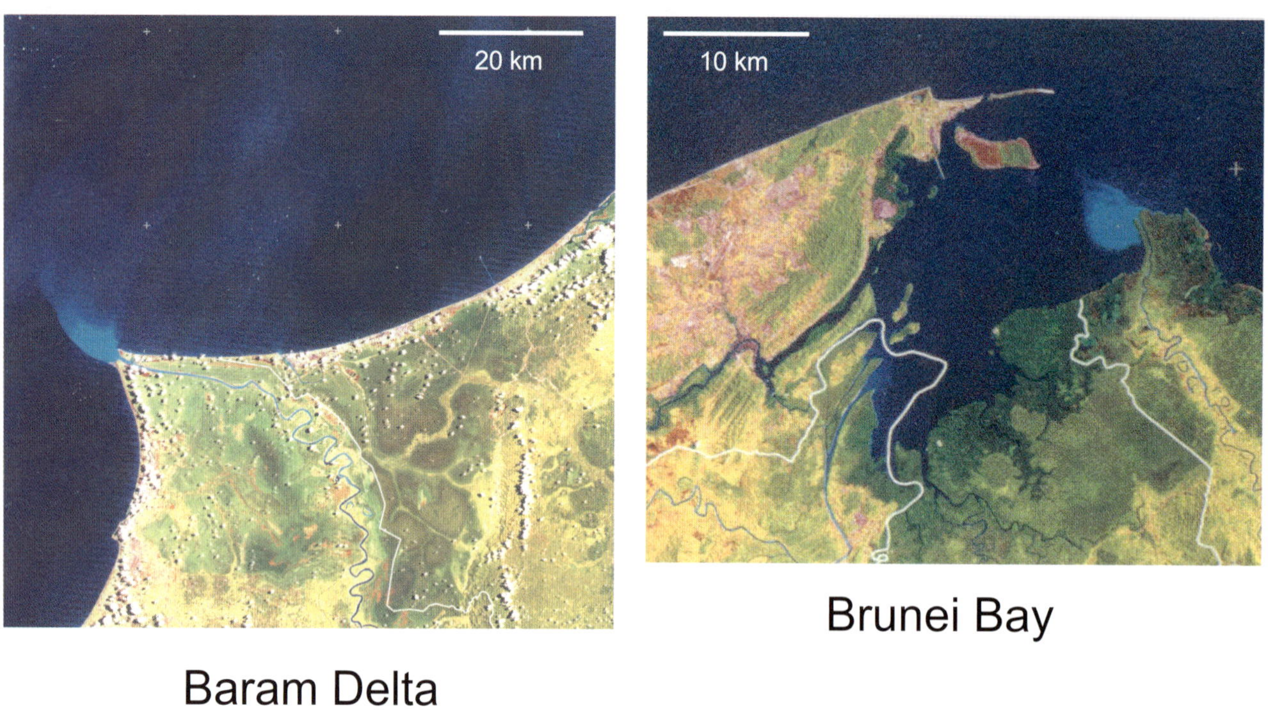

FIG. 4.—Satellite photograph of the modern wave-dominated Baram delta, and Brunei Bay with tide- and river-dominated deltas.

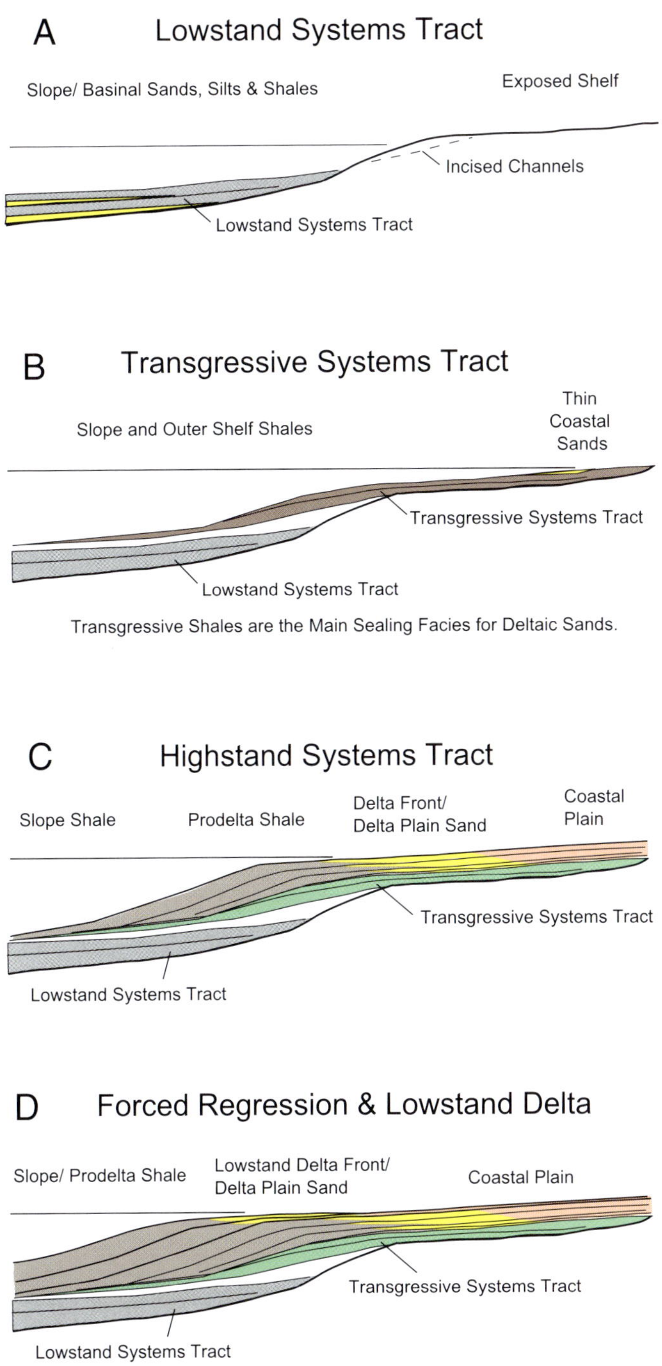

FIG. 5.—Stratigraphic model for Tertiary sequences of offshore Brunei.

represent "forced regressions", but a downstepping of the depositional system cannot be demonstrated in seismic geometries. Thick packages of slope strata generally lack a single onlap surface; rather, those slope strata are apparently foresets of the most basinward shelf margins. This suggests that many of the basinward shelf margin and delta complexes formed during the times of lowest relative sea level (Fig. 5D). The entire shelf often experienced continuous subsidence, causing substantial coastal-plain deposition on the modern inner shelf when progradation (probably forced regressions) caused the coastline to be at the modern middle to outer shelf (Fig. 5D). A significant depositional slope across the coastal plain and shelf also hindered sea level from dropping below the shelf margin for long periods of time.

Wireline Log Patterns

Deltaic sequences are clearly expressed by the gamma-ray curves, with high gamma-ray readings in the lower shaly parts of sequences, and gamma-ray radiation decreasing upward as silt and then sand increase upward in association with progradation and shallowing (Fig. 7). Coarsening-upward sequences that end basinward of the shelf margin (topset–foreset inflection) are still generally shaly. Coarsening-upward sequences that end at or slightly landward of the shelf margin (topset-foreset inflection) are still largely siltstones (Figs. 6 and 7). Reservoir-quality sands begin to cap coarsening-upward sequences approximately 1–3 km landward of the shelf margin (Figs. 6 and 7).

Biotic Pattern

Depositional environments and sequences can also be defined using biotic indices such as benthic foraminiferal biofacies, planktonic foraminiferal abundance, nannofossil abundance, and variations in benthic foraminiferal assemblages. In outer-shelf well #3, the intervals at 1670–1800 m and 1010–1245 m are dominated by the genera characteristic of deltaic and inner-shelf environments and lack, especially in the upper interval, deeper-water genera (Fig. 8). Hence, those intervals are interpreted as representing generally shallow and deltaic environments with periodic deepenings related to fourth-order sequence-scale transgressions. In those intervals (1670–1800 m; 1010–1245 m), sequences are defined by biotic patterns that are approximately coincident with coarsening-upward trends in gamma ray. The biotic characteristics of shallowing environments within sequences include: (1) decrease in faunal abundance, (2) decrease in diversity of both benthic and planktonic foraminifera, (3) decrease in relative percentage of arenaceous foraminiferal assemblages, (4) decrease in middle-shelf, outer-shelf, and slope species, and (5) increase in deltaic species (Fig. 8). The benthic foraminiferal biofacies show an inverse cyclic relationship between deltaic species and mid-shelf and outer-shelf groups (Fig. 8). These biotic sequences are commonly 100–150 m thick in the Pliocene at 1670–1800 m and 20 to 80 m thick in the Pleistocene at 1010–1245 m (Fig. 8). The Pliocene interval is in a generally deeper depositional environment in well #3, probably indicating greater accommodation space to fill during deposition of a sequence.

Higher gamma-ray values at 1245–1670 m indicate dominantly prodelta or upper-slope environments, and this is supported by (1) increased abundance and diversity of foraminiferal faunas, (2) abundant calcareous nannofossils, and (3) more abundant mid- and outer-shelf to possibly upper-slope biofacies (Fig. 8). In this environmentally deeper interval (1245–1670 m), individual sequences are not clearly defined by biotic trends (Fig. 8). Seismic correlations indicate that multiple sequences are present in the interval, but changes in relative sea level were not great enough to affect faunas in the deeper depositional environment.

Depositional Environments and Controls on Reservoir Distribution

Several observations suggest that wave-dominated deltas were common in the middle to late highstand parts of sequences in the Late Miocene and Pliocene, including (1) broad arcuate shelf-margin geometries discerned by seismic mapping in the

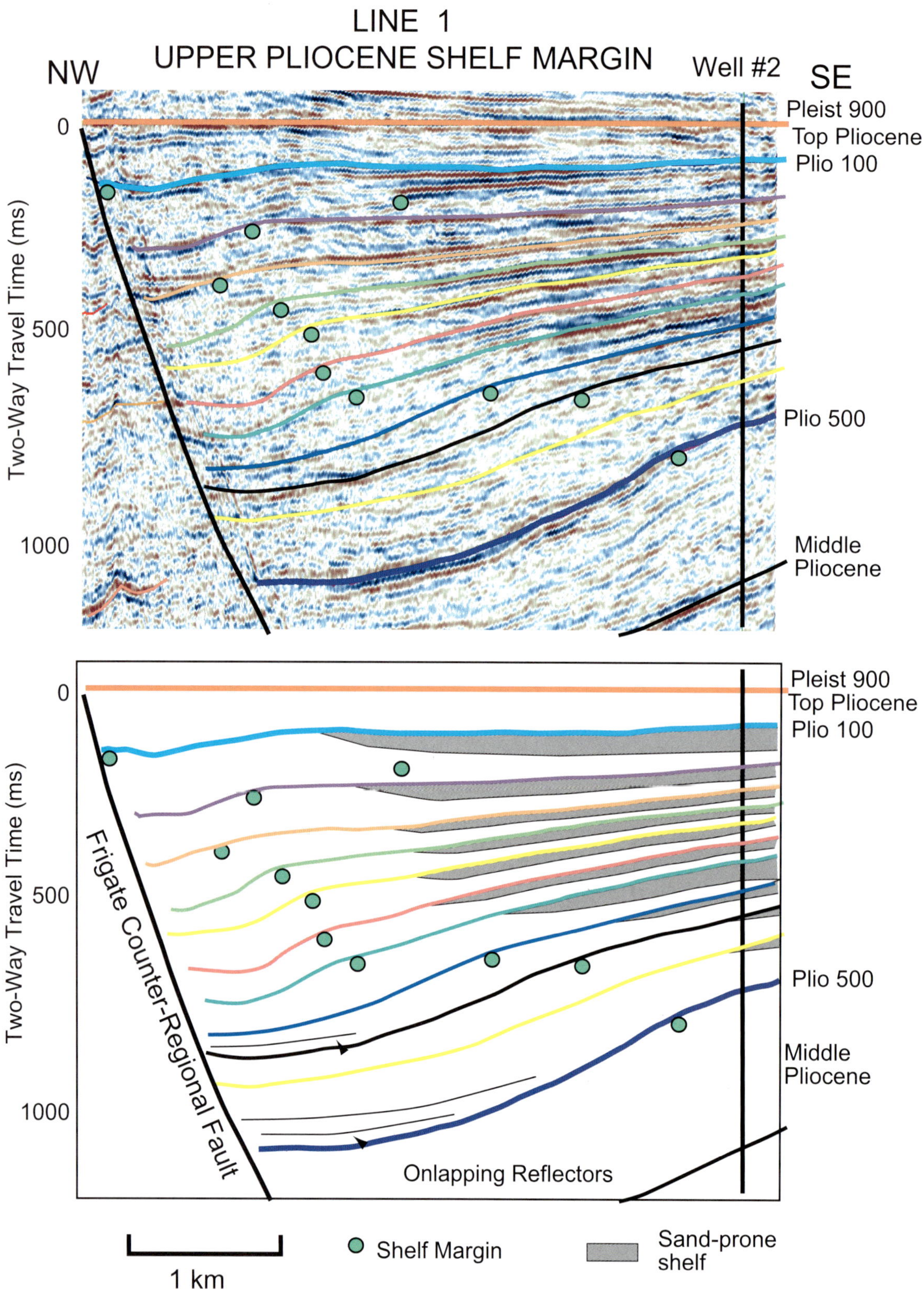

FIG. 6.—Seismic line flattened on the lower Pleistocene Pleist 900 horizon. Upper Pliocene sequences are prograding basinward (to the left) with (1) lowstand reflectors onlapping sequence boundaries on the slope, (2) prograding to aggrading shelf margins (inflection points from topsets to foresets), and (3) higher-amplitude, sand-prone topset reflectors starting approximately 1–3 km landward of the shelf margins. Gamma-ray curve for well 2 is shown in Figure 7. A transgressive backstepping is distinct in the sequence below Plio 100.

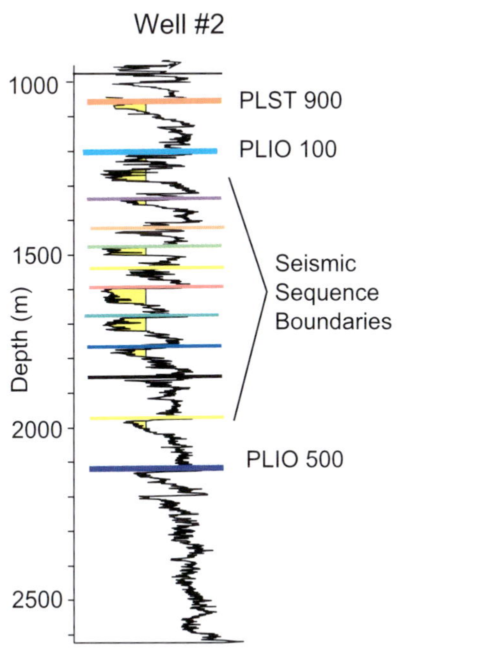

←

FIG. 7.—Well-log expression of upper Pliocene sequences in well 2 shown on Figure 6. Colored lines are sequence boundaries that correspond to sequence boundaries on seismic line in Figure 6. Lower parts of sequences are shaly with high gamma ray, and the upper parts of sequences have decreasing gamma ray related to upward coarsening. Yellow shading indicates sand.

FIG. 8.—Trends of microfossils in well #3. Gamma ray curve at left has been "butterflied". Sandy and silty intervals are highlighted in yellow. Paleoenvironments of sequences are shown. Deeper-water, more transgressive intervals are characterized by higher total diversity, higher planktonic and benthic diversity, and more outer-shelf forms. Shallower, more deltaic intervals are characterized by lower total, planktonic, and benthic foram diversity, and more deltaic forms.

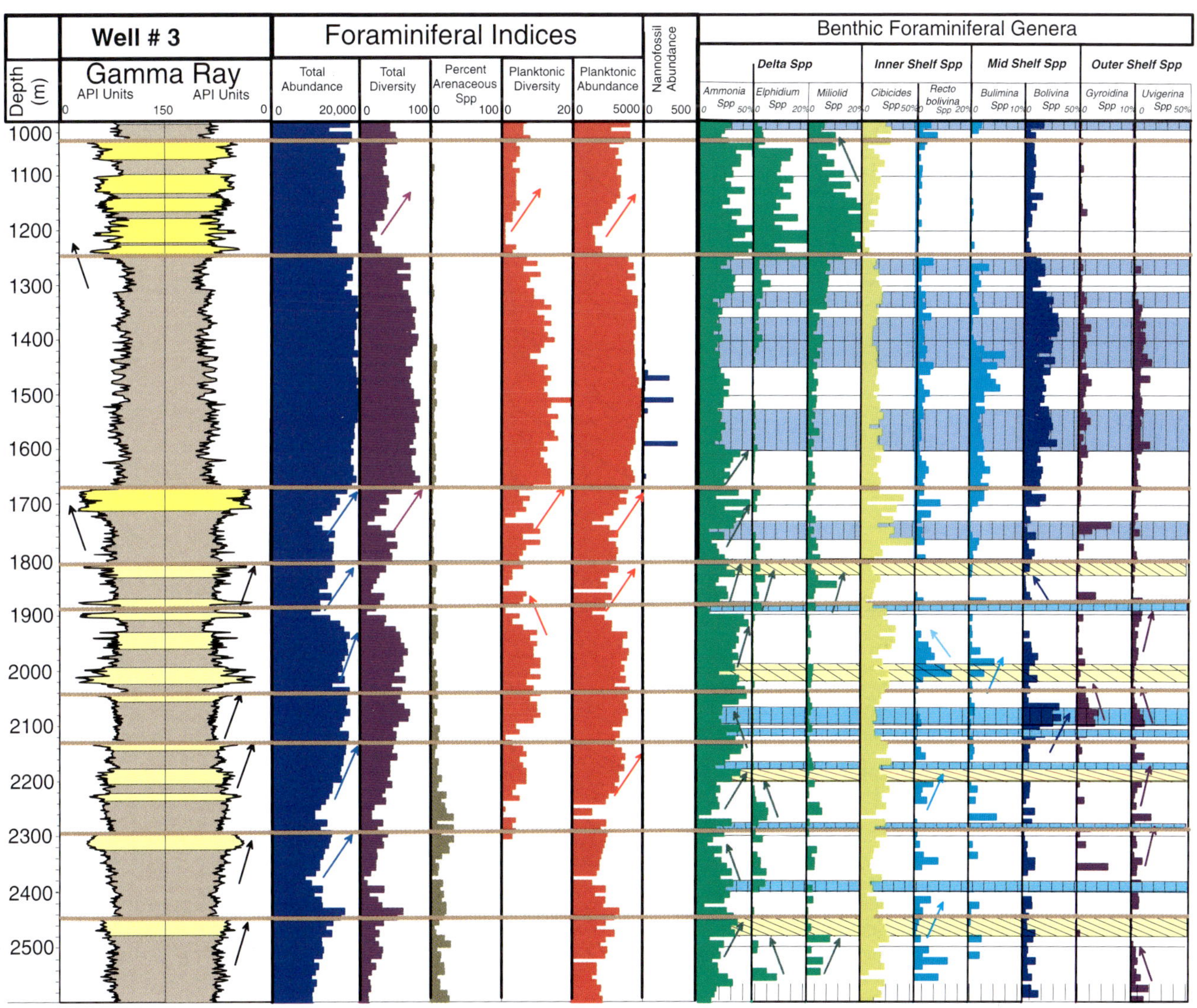

late highstand parts of sequences (Fig. 9) that are characteristic of wave-dominated deltas (Galloway, 1975); (2) seismic lines that show only gradual thickening and thinning along paleo-shorelines in contrast to rapid thickening and thinning of lobes that are characteristic of fluvial-dominated deltas (Shlemon, 1975), (3) widespread occurrence of sand along depositional strike in middle to late highstand parts of sequences, which is characteristic of wave-dominated deltas (Galloway, 1975), and (4) paleogeographic reconstructions that suggest a broad open ocean to the north from the Late Miocene to the present (Hall, 1996; Wilson and Moss, 1999). Compared to river- or tide-dominated deltas (Miall, 1976; Miall, 1984), the distribution of sand on the modern middle to outer Brunei shelf is relatively consistent along depositional strike. In summary, the highstand parts of Brunei deltaic sequences probably start with tide- and fluvial-dominated deltas as described by Caline and Huong, (1992) and Lambiase et al. (this volume) and pass upward to wave-dominated deltas during the later parts of the highstand (Fig. 10).

Sands are well developed in inner-shelf environments, but outer-shelf and upper-slope environments were generally shaly. An exception to this is where incised upper slope channels were infilled with sand (generally associated with sea-level lowstands)(Sandal, 1996, p. 161). Reservoir-quality sands occur 1 km or more landward of the contemporaneous shelf margin and are apparently associated with wave-dominated deltas with sand redistributed along paleo-shorelines. Landward of the paleo-shorelines, coastal-plain environments with fluvial channel sands and overbank shales are present (Fig. 5; Sandal, 1996). Substantial thicknesses of outer-shelf shales suggest that the Pliocene deltaic systems of Brunei were mud-dominated. Similarly, sands are best developed in coastal and coastal-plain environments in prograding Miocene and Pliocene strata in adjacent areas of Sabah (Rice-Oxley, 1991).

LONG-TERM DEPOSITIONAL TRENDS

Deltaic sediments of the Brunei shelf show distinct variations in the Late Miocene, Pliocene, and Pleistocene. At the end of the Middle Miocene (11 Ma), the shelf margin was approximately at the modern Brunei coastline except in the northeast near Sabah, where the shelf margin of the early Champion delta was approximately 30 km basinward of the modern shoreline (Sandal, 1996)(Fig. 9). The shelf margin on the northeast part of the Brunei shelf (Champion delta) prograded 40 km during the Late Miocene (11 to 5.3 Ma [Mio 250]; TB3.1), giving an average rate of

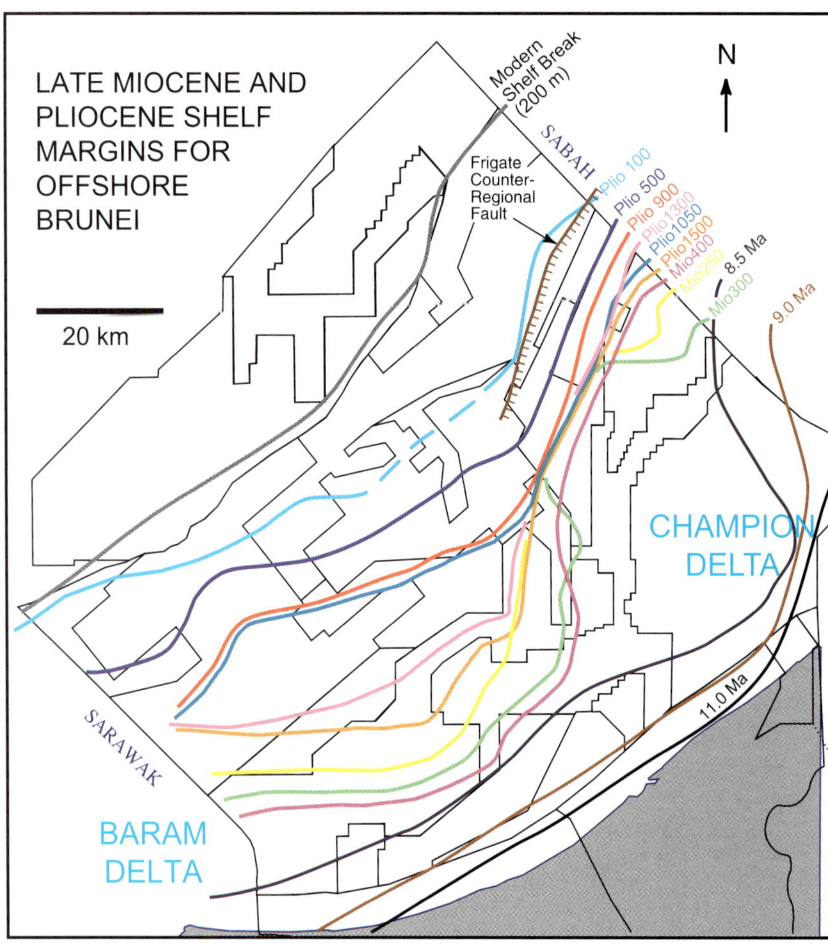

FIG. 9.—Map showing upper Miocene and Pliocene shelf margins which generally prograde basinward (to the northwest). The shelf margins were mapped from inflections between shelfal topsets and slope foresets. The modern shelf break is approximately the latest Pleistocene shelf margin. See Figure 3 for ages of seismic horizons. Shelf margins in millions of years (Ma) are from Sandal (1996).

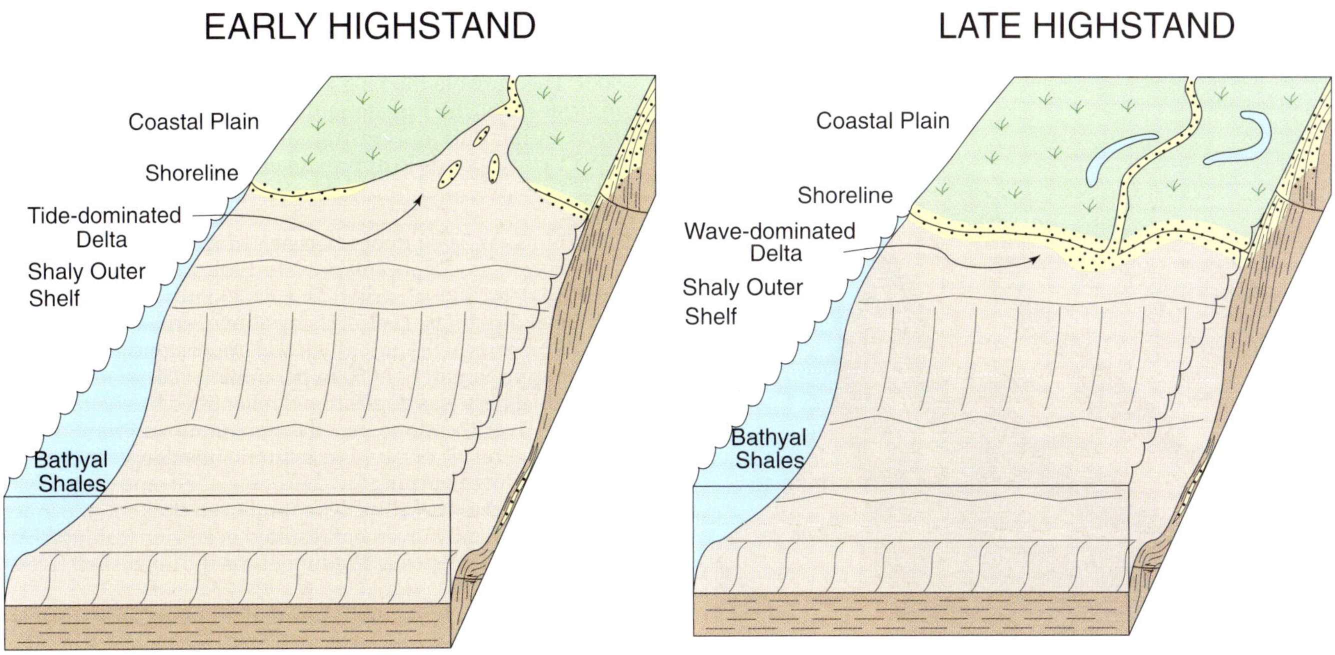

FIG. 10.—Schematic model of delta evolution during sea-level highstands.

progradation of approximately 4 km/My. In contrast, shelf-margin progradation was only 20 km on the southwest part of the Brunei shelf during the same period of time (Fig. 9). Structural deformation and lack of data have not allowed us to conduct a detailed study on most of lower Upper Miocene, Middle Miocene, and Lower Miocene.

Isochron maps show that the locus of deposition shifted to the Baram delta area during the latest Miocene (Fig. 11). The shelf margin in the Baram delta area prograded approximately 30–40 km basinward during the Pliocene (approximately 10 km/My), while the Champion delta was aggradational during the early to middle Pliocene, and modestly progradational during the late Pliocene with only 12 km of total Pliocene progradation (Figs. 9, 12–14, 16, 17). It is thought that decreased sediment supply coupled with counter-regional faulting on the shelf and upper slope contributed to slow Pliocene progradation in the Champion area. The decreased sediment supply probably resulted from a capturing of substantial parts of the Champion hinterland drainage by the Baram River.

Approximately 21 sequences are recognized in the Pliocene (approximately 3.7 My in duration; Berggren et al., 1995) suggesting they have an average duration of approximately 180,000 years. We are probably missing some sequences, and hence Pliocene sequences may be closer to 120,000 years in duration, and be consistent with a major Milankovich frequency.

Deltaic strata offshore Brunei show a consistent basinward shift in depocenters from the Late Miocene, through the Pliocene and Pleistocene (Figs. 11–15). Although we cannot precisely determine the time represented by each of the stratigraphic intervals in Figures 11–15, we estimate that each represents 1–2

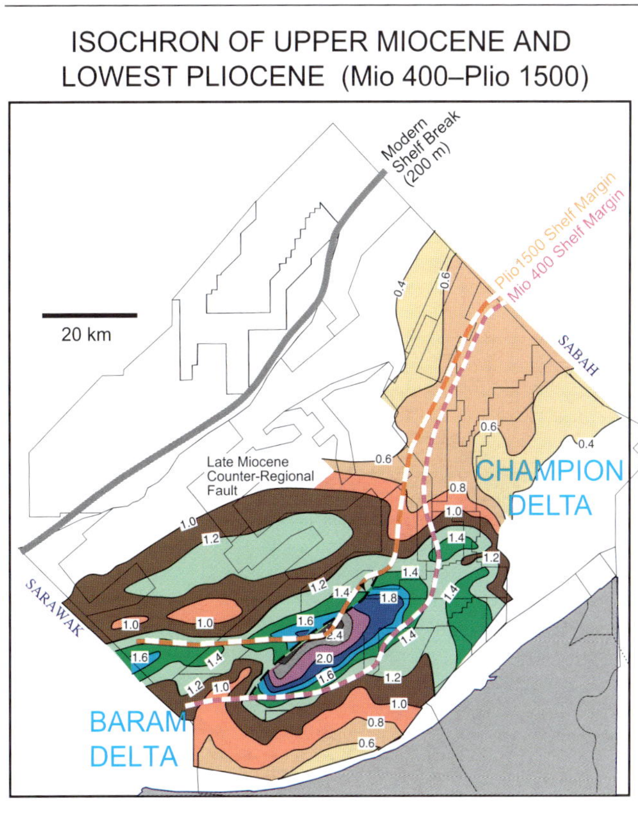

→

FIG. 11.—Isochron map of upper Miocene and lowest Pliocene strata (Mio 400 to Plio 1500). Depocenter for the Baram delta roughly corresponds with a contemporaneous counter-regional fault and the location of the modern Ampa shale ridge. Thicknesses are from seismic data and are in seconds of two-way travel time (TWTT). One second of TWTT corresponds to approximately 1000 m of strata.

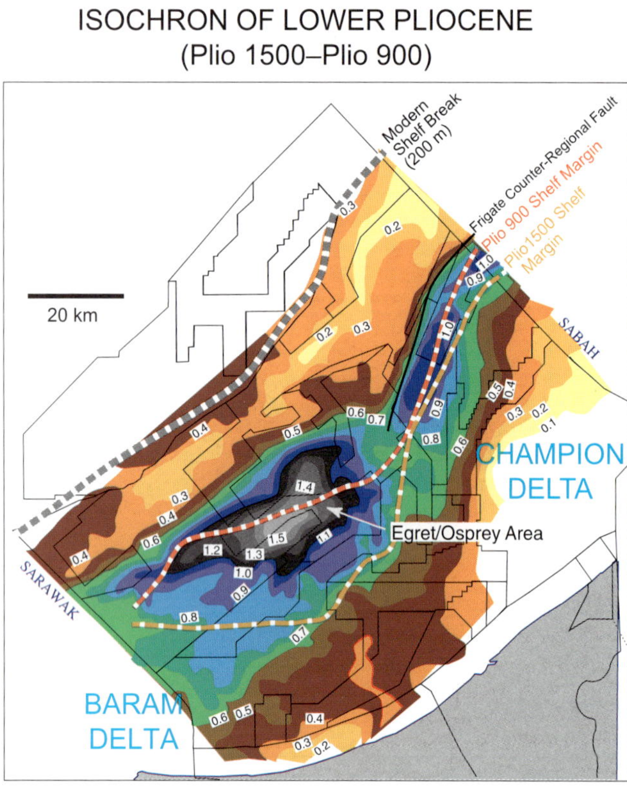

←

FIG. 12.—Isochron map of lower Pliocene strata (Plio 1500 to Plio 900). The variable thickness of strata in the Champion area indicates that the Frigate fault was active. Thicknesses are from seismic data and are in seconds of TWTT. One second of TWTT corresponds to approximately 1000 m of strata.

million years (i.e., Figure 11 is approximately the top third of the Upper Miocene [5.2 My / 3 = 1.7 My]; ~ 3 subequal periods of the Pliocene [3.7 My / 3 = 1.2 My] for Figures 12-14; Pleistocene ~ 1.7 My for Figure 15). In general, maximum sediment thickness occurs near and basinward of the contemporaneous shelf margins. As much as 3,000 m of Pliocene sediment accumulated in some depocenters. Depocenters were typically 40–60 km long (parallel to depositional strike) and 10–30 km wide (parallel to depositional dip) (Figs. 11–15). Using above estimates, sedimentation rate on parts of the outer shelf and upper slope of the Baram delta was approximately 1,000 m per million years. Perhaps 200 m of that thickness was depositional relief filled by shelfal strata prograding over the slope, but the remaining accommodation space was the result of rapid structural subsidence. In several areas, rapid downwarping of the outer shelf and slope was coupled with very slow subsidence and even uplift on the inner shelf. This rotational movement resulted in erosion (cannibalization) of previously deposited sediments on the inner shelf (Pleistocene in Figure 16).

Shelf-margin progradation during the Pleistocene was relatively rapid and uniform (approximately 20 km; 11 km/My) across the Brunei shelf. However, the Pleistocene is much thicker in the southwest because of large throw along some outer-shelf growth faults. Much of the Pleistocene progradation was probably associated with lowstand deltas.

Stratigraphic patterns in these deltaic strata are difficult to relate to global sea-level changes. Two major transgressive events have been recognized: (1) near the Miocene–Pliocene boundary and (2) near the Pliocene–Pleistocene boundary. Owing to lack of precise age dates, these cannot be clearly related to global eustatic events. In addition, most slope strata cannot be accurately correlated to lowstand surfaces on the shelf because of lack of biostratigraphic precision and faults that cause difficulty in correlating seismically between the shelf and slope. However, erosional channels in the lower to middle Pliocene of the Baram delta area (Sandal, 1996, figure 5.61, p. 190) indicate major drops in relative sea level.

RELATION OF SEDIMENTATION TO STRUCTURAL MOVEMENT

As indicated above, syndepositional tectonics on the Brunei shelf affected deltaic deposition, and vice versa. The Brunei shelf is underlain by a thick interval of "Setap" shale, interpreted as deep marine (Sandal, 1996). The base of the Setap shale is apparently very deep and has not been recognized in wells or seismic data on the shelf. The "Setap" shale represents a relatively plastic substrate over which deltaic sands, silts, and shales prograded. McClay et al. (1998) have shown that features similar to those offshore Brunei can be produced by sediment loading of a two-layer plastic (shale) substrate.

The Brunei shelf and slope have a number of related structures. The inner shelf contains diapiric (?) shale ridges, adjacent shale-evacuation troughs, and a variety of other complex fault

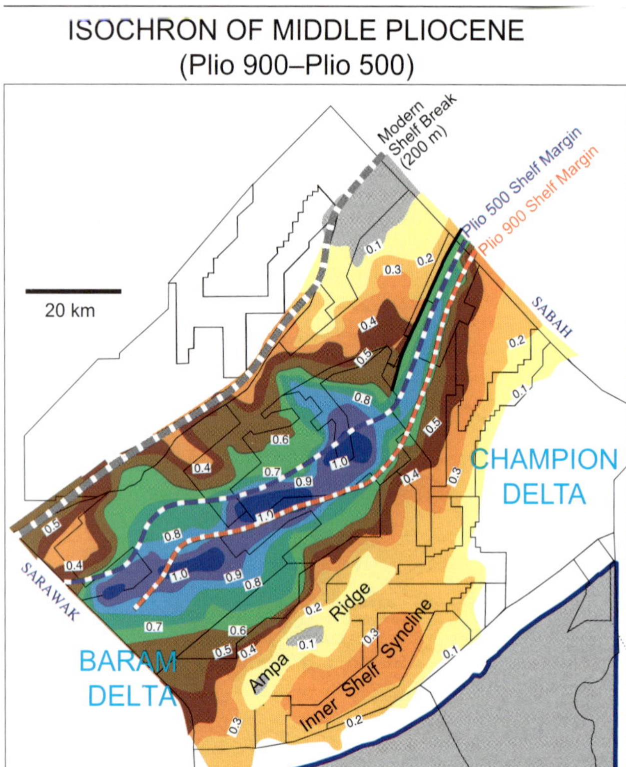

←

FIG. 13.—Isochron map of middle Pliocene strata (Plio 900 to Plio 500). Thinning in the vicinity of the Ampa ridge indicates active diapiric movement on that ridge. Thicknesses are from seismic data and are in seconds of TWTT. One second of TWTT corresponds to approximately 1000 m of strata.

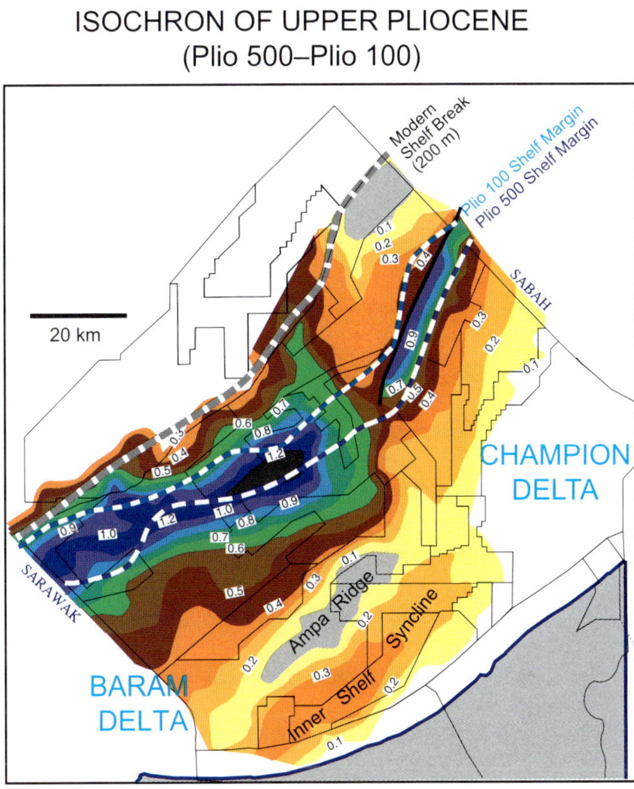

FIG. 14.—Isochron map of upper Pliocene strata (Plio 500 to Plio 100). The Frigate fault was still active, resulting in a hinged subsidence between the Frigate fault and the Champion inner shelf. Thicknesses are from seismic data and are in seconds of TWTT. One second of TWTT corresponds to approximately 1000 m of strata.

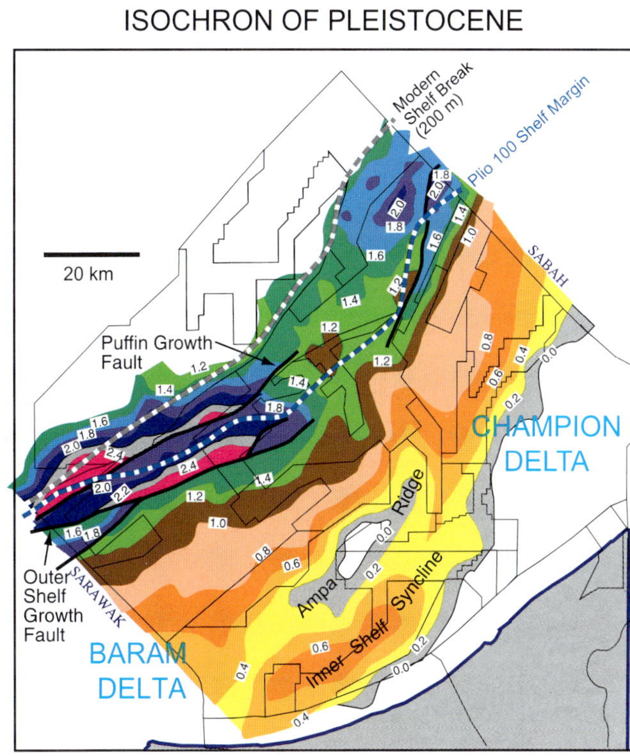

FIG. 15.—Isochron map of Pleistocene strata (Plio 100 to modern mud line). The outer-shelf down-to-the-basin growth faults were active. Thinning in the vicinity of the Ampa ridge indicates active diapiric movement on that ridge continued. Thicknesses are from seismic data and are in seconds of TWTT. One second of TWTT corresponds to approximately 1000 m of strata.

structures. Down-to-the-basin and counter-regional normal growth faults are common on the outer shelf and upper slope. Toe-thrust anticlines are common on the middle and lower slope (Fig. 2). A "central graben" area with shale diapirs is present on the upper slope between growth faults and toe-thrust anticlines (Fig. 2; McClay et al., 1998). In general, counter-regional growth faults initially developed at or basinward of the shelf margin during early progradation (Fig. 18B). In the northeast shelf (Champion area), the large Frigate counter-regional fault continued to move after the shelf margin prograded over it (Figs. 15, 16, 18C, E). In outer-shelf environments of the Baram area (to the southwest), early counter-regional faults were succeeded by down-to-the-basin normal growth faults (Fig. 18D). Those down-to-the-basin normal growth faults developed approximately 0.5 km or more landward of the contemporaneous shelf margin (Figs. 17 and 18D), and hence appear during later stages of progradation.

The late Miocene was the time when the locus of major deltaic sedimentation shifted from the Champion delta to the Baram delta (Figs. 11–15). The depocenter for the uppermost Miocene section occurs in an area with a syndepositional counter-regional fault (Fig. 11). This coincides with the location of the Ampa shale ridge formed in the late Pliocene and Pleistocene (Fig. 15). The uppermost Miocene section has a fairly constant thickness in the Champion area (Fig. 11), indicating that the Frigate counter-regional fault and Champion anticline were not actively moving then.

Thickening of the lower Pliocene section on the downthrown side of the Frigate counter-regional fault indicates that the Frigate fault was active (Fig. 12). A gradual thinning of sediments from the Frigate fault to the Champion Field anticline indicates hinged syndepositional subsidence with subsidence steadily decreasing to the Champion Field anticline. Lower Pliocene strata thin over toe-thrust anticlines on the northeastern (Champion) side of the present Brunei lower slope, but not the southwest side of the Brunei lower slope, suggesting toe thrusting on the lower slope contemporaneous with and probably caused by counter-regional faulting on the adjacent outer shelf and upper slope. Relatively uniform lower Pliocene sediment thicknesses in the vicinity of the Ampa Ridge and Inner Shelf Syncline (Fig. 12) indicate that those structures had not started moving. Deposition of approximately 1500 m of slope and outer-shelf strata in the Egret–Osprey area (Figs. 2 and 12) indicate substantial early Pliocene downbowing in that area.

Middle Pliocene strata are quite variable in thickness on the Brunei shelf, indicating that some early Pliocene structural features continued to move and new ones became active. Middle Pliocene strata thicken greatly on the downthrown side of the Frigate counter-regional fault and thin toward Champion Field, indicating continued hinged subsidence and uplift in that area (Fig. 13). Lower-slope strata on the northeast side of Brunei show continued thinning over toe-thrust anticlines, indicating that those structures continued to be active. The main locus of shelf

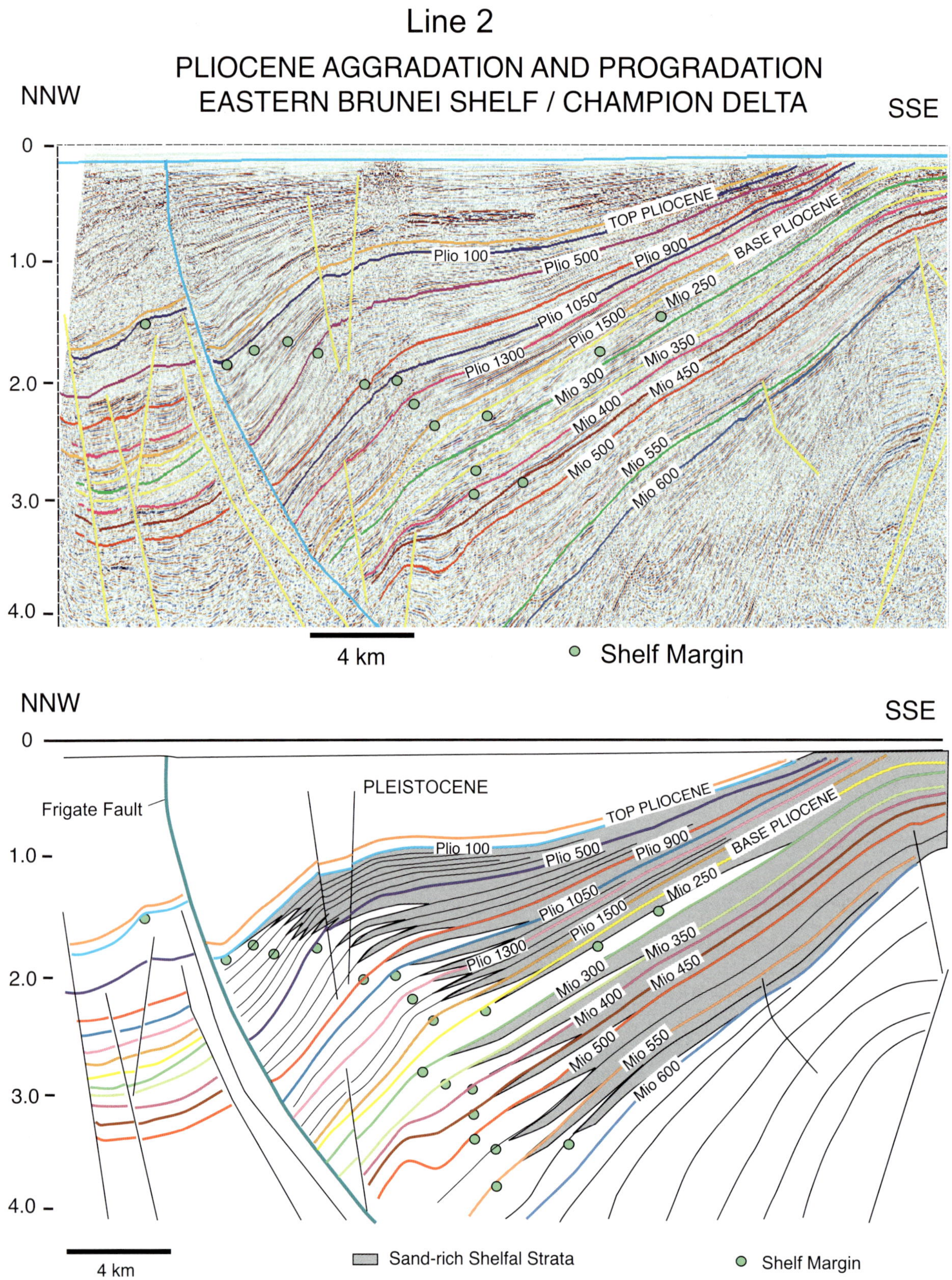

FIG. 16.—Seismic line and interpretation of stratigraphy of the northeast (Champion) side of the Brunei shelf. The Frigate fault was actively moving during the Pliocene and Pleistocene. The shelf margin was generally aggradational during the latest Miocene and Pliocene, and then began prograding in the latest Pliocene and Pleistocene. See Figure 2 for approximate location. Vertical scale is seconds of two-way travel time (TWTT).

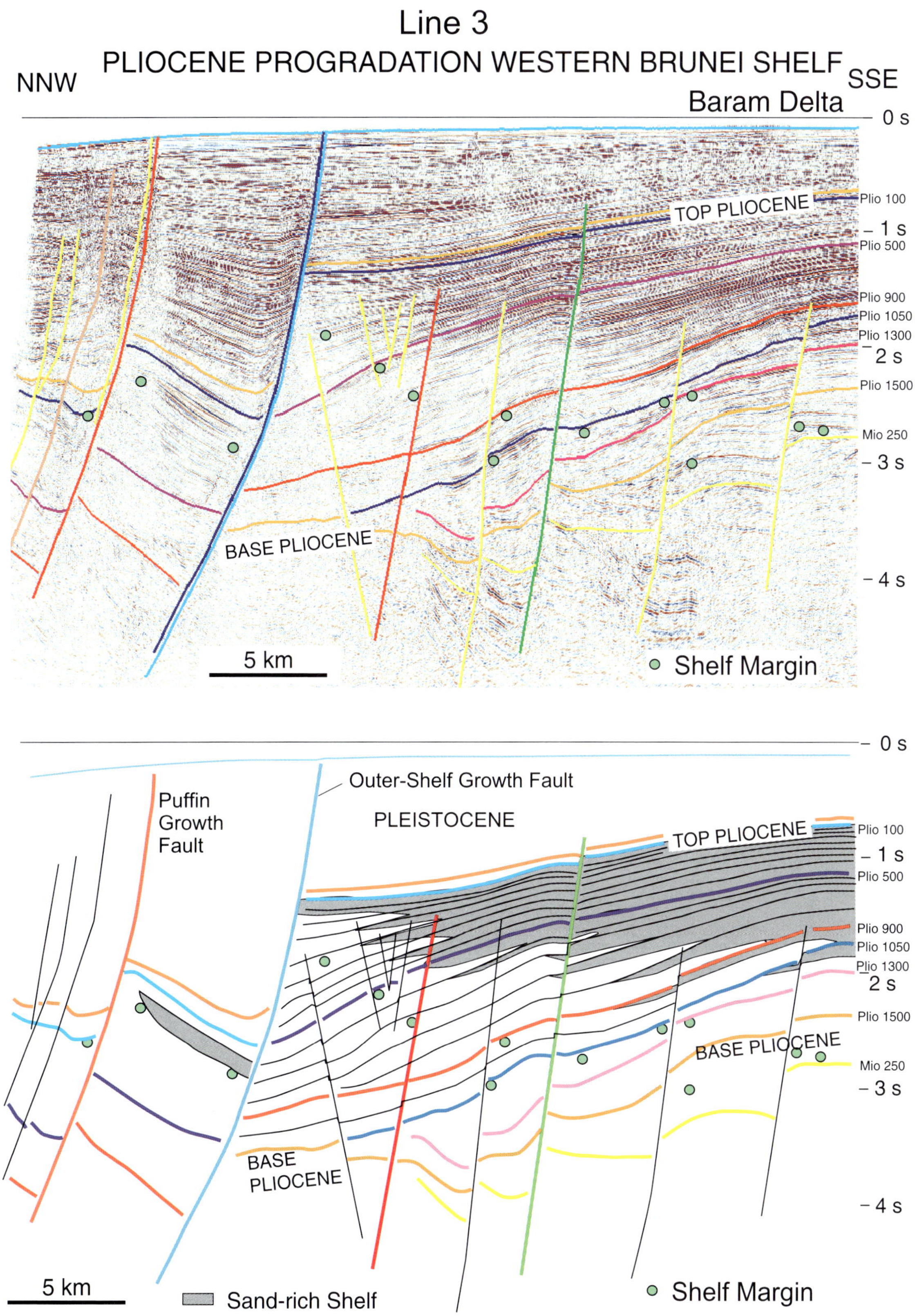

FIG. 17.—Seismic line and interpretation of stratigraphy of the southwest (Baram) side of shelf. The outer-shelf down-to-the-basin faults were actively moving during the Pleistocene. The shelf margin was generally progradational during the Pliocene and Pleistocene. A counter-regional fault just to the right of the "Outer Shelf Growth Fault" shows offset of lower Pliocene reflectors, and preceded the major Pleistocene "down-to-the-basin" growth faults. See Figure 2 for approximate location.

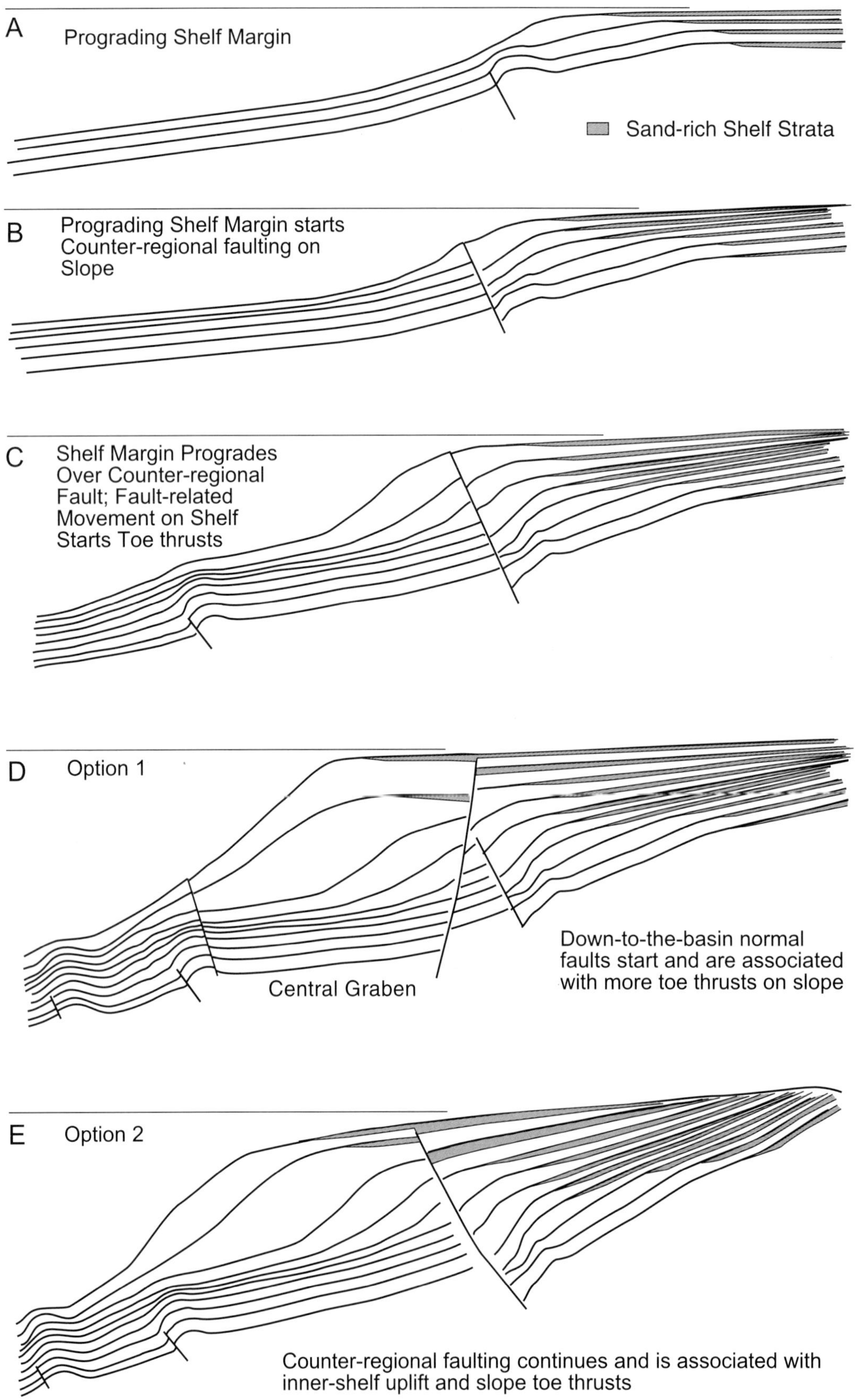

FIG. 18.—Typical sequence of syndepositional structures associated with the prograding Champion and Brunei deltas.

sediment deposition was an elongate trend on the upper slope and outer shelf in the Baram area (modern middle to outer shelf; Figure 13). A distinct thickening of sediments on the downthrown side on the Fairley fault indicates that the normal, down-to-the-basin growth faults were active in that area. Thinning over the Ampa Ridge and thickening in the adjacent Inner Shelf Syncline (Fig. 13) indicate that those structures started moving in the middle Pliocene.

Most middle Pliocene structure and sedimentation trends continued into the late Pliocene. The main elongate Baram depocenter shifted basinward with the prograding shelf margin and slope (Fig. 14). Some minor counter-regional and down-to-the-basin normal faults were active in the Baram area, but the large Outer Shelf and Puffin down-to-the basin faults had not started moving. Variations in sediment thickness indicate continued movement on the Frigate counter-regional fault, northeast lower-slope toe thrusts, Ampa shale ridge, and adjacent Inner Shelf Syncline (Fig. 14).

The most striking changes in depositional structure during the Pleistocene occurred on the southwest (Baram) part of the outer shelf and lower slope. Two large down-to-the-basin normal growth faults, Outer Shelf and Puffin, developed on the southwest part of the outer Brunei shelf (Figs. 15 and 17). The Outer Shelf Growth fault is older and more landward than the Puffin fault, has throws in excess of 1000 m, and is apparently still moving. The more basinward fault, the Puffin fault, is young, is still actively moving, and has as much as 400 m of throw. As Figure 15 shows, Pleistocene sediments thicken considerably on the downthrown side of these growth faults. Substantial subsidence has occurred along the entire outer shelf of Brunei during the Pleistocene, with the entire outer shelf having in excess of 1000 m of Pleistocene sediment and 2400 m of Pleistocene sediment on the downthrown sides of the Outer Shelf Growth Fault (Fig. 15). Thinning of Pleistocene strata over toe-thrust anticlines in the west part of the lower slope indicate that toe-thrust anticlines began moving on the west part of the lower slope during the Pleistocene. Most other offshore structures continued to move during the Pleistocene, including the Frigate counter-regional fault, northeast lower-slope toe thrusts, Ampa shale ridge, and adjacent Inner Shelf Syncline (Fig. 15). Pleistocene uplift in the Champion area resulted in depositional thinning and truncation of substantial parts of the Pleistocene and Pliocene shelf section (Fig. 16).

A model for structural deformation associated with progradation of the wave-dominated deltaic system is shown in Figure 18, and is in many ways similar to ideas of Morley and Guerin (1996). Initial progradation of the wave-dominated delta initiated substantial outer-shelf and upper-slope subsidence. As deltaic sediments prograded over former toe thrusts, counter-regional faults developed on the shaly upper slope where surfaces of weakness were present (Fig. 18B). The fault surface geometries of the counter-regional faults suggest that they may be following and reversing relict toe-thrust faults. In some locations, counter-regional faults continued moving after the outer shelf had prograded over them (Fig. 18C, E). Down-to-the-basin growth faults started on the outer shelf where major counter-regional faults were no longer active (Fig. 18D). It is not clear why down-to-the-basin growth fault are dominant in the Pleistocene of the southwest Brunei outer shelf whereas counter-regional growth faults dominate on the northeast side. Sandal (1996) suggests that the eastern side of the Brunei shelf is dominantly transpressional and the western side is dominantly transtensional, hence counter-regional growth faults in the east and down-to-the-basin growth faults in the west. Large amounts of deltaic sediments have been ponded by those Pleistocene outer-shelf growth faults. Subsidence along those growth faults has slowed basinward progradation of the delta complexes.

Diapiric shale ridges formed on the inner shelf landward of massive shelf-margin subsidence and counter-regional faults (Figs. 13–15). The ridge-like morphology of diapiric shales on the inner shelf may be related to widespread sediment loading associated with wave-dominated deltaic deposition; in contrast, fluvial-dominated deltas have more localized depocenters, resulting in more cylindrical diapiric shales.

Toe-thrust anticlines develop on the middle and lower slope in coincidence with times of major movement on upper-slope and outer-shelf growth faults (Fig. 18D, E). Growth faults extend down into a thick massive shale zone, and distinct decollement fault surfaces connecting them with slope toe thrusts could not be recognized in seismic data. Toe thrusts also extend down into the thick massive shales on the slope. A "central graben" (as per McClay et al., 1998) with narrow shale ridges is present between active toe-thrust faults and counter-regional or down-to-the-basin growth faults (Figs. 2 and 18D).

CONCLUSIONS

The Brunei shelf has been a dynamic deltaic system during the Late Miocene, Pliocene, and Pleistocene. Mapping of Late Miocene and Pliocene shelf margins shows that highstand deltaic deposition on the Brunei shelf generally occurred in association with wave-dominated deltas. Fourth-order sequences are commonly 50–100 m thick on the middle to outer shelf, with shaly transgressive bases passing up to sandy progradational tops. The Champion delta prograded 40 km during the Late Miocene but was largely aggradational during the Pliocene because of decreased sediment supply and counter-regional faulting. The Baram area was aggradational to slightly progradational during the latest Miocene and then prograded approximately 40 km during the Pliocene. Isopach maps show that near the shelf margin, sediment loading associated with rapid deltaic accumulation caused counter-regional faults to start on the upper slope. To the northeast, counter-regional movement continued after the outer shelf prograded over the fault. Down-to-the-basin growth faults developed on the outer shelf to the west in an area where counter-regional faults stopped moving. Currently, down-to-the-basin faults dominate the outer shelf in the Baram area (to the west), at the same time when counter-regional faults dominate in the Champion area (to the northeast). Both fault systems have trapped large amounts of sands and shale deposited in wave-dominated deltaic systems.

ACKNOWLEDGMENTS

This paper is part of a regional study of the shelf of Negara Brunei Darussalam. We benefited greatly from discussions with other team members, including Gregor Dixon, John Baines, Bill Mueller, Sherman Smith, Chip Morgan, Bren Blackburn, Paul Wagenhofer, and Nursyafrin Arbi. Reviews by Shirley E. van Heck, Dag Nummedal, and three anonymous reviews greatly improved this paper. We thank the Petroleum Unit of the Government of Negara Brunei Darussalam, Unocal Corporation, and Fletcher Challenge Energy for permission to publish this paper.

REFERENCES

Berggren, W.A., Kent, D.V., Swisher, C.C., and Aubry, M.-P., 1995, A revised Cenozoic geochronology and chronostratigraphy, *in* Berggren, W.A., Kent, D.V., and Hardenbol, J., eds., Geochronology, Time

Scales and Global Stratigraphic Correlation; A Unified Temporal Framework for Historical Geology: SEPM, Special Publication 54, p. 129–212.

CALINE, B. ,AND HUONG, J., 1992, New insight into the recent evolution of the Baram delta from satellite imagery: Geological Society of Malaysia, Bulletin, v. 32, p. 1–13.

GALLOWAY, W.E., 1975, Process framework for describing the morphologic and stratigraphic evolution of deltaic depositional systems, *in* Broussard, M.L., ed., Deltas: Models for Exploration: Houston Geological Society, p. 87–98.

HALL, R., 1996, Reconstructing SE Asia, *in* Hall, R., and Blundell, D.J., eds., Tectonic evolution of Southeast Asia: Geological Society of London, Special Publication 106, p. 153–184.

JOHNSON, H.D., KUUD, T., AND DUNDANG, A., 1989, Sedimentology and reservoir geology of the Betty field, Baram delta province, offshore Sarawak, NW Borneo: Geological Society of Malaysia, Bulletin, v. 25, p. 119–161.

MCCLAY, K.R., DOOLEY, T., AND LEWIS, G., 1998, Analog modeling of progradational delta systems: Geology, v. 26, p. 771–774.

MORLEY, C.K., AND GUERIN, G., 1996, Comparison of gravity-driven deformation styles and behavior associated with mobile shales and salt: Tectonics, v. 15, p. 1154–1179.

MIALL, A.D., 1976, The Triassic sediments of Sturgeon Lake South and adjacent areas, *in* Lerand, M., ed., The Sedimentology of Selected Clastic Oil and Gas Reservoirs in Alberta: Canadian Society of Petroleum Geologists, p. 25–43.

MIALL, A.D., 1984, Deltas, *in* Walker, R.G., ed., Facies Models, Second Edition: Geological Society of Canada, p. 105–118.

PLINT, A.G., EYLES, N., EYLES, C.H., AND WALKER, R.G., 1992, Control of sea level change, *in* Walker, R.G., and James, N.P., eds., Facies Models: Response to Sea Level Change: Geological Association of Canada, p. 15–25.

POSAMENTIER, H.W., JERVEY, M.T., AND VAIL, P.R., 1988, Eustatic controls on clastic deposition I—conceptual framework, *in* Wilgus, C.K., Hastings, B.S., Kendall, C.G.St.C., Posamentier, H.W., Ross, C.A., and Van Wagoner, J.C., eds., Sea-Level Changes: An Integrated Approach: SEPM, Special Publication 42, p. 111–124.

RANGIN, C., BELLON, H., BERNARD, F., LETOUZEY, J., MULLER, C., AND SANUDIN, T., 1990, Neogene arc–continent collision in Sabah, northern Borneo (Malaysia): Tectonophysics, v. 183, p. 305–319.

RICE-OXLEY, E.D., 1991, Palaeoenvironments of the Lower Miocene to Pliocene sediments in offshore NW Sabah area: Geological Society of Malaysia, Bulletin, v. 28, p. 165–194.

SANDAL, S.T., ed., 1996, The Geology and Hydrocarbon Resources of Negara Brunei Darassalam, 1996 Revision: Syabas, Bandar Seri Begawan, Brunei Darassalam, ISBN 99917-900-0-4, 243 p.

SHLEMON, R.J., 1975, Subaqueous delta formation—Atchafalaya Bay, Louisiana, *in* Broussard, M.L., ed., Deltas: Models for Exploration: Houston Geological Society, p. 209–221.

TAN, D.N.K., AND LAMY, J.M., 1990, Tectonic evolution of the NW Sabah continental margin since the Late Eocene: Geological Society of Malaysia, Bulletin, v. 27, p. 241–260.

WILSON, M.E.J., AND MOSS, S.J., 1999, Cenozoic paleogeographic evolution of Sulawesi and Borneo: Palaeogeography, Palaeoclimatology, Palaeoecology, v. 145, p. 303–337.

RESERVOIR HETEROGENEITY OF MIOCENE–PLIOCENE DELTAIC SANDSTONES AT ATTAKA AND SERANG FIELDS, KUTEI BASIN, OFFSHORE EAST KALIMANTAN, INDONESIA

ARTHUR S. TREVENA
Unocal Exploration and Production Technology, 14141 Southwest Freeway. Sugar Land, Texas 77478 U.S.A.
e-mail: atrevena@unocal.com
YOSEPH J. PARTONO
Unocal Indonesia Co., Pasir Ridge, P. O. Box 276, Balikpapan, East Kalimantan, Indonesia
e-mail: yjpartono@unocal.com
AND
TOM CLARK
Unocal Gulf of Mexico Region, 14141 Southwest Freeway. Sugar Land, Texas 77478 U.S.A.
e-mail: tom.clark@unocal.com

ABSTRACT: Heterogeneous sandstones in upper to middle Miocene lobes of the Mahakam Delta are prolific hydrocarbon reservoirs in the lower Kutei Basin, offshore East Kalimantan, Indonesia. At Attaka and Serang fields, more than 99 million cubic meters (627 million barrels) of oil and 37 billion cubic meters (1.3 trillion cubic ft) of gas have been produced from sandstones that we interpret as delta-front bars and tidal / fluvial distributary channels. Sand bodies of the modern Mahakam delta are analogs for many of these reservoirs.

Delta-front bars are burrowed to laminated, fine-grained sandstones that form equant to somewhat dip-elongate bodies that range in thickness from < 1 m to 5 m and may exceed several kilometers in width. Cross-stratified, coarse- to fine-grained tidal / fluvial distributary channel sandstones are 3 to 17 m thick and narrow (< 1.5 km wide).

Distributary-channel sandstones are typically highly porous (20–35%) and permeable (100–10,000 md), although tidal distributaries exhibit permeability heterogeneity, due to mud drapes and local burrows. Delta-front sandstones, although areally extensive, have generally poorer reservoir quality than the distributary channel sandstones (k = < 0.1–1000 md; porosity = 10–25%). Also, the delta-front sands exhibit major centimeter- to decimeter-scale variations in permeability, which are related to variations in clay content and intensity of burrowing.

Stacked, distributary-channel reservoirs are especially well developed in the Upper Miocene "Main Deltaics" interval, which we interpret as a succession of lowstand deltaic lobes.

The coarsest-grained and thickest deltaic sandstones typically accumulate during relative lowstands, times when deltas have prograded long distances across marine shelves. Such lowstand sandstones also form the most porous and permeable hydrocarbon reservoirs. Thin-bedded sandstones and burrowed sandstones are common in distal deltaic deposits, which were probably deposited during times of somewhat higher relative sea level. Such sandstones may form low-resistivity reservoirs, which can, if recognized, contribute substantially to hydrocarbon production.

INTRODUCTION

Sandstone reservoirs at Attaka and Serang fields, offshore East Kalimantan, Indonesia, are interpreted as delta-front bars and distributary channels in upper Miocene to Pliocene lobes of the ancestral Mahakam delta. Attaka and Serang fields are located on the shelf of the lower Kutei Basin, approximately 30 kilometers northeast of the present Mahakam delta (Fig. 1). Since 1972, deltaic reservoirs at Attaka have produced more than 93 million cubic meters (586 million barrels) of oil and 34 billion cubic meters (1.2 trillion cubic ft) of gas. Serang reservoirs have produced more than 6.5 million cubic meters (40.9 million barrels) of oil and 2.9 billion cubic meters (104 billion cubic ft) of gas since 1994. Attaka field is a large fold that is partitioned by northwest-trending normal faults that step down to the northeast, the basinward direction (Fig. 2). At Serang field, structural closures occur in both the upthrown and downthrown blocks of the Serang Fault, a major extensional fault on the lower Kutei shelf (Figs. 3 and 4). Production at Attaka derives from a 2000$^+$ m thick section of upper Miocene to Pliocene strata, which have been informally subdivided into intervals designated as "Shallow", "Main Deltaics", "Intermediate", "Deep", and "Overpressured" (Fig 5). At Serang, hydrocarbons are produced from deltaic sandstone reservoirs of the "Shallow" and "Main Deltaics" intervals of the upper fault block, but only from the "Shallow" interval of the lower fault block. Attaka reservoirs in the "Intermediate", "Deep", and "Overpressured" intervals are mostly thin-bedded sandstones (< 2 m thick), which we interpret as delta-front bars. Reservoirs in the "Shallow" and "Main Deltaics" intervals are diverse; we interpret them as deposits of river- and tide-influenced distributary channels and as delta-front bars.

ATTAKA–SERANG "SHALLOW" RESERVOIRS

Upper Miocene to Pliocene strata of the Attaka–Serang "Shallow" interval overlie the X0 Sequence Boundary, also known as the RT marker. The RT marker corresponds to a major discontinuity in resistivity. Shale base-line resistivity decreases markedly above the RT maker; this shift is probably caused by changes in clay mineralogy (discussed later). "Shallow" interval strata at Attaka field include sandstones, shales, siltstones, and coals. Representative facies from the "Shallow" interval are illustrated by core No. 1 from Attaka B-15 (Fig. 7, refer to Fig. 6 for explanation of symbols). A sharp-based, cross-bedded sandstone, interpreted as a river-dominated distributary deposit, overlies burrowed and rooted mudstone, a possible bay deposit at the base of core 1. The cross-bedded sandstone is in turn overlain by rooted mudstone and coal, interpreted as delta-plain deposits. The coal is overlain by an upward-coarsening succession of laminated to burrowed shales, siltstones, and silty

Tropical Deltas of Southeast Asia—Sedimentology, Stratigraphy, and Petroleum Geology
SEPM Special Publication No. 76, Copyright © 2003
SEPM (Society for Sedimentary Geology), ISBN 1-56576-086-7, p. 235–254.

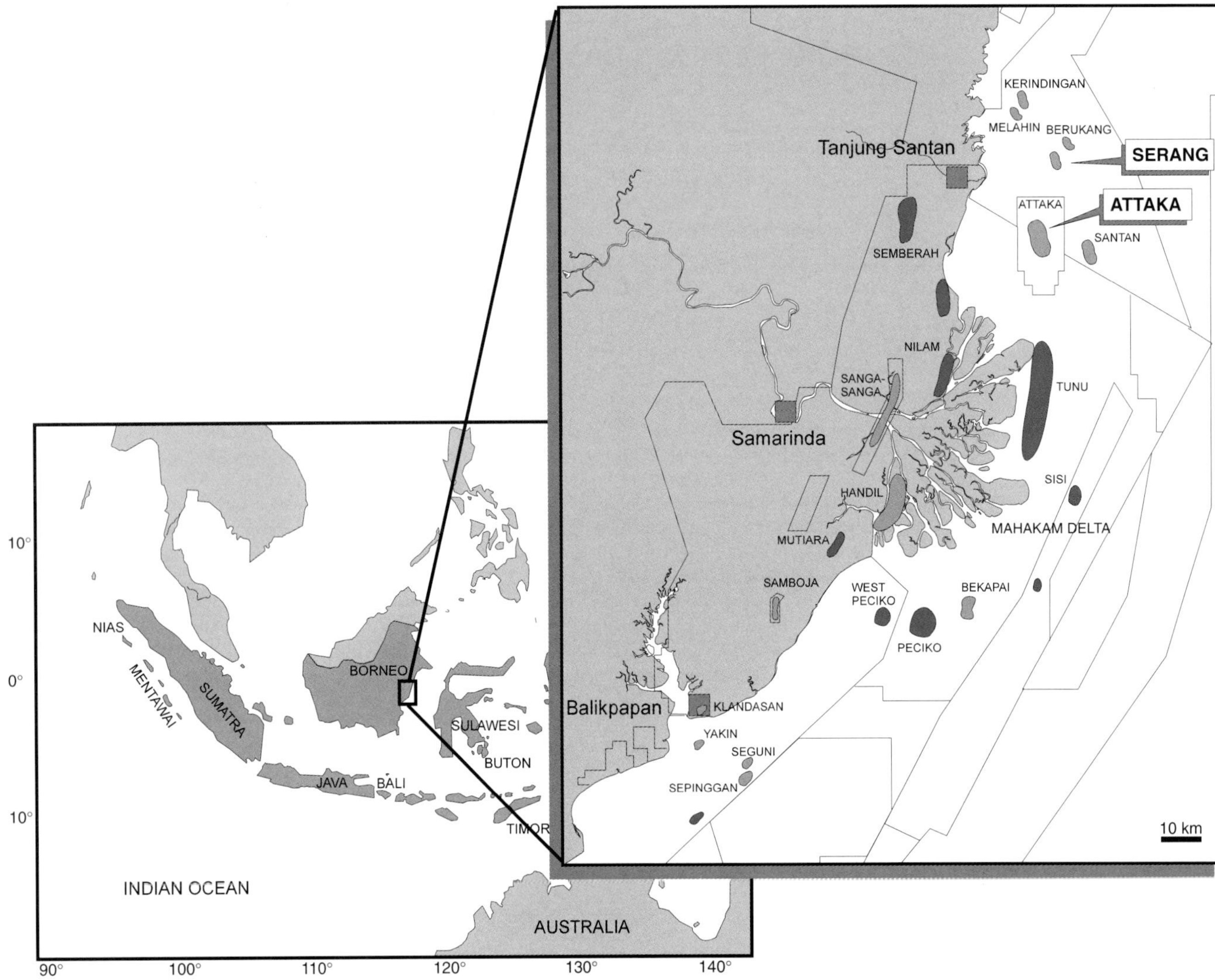

FIG. 1.—Location map for Attaka and Serang Fields and the modern Mahakam Delta, E. Kalimantan, Indonesia.

sandstones that contain brackish-water foraminifera and pelecypods. We interpret that a flooding surface is present at the top of the coal at 1332.6 m, and that the overlying upward-coarsening succession records the progradation of a delta-front bar over bay or prodelta muds. Laminated to burrowed mudstones near the top of core 1 probably represent bay or prodelta deposits that overlie another flooding surface (at 1325.0 m).

DEPOSITS OF THE "MAIN DELTAICS"

The "Main Deltaics" interval at Attaka and Serang fields includes strata between the X-0 ("RT marker") and X-2 sequence boundaries (Fig. 5). We interpret that much of the 500-m-thick "Main Deltaics" interval was deposited during periods of generally low relative sea level, when the ancestral Mahakam delta was situated near the Late Miocene shelf edge. "Main Deltaics" strata include sandstones (here interpreted as deposits of river- and tide-influenced distributaries and delta-front bars), mudstones (interpreted as estuary, bay, prodelta, and tidal-flat deposits), coals (delta-plain marsh and peat-ridge accumulations), and local carbonate buildups (associated with marine transgressions). Many of these Miocene facies are analogous to sediments from the modern Mahakam delta (Allen et al., 1976, 1979; Allen and Chambers, 1998).

Serang 68-4 Reservoir

The Serang 68-4 reservoir is typical of sandstones that overlie a regional erosional surface within the "Main Deltaics", designated as the X-1 Sequence Boundary (Figs. 8, 9, 10). The 68-4 reservoir lies at the top of an alternating mudstone–sandstone succession that overlies and probably downlaps onto transgressive marine carbonates (Fig. 9). We interpret these rocks as representing prograding prodelta and delta-front deposits, on the basis of core descriptions and log response (Fig. 9). Reservoirs overlying the X-1 Sequence Boundary consist of sharp-based, upward-fining sandstones that range up to 17 m in thickness. Locally, thin, low-resistivity shales and upward-coarsening sandstones are also present. In places, several sandstone bodies may be stacked vertically to form compound reservoirs as thick as 40 m (Fig. 8). In Serang SA-2RD, a 28-m-long conventional core was cut through the X-1 Sequence Bound-

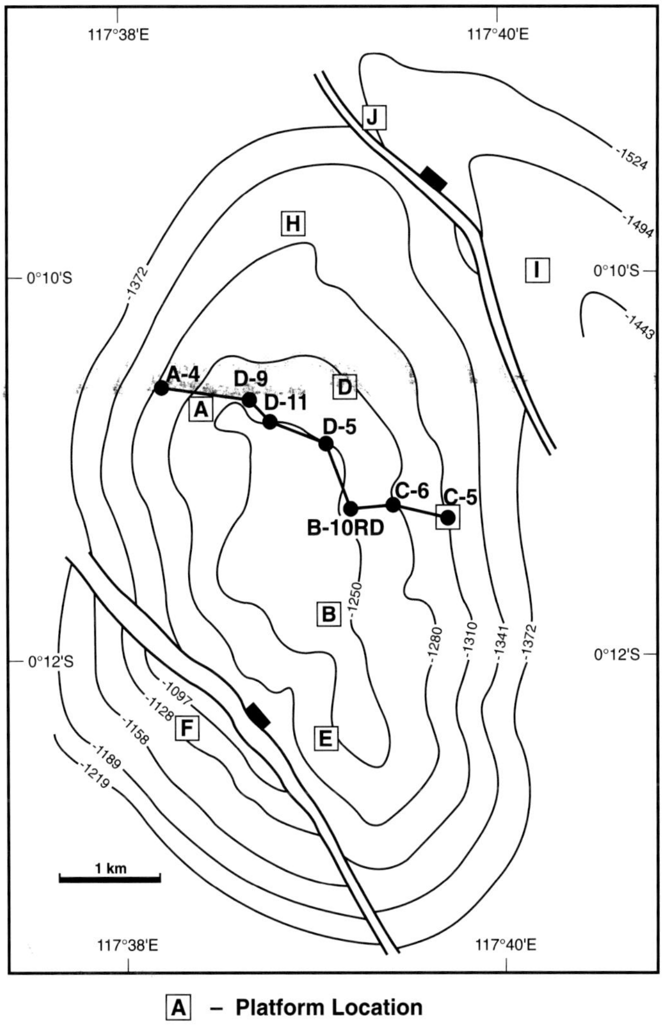

FIG. 2.—Attaka field structure map for top of "Main Deltaics", showing northwest-trending normal faults. Structure contours in meters (subsea). Note location of cross section depicted in Figure 13. Datum is X0 or "RT" marker.

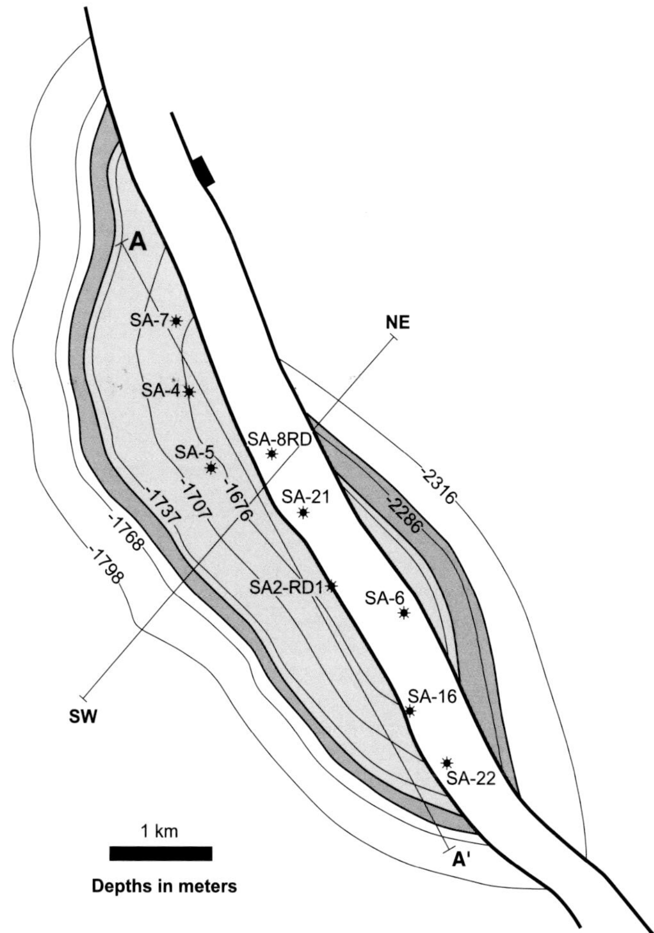

FIG. 3.—Serang field structure map, showing Serang fault, gas accumulation (light gray), oil accumulation (dark gray), and location of cross section A–A' shown in Figure 8 and seismic line shown in Figure 4. Datum is 59-6 marker (slightly above X0 or "RT" marker).

ary (Fig. 10; refer to Fig. 6 for explanation of symbols). In this well, the X-1 Sequence Boundary is at the base of a 10-m-thick sandstone bed. Strata immediately below the X-1 Sequence Boundary consist of laminated to burrowed, pelecypod-bearing mudstones and silty, very fine-grained sandstones, which we interpret as distal delta-front and prodelta deposits. The overlying, sharp-based sandstone grades from conglomeratic, coarse-grained sandstone at the base to very fine-grained sandstone at the top. It is dominated by cross lamination in its lower part and ripple lamination in its upper part. Locally, thin silt interbeds and burrows are present in the upper part (Fig. 10). We interpret this sandstone as a tidally influenced distributary, similar to modern Mahakam distributary sands described by Allen and Chambers (1998), though perhaps coarser at the base (compare Figs. 10 and 11). The abrupt vertical shift from distal delta-front to distributary-channel facies and the conglomeratic character of sandstone above the X-1 indicate a significant basinward shift in facies, and consequently, a sequence boundary. This sandstone is overlain by a thin, laminated, coaly shale (2648.0–2648.4 m in Figure 10) and an upward-thickening, ripple-laminated to cross-laminated sandstone, which may represent a bay or estuary and a prograding bay-head delta (or splay), respectively (Fig. 10). The upper part of the core consists of cross-bedded, conglomeratic, very coarse-grained to medium-grained sandstones, which are probably deposits of river-dominated distributaries or fluvial channels characterized by high current velocities. We offer a sequence stratigraphic interpretation for the Serang SA-2RD core in Figure 10.

Strata underlying the X-1 Sequence Boundary are interpreted to represent prodelta and distal delta-front facies, and strata above this surface represent delta-plain facies. The X-1 Sequence Boundary is associated with erosional relief on the northern Kutei shelf, judging from regional seismic interpretations. Unfortunately, fault shadow and shallow gas effects serve to obscure this relationship within the productive areas of Serang and Attaka fields.

It is probable that lowstand deltaic deposits, like the strata associated with the X-1 Sequence Boundary, have characteristics similar to those documented by Sydow (1996; refer to Fig. 12 of this paper) from Quaternary lowstand deltas of the Mahakam shelf. In these Quaternary lowstand deltas, prograding, muddy clinoforms of the delta front are truncated by erosional valleys

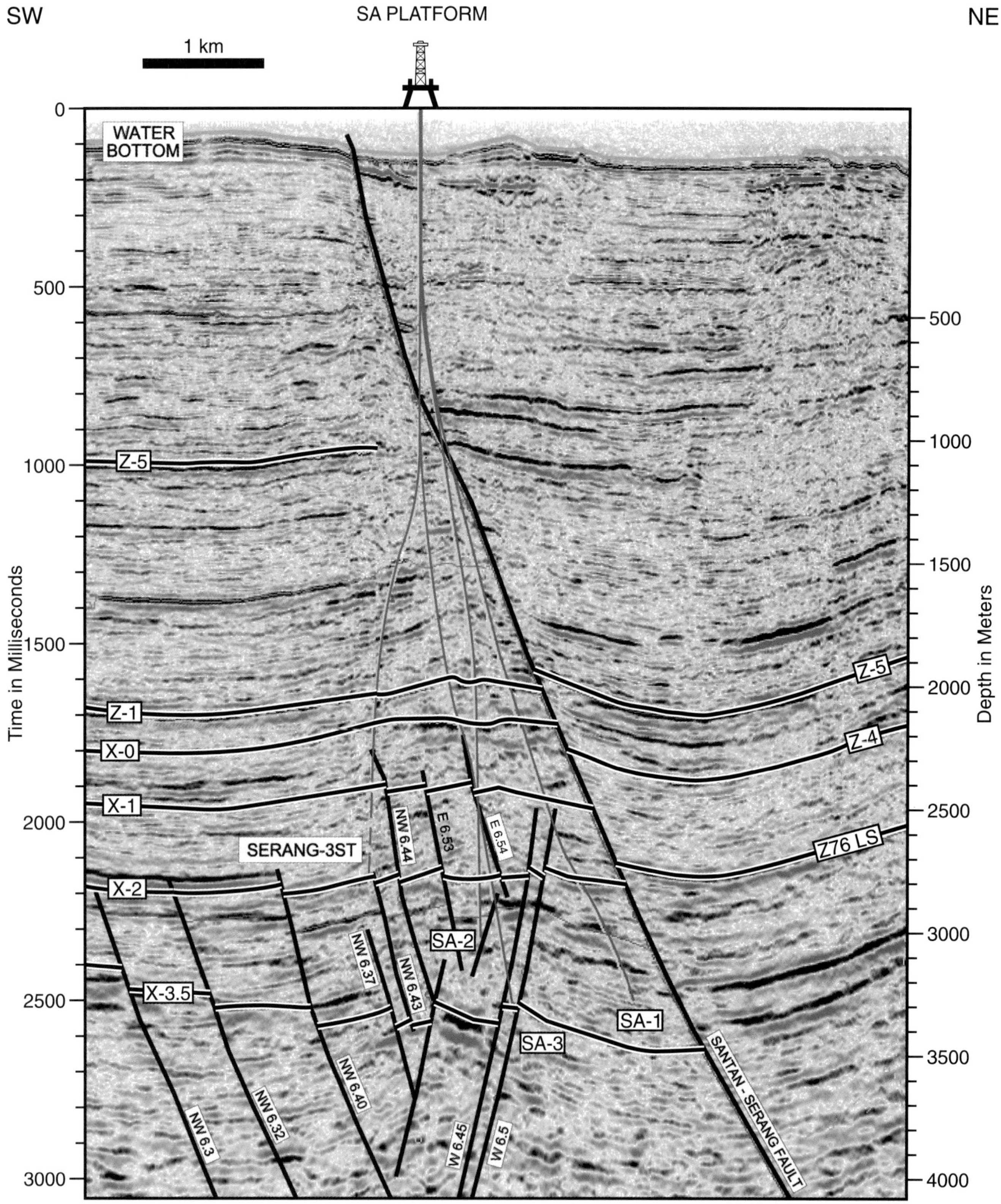

FIG. 4.—Dip-oriented seismic line through Serang field, showing Serang fault, several smaller faults, and stratigraphic markers. Z5 = top "Shallow", X0 = top "Main Deltaics", and X2 = top "Intermediate."

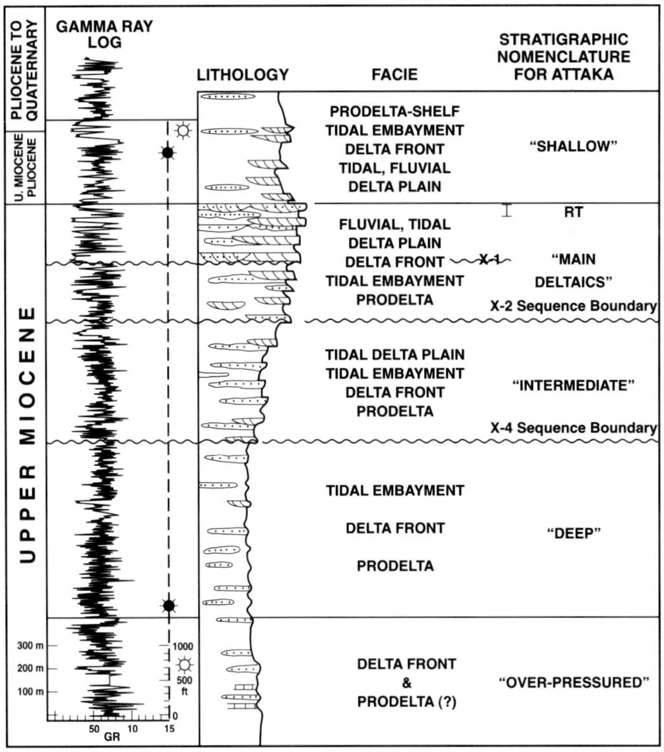

FIG. 5.—Composite stratigraphic column for Attaka field.

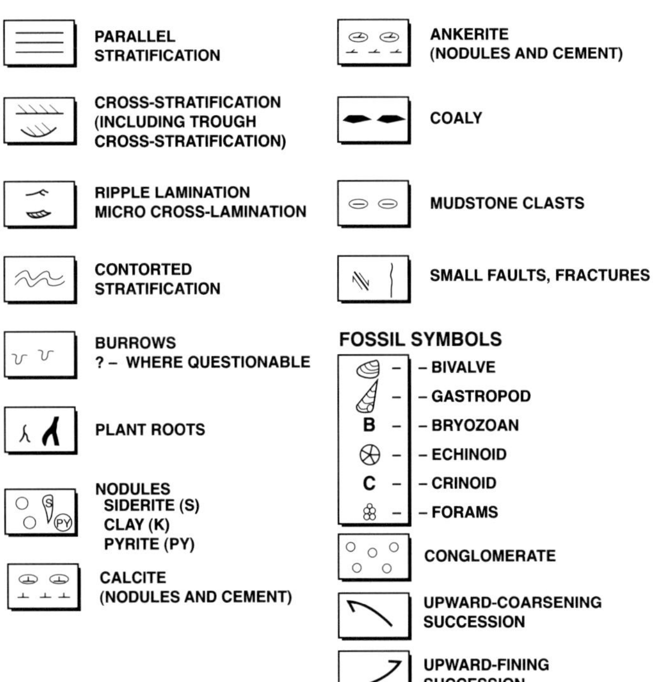

FIG. 6.—Explanation of symbols used in core graphs.

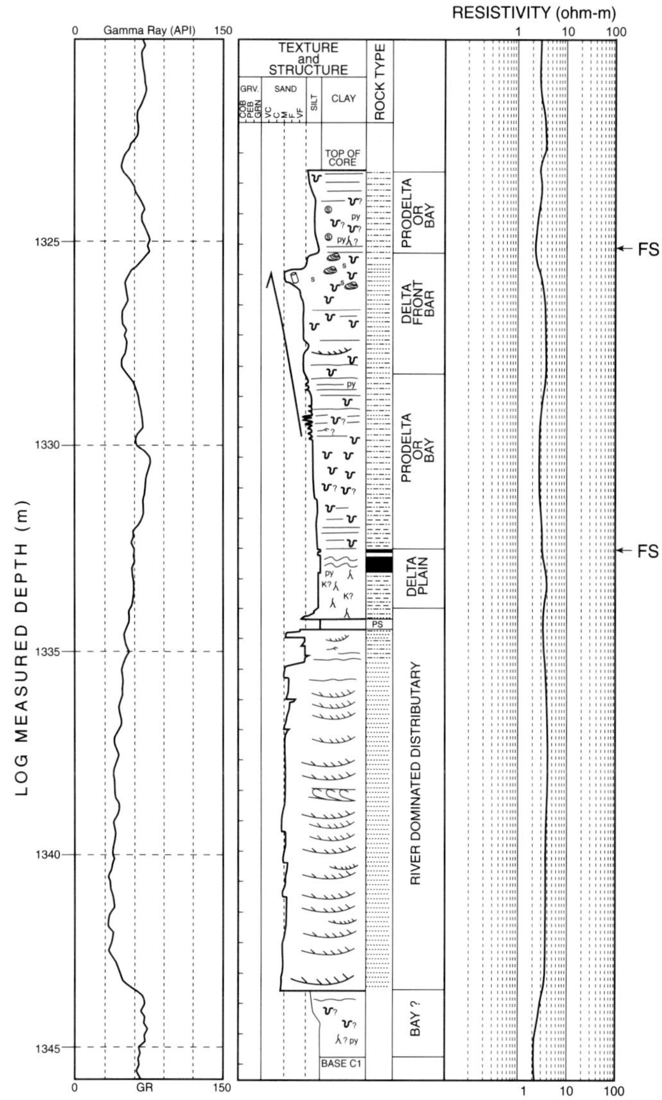

FIG. 7.—Core graph of representative reservoir facies from Attaka "Shallow" interval. Note low resistivity of the thick-bedded sandstone, which is an oil reservoir (interpreted as a river-dominated distributary-channel deposit).

that are probably filled by coarse distributary and fluvial-channel sands.

Attaka RT Reservoir

The Attaka RT reservoir, near the top of the "Main Deltaics" interval, includes many interesting features, such as sandstones interpreted as river- and tide-influenced distributary and delta-front deposits, coals, an erosional surface, and flooding surfaces overlain by organic-rich, transgressive shales (Figs. 13–16). Sharp-based sandstones, which we interpret as distributary channels, are typically narrow and elongate (Fig. 16). Cores and logs illustrate cyclicity of lithologies and facies in the RT reservoir (Figs. 13 and 14). Cores from Attaka D-9 are cross-bedded, coarse- to fine-grained sandstones, which we interpret as deposits of river-dominated, distributary channels. Judging from log correlations of D-9 and adjacent wells (Figs. 13 and 14A), the

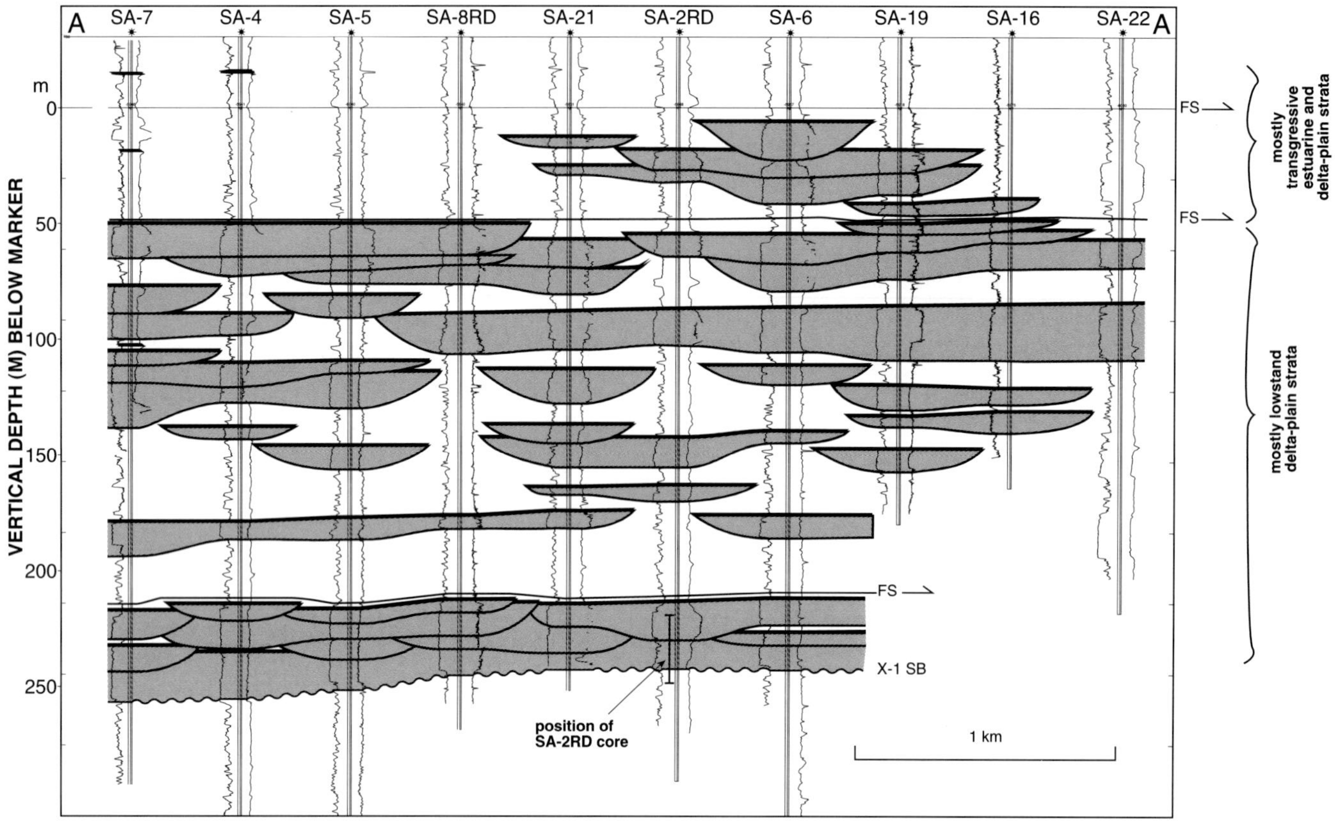

FIG. 8.—Cross section through the upper part of the "Main Deltaics" interval at Serang field, showing the "X-1" Sequence Boundary and position of core in 68-4 reservoir (see Fig. 10). Line of section is shown in Figure 3.

lower sandstone in core 2 probably rests on an erosional surface with significant relief (a probable sequence boundary). The upward-fining, mostly cross-stratified sandstone in core no. 3 from the D-11 well (Fig. 14B) is interpreted as a tidally influenced distributary deposit, on the basis of siltstone interbeds and burrows in its upper part, similar to tidally influenced distributaries described by Allen et al. (1976) from the modern Mahakam (Fig. 11). This distributary sandstone is overlain by a clean, burrowed to laminated sandstone (transgressive delta-front "roll-back" or estuarine sand?) and laminated, pelecypod-bearing, organic-rich (TOC = 3.5–6.1%) shale. This dark, organic-rich shale is interpreted as a low-energy, estuarine deposit (on the basis of the occurrence of brackish-water foraminifera); it contains abundant, hydrogen-poor (HI < 120) organic matter that is isotopically light ($\delta^{13}C$ = -29%). The overlying, lighter-colored, burrowed to laminated shale is interpreted as a prodelta deposit (on the basis of the occurrence of marine foraminifera) that contains organic matter that is more hydrogen rich (HI > 130) and somewhat heavier ($\delta^{13}C$ = -28%), probably a mixture of terrestrial and marine-algal types. The overlying, upward-coarsening, burrowed and laminated siltstone to medium-grained sandstone (1603–1609 m) is interpreted as a prograding delta front. Burrowed, silty sandstones and siltstones (1605–1609 m) probably represent bar-margin or distal delta front deposits. The overlying, laminated, clean fine-grained sandstone (1603–1605 m) probably represents bar-crest deposits. The rooted, burrowed, and laminated siltstones above this sandstone are interpreted as deposits of a tidal deltaic plain.

THIN-BEDDED RESERVOIRS OF THE "INTERMEDIATE", "DEEP", AND "OVERPRESSURED" INTERVALS

Strata from the "Intermediate", "Deep", and "Overpressured" intervals are best known from Attaka field, where they include numerous thin-bedded sandstone reservoirs interstratified with thick successions of mudstone (Fig. 17). These rocks are interpreted mostly as distal delta-front, prodelta, and shelf deposits, although some distributary channels, delta-plain, and estuarine sediments appear to be present in parts of the "Intermediate" interval (Figs. 17 and 18). In Figure 18 we compare an interpretation of the M75-91 reservoir from the Attaka "Intermediate" interval with sediments described by Allen and Chambers (1998) from the modern Mahakam delta front.

Delta-front sandstones have been identified on the basis of upward-coarsening grain-size profiles and the presence of foraminfera that live in sandy, brackish-water environments. We separate delta-front sands into two subfacies. Clean, current-laminated sandstones are interpreted as high-energy, bar-crest deposits of the proximal delta front, whereas muddy, mostly bioturbated sandstones are interpreted as bar-margin or distal delta-front deposits (Fig. 17). Thin interbedding in delta-front deposits gives rise to numerous, small-scale variations in mud content, grain size, and stratification, which in turn produce major small-scale variations in permeability (Fig. 19). Such beds probably form clinoforms that dip and "shale out" in the seaward direction.

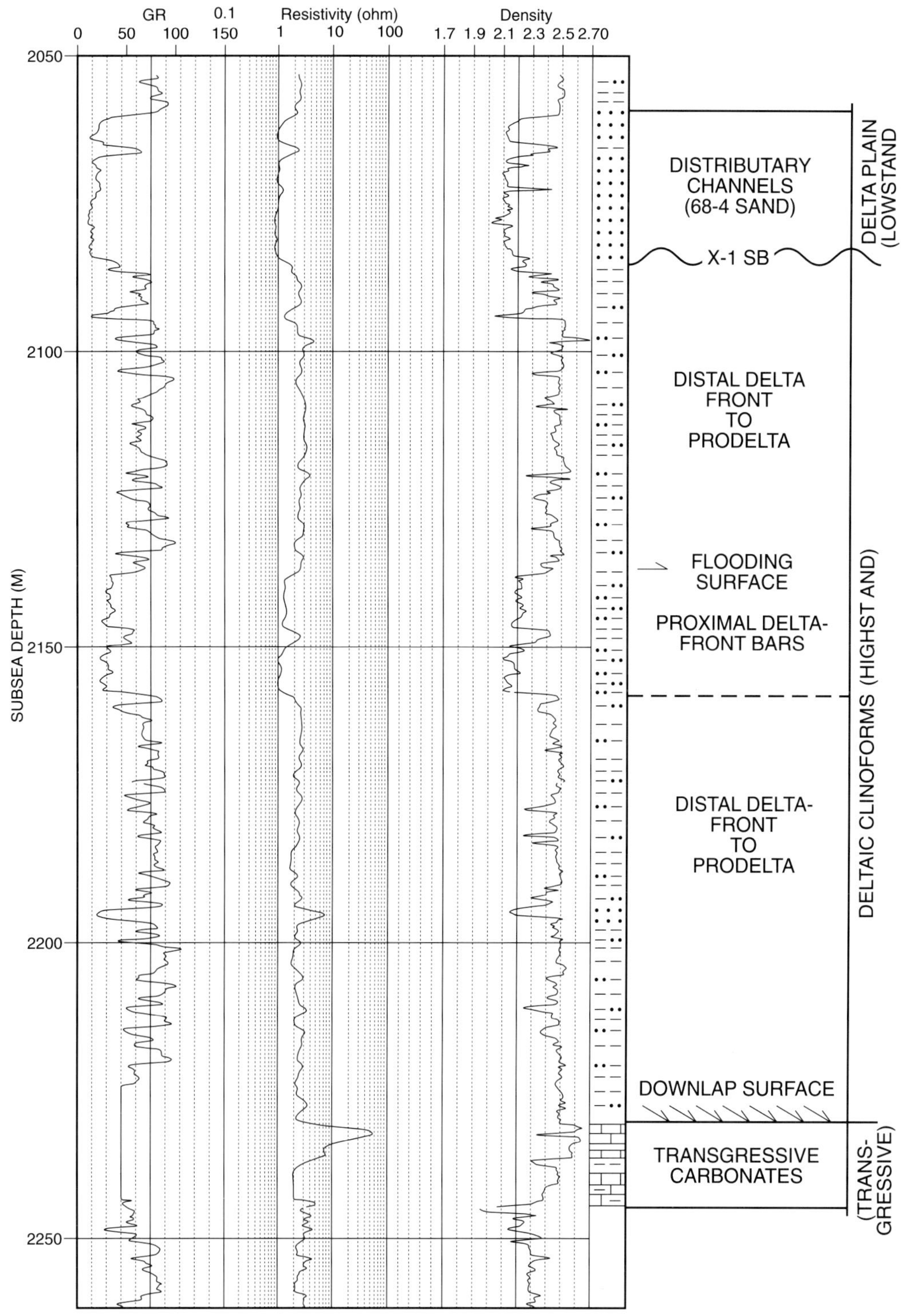

FIG. 9.—Wire-line log traces and interpretation for part of "Main Deltaics" interval, including 68-4 sand in Serang SA-3RD2.

MINERALOGY OF THE DELTAIC SANDSTONES

Upper Miocene to Pliocene Mahakam delta sediments at Attaka and Serang fields have diverse framework-grain and clay-mineral compositions.

Framework Mineralogy

Sandstones from the "Main Deltaics", "Intermediate", and "Deep" intervals contain abundant quartz with only minor to moderate amounts of feldspar and rock fragments (Fig. 20). How-

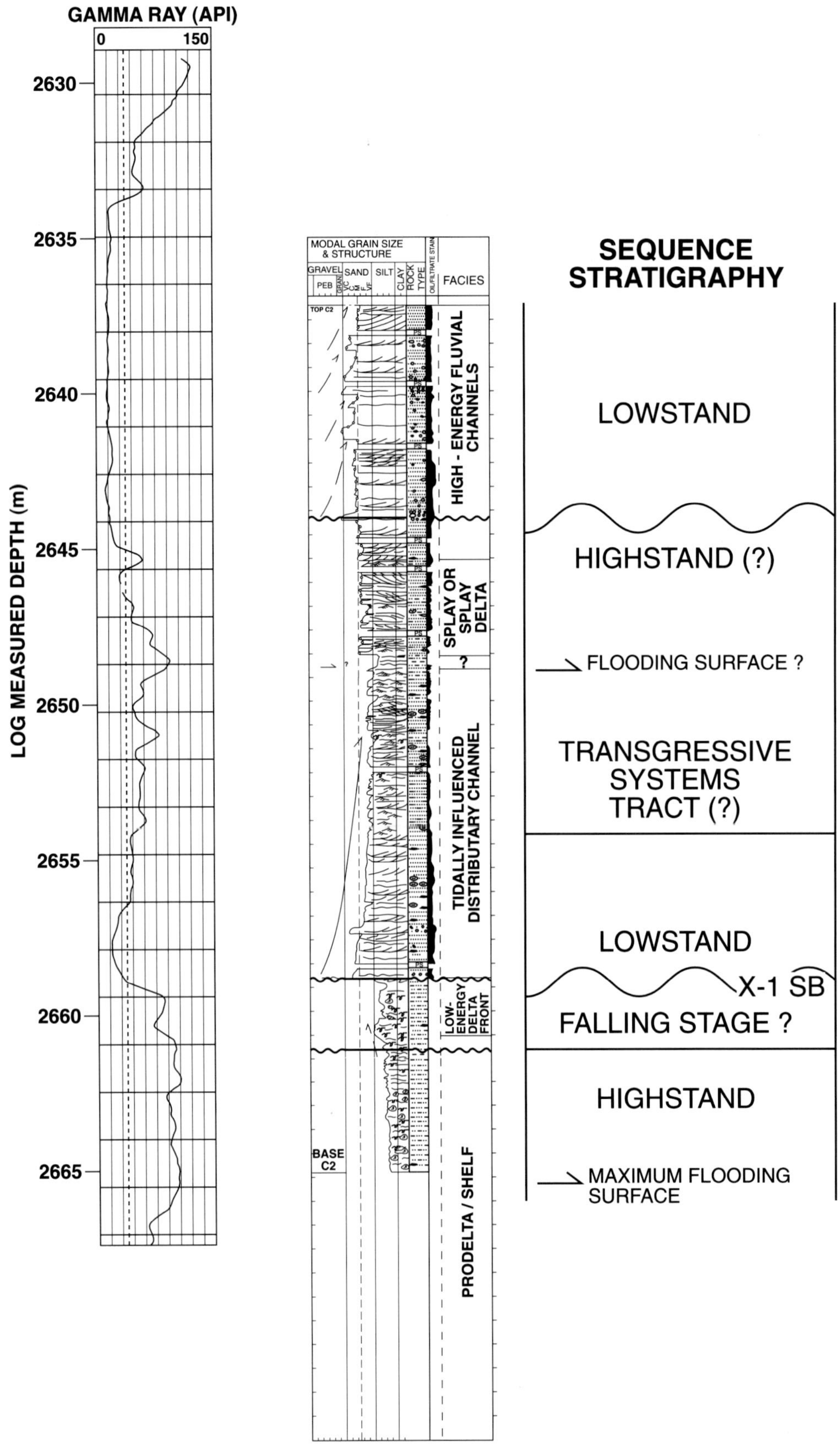

FIG. 10.—Core graph of 68-4 reservoir ("Main Deltaics") in Serang SA-2RD.

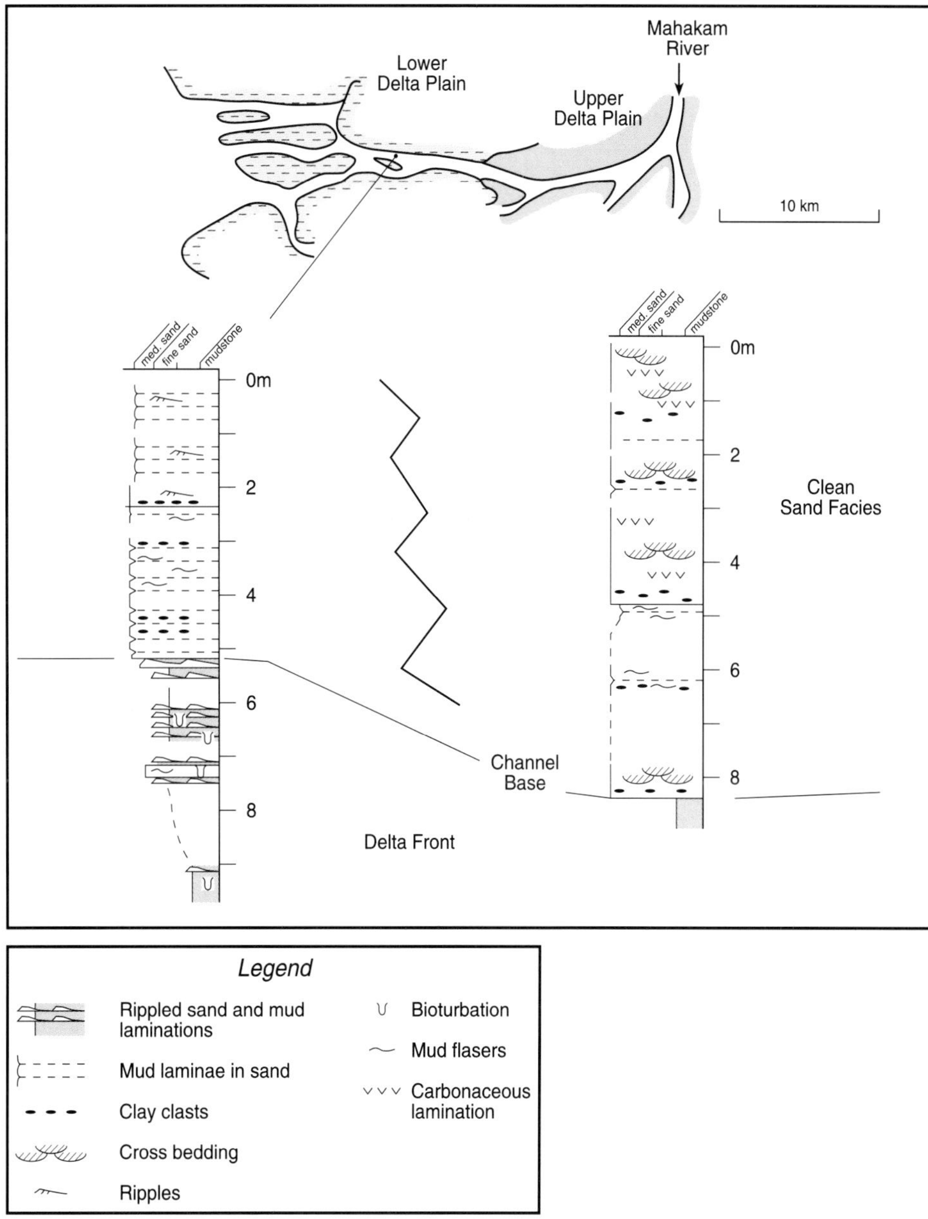

FIG. 11.—Core sections of sand in modern Mahakam distributary channel (from Allen and Chambers, 1998).

ever, many (but not all) sandstones from the "Shallow" interval (above the X0/RT maker) contain abundant, partly devitrified volcanic grains (Fig. 20). The volcanic fragments range from intermediate varieties, which contain plagioclase phenocrysts of andesine composition (42–48% Ca feldspar), to silicic varieties, which contain phenocrysts of quartz, alkali feldspar, and sodic plagioclase (10% Ca feldspar), from electron probe microanalyses. Incomplete devitrification of groundmass and the lack of albitization of plagioclase phenocrysts indicate that the volcanic rock fragments are geologically young, probably only slightly older than the sandstones in which they occur. The Mahakam River may have become choked with volcanic detritus from occasional extrusions during the latest Miocene, giving rise to successions of volcaniclastic deltaic sediments. Such extrusions may have temporarily increased stream gradients and rate of deltaic progradation.

Secondary and Clay Mineralogy

Sandstones from Attaka and Serang fields have sustained only a moderate degree of diagenetic modification, including compaction, precipitation of secondary minerals, and minor to moderate leaching of feldspars and rock fragments. Siderite, pyrite, and kaolinite are widespread as secondary minerals, whereas smectite, mixed-layer illite–smectite, chlorite, quartz, calcite, and ankerite occur in minor amounts, or they are found only locally. On the basis of clay mineralogy, deltaic sandstones from Attaka–Serang can be divided into two groups: (1) "Shallow" interval rocks that are dominated by smectite (or randomly ordered illite–smectite) and kaolinite and (2) rocks from the "Main Deltaics", "Intermediate", and "Deep" intervals, and from some zones in the "Shallow", that are domi-

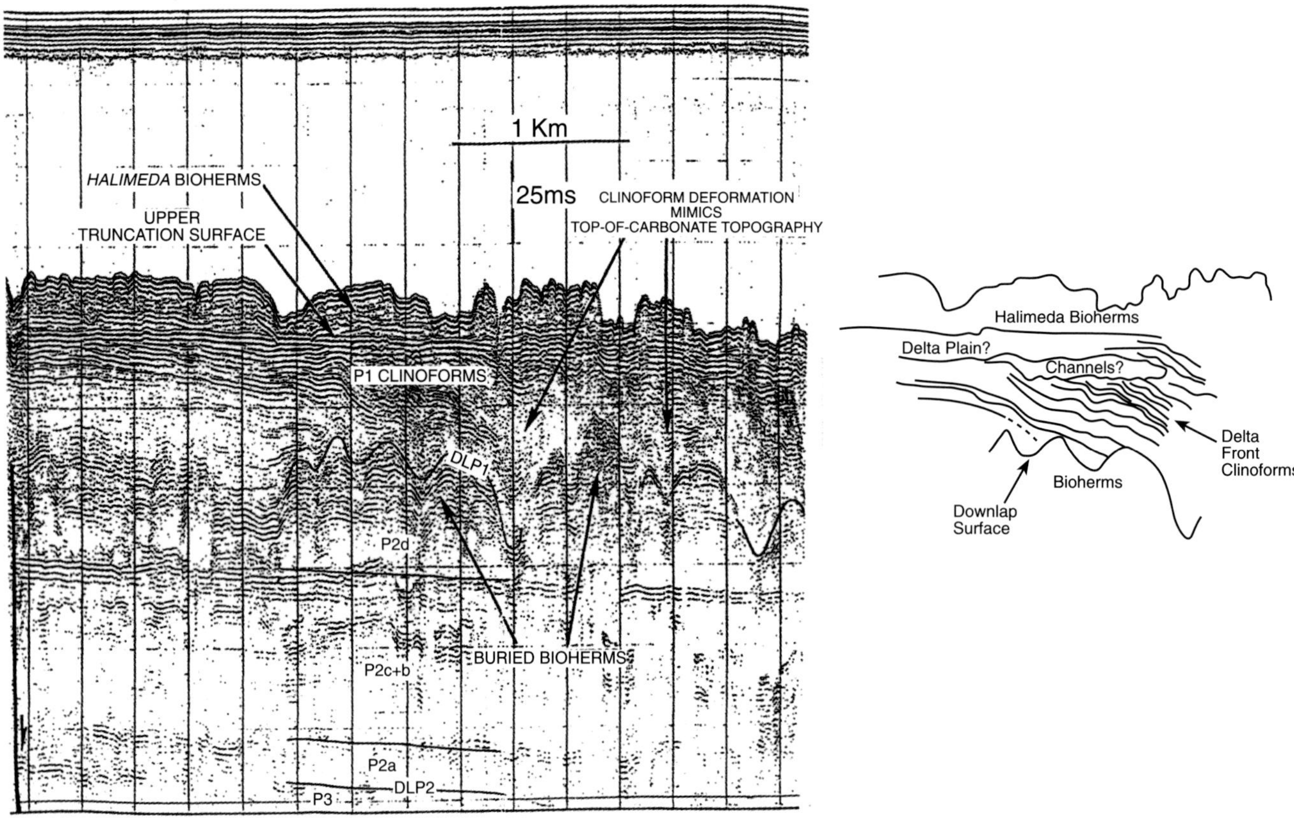

FIG. 12.—Clinoforms of Mahakam lowstand delta of Quaternary age (from Sydow, 1996) with interpretation at right.

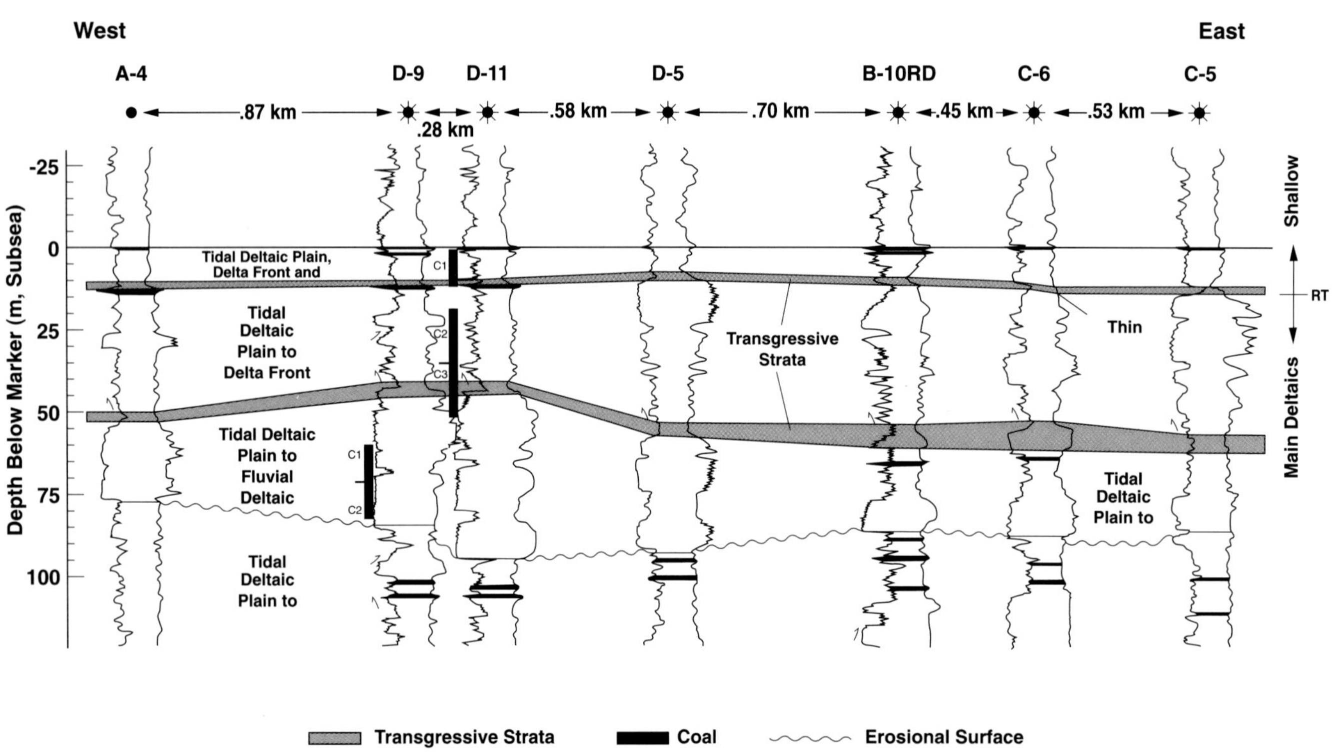

FIG. 13.—Cross section through RT Reservoir in Main Deltaics, Attaka field. Line of section is shown in Figure 2. Logs shown are gamma ray (left) and resistivity (right). Positions of cores in Attaka D-9 and D-11 are shown by dark columns.

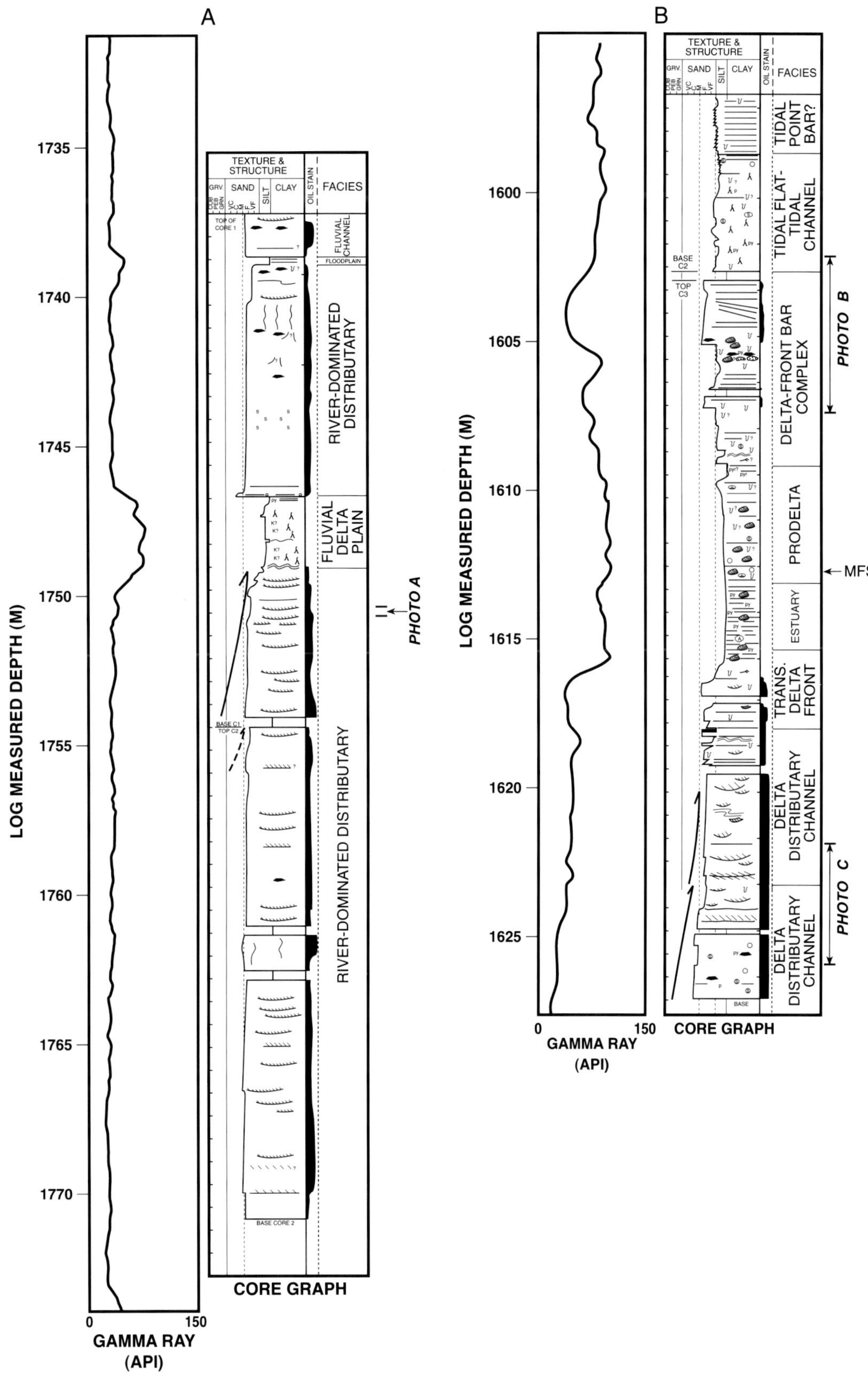

FIG. 14.—Cores from Attaka "Main Deltaics." **A)** Core graph of river-influenced distributary channels, Attaka D-9. **B)** Core graph of deltaic facies from Attaka D-11

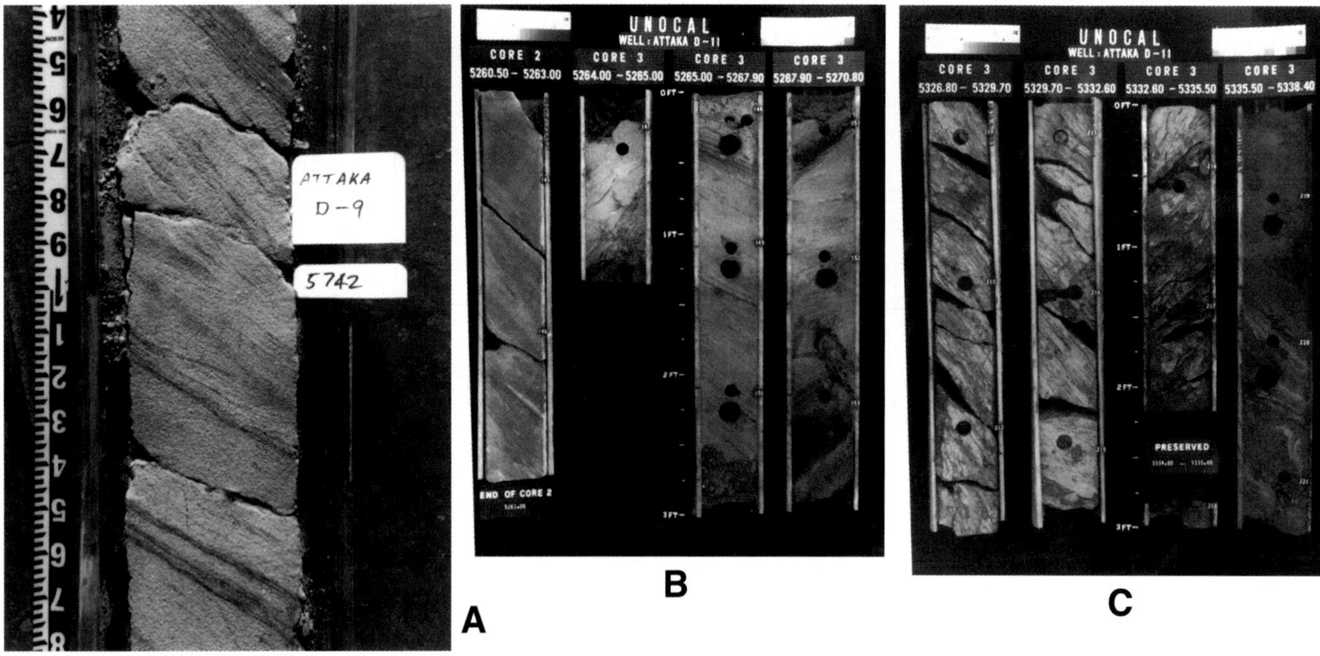

FIG. 15.—Core photographs of deltaic facies from the RT reservoir in the Attaka "Main Deltaics." Wells deviate from vertical. **A)** Cross-bedded sandstone, river-dominated distributary channel, Attaka D-9. **B)** Low-angle laminated delta-front sand with *Ophiomorpha burrow*, Attaka D-11. **C)** Cross-bedded sandstone with abundant coalified plant debris, lower part of tide-influenced distributary channel, Attaka D-11.

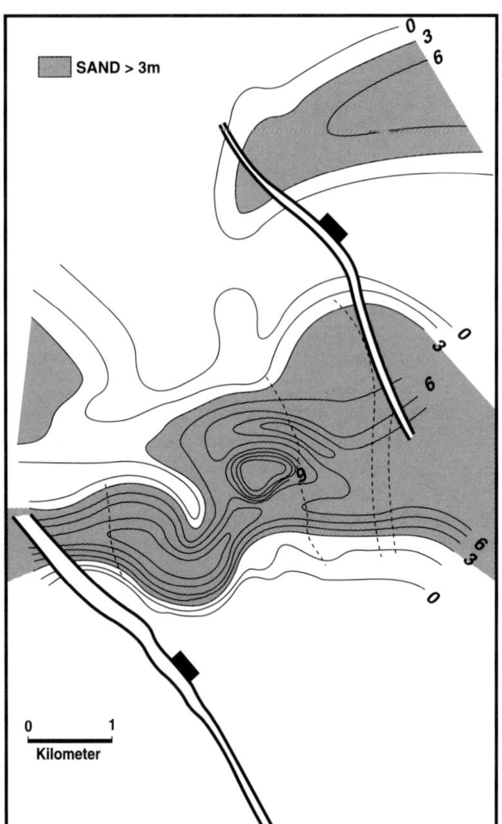

FIG. 16.—Map of net sand for Attaka 42-0A reservoir, immediately below RT marker. Contours in meters.

nated by kaolinite and ordered, mixed-layer illite–smectite but contain little or no high-expandable clay (Fig. 21). The abundant high-expandable smectite and randomly ordered illite–smectite in many "Shallow" reservoirs are associated with abundant volcanic grains, low resistivity, and severe completion challenges (Trevena et al., 1993, p. 95).

CONTROLS ON POROSITY AND PERMEABILITY

Importance of Facies

Depositional facies profoundly influence the reservoir quality of deltaic sandstones at Attaka and Serang fields (Figs. 22 and 23). River-dominated distributary-channel sandstones form the most porous and permeable reservoirs (average porosity = 28.4%, median permeability 665 md). These sandstones are coarse to medium grained, and they contain little depositional clay. Tide-dominated distributary-channel sandstones are also porous and permeable (average porosity = 26.1%; median k = 256 md) but somewhat less so than the river-dominated distributary sandstones. This is probably due to the somewhat finer grain size and local occurrence of depositional clay (as laminated, tidal drapes and dispersed mud introduced by burrowing). Delta-front sandstones have been subdivided into high-energy, "bar-crest" deposits (average porosity = 15.7%; median k = 10.3 md) and "bar-margin" deposits or distal delta front deposits (average porosity =11.8%; median k = 0.33 md). "Bar-crest" deposits are either laminated sandstones or laminated to bioturbated sandstones; they are medium to very fine grained and have widely ranging clay content. "Bar-margin" and distal delta-front deposits are typically heavily burrowed and contain abundant dispersed clay, which limits their permeability.

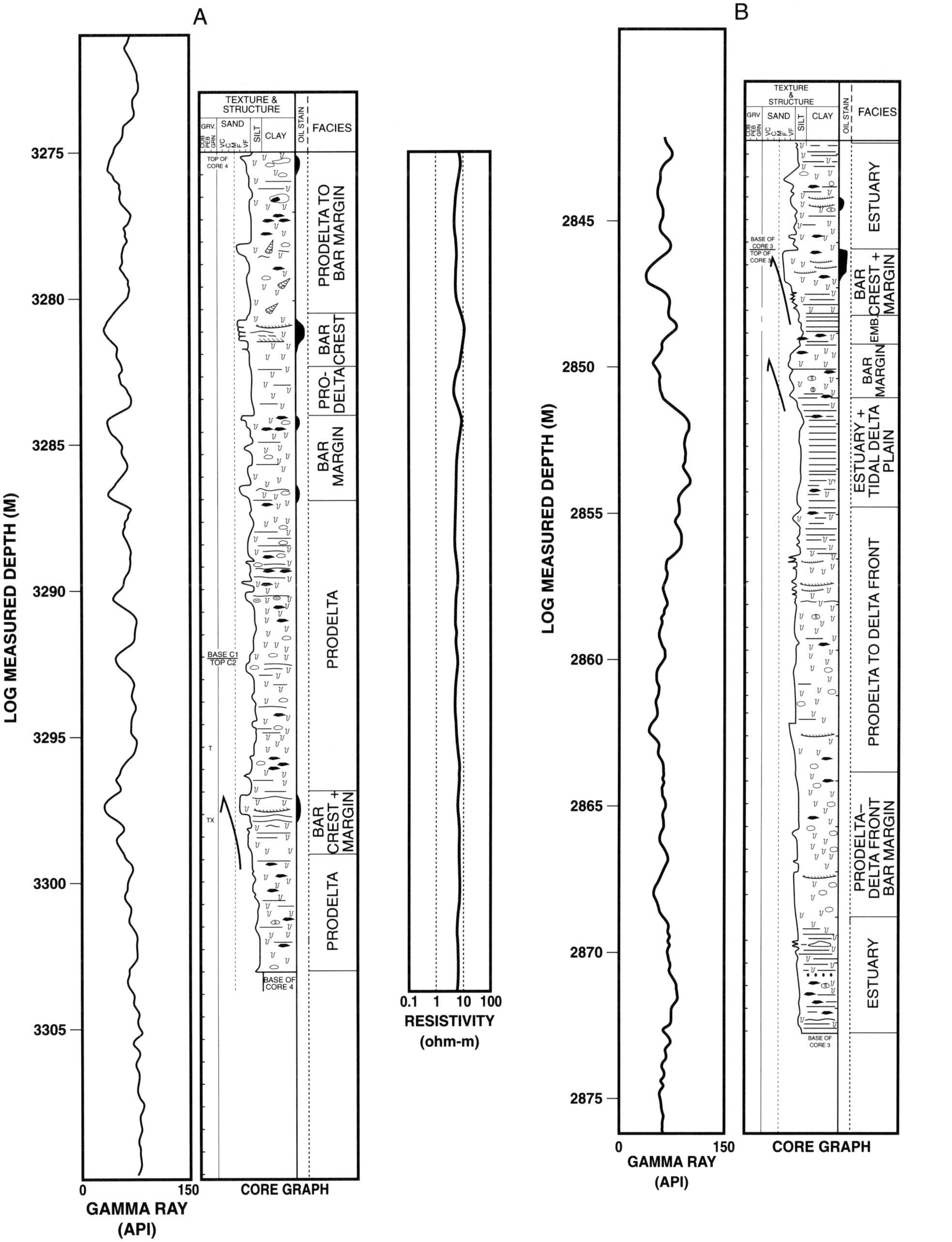

FIG. 17.—Core graphs of thin-bedded deltaic reservoirs, Attaka I-5. **A)** Core graph of deltaic facies from Attaka "Deep" interval. **B)** Core graph of deltaic facies from Attaka "Intermediate" interval.

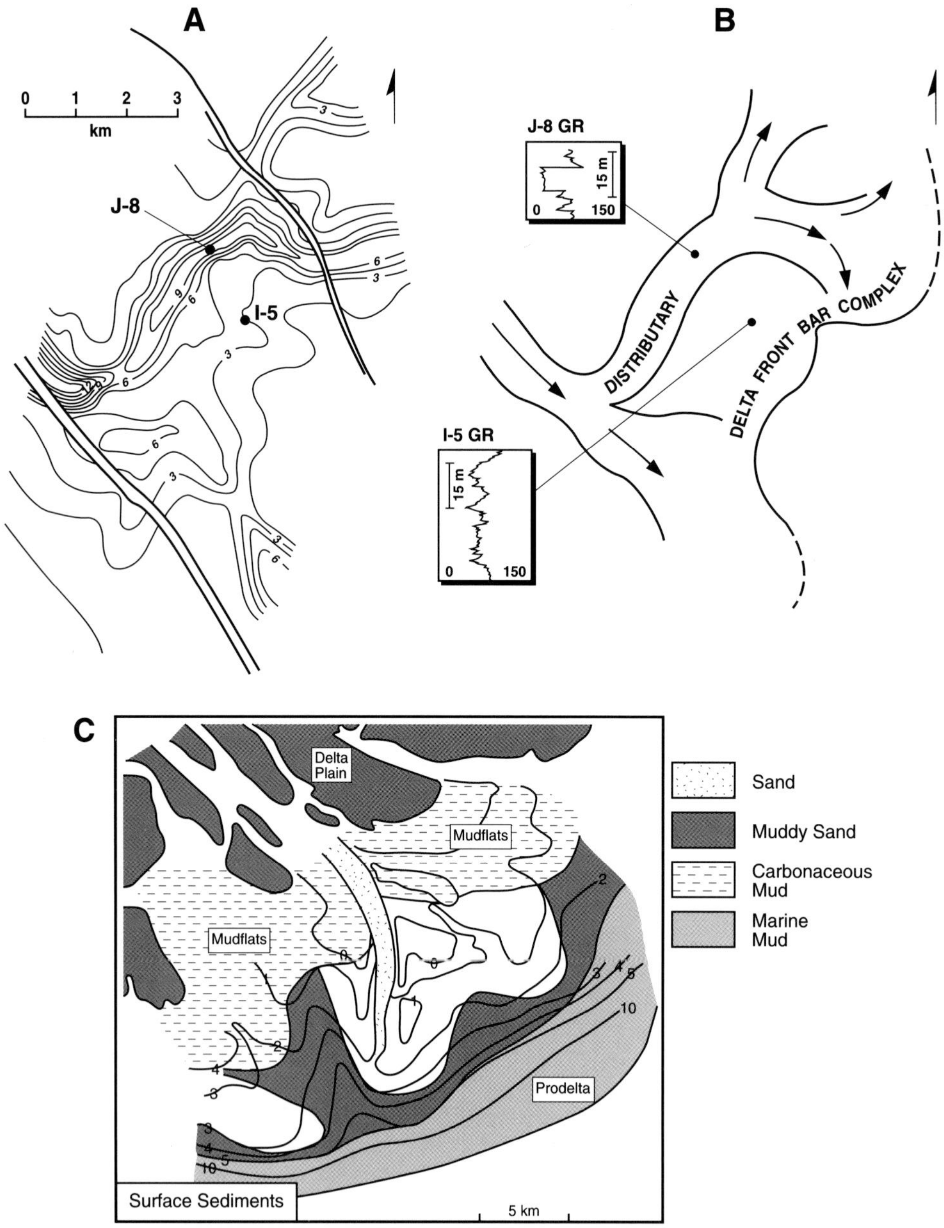

FIG. 18.—Delta-front sands of Miocene and Modern Mahakam Delta. **A)** Net sand map of M75-9A reservoir in Attaka "Intermediate" interval. Contours in meters. **B)** Interpretation of net sand map in Part A. **C)** Surface sediments of the modern Mahakam delta front (from Allen and Chambers, 1998)

Influence of Clay

The influence of clay on permeability is illustrated by high-resolution, probe-permeameter studies of conventional cores from delta-front sandstones. Laminated, relatively clay-free sandstones (Fig. 19A, 3274.75–3275.21 m) are typically highly permeable ($k > 100$ md). Sandstones that contain abundant laminated clay typically show centimeter-scale variations in permeability of more than one order of magnitude (Fig. 19A, 3275.24–3275.39 m). Highly burrowed sandstones typically have low permeability (< 1 md), owing to high dispersed clay content (Fig. 19A, 3274.48–3274.69 m; Fig. 19B, 3275.60–3275.70 m).

LOW RESISTIVITY IN DELTAIC RESERVOIRS

Low-resistivity deltaic reservoirs are hydrocarbon-producing sandstones, which exhibit resistivity so low that they are typically interpreted as water-saturated rocks, on the basis of assumptions and techniques of conventional log analysis. Currently, 40–50% of Attaka oil production and nearly 90% of Serang gas production derives from low-resistivity reservoirs. We recognize two main types of low-resistivity reservoirs at Attaka–Serang: (1) thick, cross-bedded sandstones of the "Shallow" interval (mostly river- and tide-influenced distributaries) and (2) thin-bedded, variably bur-

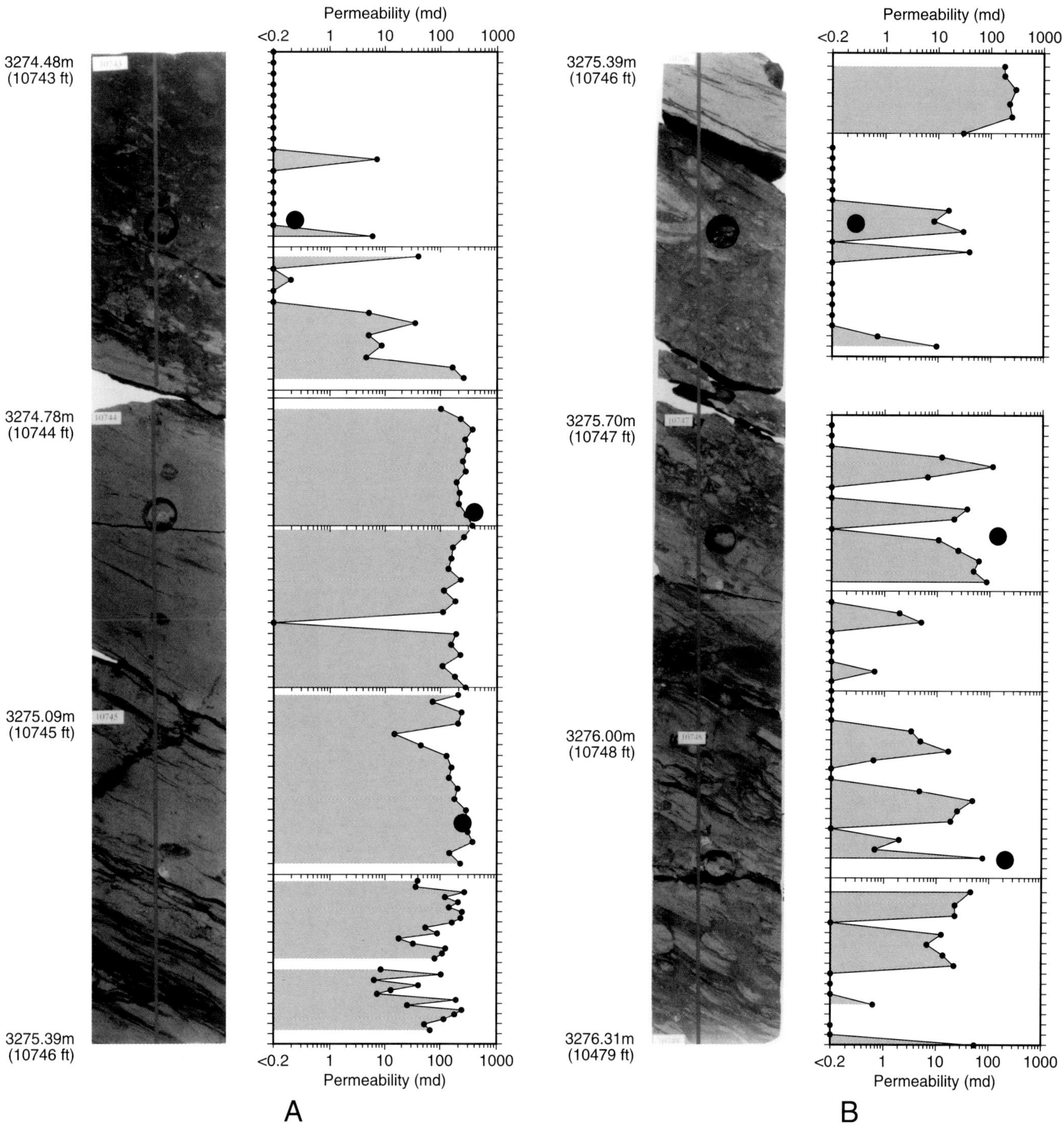

FIG. 19.—High-resolution, probe-permeability profiles through thin-bedded delta-front sandstone from Attaka "Deep" interval, Attaka I-5. Plug-based values are shown as large circles. **A)** 3274.48–3275.39 m (10,743-10,746 ft), **B)** 3275.39–3276.31 m (10,746-10,749 ft).

rowed sandstones (mostly distal and bar-margin delta-front deposits), especially those in the "Intermediate" and "Deep" intervals at Attaka field.

Low-resistivity, thick-bedded, distributary reservoirs in the "Shallow" interval at Attaka and Serang typically contain pore linings of smectitic clay (Figs. 7, 21, 24). They exhibit resistivities as low as 2 ohm-m. Pore linings of smectite and randomly ordered illite–smectite (containing 70% expandable layers) bind water and increase conductivity, thereby lowering resistivity. High conductivity and cation exchange capacity of the smectitic

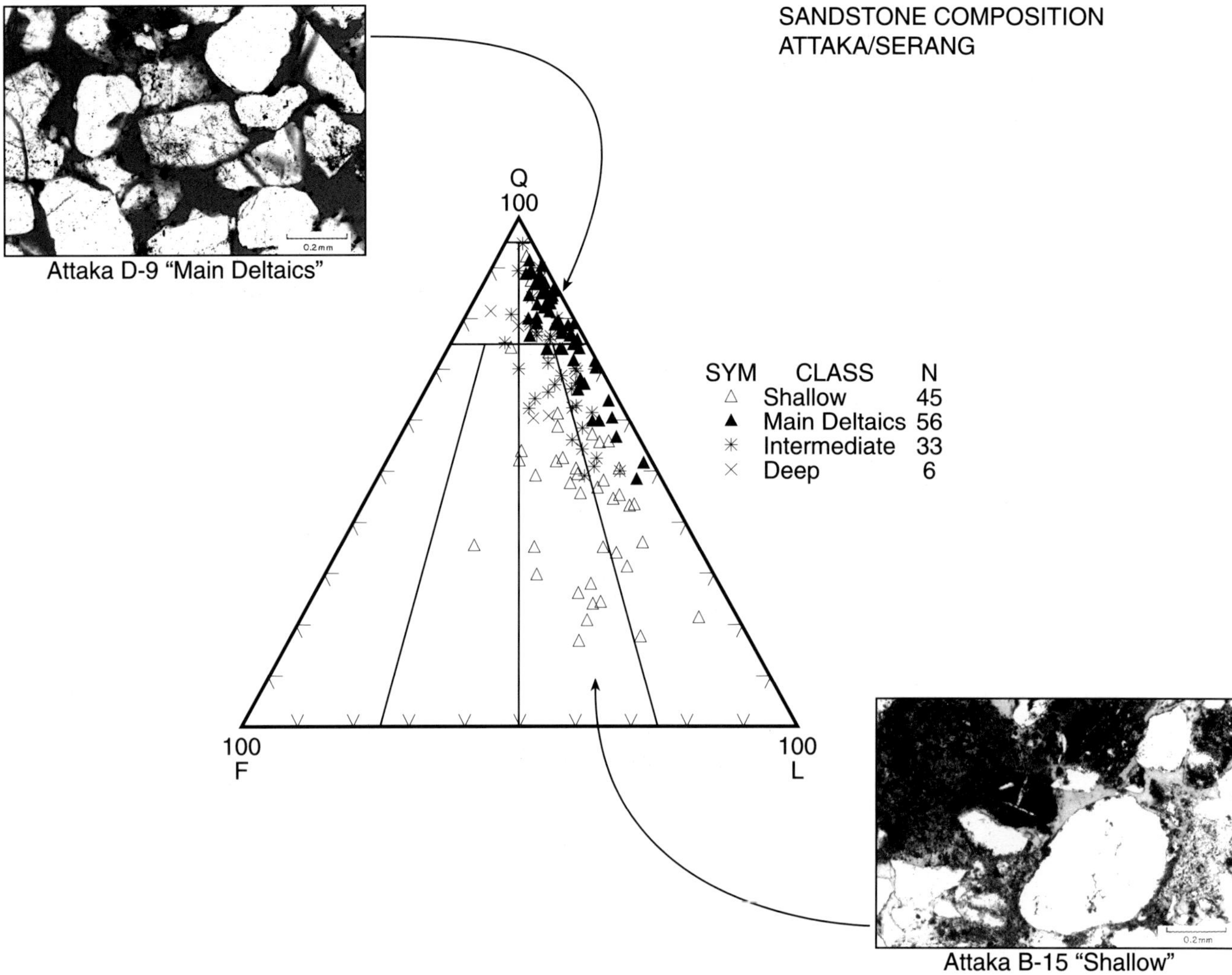

FIG. 20.—Ternary diagram with photomicrographs illustrating composition of framework grains in sandstones from Attaka and Serang fields. Q = quartz, F = feldspar + plutonic fragments, L = rock fragments. Much sandstone from the "Shallow" interval contains abundant volcanic rock fragments.

clay cause low resistivity, regardless of the fluid content of the reservoir (Partono, 1992).

Delta-front reservoirs of the "Intermediate" and "Deep" intervals are also associated with low resistivity (Fig. 17), even though they contain clays that are relatively benign, compared to the smectite of the "Shallow" low-resistivity reservoirs. Sandstone bed thickness is less than 0.5 m in many distal and bar-margin delta-front deposits. Conventional logging tools cannot adequately resolve beds thinner than about 1 meter, and consequently, the true resistivity of oil-saturated, thin-bedded sandstones cannot be determined in most cases. Also, burrowing in many delta-front deposits (Fig. 19) results in the mixing of mud and sand, ultimately lowering sandstone resistivity.

Low-resistivity sandstone reservoirs are probably present in many deltaic successions. They can be recognized through a combination of high-resolution logging tools, scrutiny of oil and gas shows, and targeted coring. Low-resistivity sandstones were recognized and developed at Attaka in the middle 1980s, thereby reversing a steep decline in oil production and greatly enhancing recoverable reserves.

CONCLUSIONS

The following conclusions drawn from Attaka and Serang may also apply to other deltaic deposits. (1) The coarsest-grained and thickest deltaic sandstones typically accumulate during relative lowstands, times when deltas have prograded long distances across marine shelves (Fig. 25A). Such lowstand deposits typically form the most porous and permeable hydrocarbon reservoirs, because they contain abundant, relatively coarse-grained distributary channel sandstones. (2) Thin-bedded sandstones and burrowed sandstones are common in distal deltaic deposits, most of which probably accumulated at times of somewhat higher relative sea level (Fig. 25B). Such sandstones may form low-resistivity reservoirs, which can, if recognized, contribute substantially to hydrocarbon production.

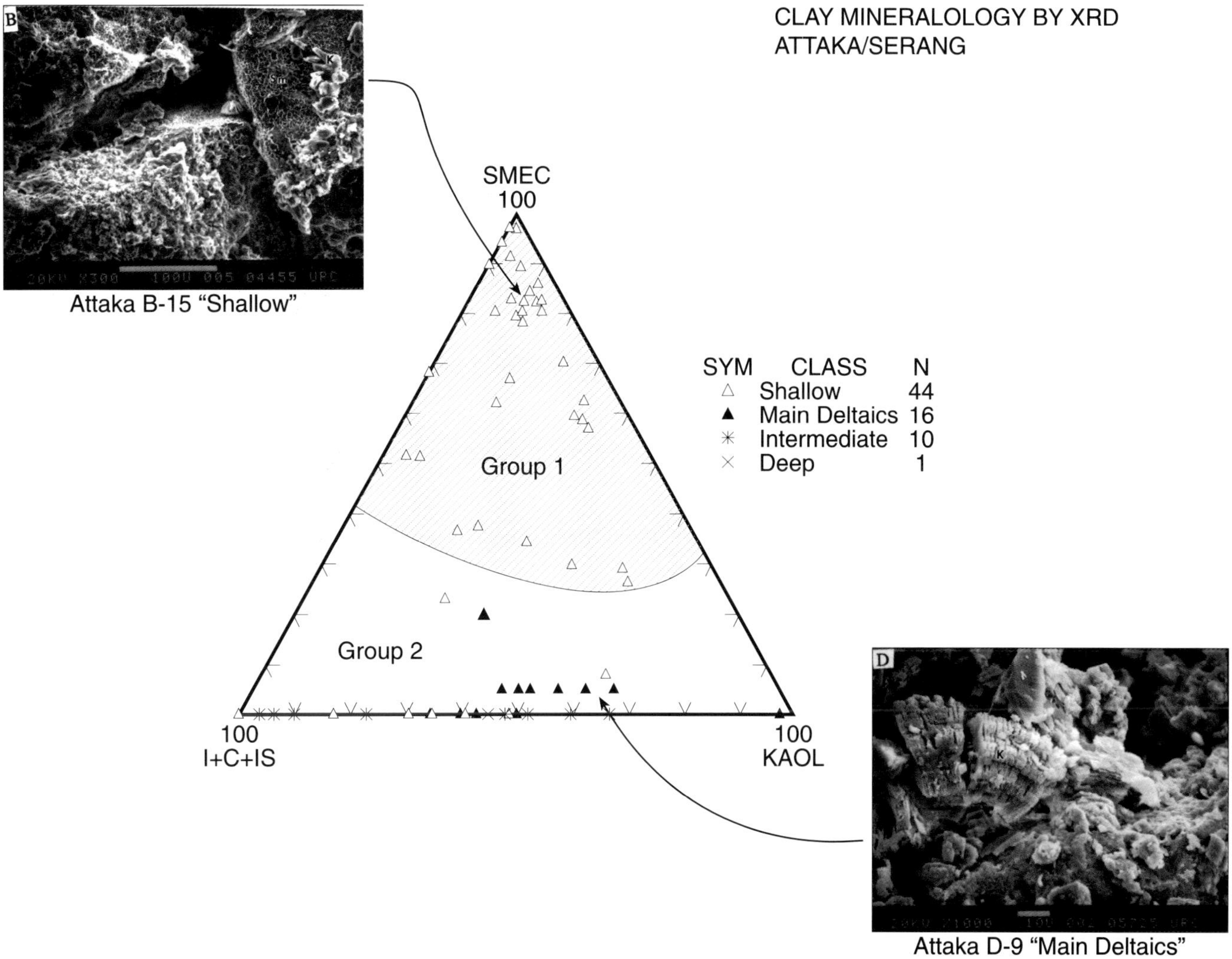

FIG. 21.—Ternary diagram illustrating XRD-based clay mineralogy of sandstones Attaka and Serang fields. SMEC = smectite + randomly ordered illite–smectite, I+C+IS = illite + chlorite + ordered illite–smectite, KAOL = kaolinite.

ACKNOWLEDGMENTS

We thank Pertamina and Unocal Indonesia Co. for permission to publish this paper. Patrice Imbert, Dag Nummedal, and an anonymous reviewer commented on an earlier draft of the manuscript and offered suggestions for its improvement. We thank Mike Kirby and Frank Wright for graphics support.

REFERENCES

ALLEN, G.P., AND CHAMBERS, J.L.C., 1998, Sedimentation in the modern and Miocene Mahakam Delta: Indonesian Petroleum Association, Field Trip Guidebook, 231 p.

ALLEN, G.P., LAURIER, D., AND THOUVENIN, J., 1976, Sediment distribution patterns in the modern Mahakam Delta: Indonesian Petroleum Association, Fifth Annual Convention, 1976, Proceedings, p. 159–167, plus figures.

ALLEN, G.P., LAURIER, D., AND THOUVENIN, J., 1979, Etude sédimentologique du delta de la Mahakam: Compagnie Française des Pétroles (TOTAL), Paris, Notes et Mémoires, No. 15, 156 p.

PARTONO, Y.J., 1992, Low-resistive reservoirs in the Attaka field: Indonesian Petroleum Association, 21st Annual Convention, 1992, Proceedings, p. 22–34.

SYDOW, J.C., 1996, Holocene to Late Pleistocene stratigraphy of the Mahakam delta, Kalimantan, Indonesia: unpublished Ph.D. dissertation, Louisiana State University, 170 p.

TREVENA, A.S., MAHADI, S., MARTINEZ, S.A., MARWOTO, PARTONO, Y.J., AND SUTIYONO, S., 1993, Characterization of Upper Miocene deltaic reservoirs at Attaka field, offshore East Kalimantan, Indonesia, *in* Atkinson, C.D., Scott, J., and Young, R., eds., Clastic Rocks and Reservoirs of Indonesia: Indonesian Petroleum Association, Core Workshop, 11 October 1993, Jakarta, p. 91–116.

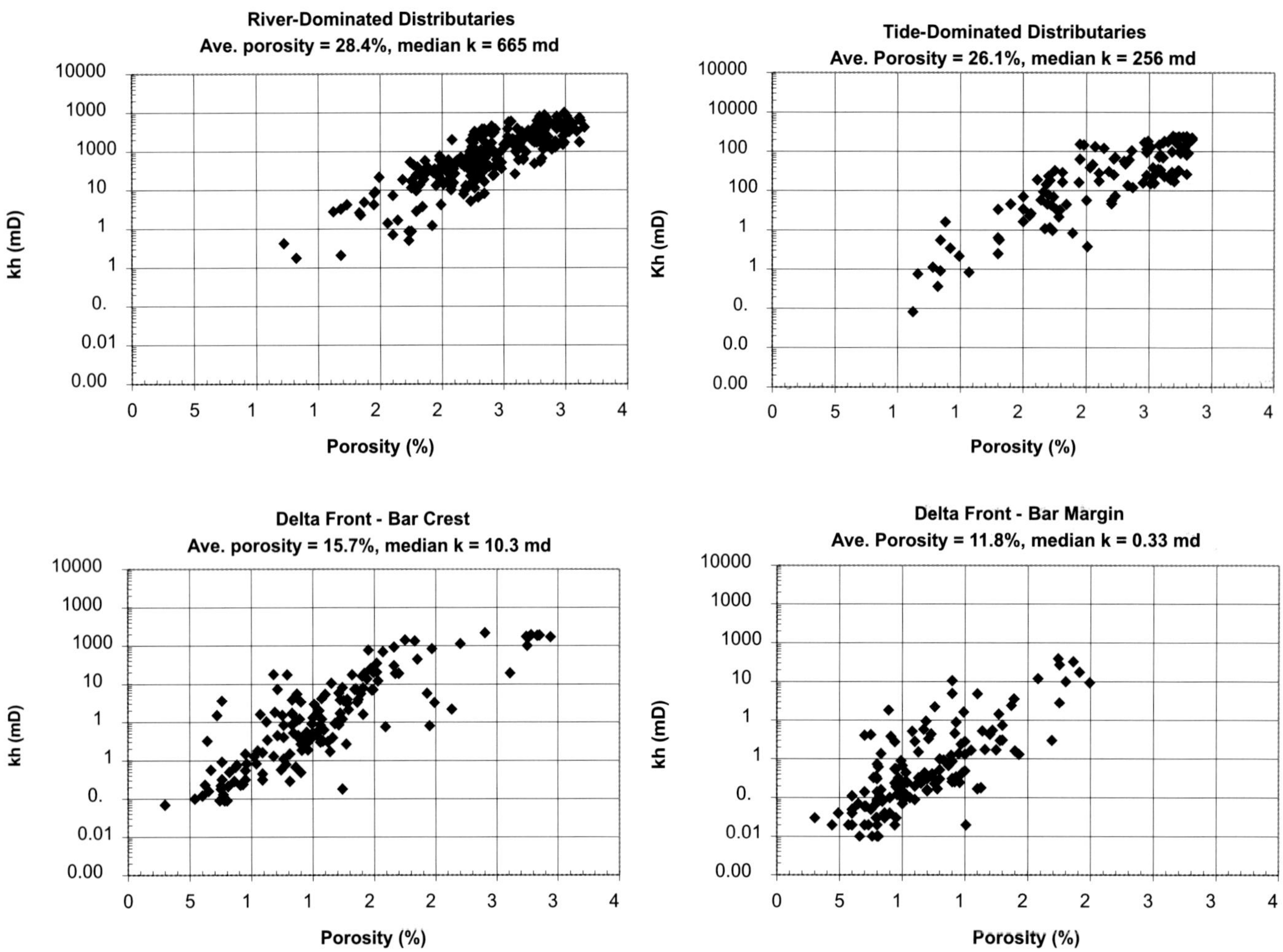

FIG. 22.—Cross-plots of porosity and permeability for river- and tide-dominated distributary sandstones and bar-crest and bar-margin delta-front deposits, Attaka and Serang fields. Based on conventional core plugs.

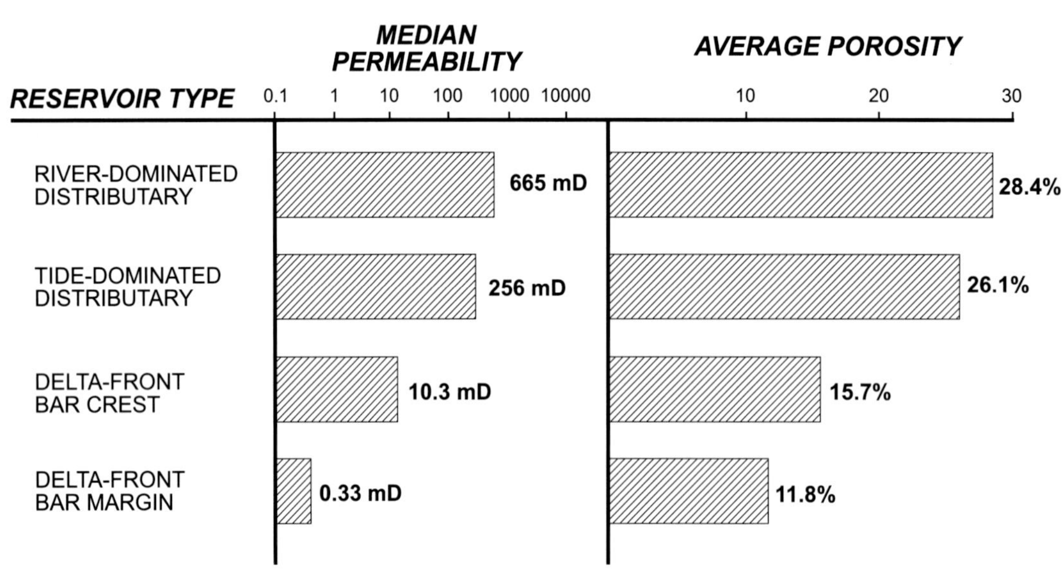

FIG. 23.—Bar chart of median permeability and average porosity for four deltaic facies from Attaka and Serang fields.

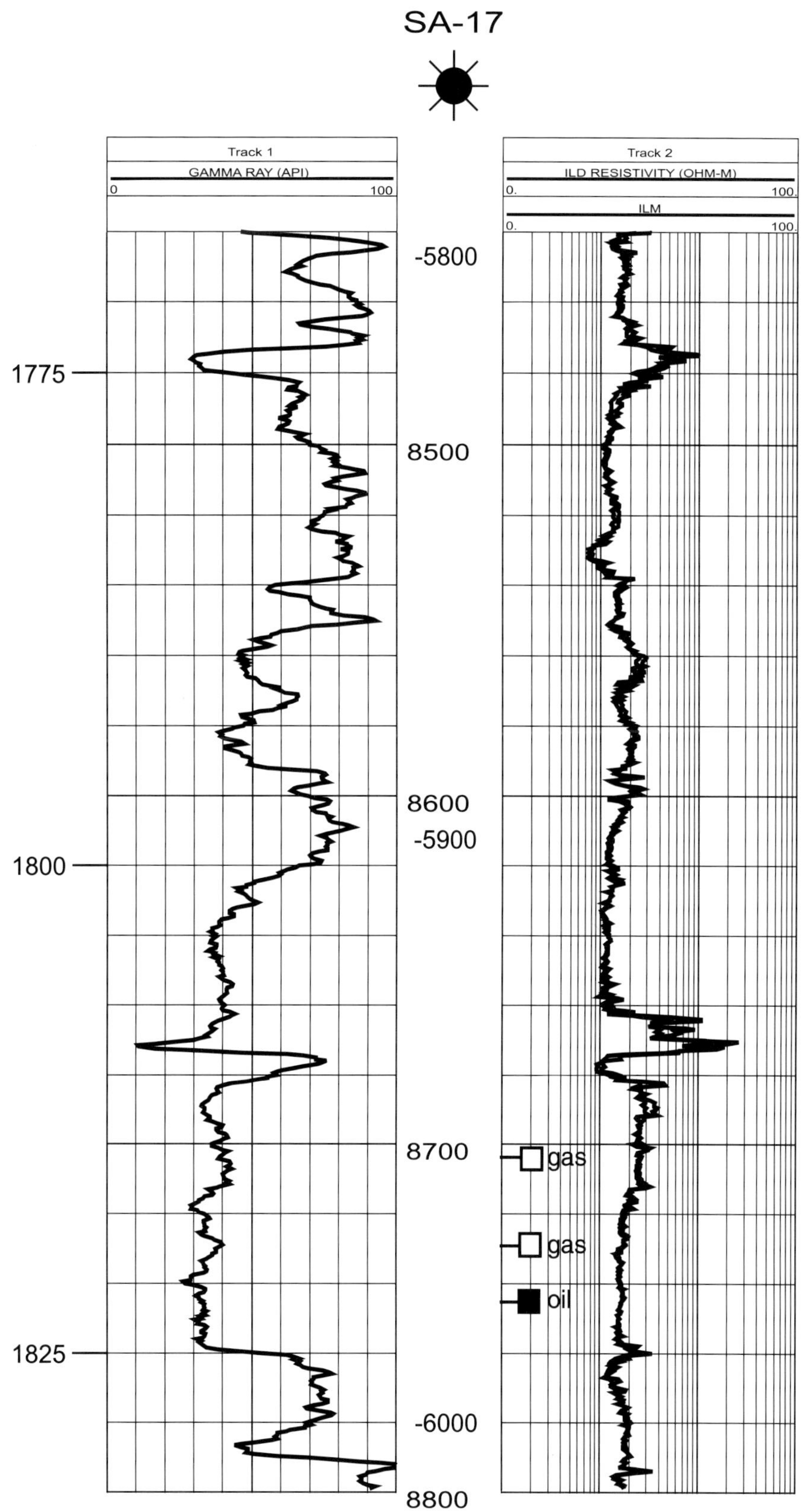

FIG. 24.—Wire-line logs from low resistive reservoir in Serang field "Shallow" interval. The reservoir is interpreted as two stacked distributary channel sandstones. Note extremely low resistivity (< 2 ohm-m). Positions of oil and gas recoveries from formation tests in lower sandstone body are indicated.

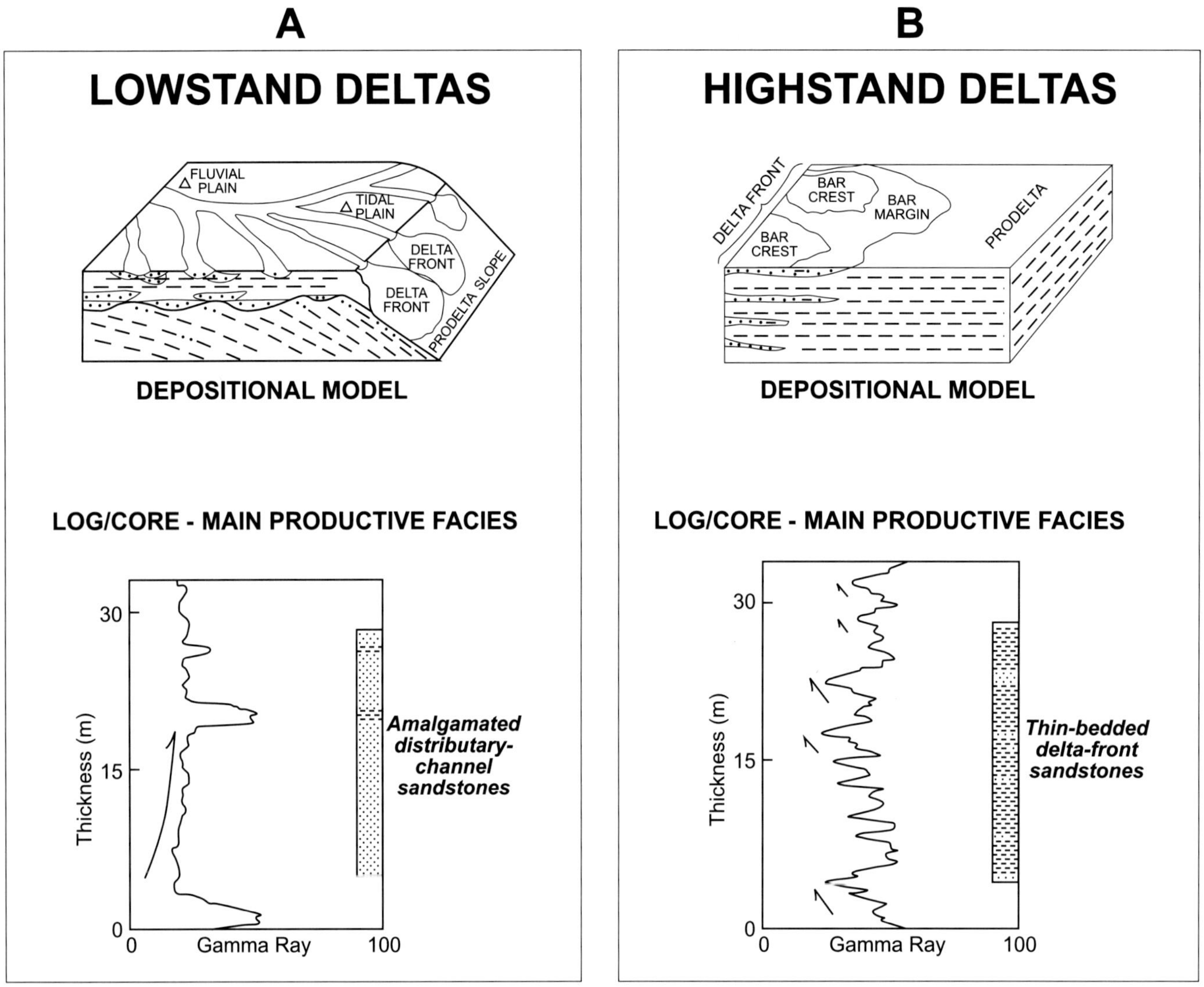

FIG. 25.—Depositional models and representative log traces for **A)** lowstand and **B)** highstand deposits of the Miocene Mahakam delta from Attaka and Serang fields.

PALEOCURRENTS AND RESERVOIR ORIENTATION OF MIDDLE MIOCENE CHANNEL DEPOSITS IN MUTIARA FIELD, KUTEI BASIN, EAST KALIMANTAN

TOBIAS H.D. PAYENBERG
National Centre for Petroleum Geology and Geophysics, The University of Adelaide, Adelaide, South Australia 5005, Australia
e-mail: tobi.payenberg@adelaide.edu.au
F. HASAN SIDI
VICO Indonesia, Jl. Rasuna Said C 11-14, Jakarta 12940, Indonesia
present address: Jason Geosystems Asia, Mulia Tower - Suite 2414, Gatot Subroto Kav 9-11, Jakarta 12930, Indonesia
e-mail: hsidi@jasongeo.com
AND
SIMON C. LANG
National Centre for Petroleum Geology and Geophysics and Australian Cooperative Research Centre,
The University of Adelaide, Adelaide, South Australia 5005, Australia

ABSTRACT: Mutiara field produces hydrocarbons from middle Miocene fluvio-deltaic successions within the Kutei Basin, East Kalimantan, Indonesia. In strata exposed over the doubly plunging Sanga-Sanga and Samboja anticlines, a large number of cores and downhole logs provide an excellent opportunity to integrate surface and subsurface data to improve reservoir characterization for exploration and development. Sedimentary facies and paleocurrent analysis were used to gain insight into the distribution of the dominant channel deposits within the succession.

Channelized sandstone bodies identified in outcrops and cores constitute the main hydrocarbon reservoirs. They commonly comprise single-story distributary-channel sandstones and occasionally multistory alluvial-channel sandstones. Paleocurrent analysis revealed the distributary channels flowing in an overall southward direction, roughly parallel to the strike of the anticlines, while two of three multistory channels trend northeast. The third multistory channel has a strong westward flow direction, which might indicate a valley incision of a previously more sinuous channel. The orientation of the single-story distributary channels can be explained as a result of active tectonism during the middle Miocene. Incision of the Mahakam River into the uplifting hinterland means that a point source of sediment supply has existed to the north of Mutiara Field since middle Miocene times. Growth of anticlines through regional inversion of older, extensional basement faults has restricted the eastward progradation of the paleo–Mahakam Delta. As a result, the delta distributary channels and delta progradation was merely towards the south and north, parallel to depositional strike, and not perpendicular, as commonly thought. In addition, low directional variance among the single-story and multistory channels suggests that the paleo–Mahakam Delta comprised low-sinuosity channels, which has strong implications for the exploration of stratigraphic traps.

Periods of dominant alluvial sedimentation produced roughly west–east striking multistory channels, which can sometimes be linked to incised-valley fills. These sandbodies might have had sediment sources to the southwest of the Mutiara field, and thus belong to a totally different fluvial system.

INTRODUCTION

The Kutei Basin is one of the largest and most important sedimentary basins, in terms of hydrocarbon exploration, in Indonesia (Verdier et al., 1980). Samuel and Muchsin (1975) estimate the basin to cover an area of 280,000 km^2, of which approximately 95% are onshore. Proven reserves exceed 3 billion (10^9) barrels of oil and 30 TCF (10^{12} ft^3) gas, making it a truly remarkable hydrocarbon accumulation. Various types of elongate sandstone deposits constitute the major hydrocarbon reservoirs within the onshore part of the Kutei Basin. The orientation of these deposits provides an invaluable insight into the overall morphology of this deltaic paleo-environment, and can lead to a better understanding of the reservoir orientation and geometry, thus improving exploration effectiveness.

The aim of this study was to use sedimentary facies and paleocurrent analysis to gain insight into the distribution of potential hydrocarbon reservoirs in the Miocene strata of the Kutei Basin by integrating data from outcrops, cores, and the downhole Formation MicroScanner tool (FMS, trademark of Schlumberger). Paleocurrents were collected from outcrops, scribe oriented and FMS oriented cores, as well as FMS log interpretations, and analyzed according to their generic nature and stratigraphic occurrence.

GEOLOGICAL SETTING

The Kutei Basin is situated on the east coast of Kalimantan, the Indonesian part of the Island of Borneo, southeast Asia (Fig. 1). Mutiara field is situated approximately 50 km south of the Equator, roughly midway between Balikpapan and Samarinda (Fig. 1). The study area is on the saddle junction of the Sanga-Sanga anticline and the Samboja anticline. Surface rocks and production zones in Mutiara field are of middle Miocene age. They range in age from around 14.3 million years (N9 biostratigraphic zone; "J" *VICO* interval) to approximately 10.2 million years (N13 biostratigraphic zone; 'D' *VICO* interval) (Fig. 2). Because of the plunge of the Sanga-Sanga anticline to the south and of the Samboja anticline to the north, the rocks cropping out in the center of the study area are youngest, with increasing age to the north and the south of the study area.

The early middle Miocene was a time of significant tectonic subsidence and little uplift. However, contactional forces associated with the Kuching uplift resulted in progressive structural uplift from west to east in this overall subsiding regime (VICO Indonesia, 1995). In the Mutiara area, the structural growth of the Sanga-Sanga anticline initiated around N12 ("E" *VICO* interval) times (Ferguson and McClay, 1997), as a part of the basin inversion that began during the Early Miocene (Allen and Chambers,

Tropical Deltas of Southeast Asia—Sedimentology, Stratigraphy, and Petroleum Geology
SEPM Special Publication No. 76, Copyright © 2003
SEPM (Society for Sedimentary Geology), ISBN 1-56576-086-7, p. 255–266.

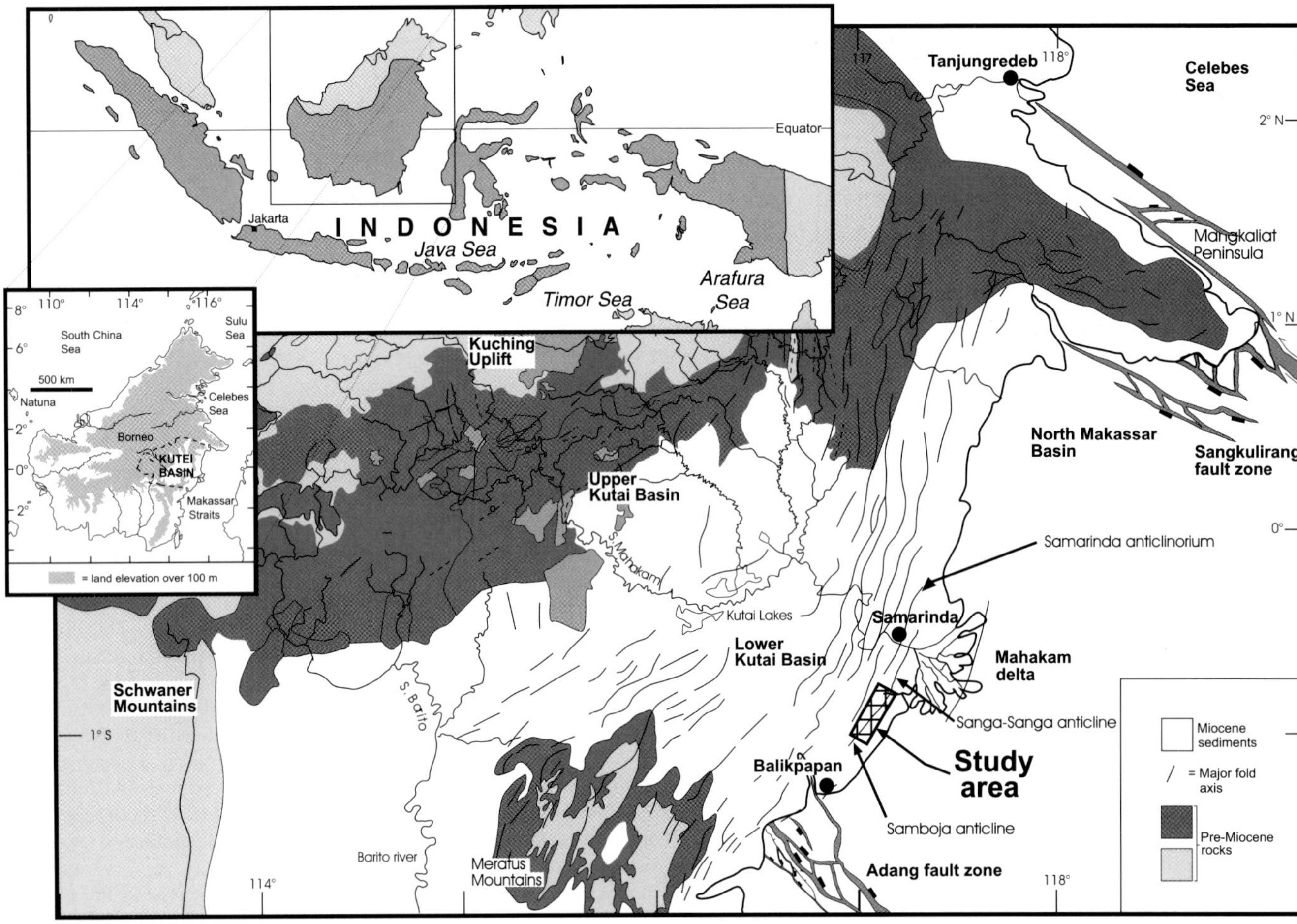

FIG. 1—Location map of the study area and simplified geological map of the Kutei Basin and surrounding margins, East Kalimantan. The anticlinal structures are clearly visible in the onshore Mahakam Delta. The rectangle outlines the study area. Geological map modified after Moss et al. (1997), Cloke (1997), and Cloke et al. (1997).

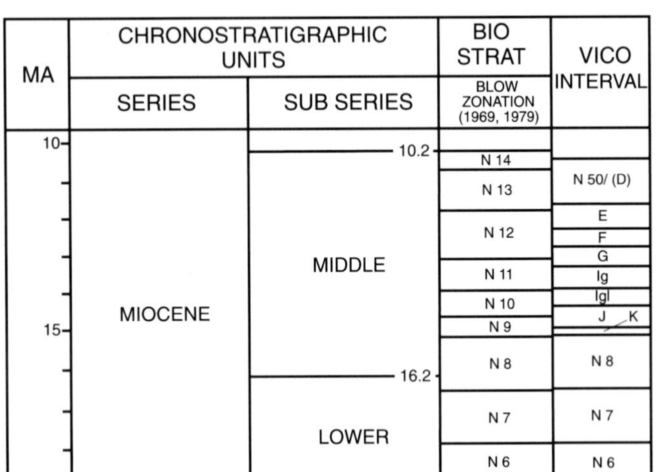

FIG. 2—Geological timetable of the Kutei Basin. While the biostratigraphic zones are basinwide, the *VICO* intervals are *VICO* internal zonations after easily recognizable markers. Modified from VICO Indonesia (1995).

1997). Trapping of sediments on the west flank of the anticlines, associated with this uplift, is evident in the Badak and Mutiara area (VICO Indonesia, 1995). It is also thought that these uplifts, at times, caused channel systems to be diverted around structural highs (Sidi, 1998), which may represent reactivation of preexisting basement lineaments (Ferguson and McClay, 1997).

The present-day Mahakam River delta is located at the current depocenter of the Kutei Basin on the east coast of Kalimantan. According to Allen and Chambers (1998), the Mahakam River incised the Samarinda anticlinorium during the middle Miocene, when uplift was progressively moving towards the east. Successive deltas have therefore prograded from a single point source near the present head of passes. The modern delta has frequently been used as an analogue to the paleo–depositional environment of the regressive rocks of the Kutei Basin (Magnier et al., 1975; Allen et al., 1976; Verdier et al., 1980; Allen and Chambers, 1998).

METHODOLOGY

The paleocurrent analysis project incorporated measurements from outcrop, as well as subsurface core and electric logs (FMS). For paleocurrents from core, a mechanical goniometer (device to measure structures in core) was custom designed and built

(Payenberg et al., 2000). A Fortran program named ROTATE (DOS) (Laycock and Payenberg, 1997) was written to correct measured paleocurrents for tectonic tilt.

The database for this study consists of eight outcrops and 425 m (1397 ft) core from the Mutiara oil and gas field. 1596 paleocurrent measurements were collected, 565 in outcrop, 522 on core, and 509 from FMS logs. The Formation MicroScanner (FMS) is a micro-conductivity subsurface well-logging tool with an average sampling density of one reading every 0.25 cm. The four-pad FMS tool is one of the highest-resolution logging tools available. Because of this high resolution, the tool can be used to investigate subsurface sedimentary structures beyond the detail of many other electrical borehole tools, such as gamma ray or sonic. Two methods for interpreting FMS data were used in this project. One method involved interpreting paleocurrent directions from the digital FMS image using a graphical workstation running FLIP (version 4.4A) from Schlumberger. The second method was to use FMS hardcopy prints, with the resistivity converted to digital borehole images across a 42-color scale, to orient conventional core using significant surfaces as described by Payenberg et al. (2000).

In outcrop, good three-dimensional exposures of the foresets of the cross-stratified bed are favored to determine the downstream direction. Slingerland and Williams (1979), DeCelles et al. (1983), and Trexler and Cashman (1990) describe other methods to measure paleocurrent directions from two-dimensional exposures. The rocks outcropping in the Mutiara area are soft, so that careful observation of the outcrop face and cutting of a "V" groove with a spade or knife was sufficient to obtain good three-dimensional exposure of the foreset surfaces of cross-stratified beds.

CHANNEL DEPOSITS

Lithofacies Analysis

Several lithofacies, and lithofacies associations, have been interpreted from detailed sedimentological analysis of cores and 3D outcrops. These lithofacies are summarized in Figure 3. The depositional environment of each lithofacies was interpreted from grainsize, grainsize changes, paleocurrent analysis, and paleoecology of the examined sections, because these criteria provide data to interpret current speed, variability, and direction.

Facies associations identified include alluvial channels, fluvial-dominated distributary channels, tide-influenced distributary channels, interdistributary mudstone, siltstone, and sandstone, distributary mouthbars, and prodelta mudstone. Evidence for the interpretation of distributary channels comes from the identification of distributary mouthbar deposits often contained within the same parasequence (Sidi, 1998). The lack of mud drapes and tidal bundles in the facies association suggests a fluvial origin. This is supported by the unidirectional paleocurrent data, discussed later. Although Willis et al. (1999) show unidirectional paleocurrent patterns from a tide-influenced delta, the lack of mud drapes and tidal bundles in the clean sandstones discussed in this paper suggests a fluvial origin.

Of major interest in terms of hydrocarbon exploration in the Mutiara area are the clean sandstones, which are interpreted to be deposited mainly in channels of distributary and fluvial origin. Channelized sandstones appear in three different styles: fluvial multistory, fluvial single-story, and tidal (Fig. 4). The single-story channels are characterized by an average sandstone thickness of 10 to 15 meters (after compaction) (Figs. 4A, 5). The channel has one major erosional surface, the channel

Lithofacies and code	Idealized Gamma-Ray log	Lithofacies description	Grain size	Character and stratification	Interpretation
Pc 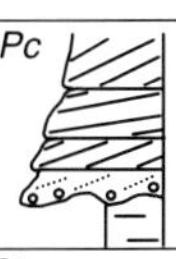		Cross-Stratified Pebbly Sandstone	Very Coarse Sand to Pebble	Beds, commonly less than 1 meter thick, stratified and graded. Contact with underlying beds is strongly erosional. Forms base of thicker (cm–meters) sandy succession.	Cross-stratified pebbly beds formed at the base of a channel or a larger barform (?flood event).
St 		Trough Cross-Stratified Sandstone	Fine to Coarse Sand	Small, 5 – 20 cm thick sets of trough cross-stratification. Often with well-rounded mudstone, siderite, or coal pebbles at the base of the trough. Common interbedded with *Sp*.	Sinuous-crested and linguoid dunes formed under varying upper and lower flow conditions formed primarily in proximal parts of a distributary channel with high flow turbulence.
Sp 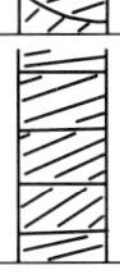		Planar Tabular Cross-Stratified Sandstone	Very Fine to Coarse Sand	Set thicknesses around 5 to 50 cm with more planar foresets and no erosional bases. Commonly interbedded with *St*. Backflow ripples present on foresets.	Migrating linguoid bars, traverse bars, and sand waves forming under low flow conditions with little flow turbulence.
Sm 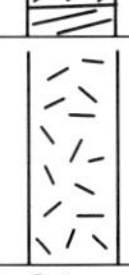		Massive Sandstone	Fine to Coarse Sand	Massive, nonstratified sandstone. Abundance of coalified plant fragments and occasional rounded siderite and mudstone pebbles. Thicknesses range from 10s of cm to 5 meters.	Sediment gravity flow in fluvial-dominated channels as a result of bank collapse and/or water escape as a result of liquefaction.
Sr & *Srb* 		Ripple Cross-Laminated (*Sr*) and Bioturbated Ripple Cross-Laminated Sandstone (*Srb*)	Very Fine to Fine Sand	Ripple cross lamination, thickness smaller 5 cm, occurs as sets, interbedded with shale on top of other bedforms.	Forms under lower-flow-regime conditions as traction deposits of 3D ripples in fine-grained sediments.
Spl 		Parallel-Stratified Sandstone	Very Fine to Fine Sand	Parallel-laminated beds with bed thicknesses smaller 4 cm. Often interbedded with *Sr* or *Srb*.	Formed at the transition of lower and upper flow regime at shallow water depth in fine-grained sandy sediment.

FIG. 3—Characteristics of the clastic lithofacies and their graphic log expressions. Facies classification scheme modified from Miall (1985).

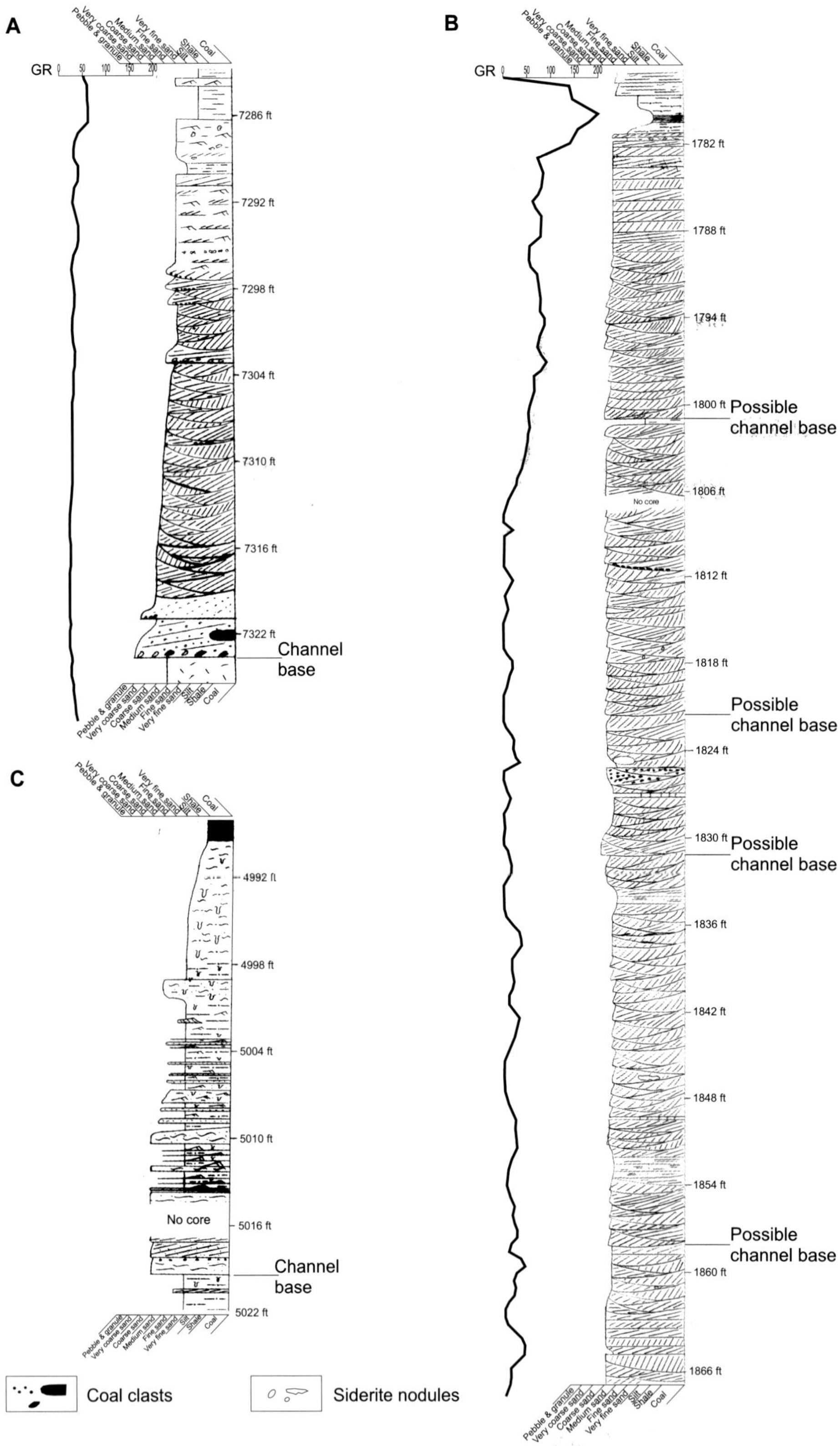

FIG. 4—Sample stratigraphic columns of **A)** a single-story distributary channel, **B)** a multistory alluvial channel, and **C)** a tidal channel. Although the base of the multistory channel is not cored, wireline logs indicate a total thickness of roughly 30 m (90 ft, B), whereas the single-story channel is only about 12 m (36 ft) thick (A). The multistory sandstone is coarser, and contains numerous surfaces that were interpreted as individual channel bases. The tidal channel clearly has a higher mud / sand ratio and is of lesser reservoir quality. See Figure 3 for lithologic symbols.

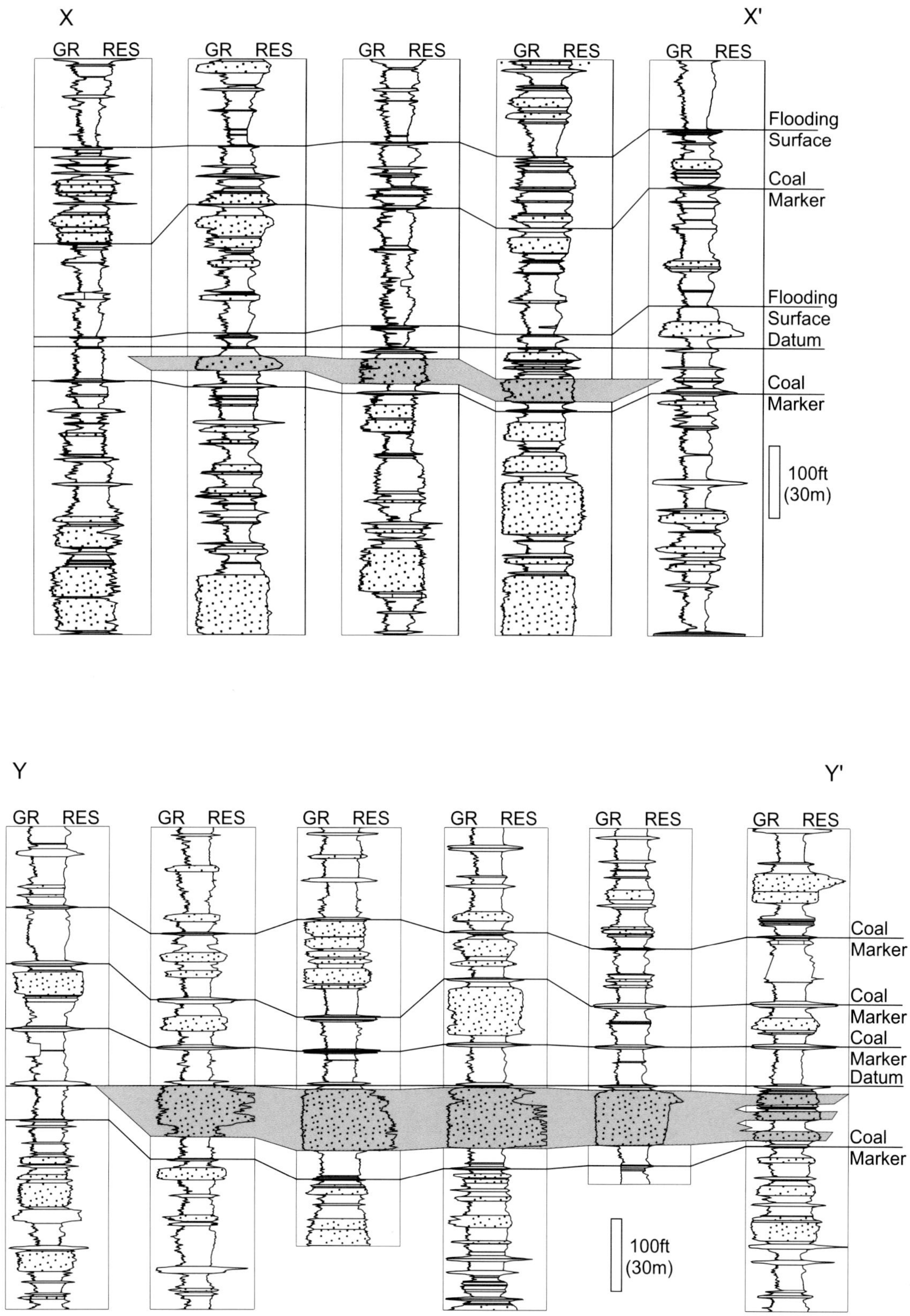

FIG. 5—Well correlation illustrating the single-story and multistory channel types. Section X–X' represents a single-story channel fill cut at an oblique angle, owing to well restrictions. Section Y–Y' is a correlation of a multistory channel cut at right angle. The multistory channel is roughly double the thickness of a single-story channel. The single-story channel is also only half as wide as multistory channel when cut at right angle to the paleoflow. Horizontal distances are not to scale.

base, overlain by a lag deposit of coarse to very coarse sandstone with rounded coal clasts and siderite nodules. The channel is filled with sandstone bars, each having a slightly erosional lower contact. This deposit represents a single, fluvial-dominated distributary channel, formed in the upper-delta-plain environment. It is clearly different in lithology from the tidal channel shown in Figure 4C. Although the depth of tidal channels is similar to that of the single-story channels, tidal channels are filled with very little sand and contain a large amount of mud drapes, tidal bundles, and interbedded, bioturbated mudstone, siltstone, and thin sandstone beds. Figure 6 shows a typical vertical succession of a single-story channel from outcrop 2. The log closely resembles the shape and size of its subsurface counterpart in Figure 4A.

The second style of fluvial channel, the multistory channel, is significantly thicker than the single-story channel, commonly exceeding 30 meters (Figs. 4B, 5). The multistory channel contains several stacked bar and channel deposits with erosional surfaces. These represent a number of crosscutting channels. Subsurface well correlations helped to identify the channel character of the sandstones encountered in core (Fig. 5).

Paleocurrent Analysis

Paleocurrent analysis was undertaken to:

1. Determine changes in channel orientation and directional variability through a stratigraphic unit as an indication for lateral and vertical changes in fluvial styles.

2. Reconstruct tributary or distributary patterns.

3. Reconstruct local and regional paleoflow patterns.

4. Determine vertical changes in flow direction through a stratigraphic section as indicators of interacting fluvial systems or in response to paleogeographic changes in the basin (after Miall, 1996).

Investigated outcrops and cores occur in a variety of stratigraphic intervals, owing to a doubly plunging fold axis in the study area. Both outcrop and subsurface data have been analyzed according to their stratigraphic occurrence and according to their channel style. In outcrop, four single-story channels were clearly identified. They are scattered across the whole field area (Fig. 7). One multistory channel in the southern part of the field area was also identified in outcrop. In the subsurface, two multistory and four single-story channels were interpreted.

Paleocurrent readings were corrected for structural deformation prior to interpretation. The regional plunge of the studied anticline (Samboja and Sanga-Sanga anticlines, Fig. 7) is very low (less than 5°), whereas maximum dip magnitudes of bedding planes were observed in outcrop to be 13°. Although not crucial, because of the low dip and plunge angles (cf. Ramsay, 1961), corrections for structural deformation were found to improve the coherency of the data slightly, and was therefore applied systematically using ROTATE (Laycock and Payenberg, 1997).

Single-Story Channels

Eight single-story channels were identified in outcrops and in the subsurface. A total of 300 paleocurrents were collected from single-story channels in outcrop, along with 468 that were measured in core and interpreted from FMS logs. Figure 8 shows paleocurrent rose plots of foresets, measured from trough and planar tabular cross-stratified beds, for all eight single-story

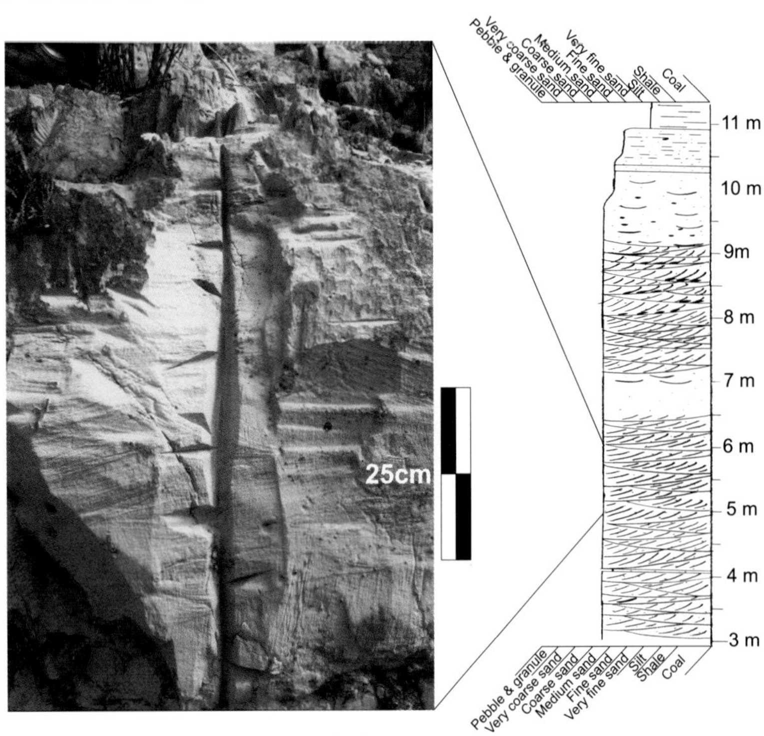

FIG. 6—Vertical log and close-up photo of a single-story channel from outcrop 2 showing facies similar to the subsurface logs (Fig. 4). The close-up photo highlights the difficult outcrop conditions: soft rock conditions and heavy weathering.

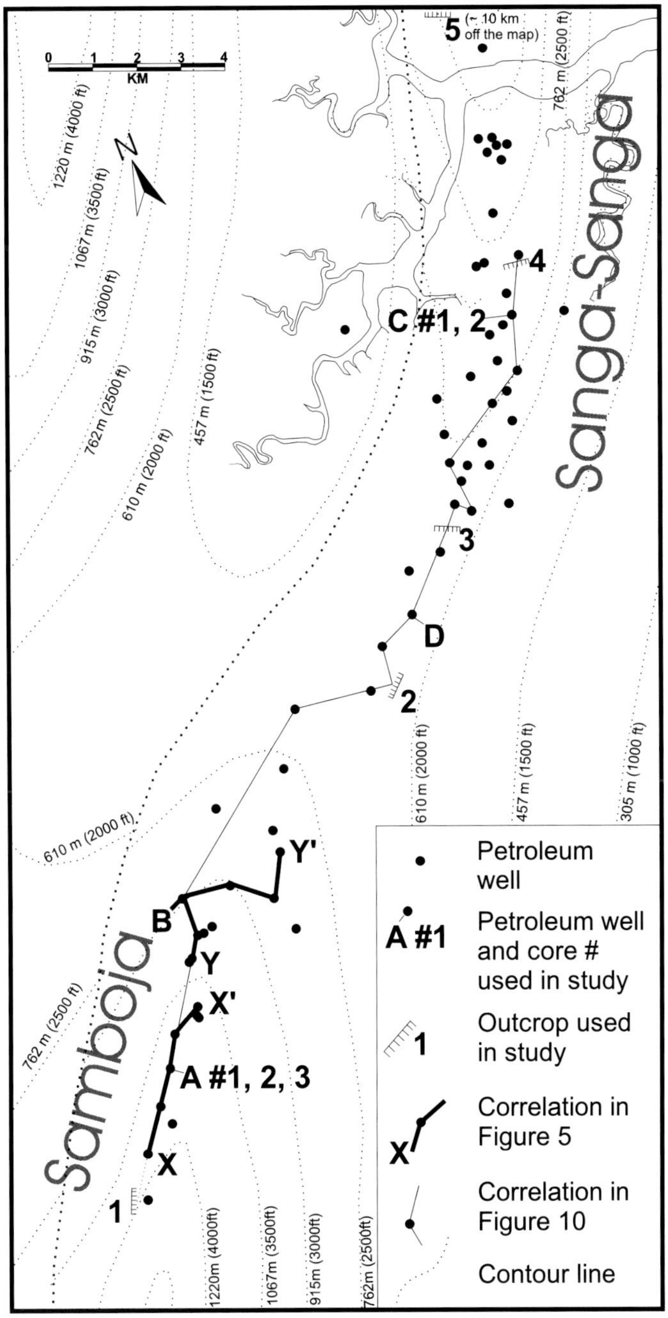

FIG. 7—Mutiara field location map showing all wells drilled prior to 1998. Outcrops numbered and shown are only those mentioned in the text. Contour lines are schematic lines of a marker horizon indicating relative amount of uplift of each contour to illustrate anticlines and synclines. The thin, solid line is the correlation line of Figure 10. Heavy lines are the well correlations in Figure 5. Note the crest-confined position of the wells, which are targeting mainly structurally closed reservoirs.

channels. Trough and tabular cross-stratification are regarded as the most reliable paleocurrent indicators, although these types of data sets can have high circular variance (High and Picard, 1974). Vector mean paleocurrent directions from all outcrops are dominantly towards the south (Fig. 8), and range from 153° in outcrop 3 to 221° in outcrop 4, a difference of only 68°.

In the subsurface, paleocurrent directions were made on FMS interpretations as well as core measurements from scribe and FMS-oriented core. Although the FMS and core data sets do not always readily match (see Payenberg, in press), all four subsurface intervals were interpreted with a southward paleocurrent direction. The variance of the paleocurrent vector mean is slightly higher but still well towards the south (111°).

Multistory Channels

Only three multistory channel complexes were encountered in this study, but because of the greater thickness of these complexes the paleocurrent data base for these channels is rather large compared to the single-story channels (Fig. 5). Figure 9 shows 157 readings from one outcrop and 291 paleocurrents from core and FMS log interpretations. Outcrop one is a multistory channel containing at least three channels, which have a mean paleocurrent direction from southwest (225°, lowest exposed channel) to west (273°, highest exposed channel). With a mean southwestward flow direction, the paleocurrent directions are very nearly 180° opposite to the subsurface counterparts. Although the evidence is not conclusive, the multistory channel might have incised a local river that was rather sinuous and was, at this location, flowing towards the west. The two subsurface channel complexes have an interpreted paleocurrent vector mean of 70° (core B#1) and 60° (core C#2), both towards the northeast.

DISCUSSION

Reservoir Orientation

The orientation and geometry of single-story and multistory channels in Mutiara field is of major importance, because they form the major hydrocarbon reservoirs. As revealed through the paleocurrent analysis, the orientation of the single-story channels is from north to south (Fig. 10). This is because the Mahakam River incised the Samarinda anticlinorium during the middle Miocene (Allen and Chambers, 1998), and the river was locked in place, acting as the source point for all successive deltas from the middle Miocene. This observation coincides with the overall isopach thickness increasing towards the east (Bob Koch, personal communication, 1997), and overall Miocene to recent deltaic progradation towards the east, observed by Gerard and Oesterle (1973) and Samuel and Muchsin (1975). Because the study area is located south of the incision, it could be argued that all distributary channels located south of the source point should have a southward flow (Fig. 11).

But tectonism also seems to have had an influence on the distributary orientation. As can be seen in Figures 1 and 11, the major fold axes in the area also have a roughly northwest–southeast strike. Synsedimentary deformation is observable in outcrop and on a larger scale. During the middle Miocene, uplift moved progressively towards the east, creating the Samboja and Sanga-Sanga anticlines (Fig. 1). Such synsedimentary structuring may have diverted channels away from the crests and forced the rivers to flow parallel to them. Because the deformation took place during a time of overall sedimentation (not erosion) and thus relative subsidence, distributary channels never existed long enough to cut into growing anticlines. Consequently, the southward orientation of the single-story distributary channels can be attributed to a combination of southward deltaic progradation from a point source located to the north of Mutiara field and forced flow parallel to anticlinal crests.

Location	Statistics for rose plots	OUTCROP x-strata		Comments
Outcrop 2	Data points: 49 Vector Mean: 179° Circular Std. Dev.: 42°	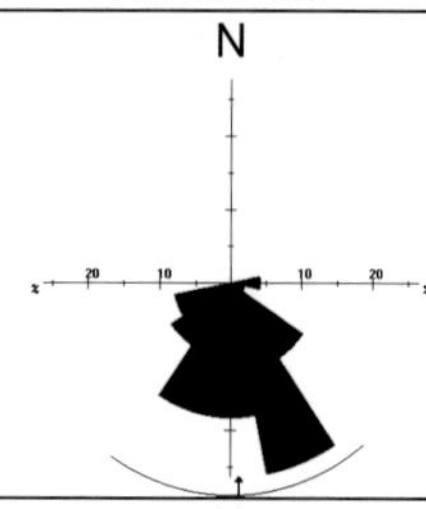		Confined flow of 3D dunes producing dominantly trough cross-stratification.
Outcrop 3	Data points: 72 Vector Mean: 153° Circular Std. Dev.: 61°	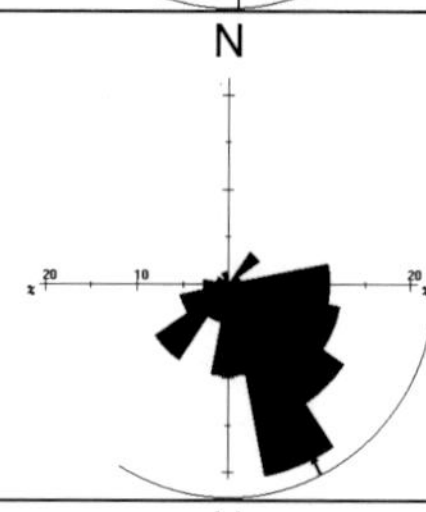		Trough cross-stratification only points towards 143°.
Outcrop 4	Data points: 52 Vector Mean: 221° Circular Std. Dev.: 42°	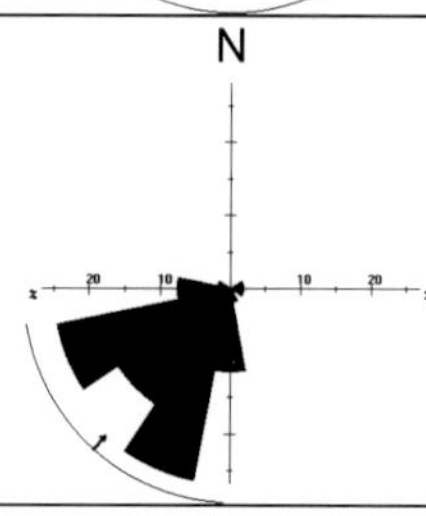		A very small outcrop face with trough and tabular cross- stratification pointing towards the same direction.
Outcrop 5	Data points: 127 Vector Mean: 187° Circular Std. Dev.: 43°	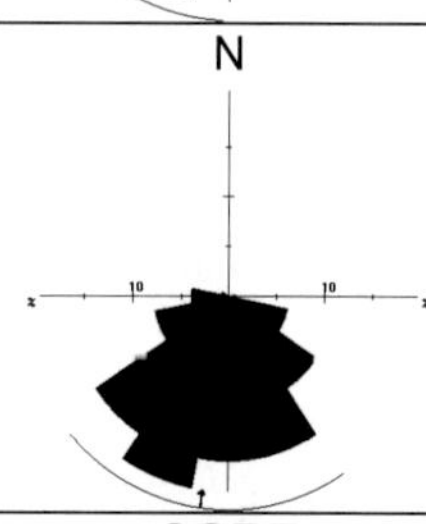		Large outcrop located to the north of Mutiara field, but within the same interval.
		CORE x-strata	**FMS x-strata**	
Core C#1	Data points: 84 Vector Mean: 158° Circ. Std. Dev.: 61°		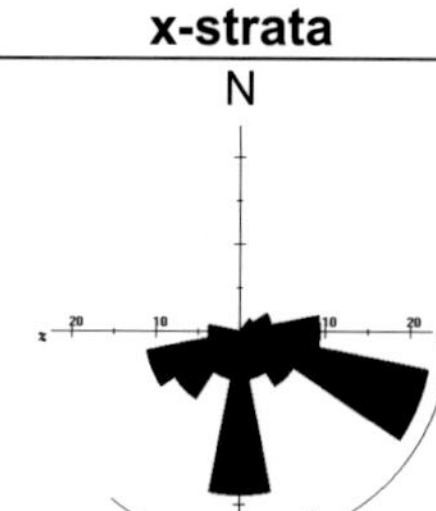	A misorientation of the scribe on the core makes the paleocurrent data from this core unreliable.
Core A#1	Data points: 26/55 Vector Mean: 197°/342° Circ. Std. Dev.: 48°/180° Core oriented using FMS logs	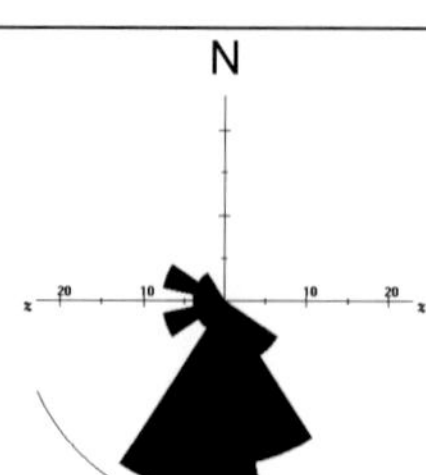	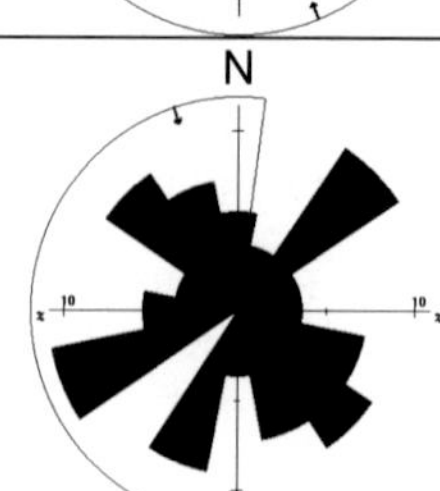	Because the poor FMS quality produced a data with a high scatter, the core paleocurrent data is used for the interpretation.

FIG. 8—Paleocurrent rose plots of the eight single-story channel deposits. The overall dominant paleoflow direction is clearly towards the south. All rose plots have a 011.25° start azimuth and a sector size of 22.5. Rose plots and statistics were generated with GEOrient (Holcombe, 1998).

Location	Statistics for rose plots	CORE x-strata		Comments
Core A#2	Data points: 57/28 Vector Mean: 220°/133° Circ. Std. Dev.: 55°/42° Core oriented using FMS logs	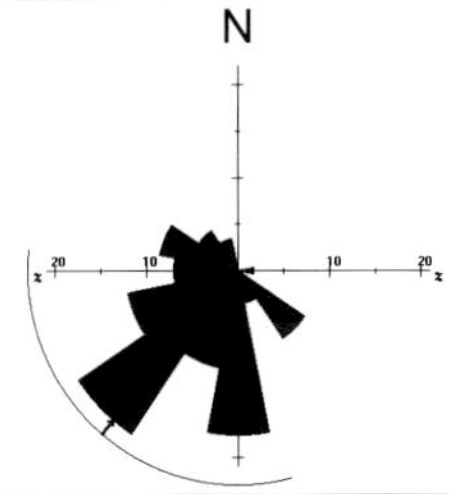	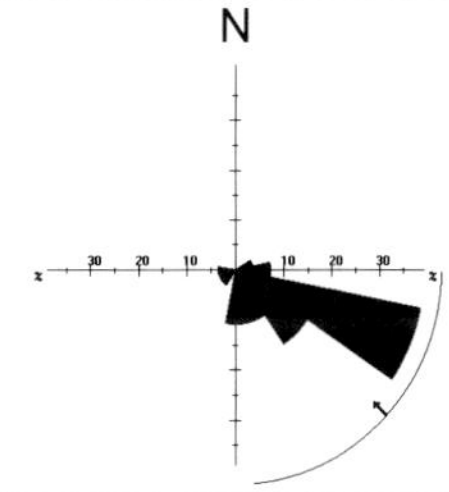	Because the core measurements have more than double the FMS measurements, a paleocurrent direction of 220° is used.
Core A#3	Data points: 45/81 Vector Mean: 241°/244° Circ. Std. Dev.: 79° Core oriented using FMS logs	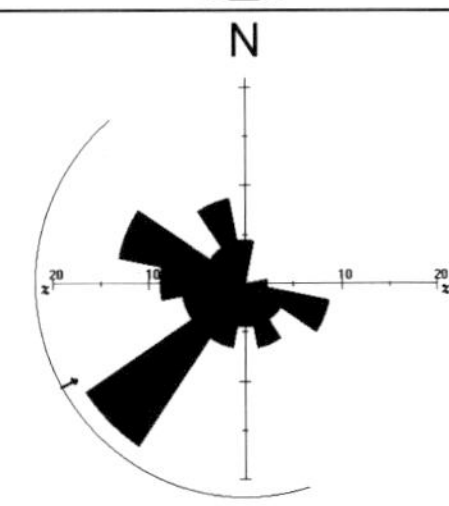	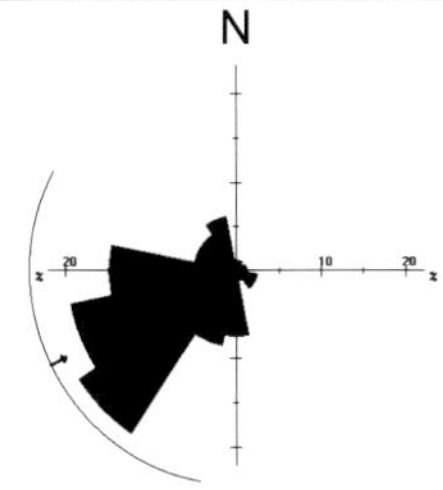	The data sets match very well, indicating a paleocurrent direction towards 243°.

FIG. 8.—(continued).

Outcrop	Statistics for "all x-strata" rose plot	OUTCROP x-strata		Comments
Outcrop 1	Data points: 157 Vector Mean: 264° Circular Std. Dev.: 60°	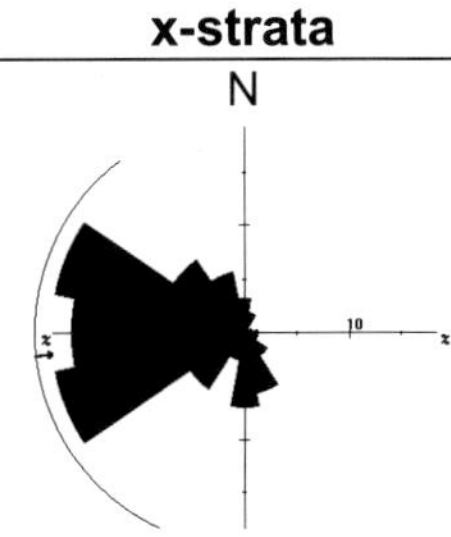		At least three, stacked channels have a confined flow towards the west.
		CORE x-strata	**FMS x-strata**	
Core B#1	Data points: 39/58 Vector Mean: 180°/070° Circ. Std. Dev.: 128°/125° Core oriented using FMS logs	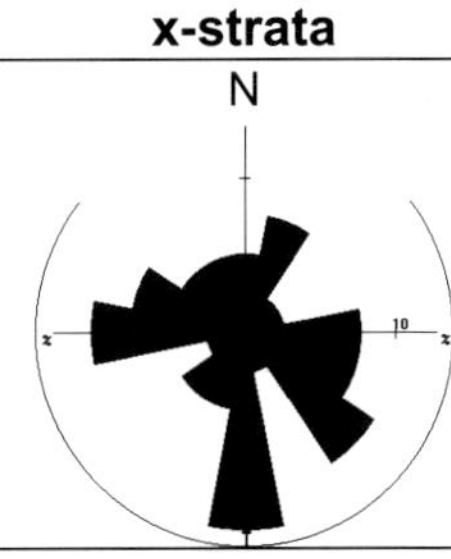	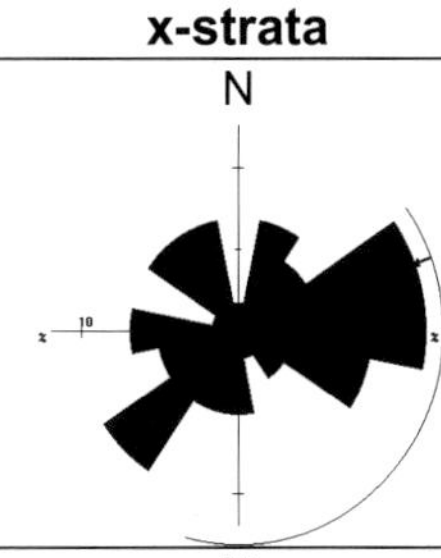	The core plot contains abundant ripple data, but troughs point towards 61°, enhancing the 70° from the FMS plot.
Core C#2	Data points: 134 Vector Mean: 093° Circ. Std. Dev.: 83°		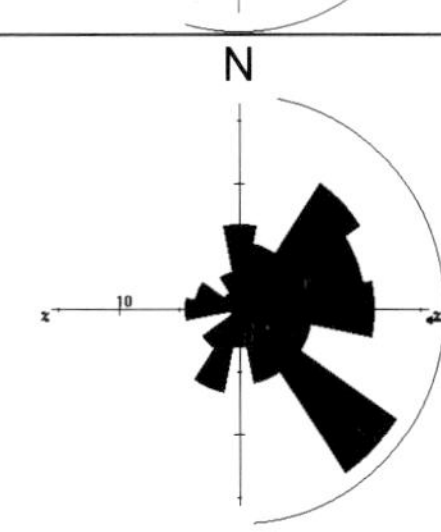	The core is again scribe oriented and thus erroneous. Paleocurrent direction is used from the trough cross-stratification of the FMS data.

FIG. 9.—Paleocurrent rose plots of the three multistory channel deposits. The orientation of two of the three channels is roughly west–east. See text for discussion. All rose plots have a 011.25° start azimuth and a sector size of 22.5. Rose plots and statistics were generated with GEOrient (Holcombe, 1998).

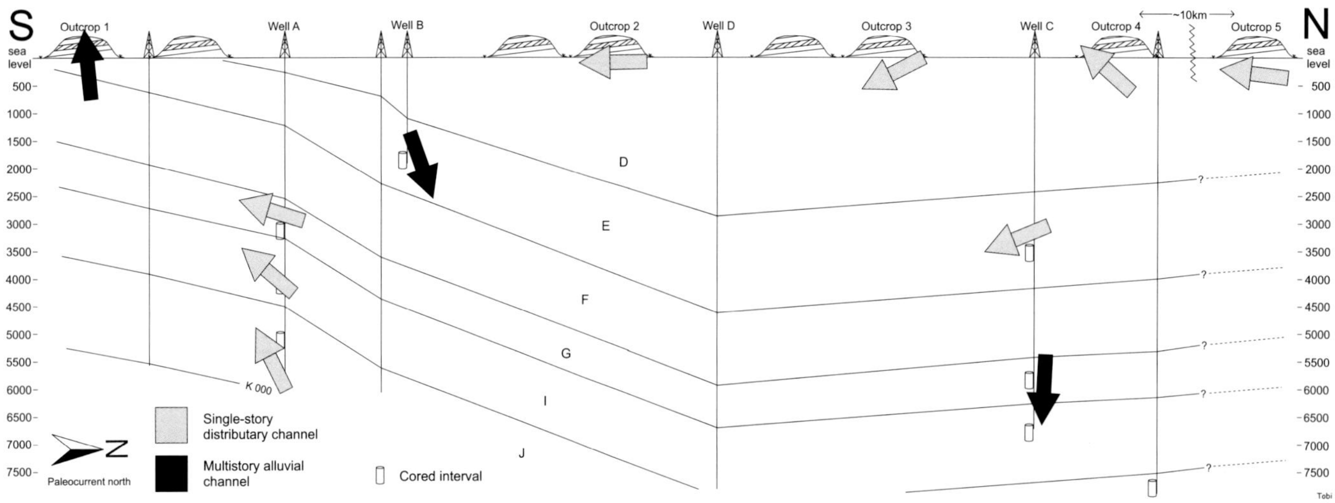

FIG. 10.—Cross section following the correlation line in Figure 7, containing paleocurrent information from Figures 8 and 9 at their appropriate depth. The arrows reflect the mean paleocurrent direction interpreted for each location. The black arrows represent multistory channels, which are west–east oriented throughout the whole subsurface succession. Gray arrows represent single-story channels, which all have a southward orientation. No change in trend of the single-story and multistory channels is evident. Their orientation origin must be structural or sedimentological in nature. Depth in feet below sea level.

The multistory fluvial channels are oriented approximately west–east, flowing at right angles to the southward direction of the distributary channels (except the one outcrop in which the flow was to the west). The fact that such channels crosscut the anticlines, together with the dimensions of the channels, suggests that clustering is likely caused by geomorphological constrictions. Some of these channels can be associated with regional sequence boundaries identified on seismic, indicating that they are incised-valley fills. Periods of hinterland uplift, producing a relative sea-level fall, could facilitate conditions suitable for valley incision. A possible source of such incised-valley fills could be reworked sediments located southwest of Balikpapan, beyond the margin of the Kutei Basin (Fig. 1, 11). Unfortunately, petrographic data were not available for this study to verify this theory.

Implications of Reservoir Orientation

The crest-parallel orientation of the north–south single-story distributary channels has significant impact on hydrocarbon exploration. So far, as can be seen in Figure 7, the main exploration in Mutiara field concentrated on the classical structural closure of the channel reservoirs close to the anticlinal crest. Potential single-story distributary channels sitting on either side of the shallowly dipping flanks of the anticlines could form excellent stratigraphic traps. Such clean sandstones sealed within interdistributary, delta-front mudstone could be easily developed, especially if they can be imaged with 3D seismic data.

A 3D seismic acquisition was run over the Mutiara area. As can be seen from Figure 5 the sediments in the Mutiara area contain an abundance of interbedded coals, shale, siltstone, and sandstone. Such highly variable lithologies in a small area make it very difficult to migrate the 3D seismic data, especially with increasing depth. Nonetheless, some preliminary interpretation confirms the suggested north–south single-story orientation, as well as the west–east multistory orientation.

Variations of Channel Style with Time

The stratigraphic location of the cored intervals is shown in a schematic correlation of wells and outcrops in Figure 10. The correlation chart is oriented north–south and shows only major marker beds illustrating the plunge of both anticlines. The zonation is that of the *VICO* intervals (Fig. 2), and it has no regional implications. As can be seen, there seems to be no stratigraphic trend on the paleocurrent distribution evident in the middle Miocene strata exposed in Mutiara field (Fig. 10). The orientation of the single-story fluvial-dominated distributary channels remains to the south throughout the whole interval, from approximately 15 Ma to 10 Ma. Despite the high variation of individual channel orientations among the intervals, no overall directional change in trend is apparent. The channel-style analysis yields more information than the stratigraphic analysis about the basin fill.

Channel Sinuosity

Apart from the overall flow direction of a channel, paleocurrent data can be used to get a sense of channel sinuosity. It is important to know whether the rivers that formed these channels were of high or low sinuosity, especially when targeting these channels as stratigraphic traps. It is argued here that both single-story and multistory channels are of very low sinuosity. The eight single-story distributary channels all have paleocurrent directions towards the south. If the river had been highly sinuous, one would expect a greater degree of scatter among the data, such as indicated by Miall (1974). Three amalgamated channels investigated in outcrop 1 (Fig. 6) have a mean paleocurrent directional variance of only 57°, again indicating a very low variance of the three consecutive channels within this multistory channel.

CONCLUSIONS

The middle Miocene succession investigated in Mutiara field contains two major reservoir styles identified in core and outcrop.

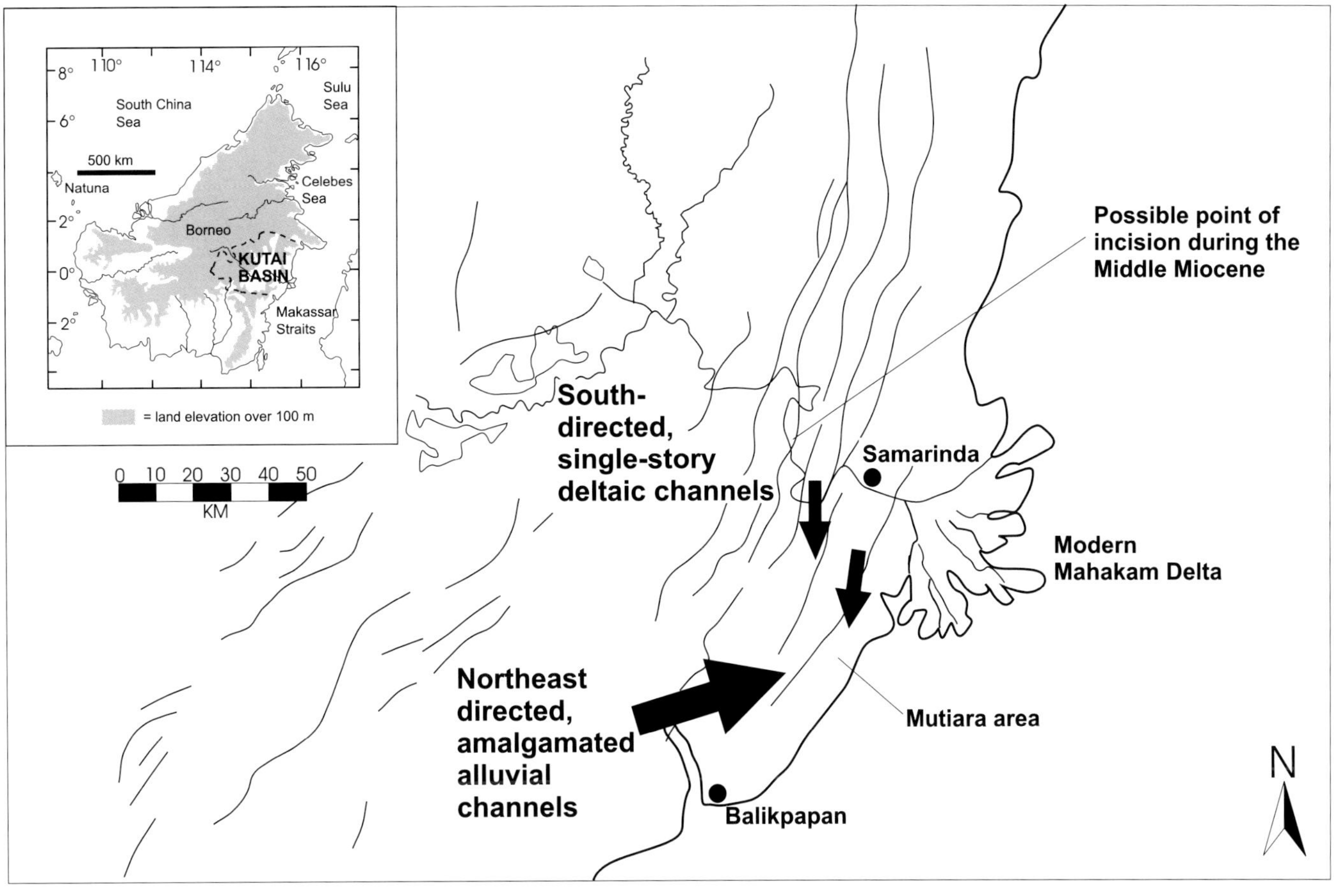

FIG. 11.—Generalized paleoflow directions of distributary and alluvial channels in Mutiara field during the middle Miocene. The point of source of the delta was to the west of the present-day point of source at the Sanga-Sanga anticline. The alluvial channels could be the result of uplift to the south. Geological map modified from Moss et al. (1997), Cloke (1997), and Cloke et al. (1997).

Both styles are based on a distinctive sedimentological character and on the sandbody orientation interpreted from paleocurrent data. These reservoirs are channel deposits and were interpreted as single-story distributary channels and multistory alluvial channels. Although the facies association of the two styles can be very similar, paleocurrent directions of the two are very different. Single-story channels had flow towards the south, whereas two of three multistory channels had flow towards the northeast.

Paleocurrent data can be directly used to interpret the orientation of channels and thus lead to a reevaluation of the hydrocarbon reserves. Furthermore, the paleocurrent data, in conjunction with facies analysis, has led to an understanding of the regional controls on channel orientations. The syndepositional initiation of synclinal and anticlinal structures, causing differential subsidence in conjunction with south-directed deltaic sedimentation, are responsible for the overall orientation of the single-story distributary channels in the Mutiara area. The flow in the single-story fluvial-dominated distributary channel systems was generally towards the south. The channels are commonly isolated single-story sandbodies, which typically form in a rapidly subsiding, deltaic environment. Multistory channels, however, have a more west–east channel orientation. They seem to have formed under alluvial-dominated depositional episodes, with a source to the southwest of Mutiara field, possibly a reworking of basin-margin deposits. The deltaic single-story channels were fed from the north or northwest of Mutiara field, where the paleo–Mahakam River incised the Samarinda anticlinorium since the Middle Miocene. High potential exists for isolated channel sandstone reservoirs on anticlinal flanks parallel to the anticlinal crests.

The paleocurrent analysis also contributed to the interpretation of the low sinuosity of the fluvial systems in the middle Miocene Mutiara field. Paleocurrents show little vertical change in paleocurrent directions of the trough cross-stratification within each channel sandbody. This has considerable implications for isopach mapping of hydrocarbon reservoirs.

ACKNOWLEDGMENTS

This paper is dedicated to the late George P. Allen, who initiated this project and who brought us all closer to the field of sedimentology and stratigraphy while all authors were at Queensland University of Technology, Brisbane, Australia. We appreciate his efforts and greatly miss his presence. Funding for this research was provided by VICO Indonesia. We thank VICO for allowing us to publish these data. Schlumberger is greatly acknowledged for providing the wireline interpretations and for their useful discussions. Special thanks go to Bob Koch, Richard Netherwood, and Peter Lloyd for their efforts in this project. Rod

Holcombe is acknowledged for providing GEOrient and modifying it to our needs. Reviewers Patrice Imbert, Mike Wilkes, and one anonymous reviewer helped to bring the paper to more focus. We are thankful to Janok Bhattacharya, Clive Jones, and Andrew Miall for critical and constructive comments on an earlier version of this manuscript.

REFERENCES CITED

ALLEN, G.P., AND CHAMBERS, J.L.C., 1997, Sedimentation in the modern and Miocene Mahakam Delta: Indonesian Petroleum Association, field trip guide book, 106 p.

ALLEN, G.P., AND CHAMBERS, J.L.C., 1998, Sedimentation in the Modern and Miocene Mahakam Delta: Indonesian Petroleum Association, Jakarta, 236 p.

ALLEN, G.P., LAURIER, D., AND THOUVENIN, J., 1976, Sediment distribution patterns in the modern Mahakam Delta: Indonesian Petroleum Association, 5th Annual Convention, Proceedings.

CLOKE, I.R., 1997, Structural controls on the basin evolution of the Kutei Basin and Makassar Straits, Indonesia: Ph.D. thesis, University of London.

CLOKE, I.R., MOSS, S.J., AND CRAIG, J., 1997, The influence of basement reactivation on the extensional in inversional history of the Kutei Basin, East Kalimantan, SE Asia: Geological Society of London, Journal, v. 154, p. 157–161.

DECELLES, P.G., LANGFORD, R.P., AND SCHWARTZ, R.K., 1983, Two new methods of paleocurrent determination from trough cross-stratification: Journal of Sedimentary Petrology, v. 53, p. 629–642.

FERGUSON, A., AND MCCLAY, K. 1997, Structural modeling within the Sanga Sanga PSC, Kutei Basin, Kalimantan: Its application to paleochannel orientation studies and timing of hydrocarbon entrapment, *in* Howes, J.V.C., and Noble, R.A., eds., Indonesian Petroleum Association, Petroleum Systems of SE Asia and Australia Conference, Proceedings, p. 727–743.

GERARD, J., AND OESTERLE, H., 1973, Facies study of the offshore Mahakam area: Indonesian Petroleum Association, 2nd Annual Convention, Proceedings, p. 187–195.

HIGH, L.R., JR., AND PICARD, M.D., 1974, Reliability of cross-stratification types as paleocurrent indicators in fluvial rocks: Journal of Sedimentary Petrology, v. 44, p. 158–168.

HOLCOMBE, R., 1998, GEOrient plots stereographic projections and rose diagrams, version 7.2: Brisbane, Australia, University of Queensland.

LAYCOCK, J., AND PAYENBERG, T.H.D., 1997, ROTATE a program to rotate structurally tilted sedimentary directional features v.1.01b: Brisbane, Australia, Queensland University of Technology.

MAGNIER, P., OKI, T., AND KARTAADIPUTRA, L., 1975, The Mahakam delta, Kalimantan, Indonesia: 9th World Petroleum Congress, Proceedings, v. 2, p. 239–250.

MIALL, A.D., 1974, Paleocurrent analysis of alluvial sediments: a directional variance and vector magnitude: Journal of Sedimentary Petrology, v. 44, p. 1174–1185.

MIALL, A.D., 1985, Architectural-element analysis: a new method of facies analysis applied to fluvial deposits: Earth-Science Reviews, v. 22, p. 261–308.

MIALL, A.D., 1996, The Geology of Fluvial Deposits; Sedimentary Facies, Basin Analysis and Petroleum Geology: Berlin, Springer-Verlag, 582 p.

MOSS, S.J., CHAMBERS, J.L.C., CLOKE, J.R., SATRIA, D., ALI, J., MILSOM, J., CARTER, A., AND BAKER, S., 1997, New observations on the sedimentary and tectonic evolution of the Tertiary Kutei Basin, East Kalimantan, *in* Fraser, A., and Matthews, S. eds., Petroleum Geology of South East Asia: Geological Society of London, Special Publication 126, p. 395–416.

PAYENBERG T.H.D., in press, Paleocurrents from FMS and scribe oriented core—a comparison, *in* Lovell, M., and Summerhays, N., eds., Geological Applications of Well Logs: American Association of Petroleum Geologists, Methods in exploration, no. 13.

PAYENBERG, T.H.D., KOCH, R., AND LANG, S.C., 2000, A simple method for orienting conventional core using microresistivity (FMS) images and a mechanical goniometer to measure directional structures on cores: Journal of Sedimentary Research, v. 70, p. 333–336.

RAMSAY, J.G., 1961, The effects of folding upon the orientation of sedimentary structures: Journal of Geology, v. 69, p. 84–100.

SAMUEL, L., AND MUCHSIN, S., 1975, Stratigraphy and sedimentation in the Kutei Basin, Indonesia: Indonesian Petroleum Association, 4th Annual Convention, Proceedings, p. 27–39.

SIDI, F.H., 1998, Sequence Stratigraphy, Depositional Environments, and Reservoir Geometry of the Middle Miocene Fluvio-Deltaic succession in Badak and Nilam fields, Kutei Basin, East Kalimantan, Indonesia [unpublished MSc thesis]: Brisbane, Australia, Queensland University of Technology, 110 p.

SLINGERLAND, R.L., AND WILLIAMS, E.G., 1979, Paleocurrent analysis in light of trough cross-stratification geometry: Journal of Geology, v. 87, p. 724–732.

TREXLER, J.H., JR., AND CASHMAN, P.H., 1990, Computer-assisted paleocurrent determination from trough cross-stratification: Computers and Geosciences, v. 16, p. 253–261.

VERDIER, A.C., OKI, T., AND SUARDY, A., 1980, Geology of the Handil field (east Kalimantan-Indonesia), *in* Halbouty, M., ed., Giant Oil and Gas Fields of the Decade 1968–1978: American Association of Petroleum Geologists, Memoir 30, p. 399–421.

VICO INDONESIA, 1995, Kutei Basin Study, an assessment of hydrocarbon potential of the Sanga-Sanga PSC: VICO Indonesia, unpublished, internal report, approx. 1000 p.

WILLIS, B.J., BHATTACHARYA, J.P., GABEL, S.L., AND WHITE, C.D., 1999, Architecture of a tide influenced river delta in the Frontier Formation of central Wyoming, USA: Sedimentology, v. 46, p. 667–688.

Index

T

W